Optimum Design and Manufacture of Wood Products

Etele Csanády · Zsolt Kovács · Endre Magoss · Jegatheswaran Ratnasingam

Optimum Design and Manufacture of Wood Products

Springer

Etele Csanády
Department of Wood Engineering
Soproni Egyetem/University of Sopron
Sopron, Hungary

Zsolt Kovács
Wood and Paper Technology
Soproni Egyetem/University of Sopron
Sopron, Hungary

Endre Magoss
Department of Wood Engineering
Soproni Egyetem/University of Sopron
Sopron, Hungary

Jegatheswaran Ratnasingam
Department of Forest Products
University Putra Malaysia
Seri Kembangan, Malaysia

ISBN 978-3-030-16687-8 ISBN 978-3-030-16688-5 (eBook)
https://doi.org/10.1007/978-3-030-16688-5

Library of Congress Control Number: 2019936293

This Springer imprint is published by the registered company Springer Nature Switzerland AG
The registered company address is: Gewerbestrasse 11, 6330 Cham, Switzerland

To the Memory of Great Woodworkers of the past and present.

Foreword

Both nature and mankind have striven for years to use natural or artificial processes to achieve an optimum or best outcome. While Nature has the inherent capability to organize its phenomena optimally, mankind has to develop his approaches based on scientific considerations and derivations supported by experimental measurements. Around 1770, Lagrange developed mathematical methods for seeking optimal solutions and originated the formulation of an "Optimal Column". At the present time, strict mathematical methods for optimization are available for nonlinear cases with constraints. At the same time, the general use of optimization procedures has not been adopted in design and manufacture.

The moderate success in optimization may have its background in the relatively complicated use of strict mathematical methods for engineers and in the lack of proper functional relationships describing the system's behaviour as a function of influencing variables. The existence of functional relationships is crucial in any optimization procedure to get reliable optimum solutions. Another problem may arise when a large and complex system should be optimized. In this case, some functional relationships are always lacking or they cannot be formulated with acceptable accuracy. A way out from this situation may be to divide the whole system into subsystems allowing an easier handling of the smaller and more definite subsystems.

The present work *Optimum Design and Manufacture of Wood Products* is a first attempt to summarize the existing scientific methods and to present new developments and results in this field. First of all, a large array of functional relationships was elaborated and described in details in Chap. 2. Besides the strict mathematical method, an engineering optimization approach is also outlined and demonstrated in several examples. The essence of this approach lies in the recognition that in many cases, the optimum is uniquely determined by a particular constraint. In this case, the optimization procedure is highly simplified with the further possible benefit of allowing us to get a general analytical solution which has a fundamental advantage over single numerical solutions because a general solution allows a quick decision in the case of conflicting optimums.

I would like to emphasize the importance of colour and gloss properties of wood as its most important aesthetic properties. Their practical use in design and manufacture of wood products is far from the real possibilities. The elaboration of more practical evaluation methods and representations using the measured reflection spectrum may help to widen the use of these important properties. Colour modification plays a distinct role in value adding of several timber species.

In order to promote the use of optimization methods as an indispensable design tool, it would be desirable to organize special courses at universities on the subject *Mathematical and Engineering Methods of Optimization* helping engineers to become more familiar with this topic.

The material presented in this book is designed for a broad circle of graduate and postgraduate students, researchers and application engineers working on optimum design and manufacture of wood products.

Sopron, Hungary György Sitkei

Preface

The world trade of furniture has grown tremendously in the last decades which requires the economic utilization of natural resources and production capacities. High-quality timber materials are becoming increasingly scarce, and therefore, optimal solutions in design and manufacture are an urgent need. Concerning mass production, the manufacturing technology has been revolutionized in the last decade where large machinery and computers have taken over craftsmanship, significantly increasing the production rate. The common technologies include computer-aided design (CAD) and manufacturing (CAM) using CNC centres or through-feed machine line.

Further attempt should be made, however, in the optimization of structural design and manufacturing processes in order to achieve a more economical use of existing resources. Moreover, the preferred use of value-adding methods for the full realization of aesthetical values of timber materials is another important way to fulfil customer demand.

Present work on *Optimum Design and Manufacture of Wood Products* provides comprehensive treatment and discussion of mathematical and engineering optimization procedures with many numerical examples. In order to facilitate the formulation of objective functions and constraints in equality and inequality forms, a wide array of functional relationships is presented based mostly on our own experimental results. A newly developed "Mechanics of Upholstering" is also included which facilitates the optimum selection of cushioning materials and support comfort design.

A detailed overview of current practices in design of furniture is also presented. The different existing methods for strength design, the interaction of tolerance and machining accuracy are also discussed and illustrated with several numerical examples. For a quick selection of furniture joints, new similarity equations are presented and elaborated.

In the final chapter, the discussion of general principles of furniture manufacturing processes is described. Interrelations among product structure, degree of automation, inventory and batch size are also explained. The description of main woodworking operations, surface coating and finishing, as well as value-adding technologies, rounds up this topic. A special care is paid to the cross-referencing of each chapter, which is presented in strong relation to each other. In Appendix, selected colour pictures demonstrate the unique ornamental properties of wood materials.

The authors are especially indebted to Prof. G. Sitkei for reading the manuscript and offering useful suggestions in working out practical examples using engineering optimisation suggested by him. Thanks also goes to Dr. Zoltán Kocsis for offering help in the preparation of illustrations. The authors are also sincerely grateful to the staff of Springer Verlag for their excellent cooperation.

Sopron, Hungary — Etele Csanády
Sopron, Hungary — Endre Magoss
Sopron, Hungary — Zsolt Kovács
Seri Kembangan, Malaysia — Jegatheswaran Ratnasingam

Contents

Chapter 1
Overview of Problems

1.1 Introduction

Every industrial production unit strives to maximize the economic utility of its productive system. This is a very complex and complicated decision-making mechanism when the company attempts to minimize the unit cost associated with transforming the incoming materials (mostly raw materials) into a completed and well-defined product at a desired level of quality, fulfilling customer requirements.

This decision-making is always associated with uncertainties and risks which cannot fully be eliminated but can be limited to an acceptable level ensuring a high enough probability of success.

This decision-making must be supported by reliable information, both in the product design and development, and also in the production phase. (Marketing issues and customer services are not treated here). This information can be derived or acquired from different sources. Some information can theoretically be derived from the general theories of material science, wood processing operations and especially from kinematic relations. Most information is the result of experimental measurements, purposefully processed to achieve more generally valid relationships as far as possible. Information from feedback of previous similar products is an integral part of the total information system.

The processing method of experimental results fundamentally influences the usefulness and possible application of the obtained results. Sadly, today a mere empirical processing is used with very limited validity. Similarly, a single finite element calculation, though it may be useful for a given solution, gives no further information for the expected effect of system variables. A carefully derived similarity equation with dimensionless numbers may have general validity and highly facilitates the prediction of the effects of system variables for arbitrary conditions. This method can be used either in the processing of experimental results or if there are differential equations having no analytical solution. In the following chapters, we give examples to show the usefulness of similarity equations which often enable us to perform an optimization procedure in a simple way.

E. Csanády et al., *Optimum Design and Manufacture of Wood Products*,
https://doi.org/10.1007/978-3-030-16688-5_1

The characteristic feature of wood products is their aesthetic value which plays an important role in their sale on the market. Aesthetic beauty is not physically measurable quantity, it is much more a subjective value and in some cases it is matter of taste. Therefore, a furniture designer should have some artistic talent and intuition to be crowned with success in his work.

There may be different motivations and reasons to start a new design and development:

- perceptible need of customers for a defined new product,
- the life cycle of an existing product is over its maturity phase,
- there are technical advances necessitate changes in product design,
- the company will widen its product variety in order to provide a stable manufacturing volume,
- changes in the market situation force the company to consider new product development.

1.2 Product Design and Development

The design procedure of a new product may be quite different depending on many factors including the various kinds, purposes, aesthetic and structural forms of wood products. The basic forms of furniture and woodworks have evolved during the last thousands of years, and sometimes surprising similarities can be found between chairs with plaiting made some 4000 years ago and in our time. Therefore, in the basic form design, there are almost always similarities with existing forms at one time or today. In the fine details, there may be many differences and possibilities for the designer to make a furniture piece more comfortable, showy, attractive and practical for the user and more economic for the producer to manufacture.

Today, the designer has modern and powerful tools at hand such as material properties, strength calculation methods, ergonomic and safety requirements, surface finishing and coating, evaluation of aesthetic properties and the general manufacturing processes. Only the increasing scarcity of precious wood may be a disturbing factor, which makes true quality products ever more expensive and unaffordable for many people.

The designer may have the following tools at hand to perform his work with maximum success.

Material properties include all aspects influencing the use of a timber. Wood properties can be described both qualitatively and quantitatively. Qualitative properties include softwood and hardwood, the density classifications as "very light", "light", "moderate", "heavy" and "very heavy" timbers. Qualitative characteristics are the different classes of durability, workability, nailing properties and finish. In many cases, the aesthetic properties such as colour hue, colour saturation, colour inhomogeneity, gloss and sheen and figure pattern also belong to the qualitative properties.

Most of the important physical and mechanical properties are available quantitatively which can be used by the designer for numerical calculations for different wood species. The strength, elasticity, shrinkage and hardness properties can be used for stiffness and strength analysis of skeletal, case and upholstered furniture and other wood products. Special requirements such as durability, reliability and stability can also be determined and predicted.

Among the broad range of wood materials, the *specialty timbers* play a very distinct role in making attractive gift items and novelties such as vases, bowls, coasters, pens, trophies, jewellery, jewel boxes and also woodwind musical instruments. These mostly tropical species grow on arid fields in Latin America, Southern Africa and Western Australia (Goldfields) under extreme environmental conditions. They are slow growing with high density, rich deep colours, flecks, streaks and striking grains. They are the raw materials for fine artistry requiring only smaller amounts of wood to produce one-of-a-kind luxury products. Due to their very low green moisture content, down to 30%, they yield unusually stable unseasoned timber suitable to wood turners. Smaller furniture pieces such as an attractive coffee table can also be produced, and they are suitable for making high quality, intricate inlays, in an array of colours to add the most elegant and subtle aesthetic detail to our lives.

One of our main problems today is to consolidate the mechanical properties of solid wood raw materials, especially their swelling and shrinkage properties. A hundred years ago, a carpenter stored his sawn boards for 5–6 years exposed to year round variations of temperature and relative humidity but without rain or sun irradiation. After this conditioning, the timber had better form stability and the swelling coefficient was decreased to half compared to that without conditioning. It was possible to design and manufacture a drawer with an upper gap under 0.5 mm and side gaps around 0.2 mm, without any risk of getting stuck in the future. Today, this time-consuming method of conditioning is not practised, but some consolidation of properties may also be achieved by short-term cyclic treatments.

Calculation methods of the engineering mechanics make it possible to calculate stiffness, strength of furniture and other wood products taking the anisotropy of wood into account (Smardzewski 2015; Eckelman 1967; Eckelman and Rabiej 1985; Koroljew 1973; Askhenazi 1978; Soboljew 1979). In certain cases, numerical methods are required to calculate complex structures.

In Sect. 2.6, there is an engineering method to calculate and design the making of upholstered furniture. Furthermore, joints are one of the most vulnerable parts in furniture, and it is very important to calculate and predict their load-bearing capacity and expected reliability. In order to facilitate the selection of the correct type of geometry for a given purpose, dimensionless expressions were developed based on experimental results (see in Sect. 4.6).

Functional relationships establish a connection among the process variables which may be needed in the planning of optimum design and manufacture of wood products. As a first attempt, we have worked out the functional relationships of the most important process variables which are described in the voluminous Chap. 2. The knowledge of functional relationships aids the selection of optimum parameters from the set of feasible ones. The use of functional relationships for optimization purposes can be found in Sect. 3.6.

There are several processing methods, termed as "value-adding methods", which make it possible to change and enhance the initial properties of raw wood. One of the oldest methods was the steaming of beech wood in Europe to enhance the stability of frame structures. Another old method is the prevention of wood deterioration by preservative treatments. In the last decades, attempts have been made to enhance surface, strength and aesthetic properties of wood species which otherwise have acceptable or even good mechanical properties (e.g. rubberwood). Further problems are the rational use of special timber species such as bamboos and oil palms which are in abundance in several countries of SE Asia.

Due to its holed stem structure, the general use of bamboo wood is considerably limited and its transform into a plane board is not easy (Bakar 2013). The wood of oil palms found in great quantity has a very loose structure with unacceptable mechanical properties. A saturation with adhesive and a compaction to the half volume can produce lumber with acceptable strength and durability (Bakar et al. 2013).

The proper use of steaming (temperature and exposure time) can be applied either for colour homogenisation or colour change. In certain cases, thermo-smoothing can be used to enhance surface properties.

Wood-based panels (particleboard, MDF) generally require to be laminated by surfacing materials (veneer, low pressure melamine, foils, laminates, coatings) for use in furniture industry. It is very important to know the life cycle of the surfacing material which is determined by its technical properties (heat strength, abrasion and scratch resistance, chemical resistance, etc.).

Furniture may be classified according to its use. Furnishing residential interiors, we may distinguish individual furniture pieces (chair), a set of furniture put together from individual pieces and suite furniture having a common style and finishing with the same overall appearance which has been designed and manufactured as a whole.

Furniture may be used in different places with the diversity of purposes. In a house, there are rooms, bathrooms, kitchens, living room furniture for inside use and garden furniture for outside use. It is obvious that an outside use imposes special requirements either in surface treating or material selection (i.e. teakwood) because it will be exposed to weathering (Csanády et al. 2015).

Another class of furniture is used in offices, public buildings (schools, hotels, hospitals, airports, entertainment and recreation) and transport equipment (trains, aircraft, ships and automobiles). One of the most important design problems is to provide comfortable seats for long-term sitting which in an aircraft or automobile may be ten hours or more. People working in offices are also exposed to long-term continuous sitting which required to develop seats of adjustability requirements.

The essential of problems in seat design can be summarized as follows. The soft tissue of the human buttocks having high water content which is practically incompressible and under hydrostatic pressure (with the same components in all directions) suffers no deformation. Therefore, the body tissue can tolerate hydrostatic pressures around 100 N/cm^2 without consequences. Vertical pressures, however,

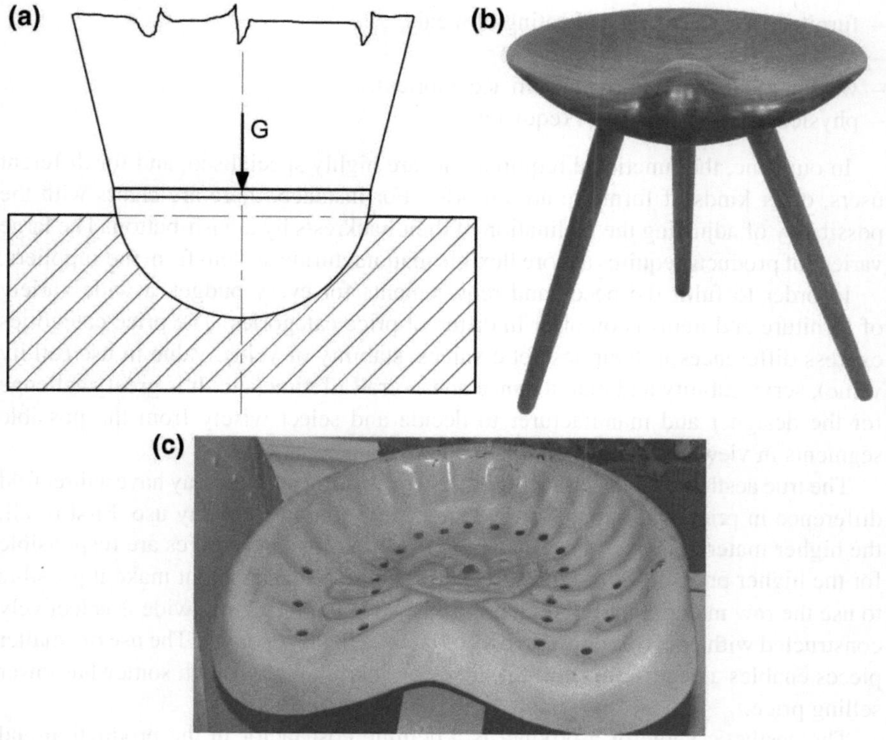

Fig. 1.1 Supporting of a bridge (**a**), a shoemaker's stool (**b**) and an early tractor seat (**c**)

cause soft tissue deformations with shear stresses, and the tolerable pressure is only around 1 N/cm^2. The main task of the designer would be to make a seat contact profile in such a way that it would transfer mostly hydrostatic pressure to the body tissue. This problem in bridge building is elegantly solved. The enormous weight of a bridge is supported on a surprisingly small contact area in the form of a hemispherical seat, Fig. 1.1, using the above principle. Due to varying radii of the human buttocks, an adjustable and accurate hemispherical seat profile seems to be a dream. Nevertheless, leading automobile manufacturers have used concave seat profiles for 50 years us allowing to drive the whole day without fatigue or pains in the hip bones (more details in Sect. 2.7). Furthermore, a century ago the three-leg shoemaker stool was carved to spherical profile making permanent sitting endurable. Early tractor seats (after 1920) have been made of steel sheets also pressed to the profile of buttocks (see Fig. 1.1). The rigid seat surface was generally covered with sheepskin.

An important property of furniture is their *functionality*. The main functions are the following:

- sitting and lounging,
- reclining and resting,
- storage for goods and items,

- furniture for preparing and eating of meals,
- mental activity (e.g. writing desk),
- washing and cleaning (bathroom accessories),
- physical training and sport requisite.

In our time, the functional requirements are highly specialised, and for different users, other kinds of furniture are on offer. For instance, there are chairs with the possibility of adjusting the inclination of their backrests by a push-button. The large variety of products requires a more flexible manufacturing system from the suppliers.

In order to fulfil the needs and requirements for every budget, a wide variety of furniture and items is on offer in different price categories. The price categories express differences in their aesthetic values, stability of value, value in use (utility value), serviceability and maintenance and overall lifetime, etc. It is great challenge for the designer and manufacturer to decide and select wisely from the possible segments in view of the competing suppliers.

The true aesthetic value has its price. A strip flooring parquet may have a threefold difference in price without any difference of its value in everyday use. First of all, the higher material cost and the matching of colour hue and figures are responsible for the higher price. A careful design and production management make it possible to use the raw material more efficiently if the strip floor 192 mm wide is selectively constructed with one, two or three rows of lamella in the top layer. The use of smaller pieces enables a better utilization of the raw material, though with somewhat lower selling price.

The aesthetic value of a product is a definite cost factor in the production and also in the decision-making. We distinguish the following classes characterizing the aesthetic value of wood product:

- unique pieces made of rare and very special or distinctly coloured wood species such as big burl, bird's eye, curly or stripped and beautifully coloured timbers. Their rarity, their eternal beauty and high-quality manual finish endow them with the highest value. The designer's main problem is what to make from the precious wood material. They are the connoisseur pieces (e.g. a round table from one piece of burl, a tiger stripped Jarrah table, a jewellery box made of bird's eye Huon pine) and can mainly be found in SE Asia and SW Australia.
- attractive gift items and novelties, music instruments and high-quality inlays made of specialty wood as fine artistry pieces.
- High-quality furniture made of precious but generally available wood with selected high-quality grades in texture and colour. In Europe, walnut has been extensively used in various countries since the thirteenth century. In other parts of the world, there are many precious wood species, especially in SE Asia, Australia, Tasmania, New Zealand and also in central and southern Africa, South America.
- quality furniture made of generally available wood species as solid wood or with veneered surfaces. This furniture is generally acquired for a lifetime, in classical styles with good value for money and stability of their value.
- furniture in the cheaper section generally made of wood composite materials with different surface coatings and manufactured by mass production. Some type of

furniture is made in elements which gives the possibility to assemble various combinations according to customer requirements (e.g. kitchen furniture).

In order to keep a given position on market, a far-sighted company carries on continuous trials and experiments in its prototype workshop to seek and establish new ideas and product improvements. The utilization and enhancement of the inherent natural beauty of wood materials requires devotion, patience and responsiveness from the designer and researcher. It also requires the combined thought of an engineer and an artist to achieve a new product with usefulness and greatest delight for the customer. The study of the magnificent works of our ancestors and predecessors helps to fulfil this task (Cescinsky and Gribble 1922; Richter 1926; Feulner 1927; Schmitz 1926). There is insufficient literature on the artistic design of furniture (Janow and Below 1971).

1.3 Manufacturing of Products

A new product dreamed up by a designer will be turned into reality by the manufacturing process. Although the product design may fundamentally govern the production process, the latter has enormous possibilities in the organisation of subsequent operations to complete the product as economically and quickly as possible. The production management profession plays a fundamental role in rational production planning and control (Anderson 1994). The production managers take part in strategy formulations, in the technical design of manufacturing, in product quality assurance and in the management of workers.

Delivery of products to the customers depends on whether they are *made for stock* or *made to order*. In the first case, the finished products are kept in stock and customers can be supplied immediately provided that stocks are not empty. In the second case, a customer must wait for his delivery time agreed upon in the order. Companies may have a mixed strategy partly making to stock and also to customer order. Special needs can generally be met by a customer's order.

In order to meet a wide variety of customer needs, the company must have a broad variety of semi-finished products allowing a short-time delivery. In other cases, longer delays in delivery should be reckoned with.

The flow of materials in a production system also has some characteristic features. In a *serial production system*, the raw material starts at one end of series of operations and will be made into a finished product. In another case, an assembly operation is needed taking several parts into an assembly (e.g. a strip flooring parquet is putting together from three layers). Major assemblies might be made by putting them together from components and subassemblies. Assembly operations, in general, put together a large number of component parts into a smaller number of finished products. An opposite case is possible in which smaller number of basic components are combined and put together into a large number of product varieties. It may be an economic way to meet various customer needs in the cheaper and mid-range price categories. Most commonly, in the last stage of production a variety of assemblies is put together and makes it possible to offer a basic furniture design (e.g. sofa) in a great number of varieties.

Another important question is the volume and variety characteristics of a production system. *Mass production* is characterized by manufacturing a small number of products in a great quantity. Generally, the subsequent operations are linked together in a line and the product moves from one operation to the next. In this case, the layout of the factory generally follows the production process.

Batch production is commonly used when the variety of products is greater and the number of items is smaller. In this case, a single machine can be used to carry out an operation on a whole batch of pieces. The machine is set up to make a similar operation on the whole batch. The *one-off production* is used when individually required products are manufactured. This means a production in great variety but a low volume of each.

Recently, a strong tendency can be observed for individual products in all fields of production which means an ever-increasing variety of models is available to customers. This trend also prevails in the woodworking industry. This situation forced companies to use maximum production flexibility to ensure a large variety of products and, at the same time, to keep unit costs competitive. One solution is the use of CNC processing centres performing different operations including panel cutting, drilling, trimming and edge banding. This highly developed processing technique is always supplemented with an appropriate control system and tooling.

New tooling systems allow the use of higher rotation speeds, and combined with jointing, an accurate edge running circle can be achieved. Furthermore, the time for changing tools can considerably reduce by increasing the net production time (Anderson 1994; Homeier 2005; Fritsch 2005).

Automation is another way to increase productivity. CNC machines can change tools automatically. The workpiece can be positioned so that different faces can be machined. The parts to be machined can be automatically transferred from one machine to another. Robots are also increasingly used to move parts or objects in a desired way. Robots not only make some workers unnecessary, but they take over dirty and dangerous jobs.

The design process is assisted with the use of computers (computer-aided design). Detailed design drawings are produced and stored by computer. This information can directly be used to command a CNC machine in order to machine a part (computer integrated manufacture).

The lifetime of furniture depends on the decision of the designer, the selected materials and manufacturing process. In the past, a piece of furniture was generally made to last at least for a century. After 150 years, a Biedermeier cabinet retains its full beauty and splendour. A set of furniture made in the early 1920s in neoclassical style, ornamented with inlays, is still in use today also in full beauty, the upholstering of chairs and sofa would need to be replaced after 40–50 years. Today, modern design aims for a shorter desired lifetime although the demand for longlasting furniture has not fully vanished. By developing high technology, China has become the biggest producer of furniture in the world. At the same time, the Chinese demand for high-quality furniture is mainly supplied from the neighbouring countries, where it is made by skilled cabinet makers from selected wood materials.

These trends and statements are not fully true for the entire world which is splitting into two different parts. It may best be illustrated by the wooden artefacts. In one part of the world, one can find only pieces made by a turner on offer which require a short production time using a simple lathe. In spite of this fact, these artefacts are not cheap. In the other part of the world hand-carved beautiful pieces are on offer for affordable prices.

A similar situation is to be seen concerning furniture. In several countries of SE Asia many small furniture makers manufacture magnificent solid wood furniture using mainly the original natural colour of wood without higher investments. A pleasant surprize is in the Pemberton area in SW Australia where some furniture makers produce superior quality furniture with unsurpassed beauty using native timbers of selected quality without high technology. Furthermore, the finest and cleanest surfaces are made by hand, using a Japanese plane (see Figs. 2.24 and 2.135).

A high standard of skill and artistic inspiration is required from the joiners and cabinet makers who devote all their efforts to the ultimate goal, the production of a furniture with eternal beauty and value. Sadly, the number of these joiners and cabinet makers is decreasing, especially in the advanced countries.

Chapter 2
Functional Relationships

2.1 Introduction

In order to successfully optimize the design and manufacture of wood products, the interaction of process variables must be known. These interactions may be described in the form of functional relationships or given in graphic representations. They can be derived by theoretical deductions or from experimental measurements. Most of our knowledge originates from experiments, and the range of validity of the results that are obtained fundamentally depends on the chosen processing technique.

The chapter on *Functional Relationships* is the first attempt to work out and compile the relationships of the most important process variables needed in the planning of design and manufacture of wood products. Most of the material presented has been newly developed or is an improvement of existing material. Relative coordinates or dimensionless numbers are used to extend the validity of obtained results.

The material in this section contains theoretical and experimental results on cutting forces and energy requirements, tool wear and tool life determination as well as their relation to surface quality, roughness and their relations to process parameters. The chapter describes an engineering method to design and evaluate upholstered furniture, hardness and abrasion resistance of wood, and its aesthetic properties such as colour and gloss. Finally, the most common objective functions are described.

2.2 Purpose and Role of Functional Relationships

Every engineering design procedure tries to select system components or processes that meet desired needs and requirements. It is a decision-making process, that often requires iterative approaches, in which the basic sciences, mathematics and engineering sciences are used to optimally convert wood to meet started objectives. The fundamental elements of the design process include the establishment of objectives and criteria, synthesis, analysis, construction details, testing and evaluation.

E. Csanády et al., *Optimum Design and Manufacture of Wood Products*,
https://doi.org/10.1007/978-3-030-16688-5_2

In order to use our mathematical and engineering science tools, we always need an array of different sets of data as an input. Engineering problems can generally be represented in terms of the input–system–output concept. The most common and useful problem is the *analysis* problem in which the input and system (functional relationship) are known, and so we can solve for the system response (output). At the same time, one of the major roles of engineering is to design new or improved equipment, furniture, processes and services. In our present context, the design or *synthesis* problems may be interpreted as those in which the inputs (different sources, materials properties, etc.) are available and certain outputs (furniture for a given purpose) must be designed and produced to ensure the desired appearance, function and lifetime. In this case, the engineer's job is to create a system (functional relationship) describing general regularities that will allow the choice of the best parameters from the available input data. This is an additional job for the design engineer, and it may sometimes be quite time-consuming.

It is important to properly process experimental results to use them in the design procedure. A simple graph of experimental results allows us to see a visual correspondence between input and output, but it does not produce a generally valid relationship as a function of possible variables. Concerning the efficiency of engineering design, it would always be desired to have all governing functional relationships allowing the use of the much more straightforward problem type, the *analysis* and only rarely the time-consuming *synthesis* method.

If the valid mathematical description of the system is not known but the system already exists as a physical entity, the method of *system identification* or *experimental modelling* may be useful. By applying suitable inputs, the system produces output responses. Both input and output should carefully be measured, and from these data, the system description can be discovered. Success depends on the proper planning of the experiments and the processing method. In some cases, the mathematical description of the system in the form of differential equations is known but the differential equations have no analytical solution. A numerical solution is possible in theory, but in the practice it is not always convenient. In this case, a similar method as the experimental modelling might be carried out. The differential equation can systematically be solved for selected input parameters, and the output results should be processed in the form of a similarity equation which gives the generally valid solution for the differential equation. Sadly, this practice is very rare all over the world though its use would always be possible (Langhaar 1951).

In the vast majority of practical designs, we do not employ formal optimization methods but we rather rely on an iterative process of *creative design analysis*. This circumstance shows the importance of analysis application, mainly in the form of a succession of analysis problems. To perform these analysis problems, the system description should be available.

The processing of experimental results requires a deep insight into the physical phenomena in question, and skill and experience in the various processing techniques which may ensure that the functional relationship is generally valid. Using dimensionless numbers, the obtained function generally has more wide-ranging validity. Then it is worth checking the limits of validity and the expected accuracy.

2.3 Classification of Functional Relationships

Functional relationships describe the dependence of output parameters as a function of influencing variables. Since technical problems are governed by natural laws, these functional relationships are always definite and uniquely defined. In practical cases, we possess *theoretically derived* and uniquely defined relationships only in cases when the dependent variable is the function of well-defined geometric and kinematic variables. This class of functional relationships includes the operational parameters of various woodworking machines, such as the feed per tooth, average and maximum chip thickness, the working diagram of a frame saw, surface waviness after milling and planning, etc. (Sitkei et al. 1990; Csanády and Magoss 2013). In this case, the accuracy of the calculated parameters depends on the accuracy of input data.

Another class of functional relationships is represented by the *similarity equations*. The similarity (dimensionless) numbers are obtained using the dimensional analysis method (Sitkei et al. 1979; Sitkei 2013). The success of this method is dependent on the proper selection of influencing variables which should be supported by theoretical considerations. If it is carefully done, it may be a very powerful method supplying generally valid relationships. The constants in the similarity equation are determined using experimental results. The accuracy of this method depends on the accuracy of the experimental results. The accuracy can be estimated by the bandwidth of scattering of measurement points on the similarity plot using a double-logarithmic scale. Sadly, this method is not commonly used although its wide-ranging application is obvious (Sitkei et al. 1979; Sitkei 2013).

The next class of functional relationships are the *semi-empirical* ones. In this case, some behaviour of a given phenomenon can be derived theoretically but without sufficient accuracy. It was theoretically derived that the wood cutting force behaves linearly as a function of chip thickness and has a minimum as a function of cutting angles (Sitkei et al. 1990). At the same time, the theoretically derived cutting force is not accurate enough for practical uses and should be corrected by experimental measurements. The form of a theoretical equation is supported by the measurement results (Koch 1964). Furthermore, in the processing of experimental results on the energy requirement of different woodworking machines, it has been theoretically shown that the proper independent variable is the feed speed multiplied with the cutting height (eH). It is interesting to note that the same conclusion could have been obtained using a similarity equation (Sitkei 2013). Therefore, possible theoretical considerations may contribute to the correct processing of experimental results.

In all cases when the physical behaviour of system is poorly known, the *empirical description method* can be used. Counting the influencing variables is an indispensable step. The form of a functional relationship is generally not known in advance. One possible way to obtain a result is to separately determine the influence of each independent variable on the dependent variable. Having many independent variables, skill and experience are needed to formulate the final equation in a proper fashion.

The validity of empirical relationships may be very limited due to different reasons. The common mistake made in experimental works is the use of a narrow range of possible values of the independent variables. Values of the dependent variable are assumed to beat zero/infinite (or maximum) values of the independent variables, and this may fundamentally influence the final form of the equation. An asymptotic phenomenon in a narrower range can be described by a power function leading to considerable error if it is used in broader range. Therefore, the use of a mere empirical relationship should be avoided as far as possible.

2.4 Cutting Force and Energy Requirement

The most of the woodworking operations are made by different tools which form chips. The main purposes of machining are the sizing, shaping and surfacing of the raw wood. There are two main types of woodworking, knife machining and abrasive machining (sanding). The latter is the most commonly used surfacing method preceding wood coating.

2.4.1 Knife Machining

Knife machining is an old method for sizing, shaping and surfacing of the raw material. The advantage of knife machining is its high cutting capacity with the smallest energy consumption. Its energy requirement is determined by the forces acting on the tool edges and by its cutting speed. The cutting speed is a key factor for a high production rate.

The force acting on the tool edge depends on many influencing factors which may be attributed to the following main variables:

Operational parameters

- chip thickness or its average value,
- cutting depth,
- cutting speed,
- cutting direction.

Wood properties

- strength and hardness in the principal cutting directions,
- modulus of elasticity in the principal cutting direction,
- moisture content,
- resin content.

Tool characteristics

- diameter of tool,

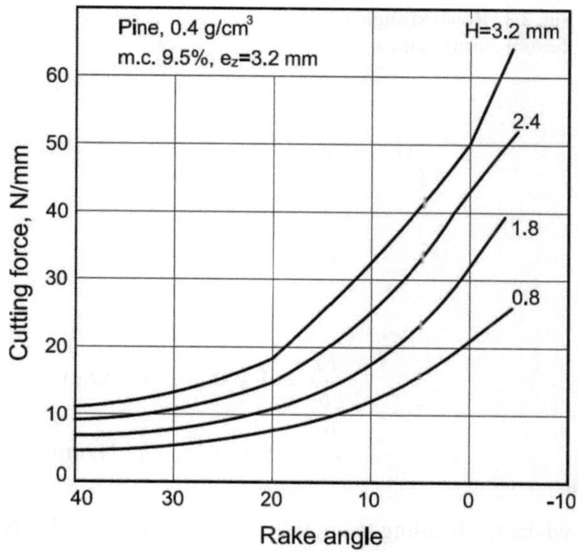

Fig. 2.1 Effect of the rake angle on the specific cutting force at various cutting depths. Tooth bite 3.2 mm

- cutting angle,
- oblique angle at slide cutting,
- sharpness of the edge,
- virtual cutting circle.

In accordance with theoretical derivations, the specific cutting force related to the unit width can be described with a linear equation with the form (Sitkei et al. 1990; Csanády and Magoss 2013):

$$\frac{P_h}{b} = A + B \cdot h \quad \mathrm{N/cm} \tag{2.1}$$

where

b is the cutting width,
h is the chip thickness,

and the A and B constants should be determined experimentally. The A and B constants depend on the mechanical properties of the wood and the cutting angle or rake angle of the tool. It is also proved theoretically that the cutting force has a minimum value as a function of the cutting angle (Sitkei et al. 1990). Measurement results depicted in Fig. 2.1 are in agreement with the theory [calculated and replotted from (Koch 1964)] and show that the cutting force increases considerably as the rake angle decreases.

It turned out that the constant B is proportional to the bending strength of the wood and the linear equation for the most common cutting angles and for sharp tools assumes the form

Fig. 2.2 Rotation angle of the tool during cutting

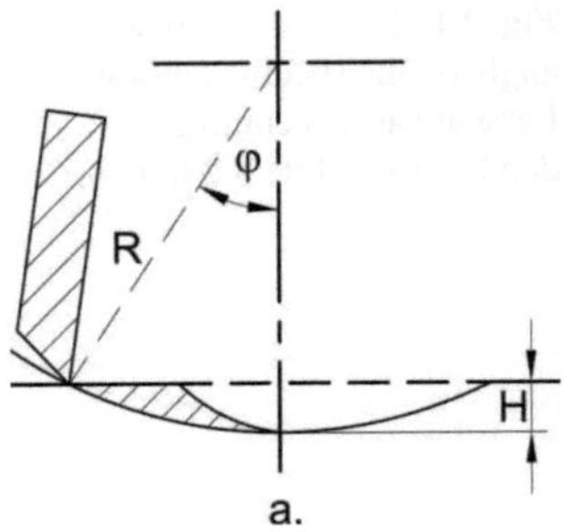

$$\frac{P_h}{b} = 4 + 0.3\sigma_b \cdot h \quad \text{N/mm} \quad (\delta = 60°) \tag{2.1a}$$

$$\frac{P_h}{b} = 5 + 0.4\sigma_b \cdot h \quad \text{N/mm} \quad (\delta = 70°) \tag{2.1b}$$

where the bending strength σ_b must be substituted in N/mm^2. The selection of appropriate bending strength may be oriented to the volume density of wood. For example, European Scotch pine may have quite different densities from 380 to 650 kg/m^3. Bending strength values belonging to the above density range vary between 60 and 90 N/mm^2. Beech and oak are much more uniform with average values of 105 and 110 N/mm^2.

In the above equations, the first term means the force component acting on the edge which is strongly dependent on the edge radius (Sitkei et al. 1990; Csanády and Magoss 2013).

Using rotating tools, the chip thickness varies as a function of the rotating angle, Fig. 2.2. The cutting force also varies in a similar way, and its average value related to the rotation angle of cutting may have different values.

The resultant force function may be composed of two parts

$$F\left(\frac{\varphi}{\varphi_0}\right) = F_{\max} \cdot f\left(\frac{\varphi}{\varphi_0}\right) \tag{2.2}$$

where $F_{\max}$ means the maximum force during cutting and $f\left(\frac{\varphi}{\varphi_0}\right)$ is the shape function. The maximum force is given by Eq. (2.1) at the maximum chip thickness, while the shape function may be approximated by the following equation

$$f\left(\frac{\varphi}{\varphi_0}\right) = C\left(\frac{\varphi}{\varphi_0}\right)^n \cdot e^{-k\left(\frac{\varphi}{\varphi_0}\right)^m} \tag{2.2a}$$

where the constants C, k, n and m are to be selected to fit the experimental force curve. The constant C adjusts the shape function to the unit height. With $R = 60$ mm, $e_z = 1$ mm, $H = 4$ mm, the cutting angle is $\varphi_0 = 21°$, the maximum force for beech wood was 29 N/mm, and the best fit was obtained for $C = 2.5$, $n = 0.5$, $k = 6$ and $m = 3$.

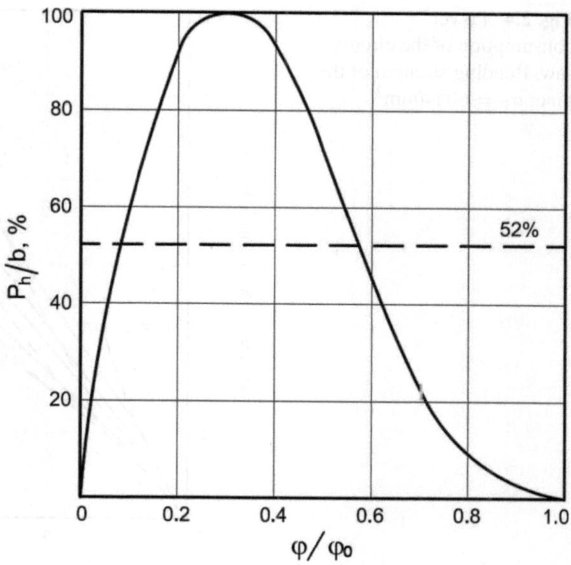

Fig. 2.3 Shape function of the cutting force

The average force integrated along the cutting angle is 52% of the maximum value, which is 15.08 N/mm distributed along the arc length $R\varphi_0 = 22$ mm. Figure 2.3 shows the shape function in relative units.

The derived calculation method greatly facilitates the correct estimation of the average cutting force required in energy consumption and friction energy calculations. The latter is the most important input data for determining the thermal load on tool edges (Sitkei et al. 1990; Csanády and Magoss 2013).

In some cases, it is interesting to know the energy required to turn 1 m^3 solid wood into chips. Using linear cutting (bandsaw), the specific cutting energy for pine and beech varies between 10 and 15 kWh/m^3 solid wood. Circular saws consume somewhat more energy due to the increase of relative cutting length. Common values fluctuate between 18 and 22 kWh/m^3 for solid wood.

In practical cases, it is more important to know the required power for a given tool and operational conditions. Earlier studies have already shown that the combined variable, the cut cross-section in the unit time *eH*, is the main influencing factor governing the power requirement in all cutting process (Sitkei et al. 1988, 1990). Figure 2.4 shows a similar representation for a circular saw cutting pine wood (Sitkei et al. 1990; Csanády and Magoss 2013).

The cutting height has a definite influence on the quantity of chips in the gullet space causing friction forces on the wall of the gap. The cutting height and the position of table also influence the power requirement due to the variation in relative arc length of cutting. This phenomenon is shown in Fig. 2.5.

The angle of cutting φ_0 is calculated by the relation

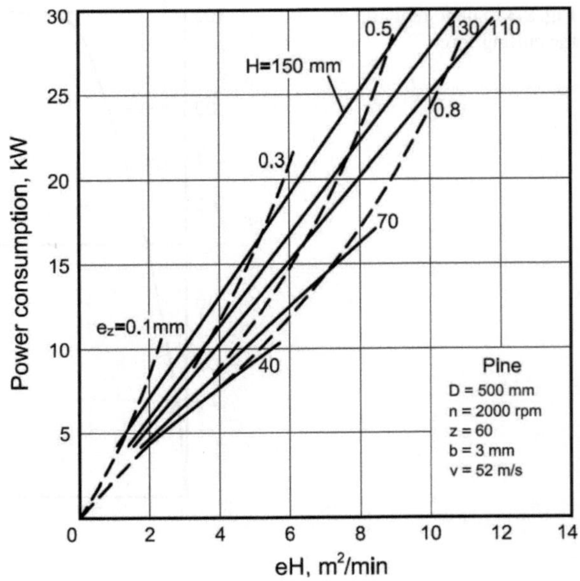

Fig. 2.4 Power consumption of the circular saw. Bending strength of the pine: $\sigma_b = 60\ \mathrm{N/mm^2}$

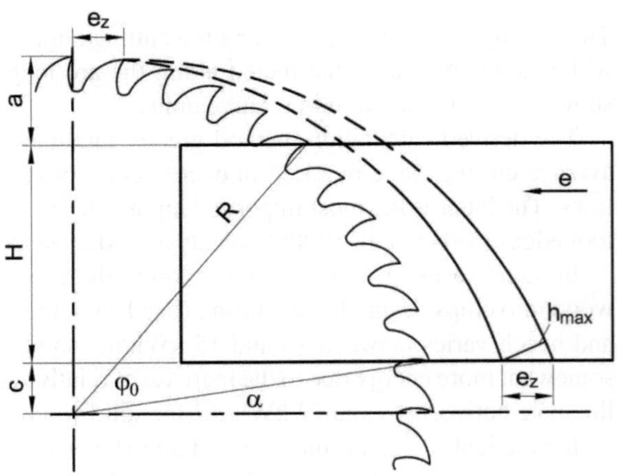

Fig. 2.5 Kinematic relations of circular saw (Sitkei et al. 1990; Csanády and Magoss 2013)

$$\varphi_0 = \arcsin\left(\frac{c + H}{R}\right) - \arcsin\left(\frac{c}{R}\right)$$

The cutting arc related to the cutting height gives the relative cutting arc as follows

$$L_0 = \frac{\varphi_0 \cdot R}{H}$$

where φ_0 must be substituted in radian. The average chip thickness is given by

$$e_{z_{av}} = \frac{e_z \cdot H}{\varphi_0 \cdot R} \tag{2.3}$$

The power requirements can be approximately estimated by the following equation

$$P = z'R \cdot b \cdot \left(A + Be_{z_{av}}\right) \cdot \frac{n}{9.55} \cdot \left(\frac{\varphi_0 \cdot R}{H}\right) \tag{2.4}$$

where the number of teeth in action z' is calculated from the cutting length divided by the pitch as

$$z' = \frac{\varphi_0 \cdot R}{t}$$

where t is the tooth pitch.

Example: A beech wood sample 10 cm high is cut with a circular saw $D = 500$ mm given in Fig. 2.5. The table position of the machine is characterized by $a = 5$ cm, $H = 10$ cm and $c = 10$ cm; the cutting angle is 0.5158 (29.55°), tooth pitch $t = 2.62$ cm and $z' = 4.93$, the average chip thickness is taken 1 mm with $e_z = 1.29$ mm. The feed speed is 154.8 m/min, and the cut cross-section in the unit time is 15.48 m^2/min. The cutting force for beech wood with these parameters is approximately 35 N/mm, and the required power is 35 kW, using Eq. (2.4). This example clearly shows that circular saws may require high power.

The power requirement of **planing and milling** can essentially be calculated in a similar way. Using the force shape function and its average force ratio Ψ as introduced in Fig. 2.3, the power consumption is given by the following equation

$$P = \Psi \cdot R \cdot z \cdot b \cdot (A + B \cdot h_{max})\frac{n}{9.55}\frac{\varphi_0}{360} \tag{2.5}$$

where z means the number of knives and b is the cutting width. The cutting angle φ_0 must be substituted here in degree.

The maximum chip thickness is calculated approximately from the equation

$$h_{max} \cong e_z \cdot \sin\varphi_0 = e_z \cdot \sin\left(1.425\sqrt{\frac{H}{R}}\right) \cong e_z 1.425\sqrt{\frac{H}{R}}$$

The average chip thickness is calculated from the chip surface divided by its length:

$$e_{z_{av}} = \frac{e_z \cdot H}{R \cdot \varphi} = \frac{e_z}{1.425}\sqrt{\frac{H}{R}}$$

or

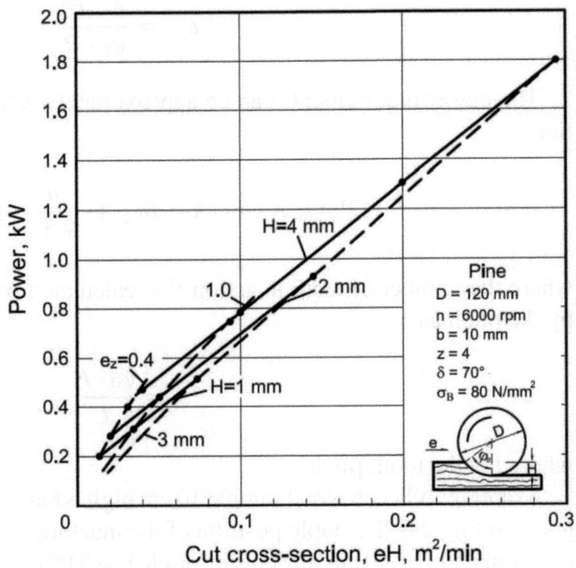

Fig. 2.6 Power requirement of milling and planing as a function of cut cross-section in the unit time

$$e_{z_{av}} \cong \frac{h_{max}}{2}$$

The relative value of cutting angle φ_0 can be expressed as

$$\frac{\varphi_0}{360} = \frac{1.425}{2\pi}\sqrt{\frac{H}{R}} = 0.2268\sqrt{\frac{H}{R}}$$

indicating that the relative duration of cutting per revolution is a function of the relative depth of the cut.

Example: Beech wood is milled with a rotating tool 120 mm in dia., 10 mm cutting width, four knives and cutting depth of 4 mm. The maximum chip thickness is $h_{max} = 1$ mm, and the rotation speed is 6000 rpm. Using these input data, the cutting angle $\varphi_0 = 0.3679$ rad or $\varphi_0 = 21°$, the tooth bite $e_z = 2.8$ mm and the feed speed is $e = 67.2$ m/min. The cut cross-section in the unit time $eH = 0.27$ m^2/min. The calculated power consumption is 1.7 kW. To turn 1 m^3 solid wood into chips requires 10.5 kWh/m^3 energy for $H = 4$ mm cutting depth, but 19 kWh/m^3 for $H = 1$ mm. Using smaller tooth bites, the specific energy consumption further increases.

Figure 2.6 gives a general overview of the energy requirement as a function of operational parameters. Cutterheads with 10–12 knives may have power requirements around 5–6 kW.

Special profile tools used in window frame production have a greater cutting depth and higher cut cross-section in the unit time. Accordingly, they require more driving power compared to the calculated value in the above example.

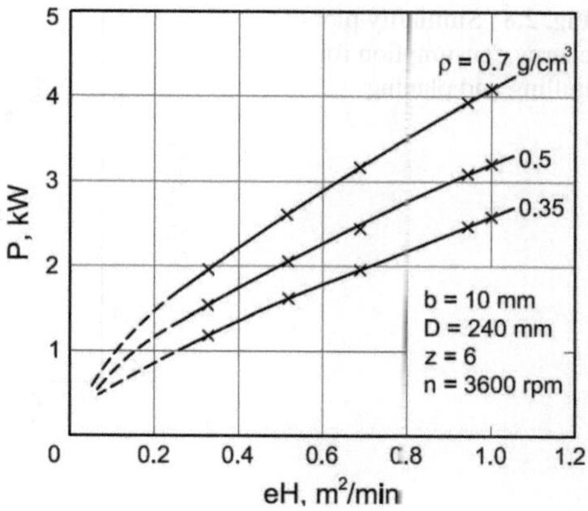

Fig. 2.7 Power requirement for planing as a function of cut cross-section in the unit time for different wood densities (recalculated from Koch 1964)

Figure 2.6 shows clearly that using smaller tooth bite, the power requirement for the same cut cross-section in the time unit increases. Because the effect of cutting depth on the power requirement is not decisive, therefore, for a quick estimate, the following simple equation may be used for pinewood:

$$P = 0.2 + 5.62(eH)\ \ \text{kW} \tag{2.5a}$$

Wood species as a function of their **density** also require different cutting energy. Figure 2.7 shows experimental results for different volume densities as a function of cut cross-section in the unit time. The curves may be described by the following equation:

$$P = 0.7 + 4.75\rho_v^{0.85} \cdot (eH)^{0.75}\ \ \text{kW}$$

where the volume density ρ_v must be substituted in g/cm³. With smaller cut cross-sections, the tooth bite e_z is generally smaller, which increases the power requirement to a certain extent. The exponent of (eH) less than one combines this effect implicitly.

A more generally valid relationship may be found using a similarity equation. For planing and milling, the following similarity equation has been derived

$$\frac{P}{\sigma_b ebH} = 0.172\left(\frac{e_z}{R}\right)^{-0.35} \tag{2.6}$$

where σ_b is the bending strength of wood characterizing the cutting resistance [see Eqs. (2.1a) and (2.1b)]. Equation (2.6) is demonstrated in Fig. 2.8 using calculated and experimentally obtained results.

Although the kinematic relations of different sawing machines (frame, band and circular saws) differ substantially, their cutting mechanism and chip formation are similar. Therefore, a successful attempt has been made to derive a similarity rela-

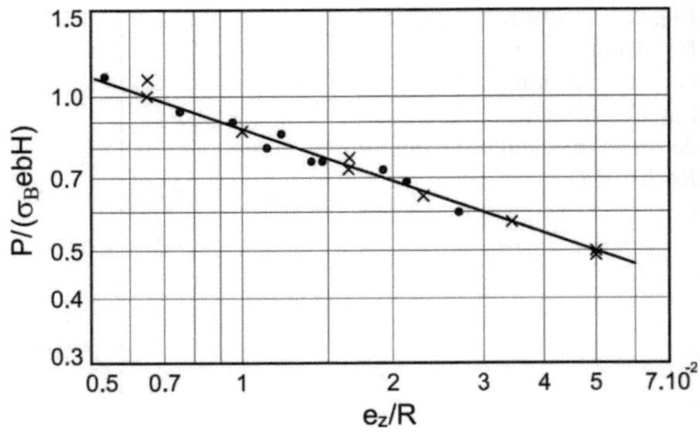

Fig. 2.8 Similarity plot of energy consumption for milling and planing

tionship to describe the average energy consumption of different sawing machines (Sitkei 2013).

Using the standard dimensional analysis method (Langhaar 1951), the following similarity equation is obtained:

$$\frac{P}{\sigma_B \cdot e \cdot b^2} = \mathrm{const}\left(\frac{H}{b}\right)^n \cdot \left(\frac{\varphi_0 \cdot R}{H}\right)^m \tag{2.7}$$

where

P is the power consumption, Nm/s,
σ_B is the bending strength of the wood, N/m^2,
e is the feed speed, m/s,
b is the width of cutting (kerf), m,
H is the sawn height, m.

The relative cutting arc, $\varphi_0 R/H$, for saws with a linear motion (frame and bandsaws) should be taken as one. The processing of experimental result with different machines and wood resulted in Fig. 2.9 (Sitkei 2013).

The diagram shows two curves lying very close to each other. The lower part represents the measurement data for circular saws and the upper part for frame and bandsaws. The tangents of curves are 1.08 and 1.04, respectively, and, for the sake of simplicity, both may be taken as $n = m = 1.08$.

With this small simplification, Eq. (2.7) has the final form:

$$\frac{P}{\sigma_B \cdot e \cdot b^2} = 0.5 \cdot \left(\frac{H}{b} \cdot \frac{\varphi_0 \cdot R}{H}\right)^{1.08} \tag{2.7a}$$

For frame and bandsaws, for which the relative cutting arc is one, the similarity equation can be rearranged to this simplified form

$$\frac{P}{\sigma_B \cdot e \cdot b \cdot H} = \mathrm{constant} \tag{2.7b}$$

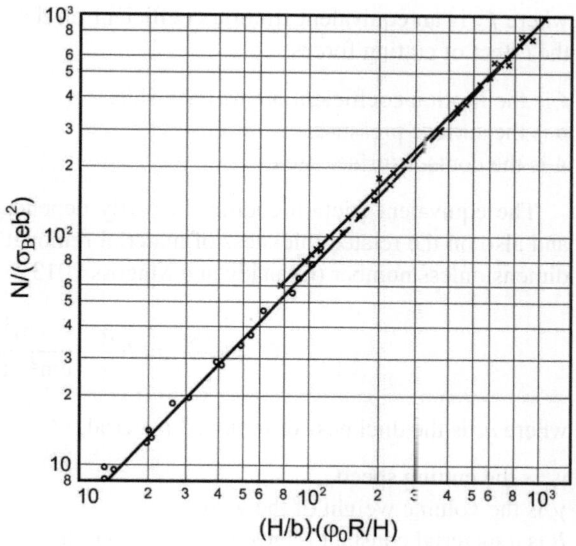

Fig. 2.9 Similarity relationship for frame, band and circular saws

indicating that this dimensionless number is invariant for these sawing machines. For a given saw and wood species, σ_B and b are constant and Eq. (2.7b) yields the simple relationship

$$P = \text{const} \cdot eH \tag{2.7c}$$

which was deduced earlier in another way (Sitkei et al. 1988). In practice, the use of Eq. (2.7c) or a similar plot is more convenient. It is valid for a given tool, and a change in the tool with different cutting width requires another constant in Eq. (2.7c).

It should be noted that Eq. (2.7) or (2.7a) gives reliable results if the tooth bite is not too small. This condition is generally fulfilled for saws.

2.4.2 Sanding

Sanding is a common woodworking operation to produce a high-quality surface. Sanding is an abrasive cutting process with very complicated mechanics and, therefore, an accurate calculation of its energy requirement is almost impossible. The following calculation gives an approximate estimation of its energy requirement.

The resultant sanding force consists of true cutting forces, but the dominant part of these forces originates from various friction components. Therefore, we use an equation based on friction forces in the following form:

$$F = (f_e + f)p \cdot A \quad \text{(N)} \tag{2.8}$$

where f_e is an equivalent friction coefficient on the contact surface including also the effect of cutting forces,

f is the friction coefficient on the back side,
p is the surface pressure,
A is the contact surface area.

The equivalent friction coefficient partly depends on the strength of the wood and also on the related thickness of material removal which is characterized with a dimensionless number (Csanády and Magoss 2013):

$$C = \frac{a_s}{v_c \cdot t} = B\frac{p}{\gamma} \quad \frac{\mathrm{cm}^3}{\mathrm{cm}^2 \cdot \mathrm{cm}} \tag{2.9}$$

where a_s is the thickness of removed material,

v_c is the cutting speed,
γ is the volume weight of the wood,
B is a material constant related to the unit width,
t is the elapsed time.

For moving workpieces with a feed speed v_f the above equation has the form

$$C = \frac{a_s}{L_c} = B \cdot \frac{p}{\gamma} \cdot \frac{v_c}{v_f} \quad \frac{\mathrm{cm}^3}{\mathrm{cm}^2 \cdot \mathrm{cm}} \tag{2.9a}$$

Because the equivalent friction coefficient contains true friction, its value is put together from two parts

$$f_e = \frac{2.3 \cdot \sigma_b \cdot C}{p} + f_1 = \frac{2.3 \cdot \sigma_b \cdot B}{\gamma} + f_1 \tag{2.10}$$

where f_1 is the true frictional part of the total resistance coefficient which can be taken as $f_1 = 0.4$ based on experimental results. The first component which substitutes the cutting force is proportional to the bending strength of the wood and the amount of specific material removal. The constant is given from experimental measures.

The power consumption is obtained by multiplying the force with the cutting velocity:

$$P = F \cdot v_c = \left(\frac{2.3 \cdot \sigma_B \cdot B}{\gamma} + f_1 + f\right) \cdot p \cdot b \cdot L_c \cdot v_c \quad \text{Watt} \tag{2.11}$$

where the friction coefficient f on the back side has values between 0.3 and 0.4.

Calculations show that the contribution of cutting forces to the total resistance coefficient is relatively small (between 0.15 und 0.3) and the main influencing factors are the friction coefficients, the surface pressure, the contact area and the cutting velocity.

Table 2.1 Sanding characteristics of selected wood species

	Pine	Beech	Oak	Alder	MDF
k, (g/cm^2 · min)	0.3–0.4	0.25–0.3	0.2–0.25	0.4–0.5	0.3–0.45
$B \times 10^8$, 1/cm	5–6	4.4–5	3.5–4.5	6–8	7–8.5

Based on experimental results, the specific wood removal (g/cm^2 · min) is often known. The following table shows average values for stock removal and the corresponding B values for 1 N/cm^2 surface pressure and 10 m/s cutting velocity (Table 2.1).

Example: Beech wood is sanded with 1 N/cm^2 surface pressure and 20 m/s cutting velocity. The volume weight is 0.007 N/cm^3, bending strength 10,500 N/cm^2, friction coefficients $f_1 = 0.4$ and $f = 0.35$, material constant $B = 4.8 \times 10^{-8}$ 1/cm. The thickness of the removed material is $a_s = 8.23$ mm/min, and the specific stock removal $k = 0.576$ (g/cm^2 · min). Taking the sanded area as 100 cm wide and 30 cm long, the power consumption is 55 kW which corresponds to 183 kW/m^2. The specific stock removal is 345.6 kg/cm^2 · h or 1.89 kg/kWh. The removal of 1 m^3 of solid wood requires 370 kWh energy which is at least a magnitude higher than a knife machining.

For comparison, a pine wood is sanded with the same operational parameters, and $\sigma_b = 8000$ N/cm^2, $\gamma = 0.005$ N/cm^3. The thickness of removed materials is $a_s =$ 12.96 mm/min and $k = 0.648$ g/cm^2 · min, and the power requirement is $P = 57$ kW or 189.7 kW/m^2. The specific stock removal is 388.8 kg/m^2 · h or 2.05 kg/kWh. The removal of 1 m^3 solid wood consumes 244 kWh which is considerably less than the energy requirement of hardwoods.

Equation (2.11) can be easily rearranged to a dimensionless form. Because the contribution of cutting forces is not decisive, for rough estimation the following simple equation can be used

$$\frac{P}{p \cdot A \cdot v_c} = 0.95 \tag{2.11a}$$

where $A = b \cdot L_c$ is the sanded surface. The dimensionless number on the left side can be regarded as a similarity number which is more or less invariant for different sanding processes. Using sanding machines with a moving workpiece, Eq. (2.9a) should be used in Eqs. (2.10) and (2.11).

If the thickness of removed material a_s is prescribed, then, using Eq. (2.9) or (2.9a), first the necessary time or surface pressure should be established and thereafter in Eq. (2.11) the value of B/γ should be replaced by

$$\frac{B}{\gamma} = \frac{a_s}{p \cdot v_c \cdot t} \quad \text{or} \quad \frac{B}{\gamma} = \frac{a_s \cdot v_f}{L_c \cdot p \cdot v_c}$$

The performance characteristics of the abrasive belt considerably decrease as a function of working time due to the continuous wear process. The power consumption also decreases but to a lesser extent than the stock removal.

The decrease of stock removal as a function of working time can be taken into account by using time-dependent material constants in the following form

$$B = B_0 \cdot e^{-\beta t} \quad \text{or} \quad k = k_0 \cdot e^{-\beta t}$$

where k is the specific stock removal in g/cm^2 · min. The constant β characterizes the rate of decrease in stock removal. The specific stock removal is integrated over the time as

$$\frac{V}{b} = B_0 \cdot L_c \cdot v_c \cdot \frac{p}{\gamma} \cdot \int_0^t e^{-\beta t} \mathrm{d}t$$

and after integration we get

$$\frac{V}{b} = B_0 \cdot L_c \cdot v_c \cdot \frac{p}{\gamma \cdot \beta} \cdot \left(1 - e^{-\beta t}\right) \quad \mathrm{cm^3/cm \cdot h} \tag{2.12}$$

If we use the specific stock removal k, the above equation has the form

$$\frac{V}{b} = k_0 \cdot \frac{L_c}{\rho} \cdot \frac{1}{\beta}\left(1 - e^{-\beta t}\right) \quad \mathrm{cm^3/cm \cdot h} \tag{2.12a}$$

where

L_c is the length of sanding,
v_c is the cutting speed,
p is the platen pressure,
γ and ρ are volume weight and density.

The main problem is the selection of reliable β values for different wood species and cutting speed s as a function of working time which would require detailed experimental results. Sadly, very few such experiments have been done and the most extensive research was carried out 60 years ago (Hinken 1954). The new processing of these old experimental results is seen in Fig. 2.10.

Hard maple (Acer saccharum) was sanded with an Al-oxide abrasive belt at 0.56 N/cm^2 surface pressure in a wide range of cutting speeds. In these experiments, the time constant β has a definite minimum at 25 m/s cutting speed. Furthermore, the time constant considerably decreases during the working time which makes the picture more complicated.

New experimental results for beech wood are depicted in Fig. 2.11 on a half-logarithmic scale (Dobrint 1991).

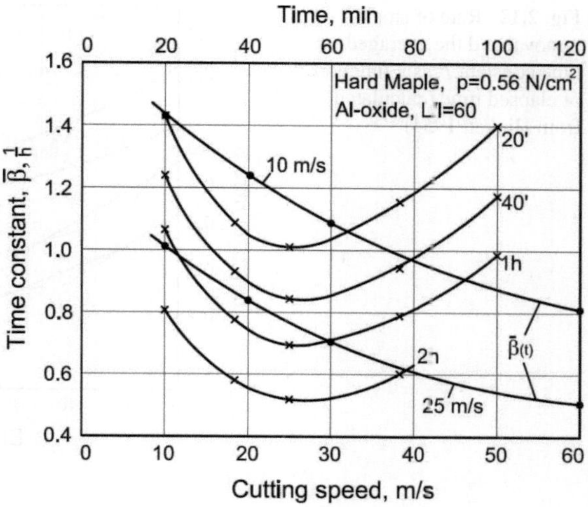

Fig. 2.10 Time constant for sanding as a function of cutting speed and working time (Hinken 1954)

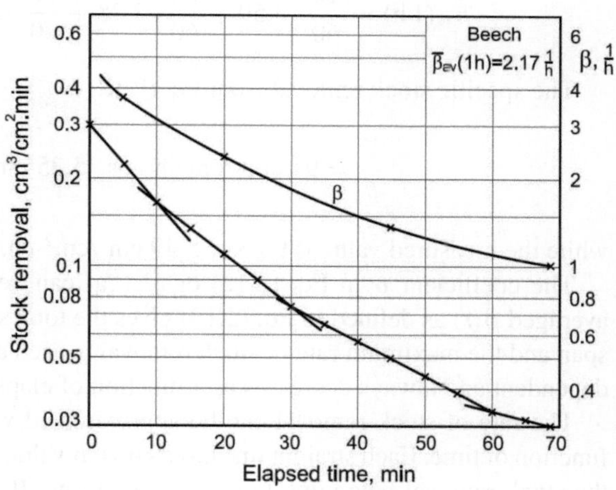

Fig. 2.11 Stock removal and time constant for beech wood as a function of working time (calculated from Dobrint 1991)

Here it is clear that the rate of decrease in stock removal, only in individual sections, may be regarded as constant. An equivalent average time constant can be calculated using the particular weighted values

$$\bar{\beta}_{\mathrm{av}}(t_0) = \sum_{i=1}^{n} \beta_i \frac{t_i}{t_0} \tag{2.13}$$

where t_0 means the total (or related) sanding time.

Example. Taking 1 h sanding time and the results shown in Fig 2.11, and with $p = 0.25$ N/cm^2, $v_c = 20$ m/s, $L_c = 17$ cm, $B_0 = 5.08 \times 10^{-8}$ 1/cm. From the figure, we can read:

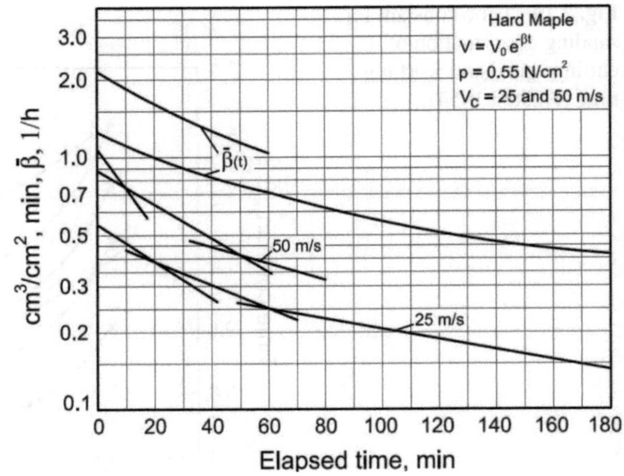

Fig. 2.12 Rate of stock removal and the averaged time constant $\bar{\beta}$ as a function of elapsed time (calculated from Hinken 1954)

$$\bar{\beta}_{\text{av}}(1\,\text{h}) = \frac{10}{60} \cdot 3.59 + \frac{20}{60} \cdot 2.28 + \frac{30}{60} \cdot 1.63 = 2.17 \quad 1/\text{h}$$

The specific stock removal from Eq. (2.12) is

$$\frac{V}{b} = 91\ \text{cm}^3/\text{cm} \cdot \text{h} \quad \text{or} \quad 5.35\ \text{cm}^3/\text{cm}^2 \cdot \text{h}$$

while the measured value is 93.3 or 5.49 $\text{cm}^3/\text{cm}^2 \cdot \text{h}$.

The coefficient β in Eq. (2.12) or (2.12a) can be used in different ways. The averaged $\bar{\beta}(t)$ as defined in Eq. (2.13) gives the total stock removal for a given time span and the maximum rate of stock removal is used at $t = 0$. In this case, $\bar{\beta}$ is time dependent and always decreases as a function of elapsed time, Fig. 2.12.

The rate of stock removal can be approximated with several straight lines as a function of time. Each straight line has its own β value which can be used to calculate the stock removal belonging to each straight line. If the first steep section is short enough, the long main section extrapolated to $t = 0$ gives the total stock removal with acceptable accuracy.

Due to the fact that researchers have different experimental conditions such as length of sanding time, length of belt, cutting velocity and platen pressure, the comparison and correct interpretation of the results is difficult. Especially the wear rate of the belts in the various investigations shows very different picture. The main reason is the different lengths of the belts, Fig. 2.13, which fundamentally alter the true contact time. The *contact time* is defined as that time during which a given area of a belt is in contact with the workpiece.

Therefore, the elapsed time t should be replaced by the reduced time t_r as

$$t_r = \frac{l}{L} \cdot t$$

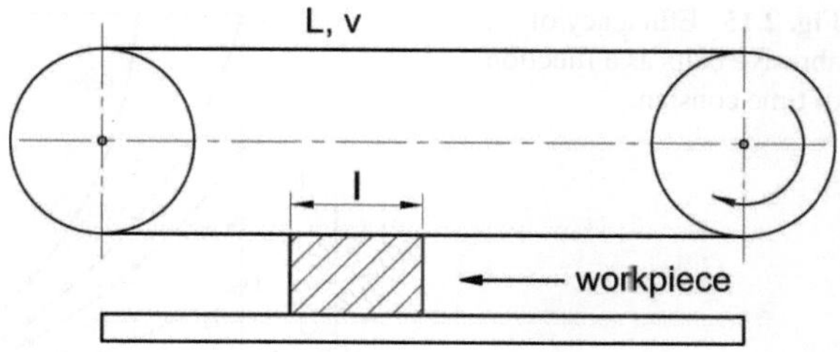

Fig. 2.13 Working principle of a belt sander

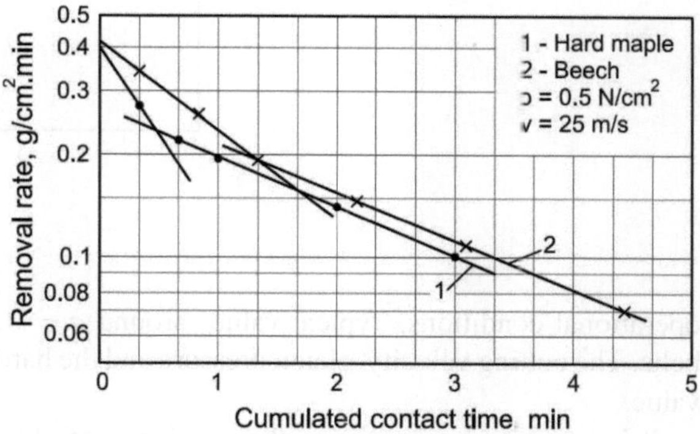

Fig. 2.14 Specific stock removal as a function of true contact time. $v_c = 25$ m/s, $p = 0.5$ N/cm^2, Al-oxide P120

where l is the length of workpiece sanded and L is the length of the belt.

Using the contact time, measurement results with very different L/l ratios (60 and 11.4) became quite comparable, Fig. 2.14 (Hinken 1954; Dobrint 1991). The material removal rate can be rewritten in the following form

$$k = k_0 \cdot e^{-a\frac{l}{L}t} = k_0 \cdot^{-\beta t} \quad \text{with} \quad \beta = a\frac{l}{L} \tag{2.14}$$

where a means the tangent of the curves. The short initial part of the curves is generally steep, with values between $a = 0.6$ and 1.1. The second steady part of the wear curve has values in a narrow range around $a = 0.33$. The beech wood in Fig. 2.11 shows $k_0 = 0.3$ g/cm$^2 \cdot$ min, $L_c = 16$ cm, $\rho = 0.72$ g/cm^3, $L/l = 11.4$ and $a = 0.33$ 1/min. Neglecting the small difference in the initial section, Eq. (2.12a) gives $V/b = 189.7$ cm^3/cm stock removal for 1 h = 60 min operational time. Please note that the true contact time is only 60/11.4 = 5.26 min!

It is important to note that to reduce belt wear and increase rate of stock removal, it is better to employ lower speeds and higher platen pressures. This removes more material with fewer cuts. As a consequence, the contact distance is reduced contributing to a longer belt life.

Practical observations show that in the initial time, the stock removal rate is proportional to the platen pressure. Later, as the belt wears, some deviations should be accounted for. The hardness of wood also influences the wear rate of the belt. The constant a in Eq. (2.14) characterizes the wear rate of a belt under the given

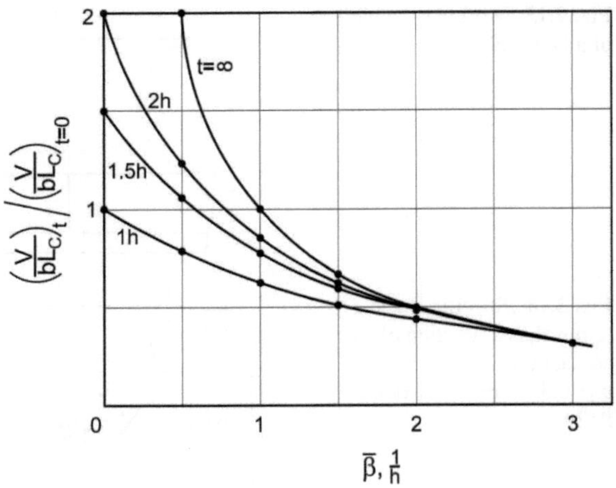

Fig. 2.15 Efficiency of abrasive belts as a function of time constant

operational conditions. Typical values around $a = 0.33$ are common for Al-oxide belts. The cutting velocity, platen pressure and the hardness of wood will modify this value.

It is interesting to examine the efficiency of abrasive belts in a given sanding operation. The main influencing factor obviously is the time constant $\bar{\beta}$. Its efficiency is measured by the effective stock removal rate at a given time related to the initial stock removal rate as a function of time constant $\bar{\beta}$. This relationship is depicted in Fig. 2.15.

If $\bar{\beta}$ is higher than 1.5, then the efficiency is not more than 60% and the practical working time is around 1 h. In the above example for beech wood, the efficiency is only 43.4% due to the high value of $\bar{\beta}$. A moderate cutting speed and higher platen pressure may ensure a better efficiency. Indeed, the time constant $\bar{\beta}$ continuously decreases with increasing platen pressures, Fig. 2.16 (Hinken 1954).

The use of aluminium oxide abrasives for sanding solid wood is becoming general and well established. Comparative experiments have shown that silicon carbide abrasives with their much higher heat conduction coefficient may have advantages for sanding MDF and particleboards containing adhesives. At high temperatures, the adhesive may soften, causing smearing and clogging (Dobrint 1991). Figure 2.17 shows experimental results for sanding MDF with P-120 silicon carbide abrasive processed and replotted from (Dobrint 1991).

Generally, Fig. 2.16 shows the same picture as Fig. 2.12, and the time constant $\bar{\beta}$ decreases considerably as a function of elapsed time. The stock removal is higher than for beech wood and has a value of 9 $cm^3/cm^2 \cdot h$. Using aluminium oxide abrasive, the same value was only 6.2 $cm^3/cm^2 \cdot h$ which is 69% compared to a SiC abrasive. Sanding particleboard in the same experiments, the specific stock removal was 6.3 and 2.4 $cm^3/cm^2 \cdot h$ for silicon carbide and aluminium oxide, respectively. With adhesives and high resin content, silicon carbide abrasives definitely surpass aluminium oxide abrasives.

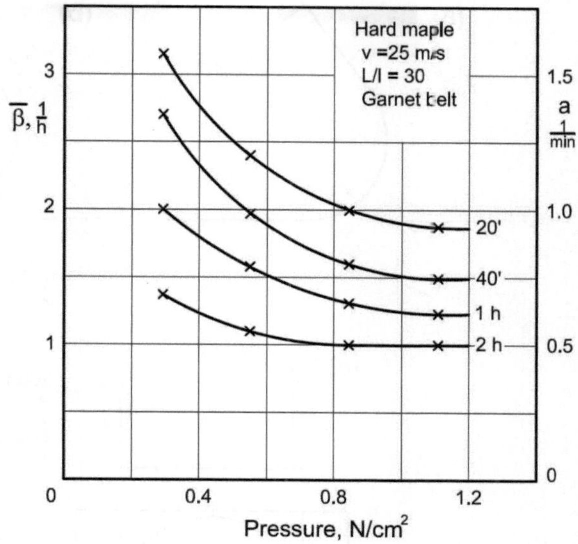

Fig. 2.16 Time constant versus platen pressure for different operational times (calculated from Hinken 1954)

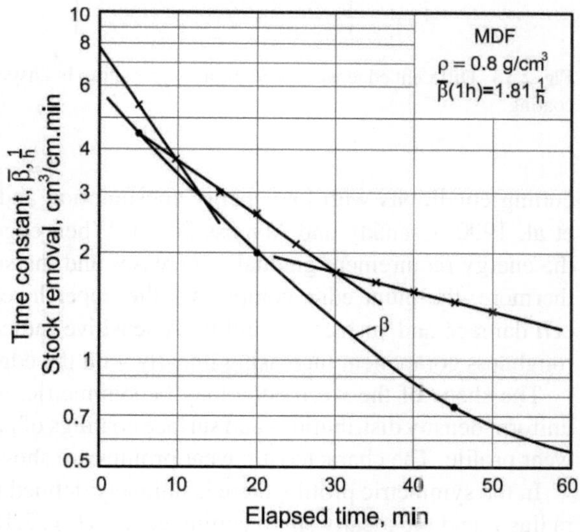

Fig. 2.17 Stock removal and time constant for MDF and using silicon carbide abrasives. Length of sanding is 16 cm (calculated from Dobrint 1991)

2.5 The Tool Life Equation

The woodworking industry generally uses high cutting speeds, and therefore, the main wear mechanisms are abrasion and high temperature corrosion. The initial edge radius of a new or newly sharpened tool generally fluctuates between 10 and 15 μm depending on the hardness of the tool material. In the initial stage of tool wear, an optimum edge shape develops with an approximate radius of 20 μm. In the following "working sharp" stage between 20 and 40 μm, the edge ensures optimum

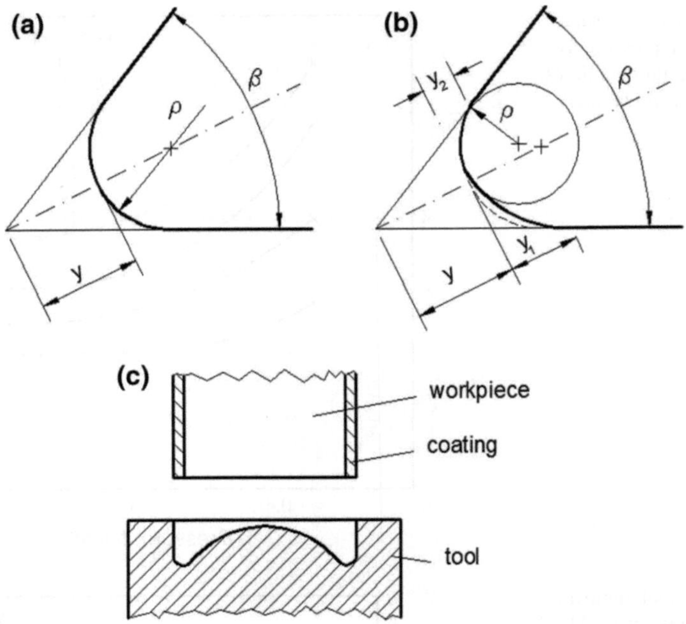

Fig. 2.18 Different edge wear profiles. **a**—symmetric, **b**—asymmetric, **c**—wear due to workpiece coating

cutting conditions with low energy consumption and good surface quality (Sitkei et al. 1990; Csanády and Magoss 2013). When edge radii increase over 40 μm, the energy requirement gradually increases and the surface quality decreases. Furthermore, the blunt edge compresses the upper layers of the cut surface causing cell damage and surface instability. A sensitive indicator for blunt edges is the R_k roughness component increasing linearly with the edge radius.

The shape of the worn edge may be symmetric, asymmetric or irregular. Non-uniform density distributions and surface coatings of particleboard cause an irregular wear profile. The characteristic wear profiles are shown in Fig. 2.18.

In the symmetric profile, there is uniquely defined relationship between the edge radius ρ and the ideally sharp cutting edge y (Fig. 2.18):

$$\rho = y\frac{\sin(\beta/2)}{1 - \sin(\beta/2)}$$

where β is the sharpening angle. For the common case with $\beta = 55°$, we get $\rho = 0.85y$.

The tool wear can be related to the feed distance (L_f) or the true cutting length (L_c) which are interrelated by the following equation

$$L_c = L_f \frac{R \cdot \varphi}{e_z \cdot z} = L_f \cdot \frac{1.42 \cdot \sqrt{H \cdot R}}{e_z \cdot z} \tag{2.15}$$

where R is the tool radius,

e_z is the tooth bite,
z is the number of teeth,
φ is the rotation angle of cutting in radian.

The edge wear mainly depends on the cutting speed v, and the true cutting length. The edge reduction can be expressed by the following empirical equation

$$y = y_0 + A \cdot v^n \cdot L_c^m \tag{2.16}$$

where y_0 is the initial edge reduction after grinding with an average value of 20 μm. The constants A, n and m should be determined experimentally for each tool material and work material. For example, using tungsten carbide tool and particleboard, the constants in Eq. (2.16) have the following values: $A = 0.0245$, $n = 0.78$ and $m = 0.65$, where the edge reduction is obtained in μm but v and L_c are substituted in m/s and m, respectively.

Considering the lifetime T of a tool between two sharpenings, the total feed distance during its lifetime is given by

$$L_{fT} = e_z \cdot n \cdot z \cdot T$$

while the true cutting distance during its lifetime for n individual tooth amounts to

$$L_{cT} = n \cdot R \cdot \varphi \cdot T = \frac{60}{2 \cdot \pi} \cdot v \cdot \varphi \cdot T \tag{2.17}$$

From Eq. (2.16), we get also an expression for L_c

$$L_c = \left(\frac{y - y_0}{A}\right)^{1/m} \cdot \frac{1}{v^{n/m}} \tag{2.16a}$$

Comparing Eqs. (2.16a) and (2.17) yields

$$T \cdot v^{\frac{n}{m}+1} = \left(\frac{y - y_0}{A}\right)^{1/m} \cdot \frac{0.1047}{\varphi} \tag{2.18}$$

which is one form of the Taylor tool life equation and for a given case the right side of Eq. (2.18) is constant. Using the value of constants in the above example, the Taylor tool life equation can be written as

$$T \cdot v^{2.2} = 91957 \quad \text{or} \quad v \cdot T^{0.4545} = 180.4$$

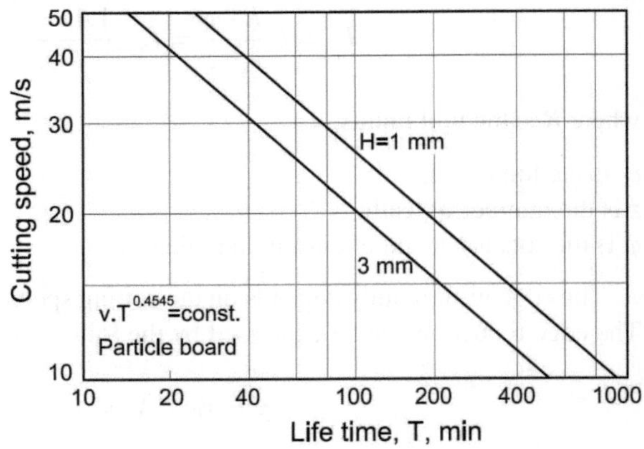

Fig. 2.19 Taylor's lifetime curve

which is plotted in Fig. 2.19.

The constant on the right side depends on the rotation angle of cutting φ which is the function of the relative cutting depth

$$\varphi = 1.42\sqrt{\frac{H}{R}} \quad \text{radian} \tag{2.19}$$

Knowing the relation between edge reduction y and edge radius ρ (appr. $\rho = 0.85y$), in Eqs. (2.16) and (2.18) ρ and ρ_0 can also be used.

It is interesting to note that the true cutting length does not depend on the depth of cut H but the feed distance does. In practice, the use of feed distance is more convenient and it can be expressed using Eq. (2.18) for T

$$L_{fT} = e_z \cdot n \cdot z\left(\frac{y - y_0}{A}\right)^{1/m} \cdot \frac{0.1047}{\varphi \cdot v^{\frac{n}{m}+1}} \tag{2.20}$$

The cutting speed v is directly proportional to the rotation speed n and, therefore, the feed distance is inversely proportional to $v^{n/m}$ which is depicted in Fig. 2.20 using $e_z = 1$ mm, $z = 4$ and $R = 60$ mm.

The depth of cut considerably influences the feed distance. Changing the rotation speed into cutting speed, Eq. (2.20) takes the form

$$L_{fT} = \frac{e_z \cdot z}{R \cdot \varphi \cdot v^{n/m}} \cdot \left(\frac{y - y_0}{A}\right)^{1/m} \quad \text{m} \tag{2.20a}$$

with

$$R \cdot \varphi = 1.42 \cdot \sqrt{H \cdot R}$$

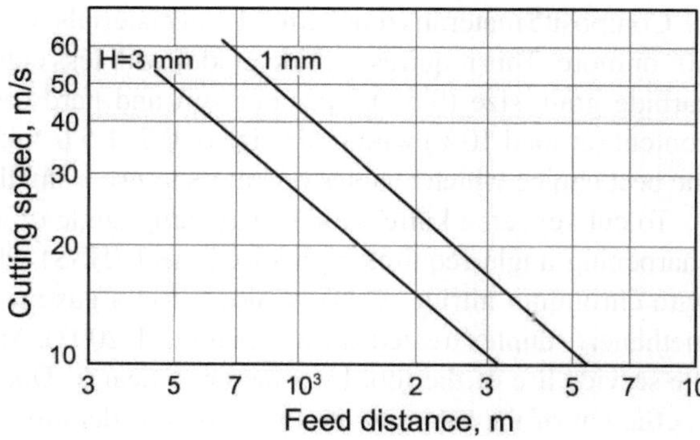

Fig. 2.20 Relationship between cutting speed and feed distance

which can easily be used in optimization too. In the above examples, the wear limit was taken as $y_0 = 100$ μm ($\rho = 85$ μm). The choice of the maximum allowable wear limit is guided by the requirements for surface quality, especially roughness parameters. There are strong correlations between tool edge radius and roughness parameters, mainly the core depth R_k (see in Figs. 2.58 and 2.59).

Excess vibrations and circle path deviation of cutting edges will considerably shorten the tool life. The resin content, the density of wood and the quantity of adhesives in composite boards alter the exponents in Eq. (2.16) and modify the relationship between cutting speed and feed distance to a certain extent. Furthermore, the tool material (carbide grain size) may also affect the wear characteristics (Sugihara et al. 1979; Salje and Dubenkropp 1983).

Tools require a careful maintenance to ensure an accurate edge geometry and running circle. Another requirement is to avoid excessive heat loads when grinding which may cause permanent tensile stresses in the surface and undesirable changes in the structure of the metal. This means an abnormal wear process is taking place rapidly blunting the edge.

In order to achieve cost-effective production, it is crucial to use cost- and edge life-optimized tools. The optimum edge life may be quite different Today the use of tungsten carbide tools is the most common practice. If a carbide-tipped circular saw in panel sizing has an edge life of one shift, it is convenient for the operator to be able to change the tool during a shift change. An extension in tool life should be doubled to ensure the same convenience and to avoid non-productive times. When the working time is not bound to a shift change, arbitrary improvements in tool life may be used.

A wide range of tungsten carbide tool materials are available to the woodworking industry for many different uses. The different carbide grades are composed of various percentages of tungsten carbide and a metallic binder (cobalt). Furthermore, these tool materials utilize a range of carbide grain sizes. The most common grain size is about 2 μ, but there are grades in the submicron range (Feld et al. 2005; Garcia 2005). Varying the carbide–binder ratio and the grain size, tool materials with different mechanical properties are made to ensure optimum performance when machining different types of wood and composites (hard and softwoods, MDF, particleboard, etc.).

Composite materials require hard tool materials, with sharpening angles of around 60° or more. This requires carbide grades with less cobalt binder (2.5–3%) and small carbide grain size (0.5–0.7 μ). For soft and hardwoods, grades with high binder content (around 10%) with grain size of 0.7–1.5 μ with good fracture resistance are the best choice which enables operators to use a smaller sharpening angle.

To cut veneer, a knife with a sharpening angle of around 20° is used. The small sharpening angle requires high-speed steel (HSS) which is generally hard coated with chromium nitride (CrN) or nitrided in a gas mixture of nitrogen, oxygen and methane, (a duplex treated knife) (Chala et al. 2003). An optimum treatment increases the service life of the tool by a factor of near 2. Due to the treatment, the friction coefficient of the tool may be lowered considerably. Furthermore, the lifetime can also be extended by using a back micro-bevel angle. A smaller sharpening angle removes less heat from the edge (Csanády and Magoss 2013). As a consequence, one should reckon with an increasing wear rate resulting in a higher exponent m in Eq. (2.16). A further condition for an adequate lifetime is a uniform running circle of the edges (Tröger and Läuter 1983). Deviations in the running circle mean different tooth bites for certain edges resulting in varying surface waviness and excessive vibrations due to self-exciting which may further enhance unequal tooth bites. A possible solution is to use jointing which minimizes the running deviation of edges. Jointing may have another beneficial effect. It makes the edge sharper and decreases the compaction of neighbouring layers due to the cut of bottom convex part of the edge. As a consequence, it reduces the surface roughness, especially the core roughness component R_k.

An interesting suggestion is to create a "self-sharpening" edge, whose principle has been used for a long time in agricultural engineering as the self-sharpening ploughshare (Itaya and Tcuchiya 2003). A thin layer of about 10 μ made of chromium nitride is laid down on the clearance face of the knife. Since coated layer is harder than the base metal, it wears more slowly and the edge will be kept sharp due to the thin hard layer. The knife of a Japanese planer is also covered with a thin layer of special steel, which can be sharpened to an extremely small edge radius.

Diamond and diamond-coated tools are also in use in the wood industry for heavy woodworking operations. They are expensive and sensitive to breakage, and therefore, their economic use requires detailed considerations.

2.6 Roughness Relationships

2.6.1 Roughness Characterization

Engineered wood surfaces have a very complex topography, and they are influenced by countless factors related to their anatomical properties and machining conditions. This surface topography cannot be described and characterized by few parameters. In the history of surface roughness measurement, many roughness parameters have been proposed and introduced in national and international standards (Csanády et al. 2015; Csanády and Magoss 2013; Dong et al. 1994; Thomas 1981; Abbott and Firestone 1933).

Roughness plays an important role in several fields of industrial technology with different requirements and importance. In the metal industry, the surface-bearing capacity and lubrication performance are greatly influenced by roughness. In the wood industry, there are no sliding surfaces and this property is much less important or negligible. The aesthetic properties of wood are very important. The surface finish and gloss fundamentally depend on the surface roughness (Csanády et al. 2015; Richter et al. 1995).

For roughness measurements, first the two-dimensional (2D) measuring and evaluating system has been developed mainly using a stylus instrument. The geometric dimensions of the stylus (needle) may be a limiting factor. Measurement results of the 2D system are well established and interpreted (Csanády et al. 2015; Csanády and Magoss 2013).

In the last decade, an increasing need has arisen for a more detailed characterization of surface topography in three dimensions (3D). Some of the 3D parameters are extended from the previously defined 2D counterparts, while others are newly defined. The proposed parameter set was mainly determined to meet the demands of the metal industry (Thomas 1981; Dong et al. 1994). Optical devices are mainly used to perform 3D measurements.

The optical 3D measurement method has undoubted advantages but in certain cases, it may supply unacceptable results, probably due to surface reflections. Figure 2.21 shows comparative results measured with a stylus and optical system using different wood species with planed and sanded surfaces. Especially, high-density species show deviations up to 300% which make it difficult to evaluate 3D optical measurements. Generally, planed surfaces are more prone to deviations due to their higher reflectivity. Until this problem is solved, it is wise to check the probable validity of optical measurement results.

Furthermore, optical devices restructure the Abbott parameters S_{pk}, S_k and S_{vk}, relative to each other as compared to the appropriate R_{pk}, R_k and R_{vk} values.

Due to its definition, S_z is dominated by extreme values and it is hardly comparable with its R_z counterpart.

The following roughness parameters will most commonly be used to evaluate wood surfaces:

Parameter	Symbol
Average peak-to-valley height	R_z
Maximum peak-to-valley height	R_{max}
Average (integrated) roughness	R_a
Root-mean-square (RMS) roughness	R_q
Skewness (the third moment of profile amplitudes)	R_{sk}
Kurtosis (the fourth moment of profile amplitudes)	R_{ku}

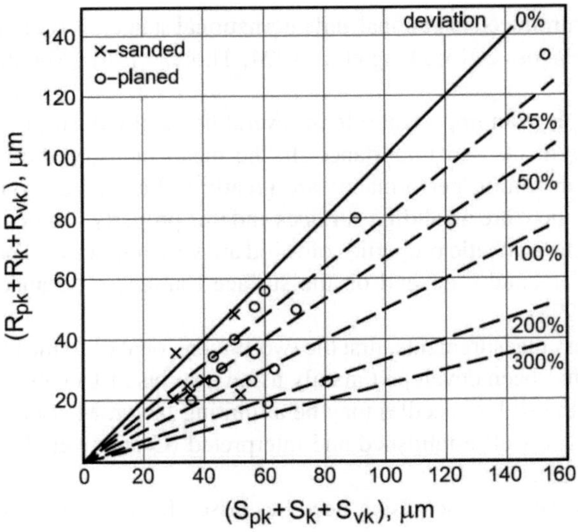

Fig. 2.21 Comparison of measurement results for a 2D Perthometer and a 3D optical system. Each point represents a given wood species

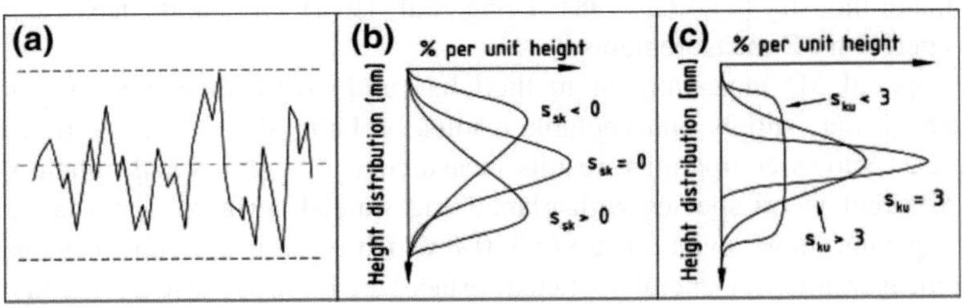

Fig. 2.22 Prifila (**a**). Height distributions for skewness (**b**) and kurtosis (**c**)

Figure 2.22 explains the possible variations of skewness and kurtosis.

Since 1933, the Abbott–Firestone parameters have been used to characterize the material and void distribution on rough surfaces.

These parameters are the following (Abbott and Firestone 1933):

Parameter	Symbol
The average height of protruding peaks	R_{pk}
Vertical difference in the core section	R_k
The averaging height of valleys	R_{vk}
Material volume above the upper core surface	M_{r1}
Void volume under the lower core surface	M_{r2}

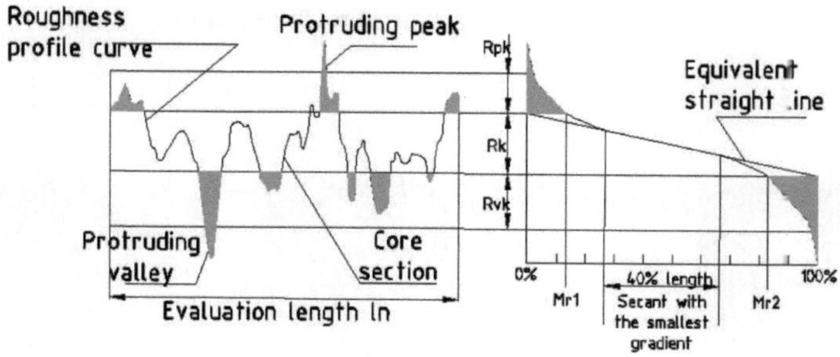

Fig. 2.23 Abbott–Firestone curve and its related parameters

Figure 2.23 shows the Abbott–Firestone curve and the meaning of individual parameters.

In contrast with other opinion, the R_q/R_a ratio is an important factor and has a strong correlation with the R_{sk} skewness. Therefore, both R_a and R_q are useful roughness parameters in the wood industry (Csanády et al. 2015).

The Japanese planer has been used for many centuries, and its ability to produce smooth surface is also well known but measurements of its effects on surface quality are not known. Samples of several wood species were planed with different chip thicknesses between 15 and 80 μm. For comparison, sanded surfaces with P-240 were also prepared.[1] A selection of roughness profiles is shown in Fig. 2.24. The Japanese planer produces extremely clean surfaces, and the repeating early and latewood are clearly seen. The same fine-sanded sample shows no signs of early or latewood.

A clean surface has excellent gloss properties, which also effect surface treatment (see in Sect. 2.12). No other woodworking operation can compete with the Japanese hand planer in this respect. It is true that curved surfaces, such as turned bowls, cannot be produced by the planer. They require different polishing techniques.

2.6.2 Anatomical Characterization of Wood

The characteristic feature of wood materials is their porous structure which is reflected in their physical properties and also in their surface topography after machining. The texture of wood is highly anisotropic and inhomogeneous with great variability within one species depending on its growing conditions. Furthermore, there is a difference between softwoods with greater uniformity and hardwoods having greater structural variations in their physical properties, colour, density and strength. Finally, all wood species have early and latewood with different density, void ratio and colour.

[1] Samples were supplied by the Woodworking Studio of Werner Weis, Pfaffenhofen, Germany

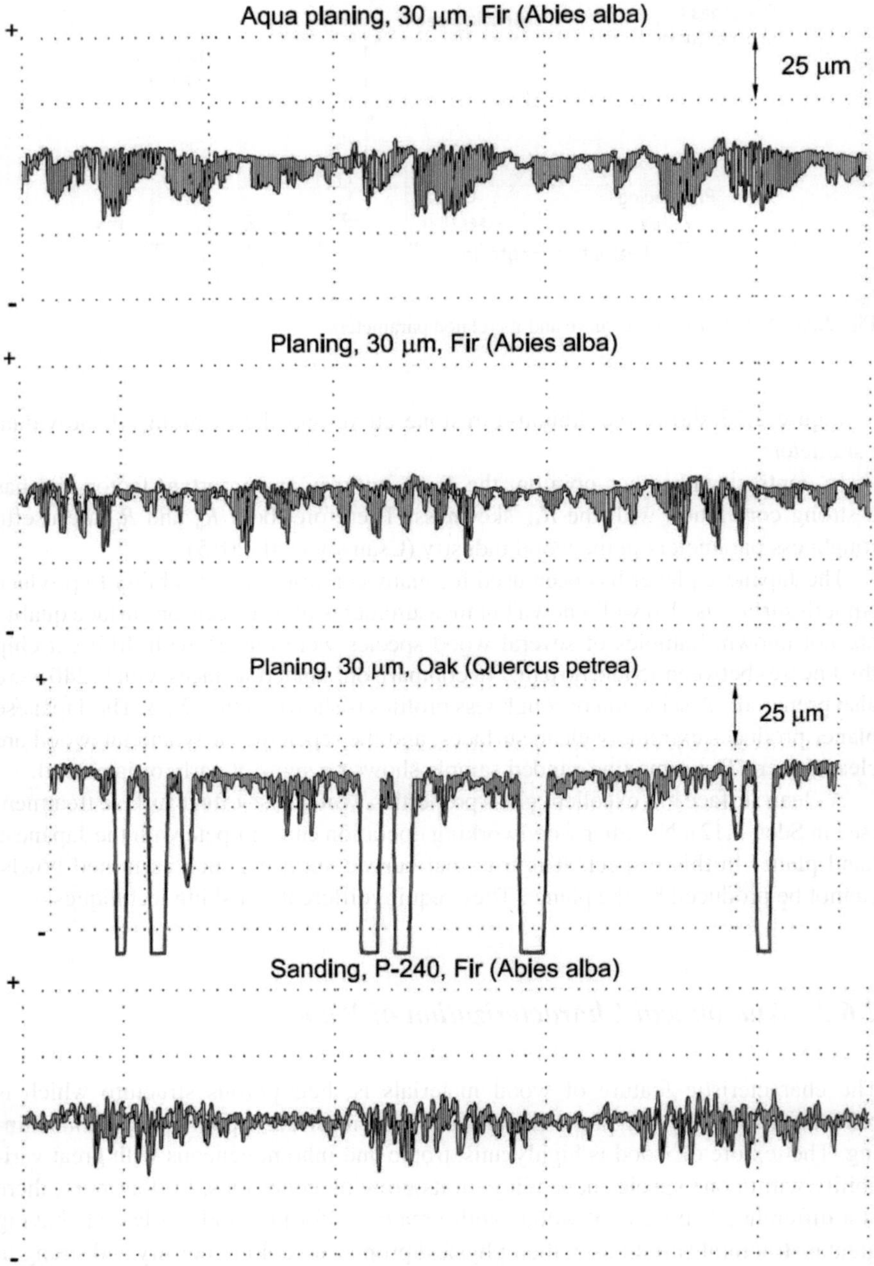

Fig. 2.24 Surface roughness profiles produced by a Japanese planer. Chip thickness is 30 μm

The most important anatomical characteristics for measuring surface roughness are:

- the diameter of vessels and tracheids in the early and latewood,
- the specific number of vessels and tracheids related to the unit cross-section in the early and latewood,
- the portion of early and latewood within an annual ring.

Using these anatomical properties, the introduction of *Structure number* ΔF proved to be very useful in recognizing general relationships, independently of wood species (Magoss and Sitkei 2000, 2001). The structure number is calculated in the following way:

$$\Delta F = \frac{\pi}{8}\left[a \cdot \left(\sqrt{n_1} \cdot d_1^2 + \sqrt{n_2} \cdot d_2^2\right) + b \cdot \left(\sqrt{n_3} \cdot d_3^2 + \sqrt{n_4} \cdot d_4^2\right)\right] \quad [\text{mm}^2/\text{cm}] \tag{2.21}$$

where

d_1, d_2 the average diameter of vessels and tracheids in the earlywood,
d_3, d_4 the average diameter of vessels and tracheids in the latewood,
n_1, n_2 the number of vessels and tracheids in the earlywood in the unit cross-section,
n_3, n_4 the number of vessels and tracheids in the latewood in the unit cross-section,
d_1–d_4 the diameter of vessels and tracheids in the early and latewood, respectively,
a, b portions of early and latewood, $(a + b = 1)$.

The value ΔF defined with the Eq. (2.21) gives an accurate definition of each wood species based on the size and specific number of cavities in the wood structure. Accordingly, surface irregularities caused by the internal wood structure are expected to have a definite correlation with their structure number.

In order to calculate the structure number ΔF, the pertinent anatomical characteristics of the given wood species must be known. The measurement of these characteristics is time-consuming and, therefore, finding some interrelation between the diameter and specific number of vessels and tracheids will facilitate the quick estimation of its structure number.

There is a good correlation between the diameter of vessels or tracheids and their specific number. This correlation for European conifers and hardwood species is depicted in Fig. 2.25. Their distribution is hyperbolic (a straight line on a double-logarithmic scale), and conifers have a slightly wider tracheid diameter for the same specific number in the unit cross-section. This diagram is also suitable to check new measurement data.

The curves are well described with the following equation:

$$\text{for hard woods } d = 3300 \cdot N^{-0.43} \ [\mu\text{m}]$$
$$\text{for conifers } d = 4230 \cdot N^{-0.43} \ [\mu\text{m}]$$

where N means the specific number of vessels or tracheids.

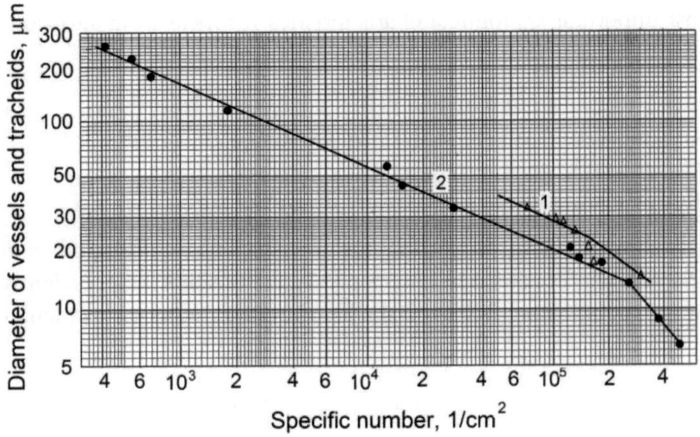

Fig. 2.25 Correlation between the diameter of vessels and tracheids and their specific number in the unit cross-section. 1—Conifers; 2—hardwoods

Using the above equation, the contribution of vessels in the early and latewood can be expressed as follows

$$\Delta F_i = a \cdot 1.574 \cdot d_i^{0.8372} \quad \left[\mathrm{mm^2/cm}\right]$$
$$\Delta F_i = b \cdot 1.574 \cdot d_i^{0.8372} \quad \left[\mathrm{mm^2/cm}\right]$$

where a and b mean the portion of early and latewood, and d_i diameter must be substituted in mm. The relationship is demonstrated in Fig. 2.26.

Example: In oakwood, the average vessel diameter is 260 µm in the earlywood and 60 µm in the latewood. The early and latewood portions are $a = 0.59$ and $b = 0.41$. Their expected contributions are $0.3007 + 0.0612 = 0.362$ mm²/cm.

Detailed calculations show that the contribution of tracheids to the structure number is from 10 to 15%. Taking the average value of 10%, oak should have a structure number of $\Delta F = 0.402$ mm²/cm.

When measuring roughness, the number of cut vessels in the trace length (generally 12.5 mm) fundamentally determines several roughness parameters (R_a, R_q, R_z, R_{vk}, etc.). On a bigger surface, the number of cut vessels may be quite different and, therefore the measurement results may also vary to an unexpected extent. Since the position of a machined surface in relation to the position of vessels is always accidental, the number and depth of cut vessels left on the surface are also accidental.

The most common values of structure number for selected wood species are summarized in Table 2.2.

The simple use of the number of cut vessels in the trace length though shows the tendency of roughness variation but with considerable scatter, Fig. 2.27.

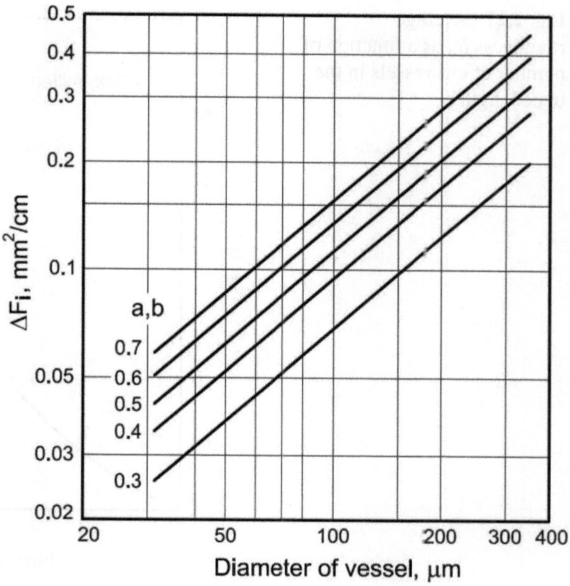

Fig. 2.26 Prediction of structure number for known vessel's diameters in the early and latewood

Table 2.2 Structure number for different wood species

Wood species	ΔF mm^2/cm
Conifers	Around 0.1
Larch	0.11–0.14
Cottonwood	0.15–0.20
Beech	0.19–0.25
Ash	0.25–0.30
Black locust	0.33–0.43
Oak	0.40–0.55

A much more accurate method is to measure the area of the cut vessels in the trace length and to plot the roughness parameter as a function of total area of cut vessels. These measurements are shown in Fig. 2.28 for milled oak. This representation shows a close correlation, and it is also suitable to read the intersection ($R_\varepsilon = 2$ μm) for zero cut vessels. This value is comparable or even better than values for conifers.

The linear relationship has the simple form:

$$R_a = 2 + 116.8 \cdot A_n \qquad (2.22)$$

where A_n means the total area of cut vessels in the roughness profile in mm^2.

An interesting comparison may be done between the average vessel cross-section (ΔF value) and the total area of cut vessels in the trace length. The expected ΔF value for oak is at least 0.4 mm^2/cm, but the maximum measured value in Fig. 2.7 is only 0.14 mm^2/1.25 cm. This finding is explained by the fact that the Perthometer

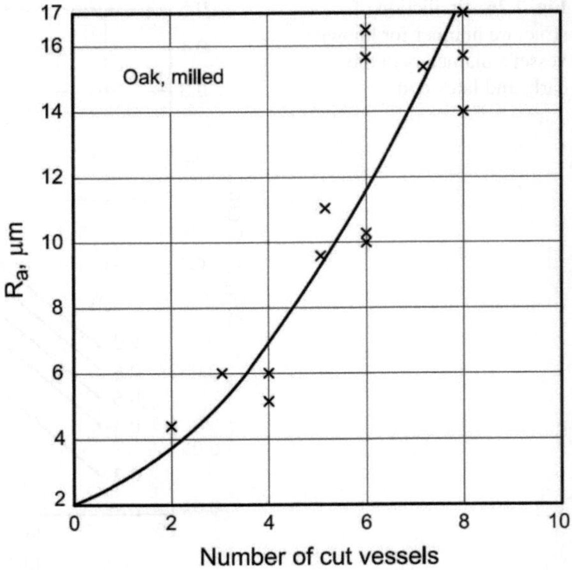

Fig. 2.27 Average roughness R_a as a function of number of cut vessels in the trace length

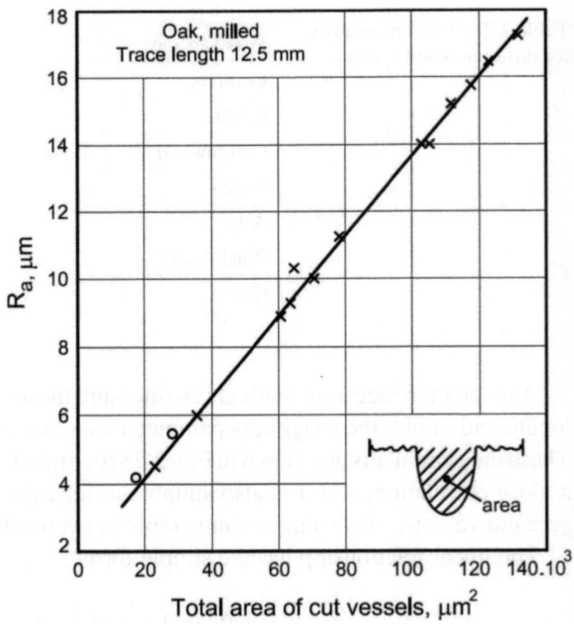

Fig. 2.28 Average roughness R_a as a function of total area of cut vessels

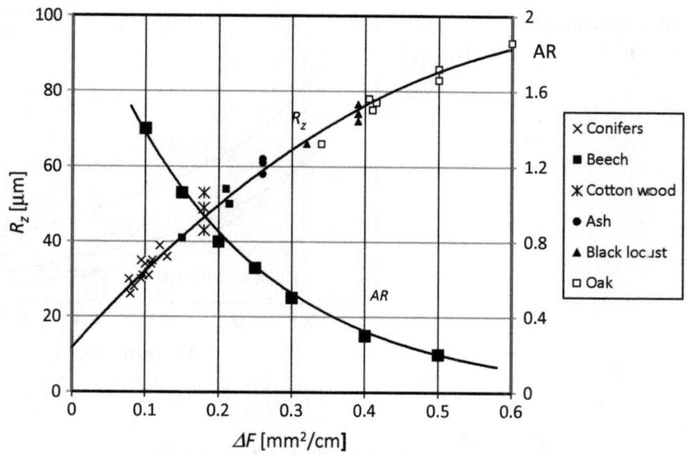

Fig. 2.29 Optimum R_z values and Abbott's ratio as a function of structure number

cannot measure the true profile of cut vessels. The average roughness is integrated from the measured profile.

The valley depth R_{vk} proportionally increases with the vessel diameter. Therefore, the *Abbott ratio* (AR) defined as $(R_{pk} + R_k)/R_{vk}$ is a further characteristic number in a dimensionless form which is distinct for each wood species. The Abbott ratio is a sensitive parameter with a wide variation. Its maximum value is around 1.6 for conifers, while its lowest value is around 0.15. It correlates well with the structure number ΔF as depicted in Fig. 2.29.

The relationship between Abbott's ratio and structure number is given by the equation

$$\frac{R_{pk} + R_k}{R_{vk}} = \frac{0.38}{\Delta F^{0.72}} - 0.4 \tag{2.23}$$

or

$$\Delta F = \left(\frac{0.38}{\frac{R_{pk}+R_K}{R_K} + 0.4} \right)^{1.39}$$

The above relationship is valid for machined surfaces using a sharp knife. Sanded surfaces show considerable deviation depending on the grit size due to surface deformation and clogging. The surface has relatively smaller valley depths R_{vk} compared to knife machining which causes higher Abbott's ratios. Figure 2.30 represents experimental results for different wood species plotted as a function of structure number. The effect of grit size is obvious, and its influence is taken into account by the following approximation:

Fig. 2.30 Effect of sanding on the Abbott ratio

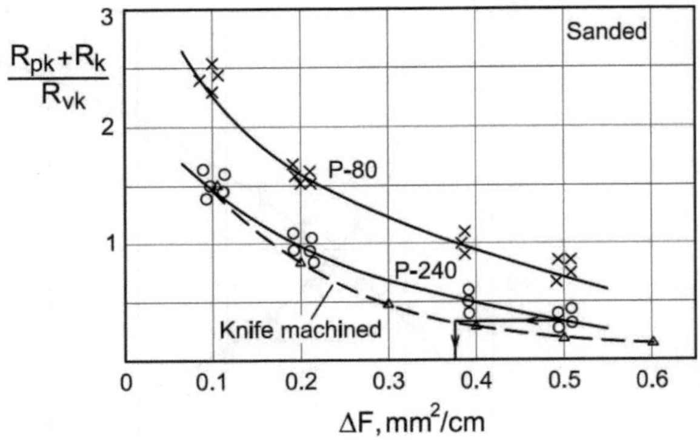

$$\frac{R_{\mathrm{pk}} + R_{\mathrm{k}}}{R_{\mathrm{vk}}} = 11.42 \cdot d^{0.4} \cdot (1 - \Delta F^{0.2}) \tag{2.23a}$$

where the average grit diameter d must be substituted in mm. The clogging effect diminishing the apparent structure number can be seen in the figure.

2.6.3 *Interrelations Among Roughness Parameters*

To generalize our knowledge of surface roughness problems, functional relationships are needed independently of wood species. Due to the high variability of wood structure, the great number of roughness parameters and possible measurement errors, this task seems to be difficult. Roughness parameters are imperfect statistical representations, and random sampling is one reason that they are prone to scatter.

Noting these difficulties, it is useful to search for possible relationships among roughness parameters using extended measurement results for many wood species. The functional variables may be single, combined or even dimensionless quantities. The obtained relationships may have different accuracy mainly depending on the relative scattering of measurement data and the proper selection of a processing technique.

The average roughness R_a and the irregularity depth R_z are linearly correlated within a given scattering zone, Fig. 2.31 (Csanády et al. 2015; Magoss 2008). Their ratio varies between 0.09 and 0.12, and the bigger ratios are generally from wood species with wider voids (vessels). It is known that R_a depends on the cross-section of voids, while R_z depends on the depth of irregularities which explains the inherent scattering of experimental results.

At the same time, their correlation is strong enough for practical uses. The average roughness R_a has a better correlation with the sum of Abbott's parameters and their correlation obeys the following empirical equation

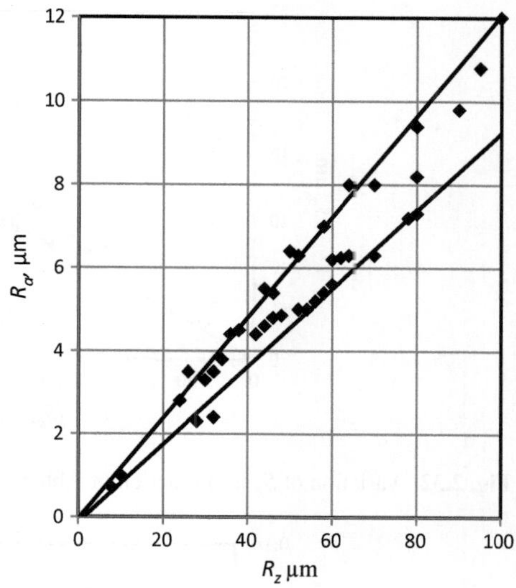

Fig. 2.31 Diagram of R_a versus R_z for different wood species

$$R_a = 0.275 \cdot \left(R_{pk} + R_k + R_{vk}\right)^{0.85} \tag{2.24}$$

The average roughness height R_z has a similar correlation function which is given by

$$R_z = 2.41 \cdot \left(R_{pk} + R_k + R_{vk}\right)^{0.85} \tag{2.25}$$

which corresponds to the averaged straight line in Fig. 2.31.

Due to the definition of S_z in the 3D system, it does not correspond to R_z and a similar relationship for S_z cannot be established. At the same time, S_a and Abbott's parameters are useful. Their correlation is depicted in Fig. 2.32 and described by the equation

$$S_a = 0.268 \cdot \left(S_{pk} + S_k + S_{vk}\right)^{0.85} \tag{2.24a}$$

which agrees well with the corresponding R_a function.

A further reason for scattering in Fig. 2.31 may be the possible variation of the core depth R_k, which is the major component of R_a. Figure 2.33 shows the variation of the R_a/R_z ratio as a function of R_k/R_z ratio for different wood species.

An increase in the R_k/R_z ratio for a given species (soft or hardwood) always enlarges the R_a/R_z ratio. Taking a given R_k/R_z ratio, hardwoods with big vessels have a higher R_a/R_z ratio compared to conifers. The curves in Fig. 2.33 can be approximated with the following empirical equation:

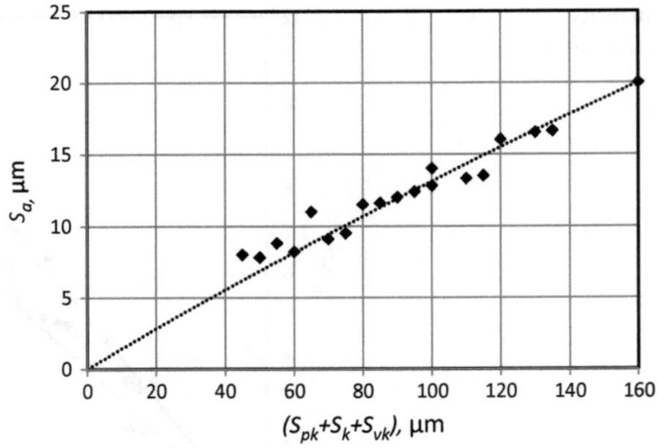

Fig. 2.32 Variation of S_a as a function of Abbott's parameters for different wood species

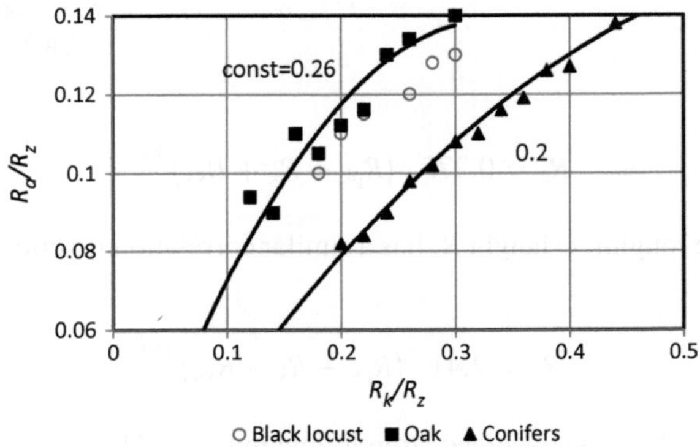

Fig. 2.33 Variation of the R_a/R_z ratio as a function of R_k/R_z ratio for different wood species

$$\frac{R_a}{R_z} = \text{const} \cdot \sqrt{\frac{R_k}{R_z}} \quad \text{or} \quad \frac{R_a}{\sqrt{R_k \cdot R_z}} = \text{const}. \tag{2.26}$$

The obtained result is a dimensionless number which may be regarded as a similarity number characterizing the machining process for a given wood species with respect to the resulting roughness. Figure 2.33 shows that the constant on the right side may vary with the presence and size of vessels.

Figure 2.34 shows a general relationship between the Abbott ratio and the related average roughness of different oak samples, other hardwood species and conifers.

Conifers lacking vessels have an almost constant-related average roughness independently of their Abbott's ratio. For hardwoods, the following correlation equation is valid:

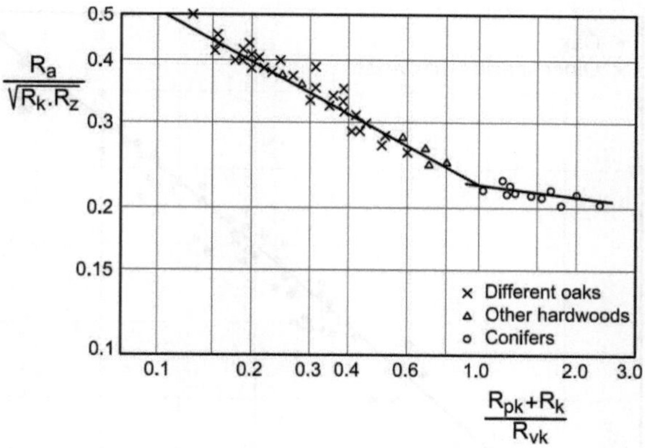

Fig. 2.34 Similarity relationship between Abbott's ratio and the related average roughness

$$\frac{R_a}{\sqrt{R_k \cdot R_z}} = \frac{0.23}{\left(\frac{R_{pk}+R_k}{R_{vk}}\right)^{0.34}} \tag{2.27}$$

Equations (2.23) and (2.27) can be combined to obtain an equation as a function of structure number ΔF. A simpler approximation may be used in the form

$$\frac{R_a}{\sqrt{R_k \cdot R_z}} = 0.61 \cdot \Delta F^{0.6} \tag{2.28}$$

For conifers, the related average roughness $R_a/\sqrt{R_k \cdot R_z}$ may be regarded as constant with an average value of 0.22.

In hardwoods with vessels, the reduced valley height R_{vk} more or less proportionally increases with R_z. Therefore, R_z may be exchanged for R_{vk} in the dimensionless number of average roughness. In this case, Eq. (2.27) will be replaced by the following relationship

$$\frac{R_a}{\sqrt{R_k \cdot R_{vk}}} = \frac{0.33}{\left(\frac{R_{pk}+R_k}{R_{vk}}\right)^{0.32}} \tag{2.29}$$

which has an even better correlation compared to Fig. 2.34. Rearranging the above equation yields

$$R_a = \frac{0.33 \cdot R_k^{0.5}}{\left(R_{pk} + R_k\right)^{0.32}} \cdot R_{vk}^{0.82}$$

and with further simplification

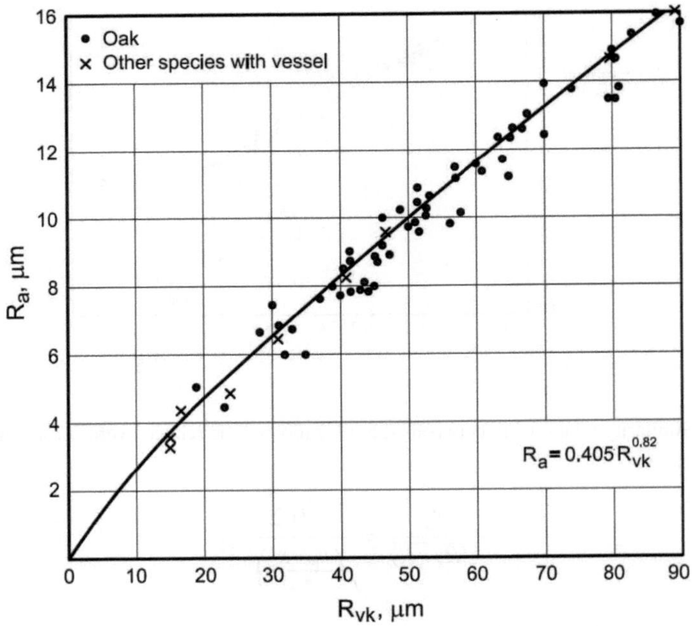

Fig. 2.35 Average roughness versus reduced valley height for oak and other hardwood species

$$R_a = 0.41 \cdot R_{vk}^{0.82} \tag{2.30}$$

The plot of experimental results and Eq. (2.30) are given in Fig. 2.35. The measured points, for oak and several other hardwood species, follow Eq. (2.30) with acceptable accuracy.

In the past, the wood industry did not commonly use height distribution parameters of *skewness* and *kurtosis*. They deserve more attention because they can provide useful information about the topography of a wood surface. As we have seen in Fig. 2.22, a negative skewness means that the peak of the height distribution is located near the surface. A kurtosis higher than 3 means a more spiked distribution.

Results with different wood species are summarized in Fig. 2.36. There is a close correlation between skewness and kurtosis, independent of wood species, which obeys a quadratic equation with the form

$$R_{ku} = 3 + 1.48 \cdot R_{sk}^2 \tag{2.31}$$

Conifers have little skewness and kurtosis and that means that the height distribution nearly corresponds to their normal distribution. With increasing hardness, both skewness and kurtosis will increase considerably.

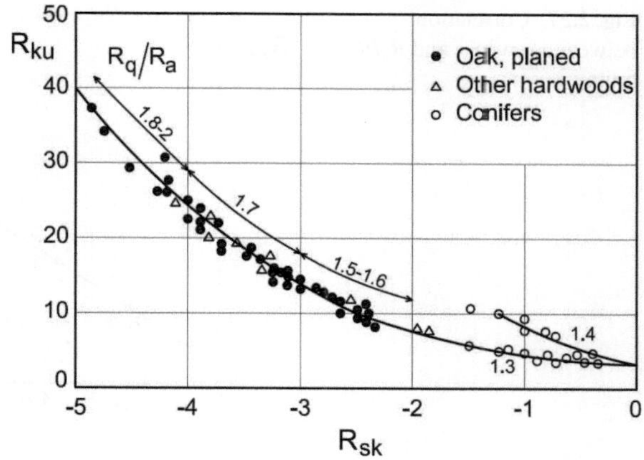

Fig. 2.36 Interrelation between skewness and kurtosis for different wood species

The skewness and kurtosis relationship is strongly connected to the R_q/R_a ratio. As skewness and kurtosis increase, the corresponding R_q/R_a ratio also increases. A detailed analysis of the R_q/R_a ratio (Csanády et al. 2015) shows a close connection to kurtosis: for more spiked height distributions, the R_q/R_a ratio uniquely increases. On a very dense (1.2 g/cm^3) rosewood (Dalbergia cochinchinensis), $R_c/R_a = 2.4$, $R_{sk} = -6$ and $R_{ku} = 45.7$ values were measured. Figure 2.36 shows that there is more kurtosis for the same skewness when the R_q/R_a ratio increases. This phenomenon was mainly observed in conifers.

The correlation between skewness and R_q/R_a ratio is depicted in Fig. 2.37. Conifers generally have R_q/R_a ratios between 1.2 and 1.5, mostly grouping around 1.3. The R_q/R_a ratio slightly increases when using a small tooth bite of 0.1 mm or less. Hardwoods may have quite high R_q/R_a ratios up to 2.4 with a corresponding high negative skewness. The correlation can be approximated by a linear function which intersects on the horizontal axis

$$-R_{sk} = 4.69 \cdot \left(R_q/R_a - 1.12\right) \tag{2.32}$$

although the curve theoretically tends to the zero skewness if R_q/R_a tends to one.

2.6.4 The Use of Structure Number

As introduced in Sect. 2.6.2, the structure number ΔF uniquely characterizes a particular wood species in relation to its internal structure which determines its inherent roughness due to its anatomical properties. The use of a structure number makes it possible to treat different wood species in a unified system and to discover general relationships valid for all wood species. The term "wood species" as a variable cannot be treated numerically and, therefore, different wood species cannot be compared.

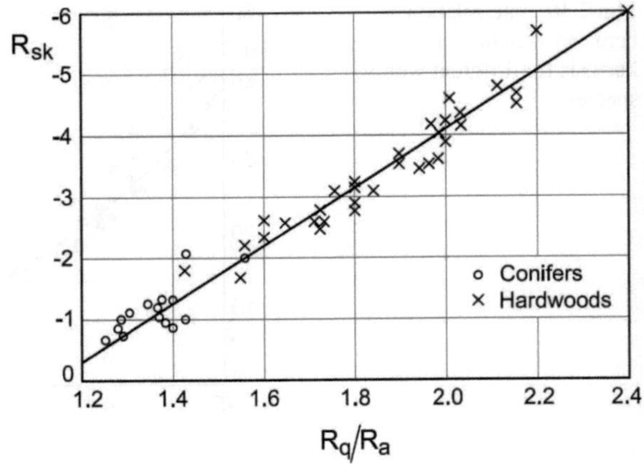

Fig. 2.37 Correlation between skewness and R_q/R_a ratio

Table 2.3 Structural properties

Wood species	Earlywood			Latewood		
	$\overline{d_i}$ (μm)	$\overline{n_i}$ (piece/cm^2)	$\overline{a}$	$\overline{d_i}$ (μm)	$\overline{n_i}$ (piece/cm^2)	$\overline{b}$
Spruce	30.0	111,335	0.8478	19.0	160,400	0.1522
Pine	28.0	125,100	0.6694	20.0	135,840	0.3306
Larch	38.0	65,490	0.6310	17.5	145,000	0.3690
Beech (vessel)	66.0	15,740	0.7000	48.0	14,020	0.3000
Beech (tracheid)	8.2	342,890		6.4	490,290	
B. locust (vessel)	230.0	546	0.5800	120.4	1500	0.4200
B. locust (tracheid)	15.0	270,000		9.6	280,000	
Oak (vessel)	260.0	400	0.5900	35.7	12,000	0.4100
Oak (tracheid)	22.5	120,000		19.6	85,000	

The structure number uses anatomical properties and differences due to growing conditions can be taken account. The structure number has a definite correlation with roughness parameters, regardless of wood species or the location where they were grown.

Table 2.3 shows typical structural properties for common European species. To use structure numbers, the structural properties must be known. The experimental determination of these properties is time-consuming which is a drawback. Nevertheless, the systematic collection of these data for different wood species in a data bank will facilitate the wide-ranging use of structure numbers. Furthermore, these anatomical properties may show some general laws which also facilitate their use.

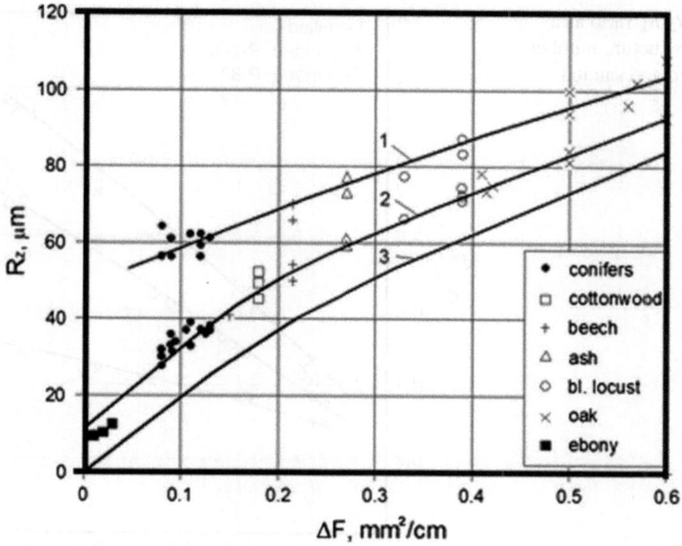

Fig. 2.38 Average roughness R_z as a function of structure number. 1—Cutting speed 10 m/s; 2—cutting speed 50 m/s; 3—anatomical roughness

The first important results were the attainable optimum roughness R_z as a function of structure number for arbitrary wood species, Fig. 2.38 (Magoss and Sitkei 2000, 2001). The same figure shows the anatomical roughness due to the internal structure (vessels, tracheids) of wood material. It clearly shows that high-speed cutting with sharp tools produces less roughness due to machining than the anatomical roughness, especially for species with big vessels. Since conifers are more sensitive to cutting speed, the roughness due to machining has the same or higher magnitude than the anatomical component, depending on cutting conditions (Csanády and Magoss 2013; Magoss and Sitkei 2003). The average roughness R_z given in Fig. 2.38 can be expressed with the following empirical equation taking into account the effect of cutting speed

$$R_z = \left(123 \cdot \Delta F^{0.75} + 35 \cdot e_z^{0.6}\right) \cdot \left(1 + \frac{50 - v_x}{50} \cdot \frac{0.1183}{\Delta F^{0.83}}\right) \quad 10\,\text{m/s} \le v_x \le 50\,\text{m/s} \tag{2.33}$$

where ΔF must be substituted in mm^2/cm, e_z in mm and v_x in m/s. The third part of this equation takes into account the lesser strength of softwoods.

Earlier experiments have shown that several roughness parameter ratios have a uniquely defined correlation with the structure number (Magoss and Sitkei 2003).

Further detailed measurements revealed that, with big-vessel species, special care must be taken in the measurement and processing procedure (the essence of these problems was already demonstrated in Fig. 2.28). Performing measurements on a

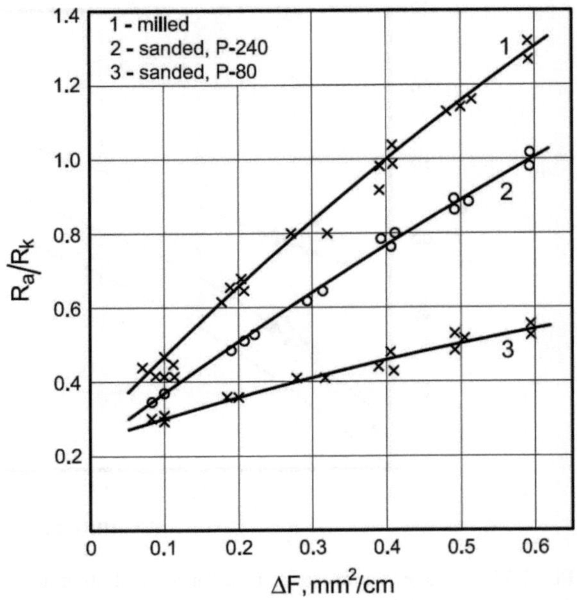

Fig. 2.39 R_a/R_k ratio as a function of structure number for machined and sanded surfaces

bigger surface, the random position of vessels on the surface can cause higher scattering in the measurement results though the average internal properties are the same throughout. This problem does not appear for conifers and much less for species with smaller vessels.

Figure 2.39 illustrates the correlation between the R_a/R_k ratio and the structure number ΔF. Variations in the anatomical structure of different species can cause a threefold variation in the R_a/R_k ratio. The curve corresponds to the case when 80% of the vessels are cut. The same sample also has smaller R_a/R_k ratios. The relationship can be described with an empirical equation

$$\frac{R_a}{R_k} = 0.25 + 1.7 \cdot \Delta F^{0.9} \tag{2.34}$$

where ΔF must be substituted in mm^2/cm.

The sanding process modifies the surface structure in considerable extent due to crushing and clogging. Especially, coarser sandpaper increases the R_k layer and decreases the R_a/R_k ratio. The effect of average grit diameter can be expressed as follows

$$\frac{R_a}{R_k} = 0.2 + 0.17 \cdot d^{-07} \cdot \Delta F^{0.85} \tag{2.34a}$$

where the grit diameter d must be substituted in mm.

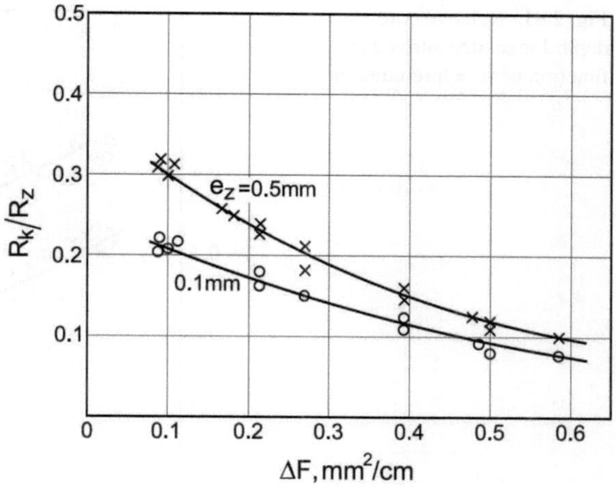

Fig. 2.40 Relative core depth for machined surfaces as function of structure number

New experiments show that the R_k/R_z ratio also depends on the tooth bite to a certain extent which is demonstrated in Fig. 2.40.

With increasing tooth bite, the radial force component increases, causing higher R_k values. In the range of 0.1 and 0.5 mm, tooth bite the correlation function has the form

$$\frac{R_k}{R_z} = 0.48 \cdot e_z^{0.25} \cdot (1 - \Delta F^{0.6}) \tag{2.35}$$

where the tooth bite e_z must be substituted in mm.

Fine sanding (P-240) creates almost the same R_k/R_z ratio as knife machining does, Fig. 2.41. Coarser sanding increases the R_k layer and the R_k/R_z ratio. The effect of average grit diameter can be given as follows

$$\frac{R_k}{R_z} = 0.78 \cdot d^{0.25} - 0.4 \cdot \Delta F^{0.7} \tag{2.35a}$$

where the grit diameter d must be substituted in mm.

The relative valley depth R_{vk}/R_z is also definitely correlated with the structure number ΔF. Figure 2.42 shows its correlation function which is described with the following empirical equation

$$\frac{R_{vk}}{R_z} = 0.15 + \Delta F^{0.8} \tag{2.36}$$

Equations (2.34) and (2.35) can be used to calculate the dimensionless number $R_a/\sqrt{R_k \cdot R_z}$, which is formally equal to

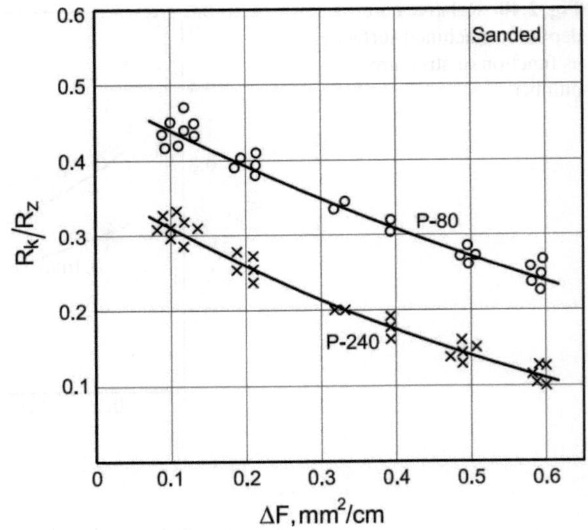

Fig. 2.41 Relative core depth for sanded surfaces as function of structure number

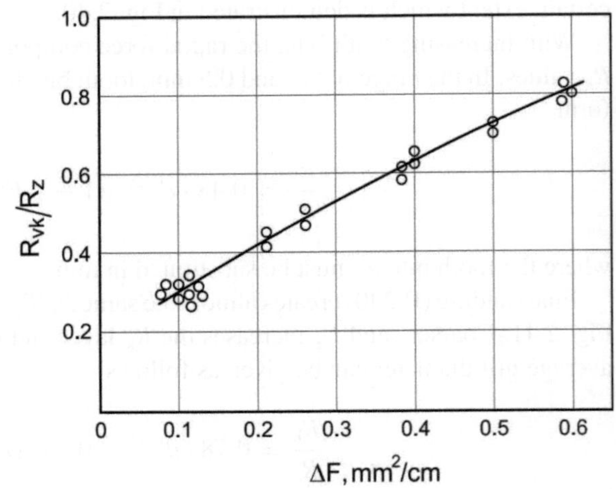

Fig. 2.42 R_{vk}/R_z ratio as a function of structure number. Knife-machined surface

$$\frac{R_a}{\sqrt{R_k \cdot R_z}} = \sqrt{\left(\frac{R_a}{R_k}\right)^2 \cdot \frac{R_k}{R_z}}$$

This method gives more accurate results compared to the simple Eq. (2.28).

The sanding operation (see Fig. 2.30) may cause distortion (deformation and clogging) in the surface layer diminishing the valley depth R_{vk}. Figure 2.43 shows experimental results with five wood species as a function of the average grit diameter. A larger grit diameter causes more deformation and clogging reduces the depths of the valleys R_{vk}.

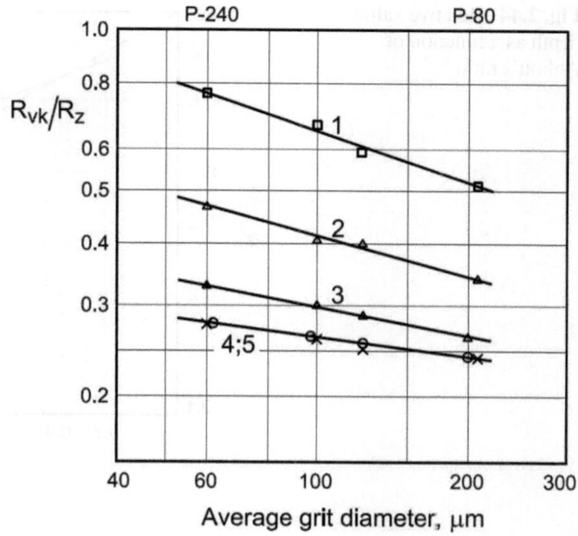

Fig. 2.43 R_{vk}/R_z ratio as a function of grit size. 1—Oak, 2—black locust, 3—larch, 4—spruce, 5—beech

The curves are described with the following equation

$$\frac{R_{vk}}{R_z} = 0.15 + \frac{0.36}{d^{0.34}} \Delta F^{0.8} \tag{2.36a}$$

which is similar to Eq. (2.36) but takes into account the effect of grit size.

The Abbott ratio provides a more accurate description of the internal structure of wood in the immediate surface layer taking into account the modifying effect of the woodworking operation. Figure 2.44 shows the relative valley depth as a function of Abbott's ratio on double-logarithmic scale using the results of Fig. 2.43. The results are well approximated with the following simple equation

$$\frac{R_{vk}}{R_z} = \frac{0.406}{\left(\frac{R_{pk}+R_k}{R_{vk}}\right)^{0.56}} \tag{2.37}$$

These results uniquely demonstrate the usefulness of the Abbott ratio in seeking general relationships which highly facilitate the processing of new experimental results or the approximate estimation of different roughness parameters.

2.6.5 *Effect of Machining on the Surface Roughness*

Machining operations may have effects on the surface roughness of any wood material. The resulting surface roughness is influenced by an array of operational parameters which have quite different importance. The main operational parameters are the following:

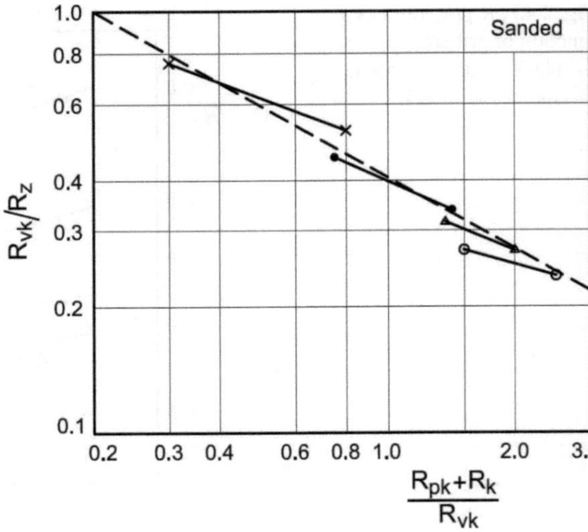

Fig. 2.44 Relative valley depth as a function of Abbott's ratio

- cutting speed,
- cutting depth,
- feed per tooth,
- cutting angle,
- workpiece vibration,
- sharpness of tools,
- climb or countercutting,
- oblique cutting,
- cutting circle radius,
- direction of cut.

The tool interacts with the wood material, and its mechanical properties influence the surface roughness. The following material properties should be mentioned:

- volume density,
- hardness of wood,
- modulus of elasticity,
- anisotropy of wood,
- early and latewood,
- inhomogeneity of wood, including irregular growth and defects.

Due to the "free cutting" mechanism, the most important influence is the cutting speed (Dobler 1972). The counterforce to the cutting edge is ensured by the strength of the material and by inertia forces. Softwoods have less strength and therefore they are more sensitive to the selection of a proper cutting speed. This is clearly seen in Fig. 2.45.

Increasing the cutting speed, the average height R_z and the valley depth R_{vk} steeply decrease, while the Abbott ratio increases.

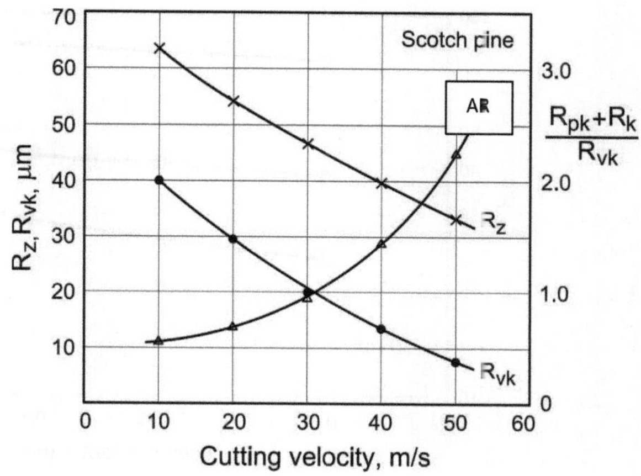

Fig. 2.45 Effect of cutting velocity on roughness parameters and the Abbott ratio (AR) for a softwood

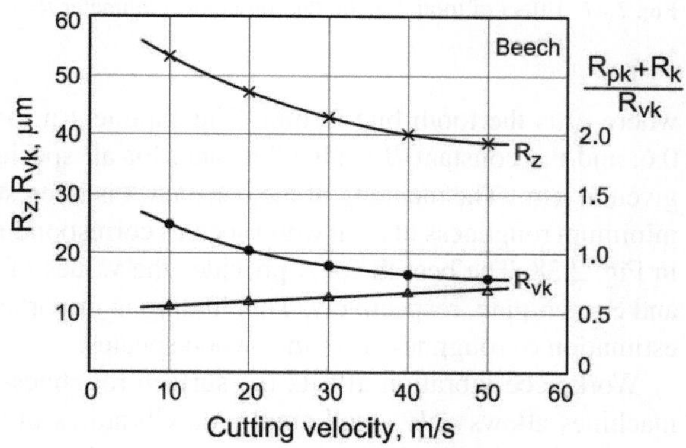

Fig. 2.46 Effect of cutting velocity on roughness for hardwood

Beech wood has more strength and therefore the inertia forces are less important. The increase of cutting speed moderately decreases the R_z and R_{vk} values, and the Abbott ratio is only slightly increased, Fig. 2.46. Around 30 m/s, the counterforce is well developed and increasing the cutting speed does not lower the roughness much (Fig. 2.46).

The tooth bite has also some influence on the surface roughness. It is well known that the essence of a smooth machining is to use a small feed per tooth as the last operation. A small tooth bite means less cutting force which causes less deformation and breakout in the deformation zone.

Figure 2.47 shows experimental results for three wood species as a function of tooth bite (Magoss 2008). The variation of R_z for all three wood species is quite similar, and the curves can be described with the empirical equation

$$R_z = A + B \cdot e_z^n \tag{2.38}$$

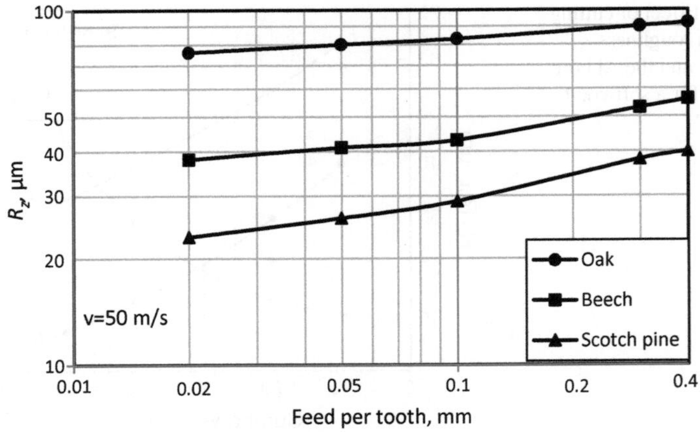

Fig. 2.47 Effect of tooth bite on the roughness parameter R_z

where e_z is the tooth bite in mm. The exponent n for all species may be taken as 0.6, and the constant B is also the same for all species with the value of 43 (R_z is given in µm). The meaning of the constant A is interesting, and it is very close to the minimum roughness of each wood species corresponding to its anatomical roughness in Fig. 2.38. The best fit for A provides the values of 69, 35 and 19 for oak, beech and Scotch pine, respectively. This finding is important and allows an approximate estimation of roughness for other wood species.

Workpiece vibration affects the surface roughness. Vacuum clamping on CNC machines allows only small-amplitude vibrations of the workpiece, not exceeding 5–10 µm amplitude. Our measurements have shown that no difference in the roughness was observed between vacuum and rigid clamping of the workpiece. If the tooth bite is higher than 1 mm, then the increased vibration amplitude may worsen the surface roughness.

Other conditions prevail when the workpiece moves during the woodworking operation. In this case, the vibration amplitude may be much higher, perceptibly worsening the surface quality (Sitkei et al. 1990).

The vibration of the workpiece mainly depends on the following factors:

- the size of the workpiece (thickness, length),
- the magnitude of exciting forces and their direction,
- the instantaneous boundary conditions, i.e. the position of pressing rolls, their prestressing and spring stiffness characteristics (Sitkei et al. 1988; Csanády and Magoss 2013).

The magnitude of exciting forces depends on the sharpness of the tool and the tooth bite. If the exciting force acts towards the machine table, the vibration amplitudes are smaller, producing less roughness. On the contrary, exciting the workpiece towards the pressing rolls, the vibration amplitudes may be high enough to considerably worsen the surface quality (Sitkei et al. 1990; Csanády and Magoss 2013).

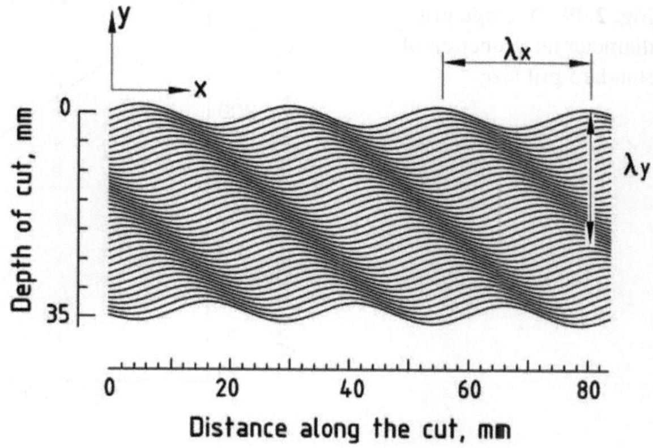

Fig. 2.48 Typical wavelike pattern of washboarding on the timber sawn with a bandsaw

A special vibration mode is "washboarding", mainly occurring with bandsaw and circular saws. Washboarding is a wavelike surface pattern appearing on the surface of sawn timber, Fig. 2.48.

The profile can be as much as 0.6–0.8 mm deep on the face of the lumber requiring larger target size in order to remove sawing deviations and to reduce the depth of washboard (Okay et al. 1995; Lehmann and Hutton 1997; Orlowski and Wasielewski 2001).

Washboarding is a self-excited resonance phenomenon occurring with high blade speed, feed speed and thinner saw blades. The frequency of the tooth passage should be slightly higher than the natural frequency of the blade (Okay et al. 1995). Washboarding occurs in a narrow frequency range, and it can be eliminated by increasing or decreasing the blade speed, or the bandsaw tension.

The cutting angle also influences surface roughness. Higher cutting angles around 65–70° exert more compressive forces into the material ensuring a smoother surface. On the contrary, the use of a small cutting angle in cutting veneer requires a pressure beam to provide the necessary compressive stresses (Sitkei et al. 1990; Csanády and Magoss 2013).

There are advantages to oblique cutting, which utilizes the combine effect of compressive and shear stresses. This cutting requires less compressive forces and produces less deformation and a better surface quality (Sitkei 1997). For the same reason, very thin veneers can only be cut by using large oblique angles. The action of pure shear stresses is utilized when cutting upholstering (foam) materials.

Sanding is important among the different woodworking operations. Sanding is an abrasive process. Due to the unusual cutting edge, with a negative rake angle and the random position of grits on the surface, it can produce a smooth surface, depending on the grit size and other operational parameters (Fischer and Schuster 1993; Scholz and Ratnasingam 2005; Siklinka and Ockajova 2001).

Earlier experiments have shown that the radius of the grits as the working edge has a distinct effect on the core depth R_k of the material ratio curve (Abbott) (Magoss 2013, 2015). Another important observation revealed that clogging may produce

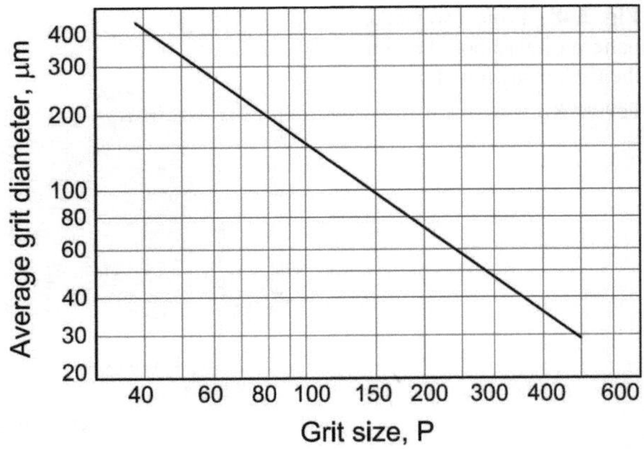

Fig. 2.49 Average grit diameter as a function of standard grit size

less roughness than the expected anatomical roughness for a given wood species. Furthermore, using general laws of contact mechanics (Bershadski and Cvetkova 1975), the effect of different working edges (spherical grit, cylindrical knife edge) on the surface roughness, especially the R_k layer, has been investigated (Magoss 2013, 2015). Another important result was that the maximum stress under grit theoretically does not depend on the grit size.

The grit size of sandpaper is determined by the number of meshes per inch of the sieves used for screening. The average grit diameter is depicted in Fig. 2.49 as a function of standard grit size notation.

The most important operational parameters of sanding are the surface pressure, feed speed, grit size and cutting speed. There are some other influencing parameters such as the hardness of grits, their heat conduction coefficient, the contact length of sanding, etc.

Due to the modifying effect of cutting edge geometry on the stress distribution under the cutting edge, a bigger radius of a sphere is equivalent to a smaller radius of a cylindrical edge concerning their effect on the R_k layer thickness, Fig. 2.50 (Magoss 2013, 2015). This important finding has been verified experimentally.

The average roughness R_a correlates with the irregularity depth R_z described by the simple equation

$$R_a = 0.12 \cdot R_z$$

which is very close to that obtained for knife-machined surfaces (Fig. 2.31).

The average roughness R_a has a better correlation as a function of the sum of Abbott's parameters ($R_{pk} + R_k + R_{vk}$) which is depicted in Fig. 2.51.

It is quite similar to Fig. 2.32 and described by Eq. (2.24). Some difference can be observed in sanding as a function of grit size but the difference is small.

An important relationship is the dependence of R_a as a function of average grit diameter which is depicted in Fig. 2.52 for spruce and beech. The relationships are almost linear and can be described as

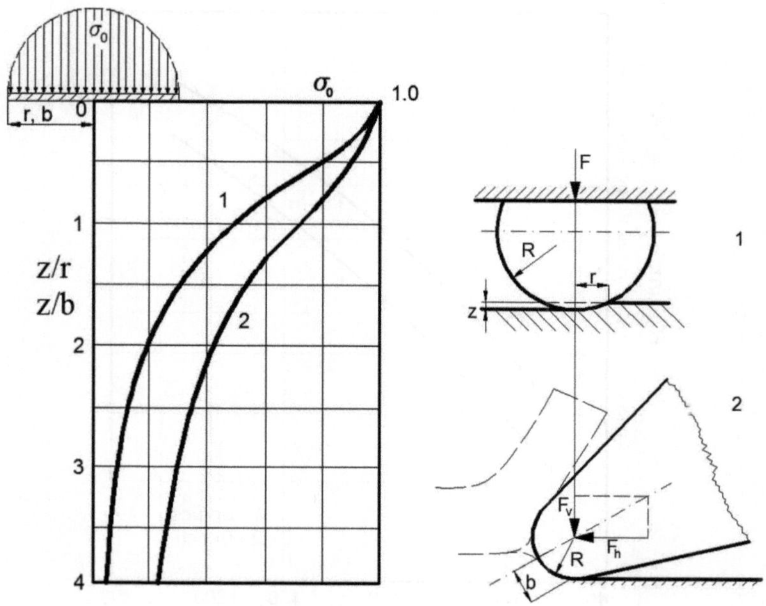

Fig. 2.50 Stress relations under a spherical (1) and cylindrical body (2) using hemispherical pressure distribution on the contact surface

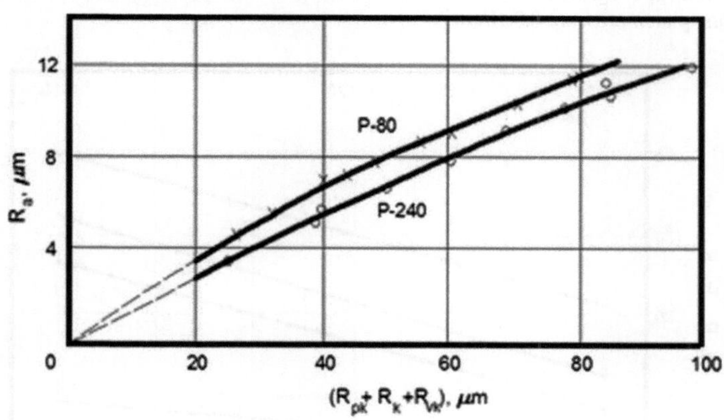

Fig. 2.51 Average roughness R_a as a function of the sum $(R_{pk} + R_k + R_{vk})$

$$R_a = 0.047 d_{av}^{0.9} \quad \text{(Beech)}$$
$$R_a = 0.065 d_{av}^{0.9} \quad \text{(Spruce)}$$

where d_{av} must be substituted in μm.

The average grit diameter mainly influences the core depth R_k as it is illustrated in Fig. 2.53.

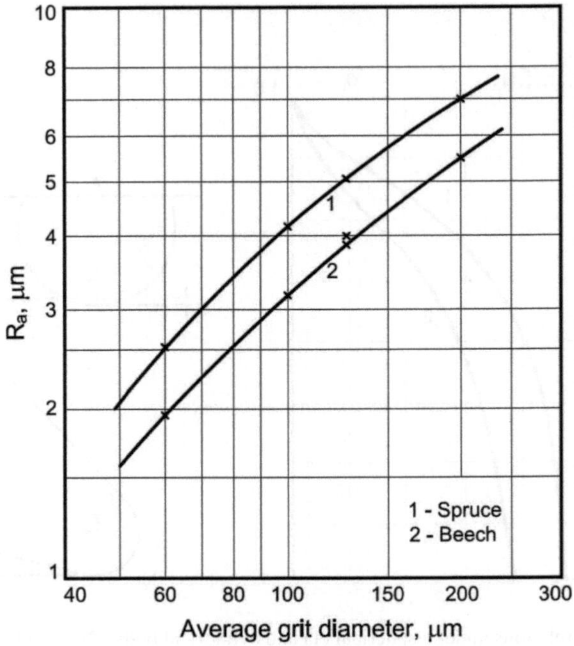

Fig. 2.52 Average roughness as a function of average grit diameter for spruce and beech. Cutting speed 25 m/s, platen pressure 0.3 N/cm^2

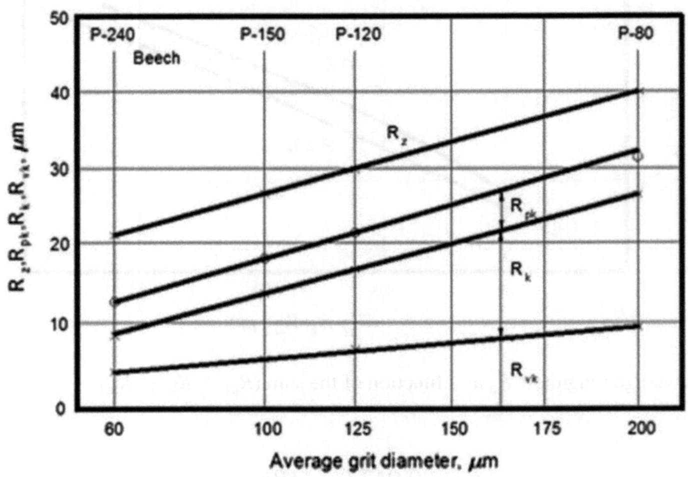

Fig. 2.53 Relationship between grit size and roughness parameters for larch with density of 700 kg/m^3

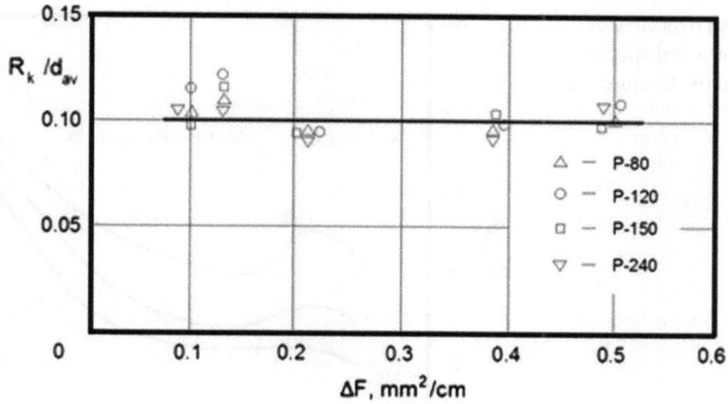

Fig. 2.54 Core depths and grit diameter ratio depending on the structure number ΔF

The dependence is linear or very close to it. Results obtained for quite different wood species revealed that the ratio of the core depth and average grit diameter R_k/d_{av} is practically constant, independent of grit size and wood species, as shown in Fig. 2.54 (Magoss 2013).

Its ratio is averaged to 0.1 which makes it possible to forecast the expected core depth R_k in advance or to select an appropriate grit size to achieve a given R_k layer thickness.

Figures 2.39, 2.41, 2.43 and 2.44 contained experimental results and relationships including sanding.

Clogging decreases the number and size of cavities in the surface. As a consequence, some roughness parameters such as R_z may be smaller than the expected anatomical roughness value. This is more explicitly demonstrated in Fig. 2.55 which gives an overview on the variation of R_z values as a function of the structure number ΔF (Magoss 2013).

Conifers are not prone to clogging. Over $\Delta F = 0.2$ (beech), the clogging effect appears and remains for all vessel species although its extent decreases for big-vessel species. Clogging can also be characterized by the Abbott ratio, as demonstrated in Fig. 2.30.

An inherent feature of milling and planing operations is the wavy surface caused by the cycloid path of the tool edge. The waviness of the surface is characterized by the wave depth t which is determined by the tool diameter D, number of teeth z, rotation speed n and the feed speed e

$$t = \frac{e_z^2}{4D} = \frac{e^2}{4 \cdot D \cdot n^2 \cdot z^2} \tag{2.39}$$

Edge machining of boards is very sensitive to edge breaking. Edge breaking can be reduced by using larger diameter tool and thin chips. This may be achieved by using conical milling cutters with a variable setting angle (Csanády et al. 2015; Krazhev 1963; Lang 1989; Tröger and Lang 1990). In this case, the cutting edge describes a

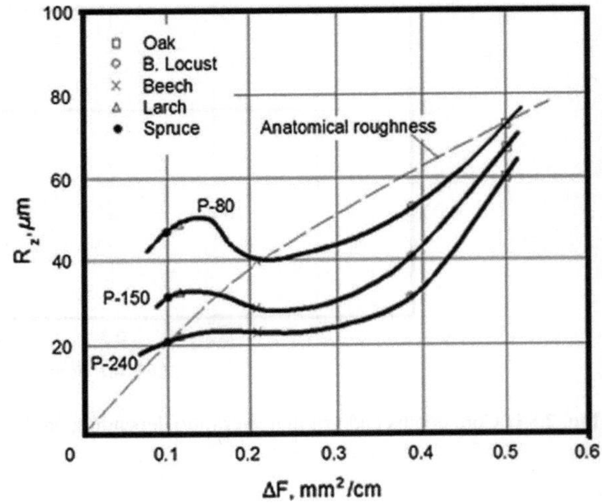

Fig. 2.55 An overview of R_Z values for wood species as a function of structure number ΔF

hyperbolic path way and its virtual cutting circle radius may have enormous values depending on the setting angle. For example, using a 5° setting angle, the virtual radius of cutting is ten times larger than the radius of the tool. Accordingly, the chip thickness will be very thin. The wave depth (geometric roughness) may also be very small due to the large cutting radius.

Tool wear affects the surface roughness. As a function of working time, the radius of a tool edge continuously increases and, according to the contact mechanics, a given pressure (stress) under the edge penetrates deeper into the material causing deformations and breakout. As a consequence, an increase in the edge radius is always associated with more surface roughness (Smardzewski 2015; Feulner 1927).

The dynamics of edge wear depend on the knife material, mechanical properties of the wood and on the operational parameters. The cutting speed is the most important parameter which determines the amount of friction power turning into heat, causing a high temperature on the edge surface (Sitkei et al. 1990; Csanády and Magoss 2013). Therefore, the Taylor tool life equation contains the cutting speed v in the following form

$$v \cdot T^n = \text{const.}$$

where T is the lifetime of the tool between two sharpenings. The exponent n in high-speed cutting has typical values around 0.5 (Sitkei et al. 1990; Csanády and Magoss 2013).

During the wear process, the edge profile generally maintains its almost semi-circle form with increasing radius. The increasing load surface is responsible for deeper crushing effects and higher core depth R_k.

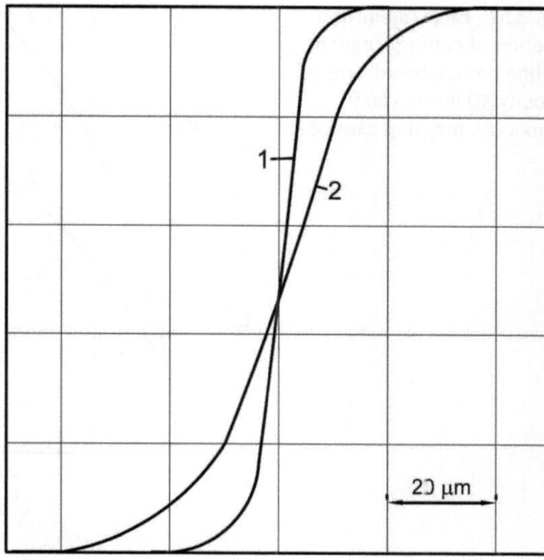

Fig. 2.56 Abbott's distribution curves for beech wood milled with a sharp tool (1) and after a feed distance of 1800 m (2)

	1		2
R_a	3.95 μm	R_a	7 64 μm (1.93 x)
R_z	37.11 μm	R_z	55.10 μm (1.48 x)
R_{pk}	5.63 μm	R_{pk}	7 71 μm (1.37 x)
R_k	10.22 μm	R_k	23.65 μm (2.3 x)
R_{vk}	10.96 μm	R_{vk}	15.81 μm (1.44 x)

The effect of edge blunting can be demonstrated by the Abbott distribution curve as seen in Fig. 2.56 for a beech sample using a sharp tool (1) and after a feed distance of 1800 m (2).

The thickness of the deformed zone doubled as the sharp edge blunted due to the feed distance.

Investigating the effect of edge blunting on roughness parameters, the tool edge radius or the feed distance can be chosen as an independent variable. The primary variable is the edge radius, which directly influences the expected surface roughness.

We distinguish the total feed distance and the true cutting length during the lifetime of a tool. The feed distance is given by the equation

$$L_{\mathrm{f}} = e_z \cdot n \cdot z \cdot T \tag{2.40}$$

While the true cutting length for a single tooth during the lifetime is expressed as

$$L_{\mathrm{c}} = n \cdot R \cdot \varphi \cdot T = \frac{60}{2 \cdot \pi} \cdot v \cdot \varphi \cdot T = L_{\mathrm{f}} \cdot \frac{R \cdot \varphi}{e_z \cdot z} \tag{2.41}$$

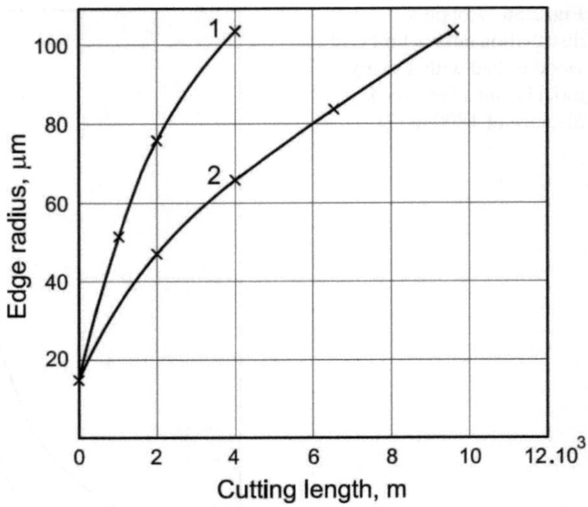

Fig. 2.57 Edge radius as a function of cutting length at milling particleboard, cutting velocity 40 m/s (1) and 20 m/s (2), tungsten carbide

where

n is the rotation speed, rpm,
R is the tool radius,
φ is the rotation angle of cutting,
z is the number of teeth,

and the angle φ must be substituted in radian. It is obvious that the wear will be determined by the true cutting length.

The edge reduction (recession) due to wear is expressed by the following empirical equation:

$$y = y_0 + A \cdot v^n \cdot L_c^m \quad (2.42)$$

For the sake of simplicity, the shape of the worn edge may be taken as a symmetrical round edge and here the radius of the edge ρ is given by

$$\rho = y \cdot \frac{\sin(\beta/2)}{1 - \sin(\beta/2)} = \frac{\sin(\beta/2)}{1 - \sin(\beta/2)} \cdot \left(y_0 + A \cdot v^n \cdot L_c^m\right) \quad (2.43)$$

where β means the sharpening angle of tool.

Figure 2.57 shows measurement results for particleboard with two different cutting speeds as a function of cutting length. The initial edge recession was approximately 20 µm which corresponds to an initial edge radius of approximately 17 µm. The exponents in Eq. (2.42) are $n = 0.78$ and $m = 0.65$, and the constant is $A = 0.0235$. Edge radius and edge recession are given in µm.

Using Eq. (2.43) and a given wear limit, the total cutting length can be calculated as

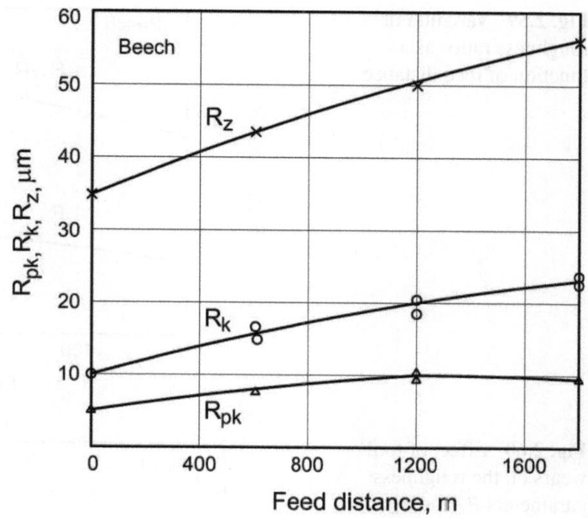

Fig. 2.58 Variation of roughness parameters as a function of feed distance

$$L_c = \left[\frac{1}{A \cdot v^n} \cdot \left(\frac{\rho}{\frac{\sin(\beta/2)}{1-\sin(\beta/2)}} - y_0\right)\right]^{1/m} \qquad (2.44)$$

Knowing the total cutting length, from Eq. (2.17) the lifetime of the tool can be estimated.

The change in the edge radius as a function of feed distance or working time generally shows a parabolic function, while the increase in edge radius influences the roughness in an almost linear fashion.

Figure 2.58 represents different roughness parameters after various feed distances. All the curves have the form of a parabolic function. The slope of the curves is different, and their ratios are not constant.

Figure 2.59 shows characteristic ratios for the same beech wood samples from Fig. 2.58. The increase of the R_k/R_z ratio indicates that the core depth R_k grows more rapidly than R_z as a function of working time.

The same is true for the R_a/R_z ratio. Although R_a is an integrated value, it is more responsive to structural changes in the surface layer due to edge wear.

The most sensitive and useful roughness parameter is the core depth R_k which uniquely indicates the change in the edge radius due to the wear process. The roughness parameter R_k is almost constant (around 10 µm) for a wide variety of wood species using sharp tools. These initial values will be at least doubled as shown in Fig. 2.58. Quite similar results were obtained for other wood species (conifers, black locust, oak). Therefore, these results may be regarded as generally valid.

The reduced peak height R_{pk} also monotonously increases with advancing wear. To produce a smooth surface, this may be a limiting factor concerning the allowable edge wear.

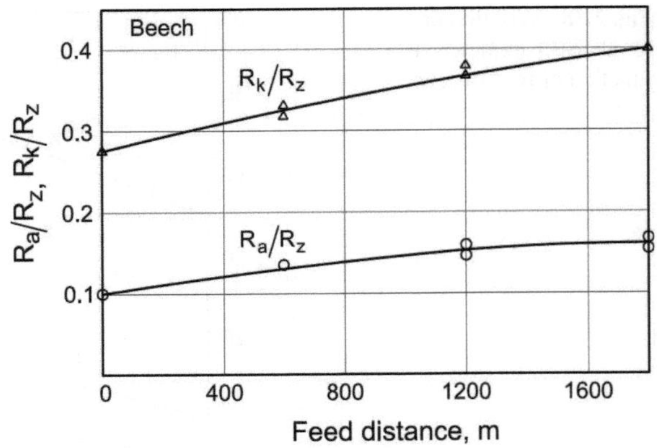

Fig. 2.59 Variation of roughness ratios as a function of feed distance

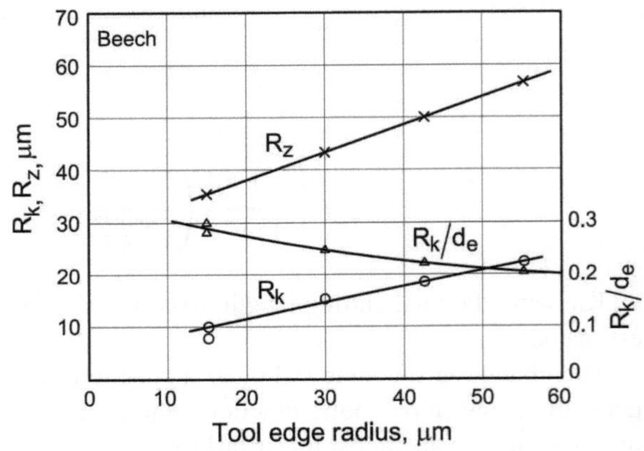

Fig. 2.60 Effect of tool wears on the roughness parameters R_z and R_k

Choosing the edge radius as the influencing variable, the roughness parameters R_z and R_k vary linearly as a function of the edge radius as depicted in Fig. 2.60.

In the figure, the R_k/d_e ratio (d_e is the equivalent edge diameter) is also given which characterizes the effect of compaction by the edge in the core zone. In sanding, this ratio was nearly constant for all wood species, see Fig. 2.54.

Using knife edges, the same ratio is higher and slightly increases towards sharp edges. This may be explained by the general theory of wood cutting (Sitkei et al. 1990). The radial component of the cutting force for sharp edges changes its sign from negative to positive. Due to tension forces in the separation zone, some fibres may be torn out from the developing surface and slightly increases the core depth. With a higher edge diameter, the radial force is always compressive which does not create any additional roughness.

Tool wear is a major factor influencing the surface quality obtainable by any machining. Furthermore, tool wear increases energy requirements. Therefore, the maintenance of tools is very important, and the sharpening cycle of tools is determined by surface quality requirements. If tools are sharpened more frequently than necessary, the higher maintenance costs will lower the economy of production.

As outlined above, the tool wear and surface roughness intensively interact. Therefore, not only the lifetime but also the allowable surface roughness is a governing factor in the proper selection of tool materials (see in Sect. 2.5).

2.6.6 *Influence of Wetting on Surface Roughness*

Wood is a capillary-porous material which always contains some water. Depending on the humidity of the surrounding air, wood tends to be in equilibrium with it. The wood surface is wetted for various purposes, e.g. aqueous coating, and a part of the liquid infiltrates into the material and the other part evaporates into the air. The movement of water from the surface into the body is the result of the moisture gradient. As a consequence, swelling and shrinkage stresses develop with subsequent residual deformations in the upper surface layer. During the wetting process, a thin surface layer moves causing distortions modifying the initial roughness of the surface.

It is well known that to coat in several subsequent layers, an intermediate sanding may be necessary, due to increased surface roughness. Due to the movements and residual deformations of a wood surface, micro-cracks may develop, accelerating diffusion and infiltrating liquids (e.g. aqueous coatings). Furthermore, the movements can open voids which would otherwise be closed to free infiltration. This may influence not only the surface roughness but also the adhesion of coating films on wood surfaces depending on the surfacing process (planing or sanding) and wood anatomy (Richter et al. 1995).

The variation of surface roughness under wetting has been investigated using different wood species and differently manufactured surfaces. Distilled water was used with a film 100 and 180 μm thick. First the water film mostly infiltrates into the upper layer. Due to continuous evaporation, the surface slowly becomes drier and the surplus water in the upper layer tends to flow back to the surface and evaporates. A transient wetting and drying process occurs with a moisture gradient causing local swelling and shrinkage.

Some experiments were conducted with hot air drying (50–60 °C) accelerating the evaporation and diminishing the infiltration. The roughness measurement was performed by an optical 3D system (GF Messtechnik-type MicroCAD) (Molnár 2018).

The thickness of an initial water film fundamentally influences the depth of action.

Figure 2.61 shows the variation of Abbott's parameters during the wetting process as a function of time. Using 180 μm film thicknesses, the water infiltrates in a deeper zone, causing temporary swelling and considerably diminishing the S_{vk} component. During this time, the other two roughness components continuously increase and they reach their maximum after half of the initial film thickness evaporates. During the drying-out, the surface will be stabilized, the roughness components tend to their end-values which are always higher than their initial values. Using a film only

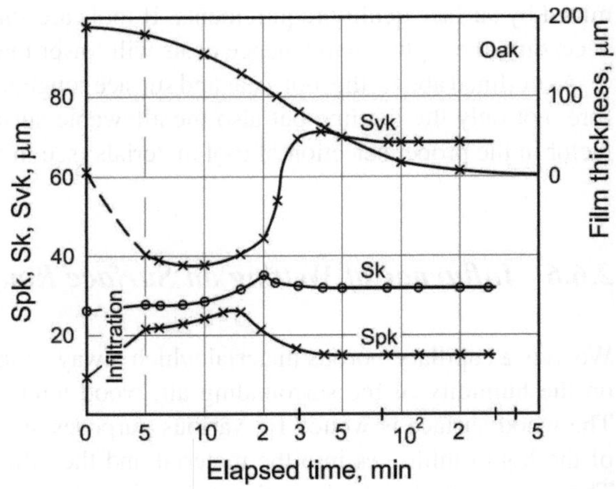

Fig. 2.61 Variation of roughness parameters during the wetting process oak planed. Initial film thickness 180 µm

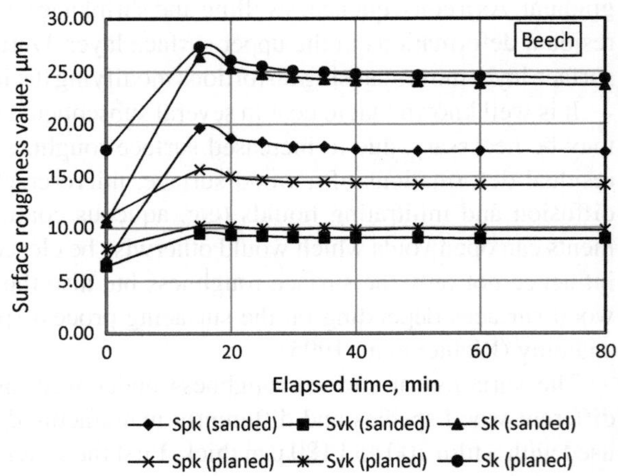

Fig. 2.62 Variation of averaged roughness parameters after wetting as a function of time. Beech samples planed and sanded

100 µm thick, the deep swelling phenomenon is absent and the S_{vk} component shows the smallest change due to wetting.

The machining of the surface may considerably influence the wetting process and the variation of roughness components. The measurements for planed and sanded surfaces are depicted in Fig. 2.62 for beech samples.

The S_{pk} and S_k components increased in both cases, but the sanded surface always had a higher variation compared to the planed surface. That means that a sanded surface can be more easily wetted. Experience has shown its advantage for coating a wood surface (Richter et al. 1995).

Figure 2.63 shows comparative measurements for different wood species when sanding with P120 grit size. The highest variation was always observed for the reduced peak height S_{pk}, especially with big-vessel species. The reduced valley height R_{vk} always shows the smallest relative increase.

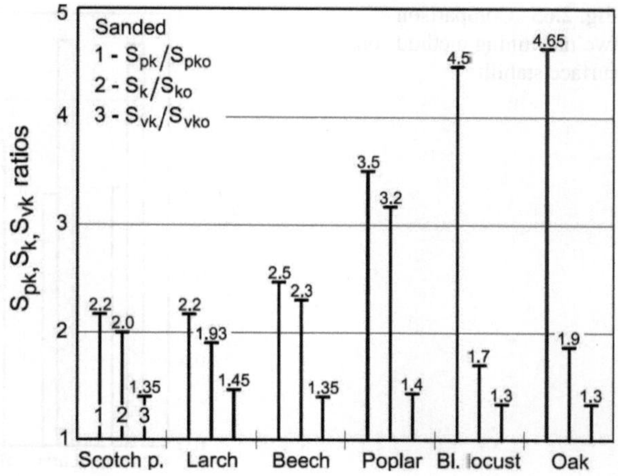

Fig. 2.63 Relative variation of Abbott's parameters for different wood species at sanding with P120 grit size

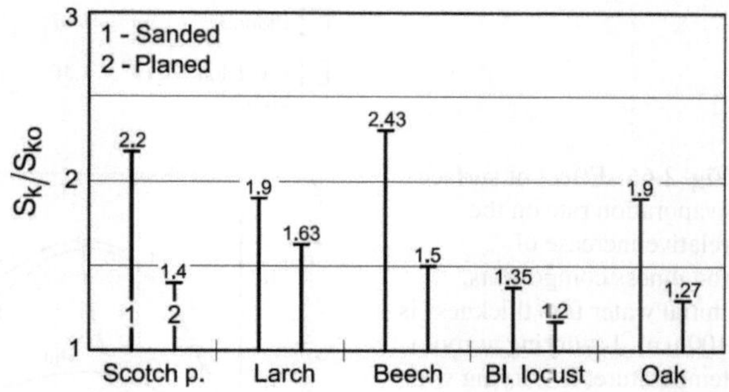

Fig. 2.64 Comparison of wetting properties of different wood species using planed and sanded surfaces

The core depth S_k (or R_k) plays the most important role in the wetting process. Its relative change in the wetting process depends on the kind of machining.

Figure 2.64 represents a comparison of changes in the S_k layer thickness for planing and sanding using different wood species. Sanding always produces a higher relative increase of S_k.

Using waterborne coating materials, the surface is subjected to wetting and it undergoes a similar distortion as in the direct wetting process. The different wood species and woodworking operations react variously to wetting and the stability of surfaces may be quite different. To evaluate wetting stability, the core depth ratio seems a good characteristic number. An interesting comparison is given in Fig. 2.65 between conventional planing and precision planing with continuous plane cutting. The latter produces better surface stability for all species, probably due to a better surface integrity (less deformation under the knife edge) resulting in a slower infiltration of water into the deeper layers.

For every thickness of a water film, the evaporation rate influences the amount of water infiltrating into the surface layer. Drying at room temperature, the halftime of evaporation fluctuates between 18 and 20 min. Using forced convection with hot

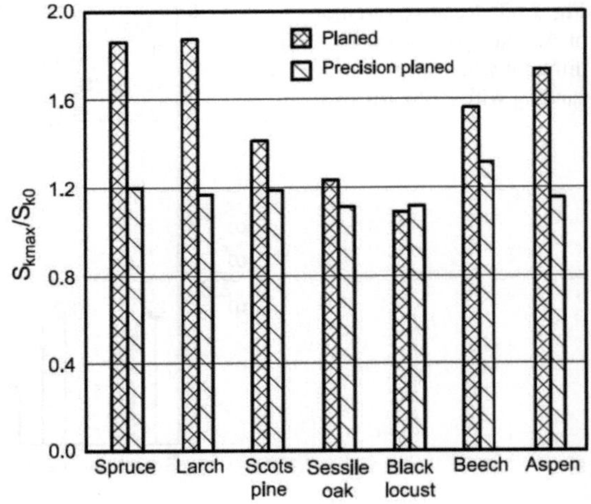

	Spruce	Larch	Scots pine	Sessile oak	Black locust	Beech	Aspen
Planed	1.86	1.87	1.40	1.22	1.08	1.55	1.73
Precision planed	1.19	1.16	1.18	1.10	1.11	1.31	1.15

Fig. 2.65 Comparison of two machining methods on surface stability

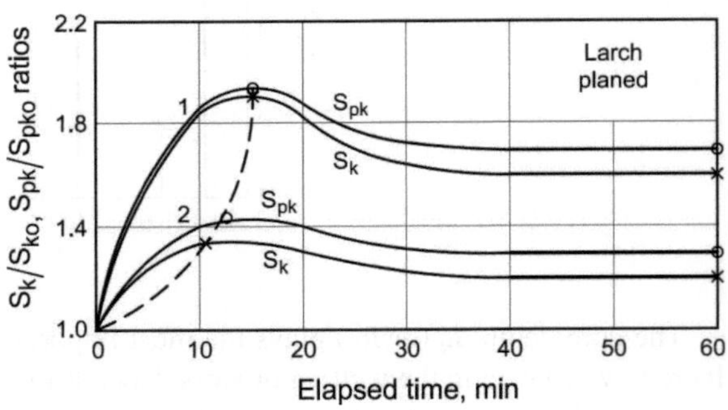

Fig. 2.66 Effect of surface evaporation rate on the relative increase of roughness components. Initial water film thickness is 100 μm, 1—drying at room temperature, 2—drying with hot air (50–60 °C)

air between 50 and 60 °C, the halftime of evaporation was reduced to 8 min which allowed to much less water penetrate into the deeper layers.

Figure 2.66 shows comparative measurements for evaporation at room temperature and hot air drying. In the latter case, the relative increase of roughness parameters is considerably less than in the case of slow evaporation due to the decreasing infiltration time.

In conclusion, the wetting of a surface has a definite influence on the surface properties and its roughness after wetting may considerably differ from its initial value. This is especially true for a sanded surface which produces a better surface for coating.

2.7 Mechanics of Upholstering

Introduction

The need for comfortable seats and beds began with the ancient Egyptians and Greeks using different stuffed pillows and mattresses covered with leather or fabric. It is interesting that the upholstered seats and couches returned only in the Renaissance in the sixteenth century. During the last centuries, spring cushions have been mainly used in furniture and also in automobiles. In the last few decades, special polyurethane foams have been widely used.

Spring cushions have an almost linear elastic behaviour which highly facilitates their testing and dimensioning. On the contrary, foam materials have a distinct non-linear mechanical behaviour due to their structure. Furthermore, when furniture is in use, the human body interacts with the cushion. This interaction is complicated by the human posture and stature. The mechanical behaviour of the human body is not well known causing further difficulties. At the same time, it is well known that pressure exerted over a long time can cause mechanical damage in tissues and the blood supply to the tissue can be cut off. An appropriate cushion distributes the pressure more uniformly over the contacting area, preventing stress concentrations.

The objective of this study is to provide an engineering method to calculate the interaction of cushions with the human body.

2.7.1 Upholstering Material Properties

Polyurethane foams are one of the basic materials used in upholstered furniture, especially seats, chairs and furniture for sleeping and relaxing. These materials have a very low volume weight generally between 20 and 40 kg/m^3, and their high elasticity permits large deformations and compactions. Due to their highly porous structure, they have distinct nonlinear behaviour under loading and unloading conditions.

In the first stage of loading, the skeleton of the foam carries the load under moderate deformation. After the skeleton collapses, the voids decrease their volume with slightly increasing loads in a wide range of deformations and finally, the material compacts with rapidly increasing loads. According to this deformation behaviour, the modulus of elasticity and the stiffness coefficient have quite different values in the course of three stages. The instantaneous modulus of elasticity is given by the tangent to the stress–strain curve.

$$\frac{\mathrm{d}\sigma}{\mathrm{d}\varepsilon} = E \quad \text{and} \quad \sigma = \int E \cdot \mathrm{d}\varepsilon$$

while the secant modulus of elasticity is given by dividing the stress with the strain

$$E_\mathrm{s} = \frac{\sigma}{\varepsilon} \text{ and } \sigma = E_\mathrm{s} \cdot \varepsilon$$

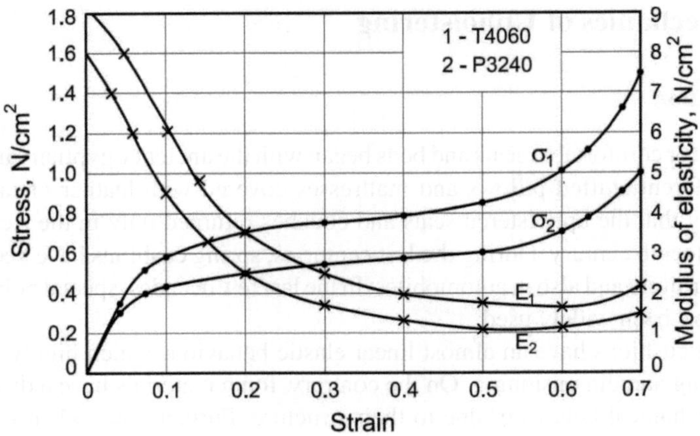

Fig. 2.67 Uniaxial loading of polyurethane foam materials (25 × 25 × 10 cm)

The stiffness coefficient is proportional to the modulus of elasticity and calculated as

$$K = \frac{\Delta F}{\Delta l} = E \cdot \frac{A_0}{L_0} \quad \text{N/cm}$$

where A_0 is the area of the material, under load,

L_0 is the thickness of the material.

Figure 2.67 shows uniaxial measurements for T 4060- and P 3240-type foams. The stress–strain relationships clearly show the three characteristic stages, while the elastic moduli strongly decrease having a minimum between the strain values of 0.5 and 0.6.

The stress–strain relationship is described by the following equation

$$\sigma = A\left(1 - e^{-\varepsilon/\varepsilon_0}\right) + B \cdot \left(\frac{\varepsilon}{0.9 - \varepsilon}\right)^n \tag{2.45}$$

where the constants A, B, ε_0 and n should be determined experimentally. The figure 0.9 in the above equation is a hypothetical maximum strain for compaction. The instantaneous modulus of elasticity is calculated as (taking $n = 2$):

$$E = \frac{d\sigma}{d\varepsilon} = \frac{A}{\varepsilon_0} \cdot e^{-\varepsilon/\varepsilon_0} + \frac{B \cdot 2 \cdot \varepsilon \cdot 0.9}{(0.9 - \varepsilon)^3} \tag{2.46}$$

The secant modulus of elasticity is simply given by

$$E_s = \frac{\sigma}{\varepsilon} = \frac{A}{\varepsilon}\left(1 - e^{-\varepsilon/\varepsilon_0}\right) + \frac{B \cdot \varepsilon^{n-1}}{(0.9 - \varepsilon)^n} \tag{2.47}$$

Table 2.4 Constants for foam materials

Foam type	A, N/cm^2	B, N/cm^2	ε_0	n
T 4060	0.8	0.057	0.07	2
P 3240	0.56	0.038	0.06	2

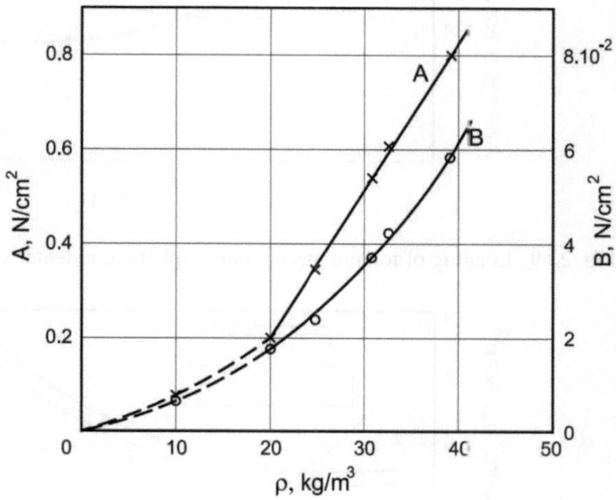

Fig. 2.68 Variation of constants A and B in Eq. (2.45) as a function of foam density

The appropriate constants for Fig. 2.67 are summarized in the following Table 2.4.

The constant A can be regarded as the average stress in a wide range of strain, and therefore, it can serve as a material characteristic. It correlates well with the volume density of the foam materials as depicted in Fig. 2.68, including constant B. The characteristic strain ε_0 and the exponent n may vary only in a narrow range.

In practice, the foam materials are subjected to loads similar to a spherical indenter (sitting person). Figures 2.69 and 2.70 show results using P 3240-type polyurethane foam and a rigid spherical indenter with radii of 90 and 110 mm radius. Contact mechanics are used to process the experimental results (force and deformation) (Timoshenko and Goodier 1951).

The contact mechanics are derived for the homogeneous elastic half space and small deformations. Since these conditions are not fulfilled, the obtained results should be properly interpreted. Due to the large deformations, we introduced smaller corrections in the calculation of the radius of contact surface.

The force–deformation relationship has the form (Timoshenko and Goodier 1951).

$$z = 0.8255\left(\frac{F^2}{R\left(\frac{E}{1-\nu^2}\right)^2}\right)^{1/3} \text{ cm} \tag{2.48}$$

where z is the maximum deformation.

F is the force,
R is the radius of the indenter,

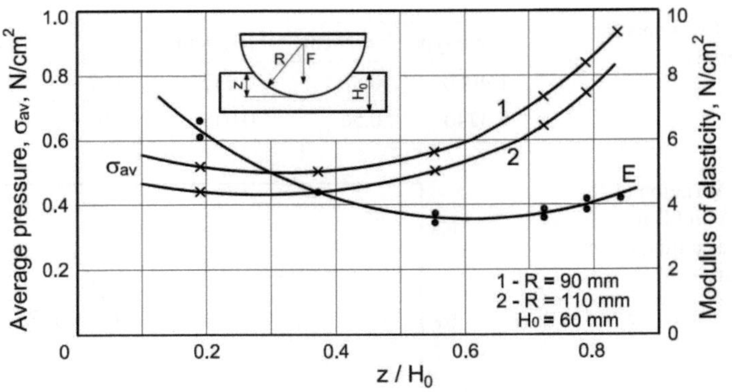

Fig. 2.69 Loading of foam material with a spherical indenter with 90 and 110 mm radii

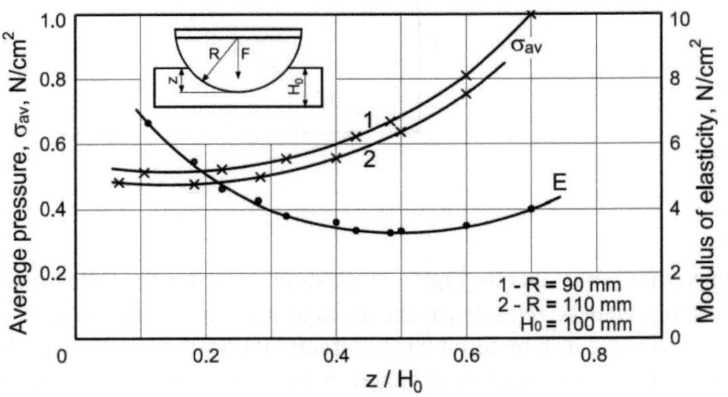

Fig. 2.70 Loading of foam material with a spherical indenter with $H_0 = 100$ mm thickness

E is the modulus of elasticity,
ν is the Poisson ratio.

The Poisson ratio for these materials is poorly known, but it has rather low values in the test material under high axial strains. Therefore, E and $E/(1-\nu^2)$ differ not much. To avoid difficulties in the proper selection of Poisson's ratio, we always determine and use the value of $E/(1-\nu^2)$.

The radius of deformation r (Fig. 2.71) is theoretically calculated as

$$r = 0.9086 \cdot \left(\frac{F \cdot R}{E_0}\right)^{1/3} \tag{2.49}$$

where $E_0 = E/\left(1-\nu^2\right)$ for shorter writing. The radius of deformation can also be calculated from the geometry in a simple way

$$r = \sqrt{2Rz - z^2} \tag{2.49a}$$

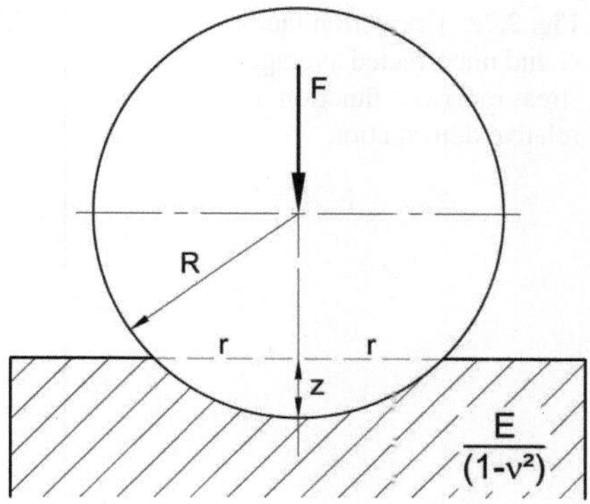

Fig. 2.71 Deformation under a rigid spherical indenter

From Eqs. (2.48) and (2.49), the force can be expressed:

$$F^z = \frac{4}{3} E_0 \sqrt{R} \cdot z^{3/2} \tag{2.50}$$

and

$$F^r = \frac{4}{3} E_0 \frac{r^3}{R} \tag{2.50a}$$

For a given case, the forces must be equal to each other and therefore

$$\frac{r^3}{R} = \sqrt{R} \cdot z^{3/2}$$

which yields

$$r = \sqrt{R \cdot z} \tag{2.49b}$$

This equation is not identical with Eq. (2.49a), and this means that the contact theory of Hertz uses an approximation in order to get a closed-form solution. Calculations show that this approximation may cause considerable errors especially with smaller deformations. Therefore, Eq. (2.49) should be corrected in the following form

$$r = C \cdot 0.9086 \left(\frac{F \cdot R}{E_0} \right)^{1/3} \tag{2.49c}$$

with

$$C = \frac{\sqrt{2Rz - z^2}}{\sqrt{Rz}} = \sqrt{2 - \frac{z}{R}}$$

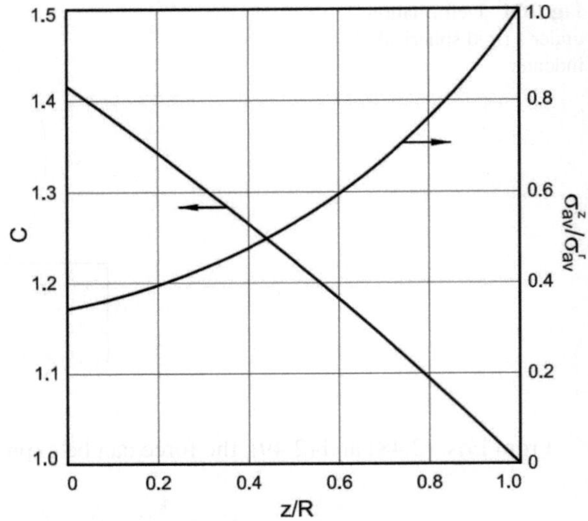

Fig. 2.72 Correction factor C and uncorrected average stress ratio as a function of relative deformation

The variation of the correction factor C is illustrated in Fig. 2.72 as a function of relative deformation z/R. At zero deformation, its limit value is $\sqrt{2} = 1.414$, if $z = R$ then $C = 1.0$.

The curve is well approximated as

$$C = 1.414 - 0.414\left(\frac{z}{R}\right)^{1.1}$$

The average stress in the contact surface is given by

$$\sigma_{\text{av}} = \frac{F}{r^2 \pi}$$

Using Eqs. (2.50) and (2.49a), the average stress has the following expression

$$\sigma_{\text{av}}^{z} = 0.4244 \cdot E_0 \frac{\sqrt{R \cdot z}}{2R - z} \tag{2.51}$$

We can use the theoretically derived Eq. (2.50a) and in this case

$$\sigma_{\text{av}}^{r} = 0.4244 \cdot E_0 \frac{r}{R} \tag{2.51a}$$

Equations (2.51) and (2.51a) should be equal it is not the case. Their ratio is given by

$$\frac{\sigma_{\text{av}}^{z}}{\sigma_{\text{av}}^{r}} = \frac{(R \cdot z)^{3/2}}{\left(2Rz - z^2\right)^{3/2}}$$

which is also illustrated in Fig. 2.72.

Without the introduced correction in Eq. (2.49c), considerable errcrs should be reckoned with.

Combining the basic Eqs. (2.48) and (2.49), further calculation equations can be obtained which may be useful in processing experimental results. For example, the average stress in the contact surface can also be derived in the following fashion

$$\sigma_{av} = \frac{0.3856}{C^2}\left(\frac{F \cdot E_0^2}{R^2}\right)^{1/3} \tag{2.51b}$$

The ratio of vertical deformation z and the radius of deformation r is easily derived as

$$\frac{z}{r} = \frac{0.9086}{C}\left(\frac{F}{R^2 \cdot E_0}\right)^{1/3} \tag{2.52}$$

or combined with Eq. (2.51b) yields

$$\frac{z}{r} = 2.356 \cdot C\frac{\sigma_{av}}{E_0} \tag{2.52a}$$

This last equation can conveniently be used in processing experimental results to obtain the modulus of elasticity E_0 (Figs. 2.69 and 2.70):

$$E_0 = 2.356 \cdot \frac{r}{z} \cdot C \cdot \sigma_{av}$$

or

$$E_0 = 2.356 \cdot \sqrt{\frac{R}{z}} \cdot C^2 \cdot \sigma_{av}$$

The maximum stress at the maximum indentation is theoretically one and half time higher than the average stress (homogeneous elastic half space). For foam materials, similar data are not known. Therefore, some experimental investigations have been conducted with a spherical intender of $R = 100$ mm equipped with a special measuring unit 40 mm in dia. for local pressure measurement. The results are depicted in Fig. 2.73. For smaller deformations, the ratio is somewhat lower than that of elastic continuum materials. The maximum stress increases rapidly over a relative deformation ($z/H_0 = 0.6$), and in practical applications, this deformation range should be avoided. Average stress values can be taken from Figs. 2.69 and 2.70. Figures 2.69 and 2.70 show that the average stress at lower deformations is nearly constant and has the lowest value. Over $z/H_0 = 0.4$ the stress increases moderately and over $z/H_0 = 0.7$ steeply. In selecting a type and thickness of cushioning material, one should strive to keep the maximum relative deformation under 0.5. In this deformation range, the maximum stress only moderately exceeds the average value, which is an advantageous property of these materials.

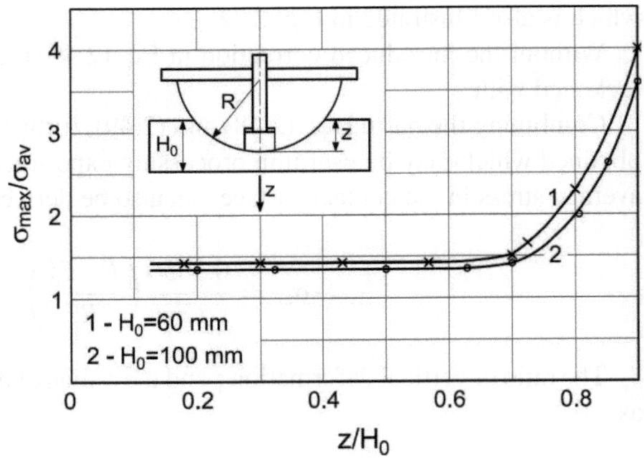

Fig. 2.73 Maximum and average pressure ratio under a spherical indenter. $R =$ 100 mm, P 3240 polyurethane foam

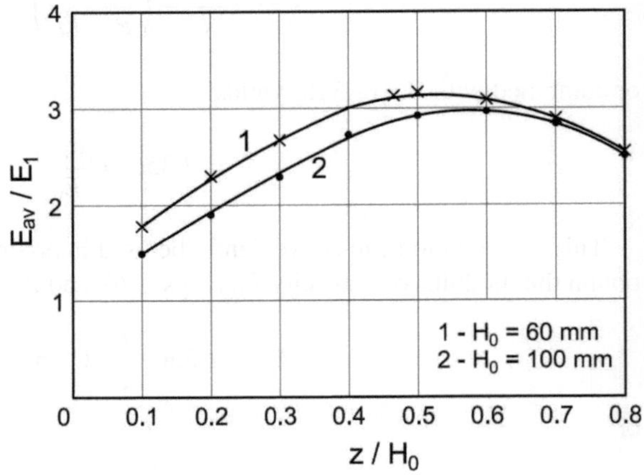

Fig. 2.74 Ratio of moduli of elasticity measured with a spherical indenter and uniaxial loading

Comparing the moduli of elasticity for uniaxial loading and those obtained with a spherical indenter for the same material shows that they fundamentally differ. Their ratio shown in Fig. 2.74 follows a definite increasing tendency due to the tension stresses in the contact layer which have not been included in the Hertz theory for small deformations.

A modified form of Eq. (2.52a) can be obtained in which the uniaxial modulus of elasticity is substituted for the relative deformation

$$\frac{z}{r} = C_0 \frac{\sigma_{av}}{E_1} \tag{2.52b}$$

where σ_{av} is the average stress under the spherical indenter, and modulus E_1 is the uniaxial modulus of elasticity for the same relative deformation. The variation of the factor C_0 is given in Fig. 2.75 as a function of relative deformation. The E_1 can be more easily determined. There is some scattering of the value of C_0 which depends on the radius of the indenter due to the dependence of E_{av}/E_1 on the radius of indenter (see Fig. 2.74).

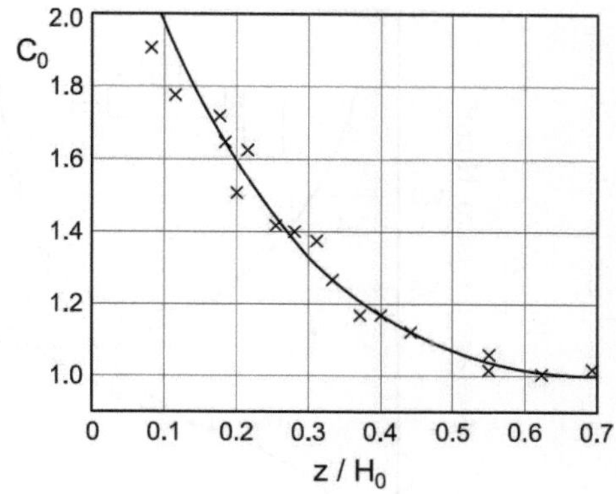

Fig. 2.75 Constant C_0 as a function of relative deformation for two different spherical indenters

Figure 2.71 shows that the lowest average pressures will be obtained at relative deformations between 0.2 and 0.5 with a flat optimum between 0.3 and 0.4. Therefore, similar relative deformations should possibly be selected for a sitting person.

A common measuring method uses circular flat plates of different diameters which are well known for the homogeneous infinite half space (Timoshenko and Goodier 1951). The relationship between the average stress and deformation is described by the following simple equation

$$\sigma_{av} = \frac{2}{\pi}\frac{E}{1-\nu^2}\frac{z}{R}$$

where z is the deformation,
R is the radius of the loading plate.

Foam materials are highly nonlinear as a function of deformation and have a finite thickness H_0 characterized by the thickness/plate radius ratio H_0/R (Yegorov 1959, 1961). The above equation demonstrates that the measurement results should be plotted as a function of relative deformation z/R. Figure 2.76 shows these measurements using two different plate radii.

The effect of the plate diameter can only be neglected for smaller relative deformations up to $z/R = 0.45$. The apparent modulus of elasticity has its minimum depending on the H_0/R ratio at a given relative deformation:

$$\frac{z}{R} \cong 0.45\frac{H_0}{R} \quad \text{or} \quad \frac{z}{H_0} = 0.45$$

As relative deformations increase, both the average pressure and the modulus of elasticity rapidly increase. These results show that the plate diameter should properly selected for the given thickness.

Using flat plates and elastic continuum materials, the smallest stress is obtained at the centre and its value is equal to half of the average pressure on the circular area. Theoretically, the pressure becomes large at the boundary of the contact area. We do

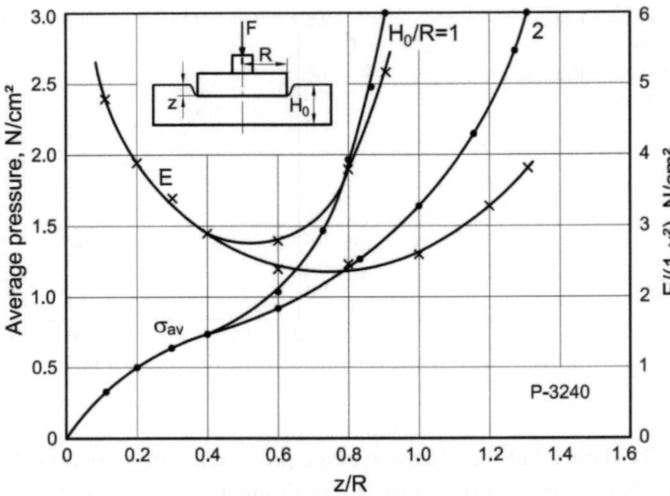

Fig. 2.76 Loading of foam material with circular flat plates. $H_0/R = 1$ ($H_0 = 6$ cm, $R = 6$ cm) $H_0/R = 2$ ($H_0 = 5$ cm, $R = 2.5$ cm)

not have any similar values for foam materials to determine the pressure distribution under rigid flat plates.

2.7.2 *Mechanical Behaviour of Soft Tissue*

When a person sits, the spherical loading body is not rigid because soft body tissue has its own mechanical behaviour. Due to its high water content, the soft body tissue is almost incompressible, with Poisson's ratios near to 0.5. Its mechanical behaviour is complicated with internal fibres and the effect of skin.

The picture is further complicated by the presence of hip bones inside the tissue which behave as a rigid core. This rigid core hinders the free deformation of the soft body tissue, and therefore, high pressures may arise on its surface (stress concentration). In extreme cases, the core can support almost the whole load acting on the soft tissue. Since soft tissue is almost incompressible, its deformation is very distinct. The bulk modulus K which is calculated as

$$K = \frac{E}{3(1 - 2\nu)}$$

has values in the range 100–150 N/cm^2 while the elastic modulus E is around 10 N/cm^2, with Poisson's ratios of 0.48 or 0.49. Hydrostatic pressures cause practically no distortion, only a slight volumetric shrinkage. At the tip of the hemisphere, especially on a rigid flat surface, the soft tissues are squeezed, which greatly increases

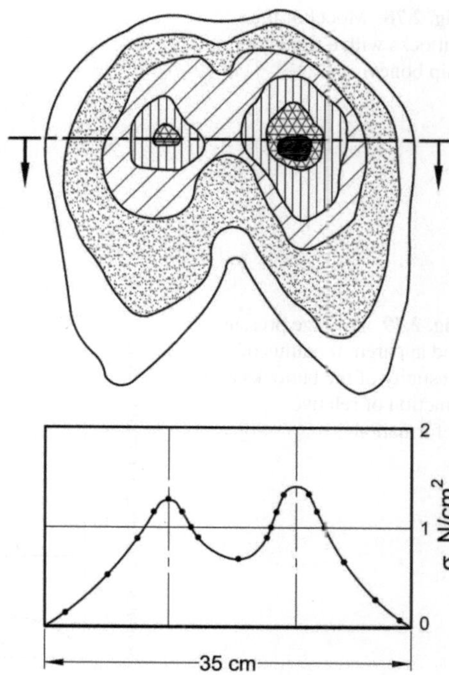

Fig. 2.77 Pressure distribution on the buttocks with almost no thigh contact

the contact area but decreases the thickness of soft tissue between the rigid surface and hip bone. As a consequence, high pressures may arise on the surface of the hip bones, supporting a high per cent of the load. Therefore, the potential damage to body tissues highly depends on the state of the stress acting on the tissue. Body tissues tolerate hydrostatic pressures as high as 100 N/cm^2 but uniaxial pressures of only around 1 N/cm^2 (Chow and Odell 1978). It is due to the fact that hydrostatic pressures cause no distortion in the tissue but uniaxial pressures with the associated shear stresses do. A concave seat profile provides a better hydrostatic pressure component and less tissue distortion. This was already used in the seat design of better automobiles 50 years ago.

The typical pressure distribution on the buttocks of a sitting person from an early measurement (Lindan et al. 1965) is demonstrated in Fig. 2.77. It is remarkable that the increase of pressure to its maximum is rather linear, giving a higher σ_0/σ_{av} ratio, which here is near 2 (a true conical distribution would give the value of 3). The sitting position almost always has some thigh contact, lowering peak pressure. Taking a position with no thigh contact gives a nearly symmetrical pressure distribution, and it was frequently used in simulations (Chow and Odell 1978; Reddy et al. 1982; Brush and Arcan 2000; Candadai and Reddy 1992).

The buttocks of a sitting person are generally modelled as spherical bodies with a rigid core as depicted in Fig. 2.78 (Chow and Odell 1978). It is assumed that the modulus of elasticity of the soft tissue is uniform throughout the body, with Poisson's

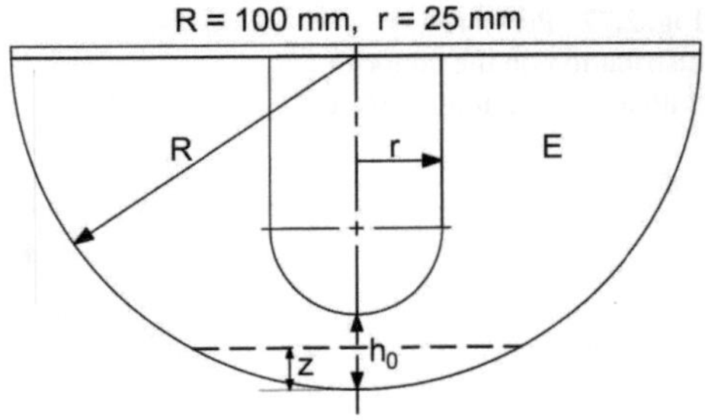

Fig. 2.78 Model of the buttocks with a rigid core (hip bones)

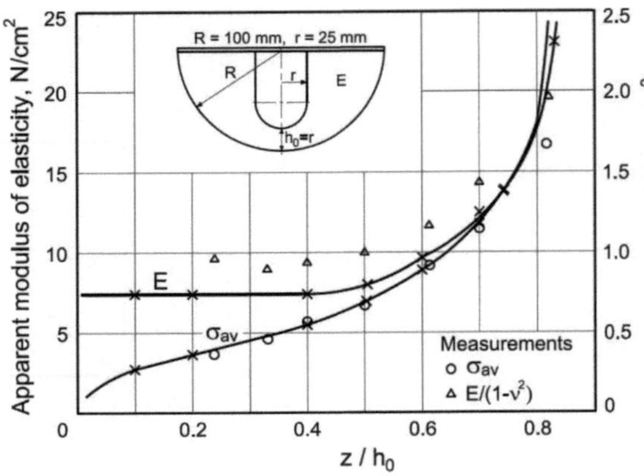

Fig. 2.79 Average pressure and apparent modulus pf elasticity of the buttocks as a function of relative deformation

ratios of 0.48 or 0.49. It is clear that the stress concentration on the core surface mostly depends on the relative deformation z/h_0.

Very little research has been done with appropriate material selection which would simulate the mechanical behaviour of human soft tissue.

One measurement done some 40 years ago is suitable for further processing to clarify some important problems (Chow and Odell 1978). The average pressure and modulus of elasticity can be calculated from the force displacement relationship with the given geometry, Fig. 2.79.

At small relative deformations, before a rigid core, the modulus of elasticity remains constant as it should be for a homogeneous body. Starting from $z/h_0 = 0.45$, the apparent value of the modulus of elasticity increases, first moderately, thereafter steeply, indicating that the deformation is hindered by the rigid core. The average pressure also changes its course with an inflexion point into a steeply rising course. Note that the maximum stress on the core (bone) surface will be much higher compared to the average stress.

New measurements on a living person (rather lanky male) sitting on a rigid flat plate produced similar results. Varying the vertical load on the buttocks, the contact surface and the vertical deformation have been measured with minimum thigh contact. The distance of the bone from the surface without a load condition was also approximately estimated. With a small vertical load of 150 N, the deformation of the

Fig. 2.80 Approximate ratio of the maximum and average stresses in the buttock as a function of relative deformation. 1—Upper boundary from model calculations on rigid surface. 2—Lower boundary on cushion

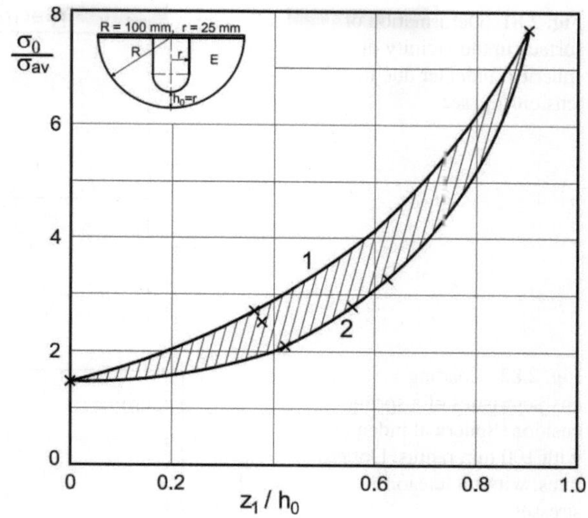

buttock was 1.25 cm which already corresponds to an approximate relative deformation of $z/h_0 = 0.5$. That means that with further loading, the deformation of the buttocks will be hindered by the bones continuously increasing the local pressure. Under a full load 600 N, the deformation was 2.1 cm and $\sigma_{av} = 1.6$ N/cm^2 which may correspond to a local pressure five to six times higher on the surface of the hip bones.

An important observation is that the radius of a buttock considerably varies as a function of loading. The initial radius of a buttock was estimated at not more than 10 cm while under full load its apparent value was 16.5 cm.

In these experiments, the modulus of elasticity of the soft tissue can be taken as $E/(1 - \nu^2) = 10$ N/cm^2 (Fig. 2.79). It is probable that this value may change from person to person. Another important variable is the initial distance of a bone h_0 from the surface of a body. A more corpulent buttock has a lower peak pressure despite the higher load.

The reliable estimate for the maximum and average stress ratio σ_0/σ_{av} is the most difficult problem due to several reasons. The measurement of internal stresses is practically impossible, cushioning materials have a complicated material laws rendering reliable calculations nearly impossible and, finally, we do not know enough about the human body tissue. The selection of an appropriate material for experimental modelling of the soft body tissue is also not very easy.

Summarizing the very few theoretical and experimental results, Fig. 2.80 shows the approximate range for the σ_0/σ_{av} ratio as a function of relative deformation of the buttocks. Using cushioning materials with appropriate thickness, a more favourable stress distribution can be achieved lowering also the σ_0/σ_{av} ratio.

We encounter relatively large deformations which may cause deviations in the processing of experimental results. Due to the large deformation, in the top layer

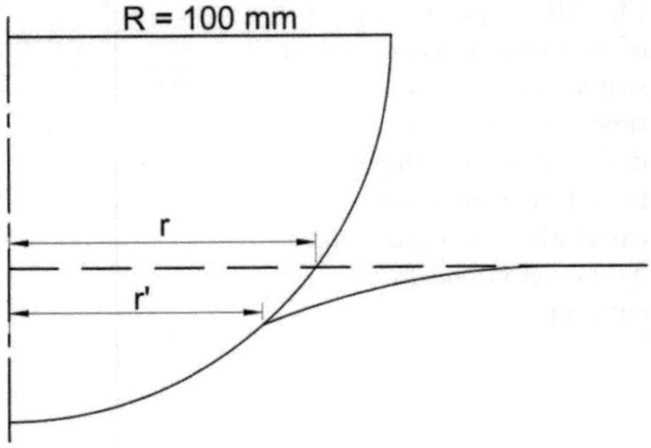

Fig. 2.81 Deformation of surface in the vicinity of a spherical indenter due to tension stresses

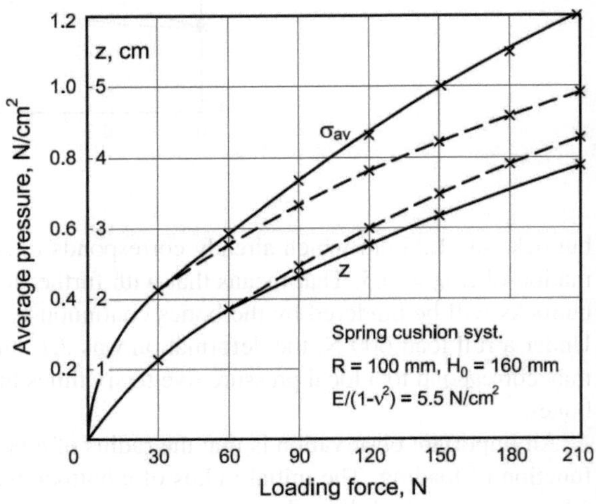

Fig. 2.82 Loading characteristics of a spring cushion. Spherical indenter with 100 mm radius. Dotted lines: without tension stresses

which is mostly a fabric, tension stresses develop which decrease the contact area, Fig. 2.81.

Furthermore, the tension stresses compensate a smaller part of the load slightly decreasing the vertical deformation.

Measurements have been made on a bed-couch upholstered with springs, cotton upper layer, and fabric cover and supported on belts. The loading was accomplished with a 100-mm-radius hemisphere. The variation of average pressure and vertical deformation as a function of loading force is depicted in Fig. 2.82.

The spring obviously behaves as a linear elastic cushion and if there are no disturbing effects, Boussinesq's theory holds (dotted lines in the figure). With larger deformations, the indentation slightly decreases but the average pressure considerably increases due to the sinking surface profile close to the indenter, Fig. 2.82.

The resultant modulus of elasticity of the cushion system is about $E/(1 - \nu^2) = 5.5\ \text{N/cm}^2$. It is interesting to note that a flat plate loading device 160 mm in diameter yielded quite similar elastic moduli indicating the elastic behaviour of the cushion system with sufficient accuracy.

The effect of tension stresses on the vertical deformation can be taken into account using the membrane analogy with large deformations (Federhofer 1936). According to this, a third power term should be included in Eq. (2.48) in the following manner

$$\frac{z}{H_0} + 1.9\left(\frac{z}{H_0}\right)^3 = \frac{0.8255}{H_0}\left(\frac{F^2}{RE^2}\right)^{1/3} \tag{2.48a}$$

where H_0 means the thickness of the cushion. Another way may be the appropriate correction (increase) of the modulus of elasticity.

Foam cushions with strong nonlinear behaviour give quite different results loaded with flat or spherical indenters which differently integrate the local values of material properties. Therefore, the correct evaluation of foam cushions requires taking the above circumstances into account. The buttocks of a sitting person are much more similar to spherical bodies, and therefore, a spherical loading device is more appropriate.

This treatment of the problems concerning upholstering materials and sitting persons is obviously a simplified one which should be useful for quick calculations and evaluations. This method does not obviate the need to conduct numerical calculations for deeper insight into the interaction of sitting or reclining persons and upholstery. Especially, the magnitude of maximum stresses is an unsolved problem but still one of the most important questions.

2.7.3 Numerical Examples

In the following, we will demonstrate the usefulness of these calculation methods and experimental results. Let us consider a sitting person on a rigid flat surface, a spring cushion system (bed-coach) and on a polyurethane foam 10 cm thick.

A. **Sitting on a rigid flat surface**. This position is practically with no thigh contact and a nearly axisymmetrical pressure distribution. The vertical load on the contact surface was 650 N (feet are supported). The contact print was measured with a maximum width and length of 170 by 280 mm. The true contact area was 405 cm^2 with an average pressure of 1.6 N/cm^2.
The total contact area corresponds to two circular prints of 160 mm diameter with the load of 325 N. The radius of each buttock is also required which requires some trials and errors. The equivalent radius depends on the posture and also the load. Its correct selection can be checked by the relation

$$R = \frac{r^2 + z^2}{2z}$$

where r is the radius of print (here $r = 80$ mm)
z is the calculated deformation.

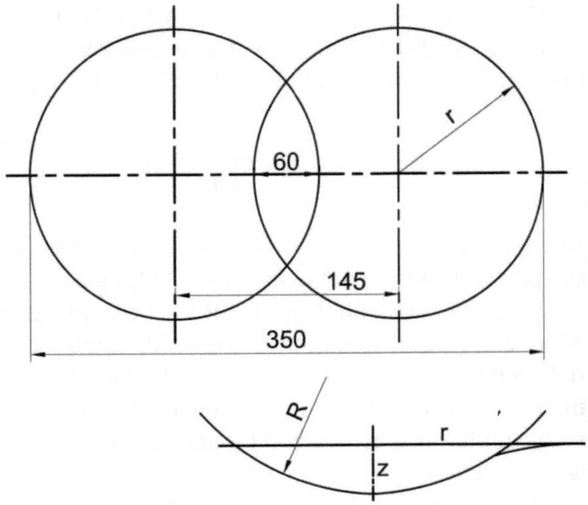

Fig. 2.83 Approximation to the calculation of buttock deformation

The equivalent radius proved to be $R = 16.5$ cm while the initial unloaded radius was around $R = 12$ cm. Now combining Eqs. (2.48) and (2.52a), we get $z = 2.1$ cm deformation and $E/(1-\nu^2) = 19.57$ N/cm^2. The distance between the hip bone and skin surface in an unloaded condition was estimated as $h_0 = 2.5$ cm and, therefore, the relative deformation is $z/h_0 = 0.84$ indicating that the bone is highly overloaded and the calculated modulus of elasticity is an apparent value due to stress concentration around the hip bone. Indeed, with a small load of 75 N and relative deformation $z/h_0 = 0.5$, the modulus of elasticity was 10.3 N/cm^2 (see Fig. 2.79). The expected maximum pressure on the hip bone is six to seven times higher than the average pressure, around 10 N/cm^2 (see Fig. 2.80).

B. **Sitting on a spring cushion system**. The cushion consists of springs, cotton layer with fabric cover supported by belts. The height of the cross-section is $H_0 = 16$ cm. The cushion was measured with a spherical indenter (see Fig. 2.82) with an elastic modulus of 5.5 N/cm^2. The print of the buttocks was also measured, and its layout is given in Fig. 2.83.
The contact area is 660 cm^2 with an average pressure of 0.985 N/cm^2. The equivalent contact radius is $r = 10.25$ cm. The radius of the buttocks is slightly smaller than in the earlier case, $R = 14$ cm. Due to the definite tension stresses in the surface layer, Eq. (2.48a) should be used which gives $z = 4.7$ cm deformation. If we use the simpler Eq. (2.48), it would require a correction in the E-modulus to 6.4 N/cm^2. Using this corrected value, Eq. (2.51b) yields $\sigma_0 = 1.267$ N/cm^2 maximum stress and 0.8477 N/cm^2 average stress. Due to the smaller effective radius (appr. 90%), the effective average stress will be higher, which is $0.8477/0.9^2 = 1.04$ N/cm^2, which is close to the measured value. Using the above calculations, we supposed a rigid spherical indenter. The actual deformations are seen in Fig. 2.84.

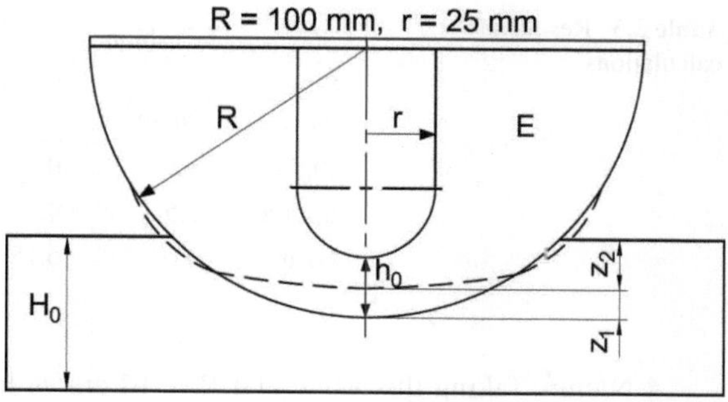

Fig. 2.84 Deformation of buttock on cushion

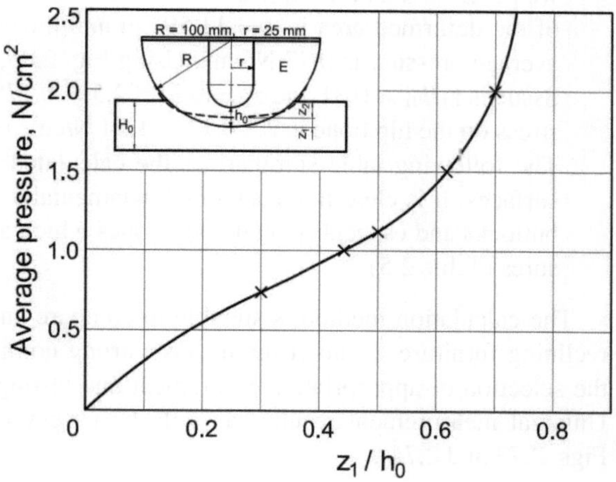

Fig. 2.85 Approximate relationship between average pressure and relative buttock deformation on cushions

In this case, the buttock remains spherical with a somewhat larger radius and therefore the results are acceptable. The deformation of a buttock can be determined in different ways. The Hertz–Boussinesq's theory gives an equation quite similar to Eq. (2.48) in which both contacting bodies may be elastic. Due to large deformations, this method is not very accurate. Using numerical results (Choww and Odell 1978) and our own calculations, the average pressure is approximated as a function of relative buttock deformation on a cushion (Fig. 2.85).
This relationship is quite similar to that of the rigid plane (Fig. 2.79), but it is higher due to the effect of cushion. Taking the calculated (and measured) 1 N/cm^2 average pressure, it corresponds to $z/h_0 = 0.43$ relative buttock deformation from which $z = 0.43 * 2.5 = 1.075$ cm. The expected maximum pressure on the hip bone is 2.25 N/cm^2 (Fig. 2.80) which is almost five time less than on the rigid surface.

C. **Sitting on polyurethane foam 10 cm thick**. The type of the foam P 3240 is the same as in Fig. 2.71 tested with a spherical loading device. In this case, Eqs. (2.48a–2.52b) may not be used in the common way due to the variation of the modulus of elasticity as a function of deformation. However, in Figs. 2.70 and 2.67 we see that the modulus of elasticity has values in a broad range around

Table 2.5 Results of calculations

	Cushion			Buttock	
Sitting on	E, N/cm^2	z, cm	σ_{av}, N/cm^2	z_1, cm	σ_{max}, N/cm^2
Rigid	∞	0	1.6	2.1	10
Spring	5.5	4.7	1.0	1.075	2.25
Foam	4.0	6.15	0.77	0.775	1.54

4 N/cm^2. Taking this value and $R = 14$ cm as formerly, then from Eq. (2.52) we get $z/r = 0.53$. Using these data, the deformation is $z = 6.15$ cm, the radius of the deformed area is $r = 11.59$ cm and the contact area $A = 422$ cm^2, the average pressure is 0.77 N/cm^2. Using Fig. 2.85, the deformation of soft body tissue is $z_1/h_0 = 0.31$ and $z_1 = 0.31 * 2.5 = 0.775$ cm. The expected maximum stress on the hip bone is $2 * 0.77 = 1.54$ N/cm^2 (Fig. 2.80).

The following table summarizes the calculated results of sitting on different surfaces. It is clear that cushions fundamentally lower the specific load on the buttocks and especially on the hip bones which are very sensitive to high pressures (Table 2.5).

The calculation method is suitable to compare and evaluate different seats and reclining furniture. Foam cushions have strong nonlinear behaviour and therefore, the selection of appropriate measurement and testing methods are very important. Uniaxial measurement results can only be used with special transformation (see Figs. 2.73 and 2.74).

Conclusions

An approximate engineering calculation method has been developed, and its applications are demonstrated. The calculations are based on experimental results performed on cushion materials and human soft tissue. The results of numerical model calculations are also taken into account which usefully complete the experimental results. Certain quantities such as internal stresses cannot be measured at all. Spring cushion systems behave almost as a linear elastic body and therefore, their modulus of elasticity can be measured using different methods such as uniaxial compression, loading with spherical indenter or a flat circular plate. With large deformations and due to tension stresses, surface deformations in the vicinity of the loading body modify the contact surface.

Polyurethane foam materials show highly nonlinear behaviour with far-reaching consequences. It is crucial to select correct measuring and testing methods to obtain useful results for system evaluation. Generally, a spherical indenter should be used but even the radius of indenter has a considerable influence on the results. More systematic measurements would be needed to achieve a further generalization of the existing results using dimensionless coordinates.

The behaviour of human soft tissue is poorly known. The possible variation in the modulus of elasticity and the position of hip bone within the tissue are still to be determined. Especially, the latter has considerable influence on the stress concentration around the hip bone surface. Nevertheless, as a first approximation, this calculation and evaluation method are suitable to dimension new seating and reclining furniture and to compare existing ones. A proper concave seat profile allows sitting for a long time and has been used for years (see Fig. 1.1).

2.8 The Hardness of Wood

Hardness is a mechanical property of wood species and a measure of its resistance to indentation by various loading devices (cone, wedge or sphere). Hardness also indicates the ability to withstand abrasion. Hence, the hardness of wood surfaces characterizes the durability and mechanical resistance against different kinds of loading such as static, dynamic, concentrated and frictional.

A hardness measurement is widely used in the metal industry and standardized methods have been developed. Due to the structural properties of wood, especially to its anisotropy, hemispherical indenters are more suitable for hardness determination in the wood industry. The diameter of the indenter may vary in a wide range between 10 and 50 mm, and earlier investigations have often stated dimensions in inches (e.g. 5/4 in.).

The hardness is an indirect mechanical property due to the complicated multiaxial stress state and the different proportion of elastic and plastic deformations. Furthermore, the same wood sample may have quite different structural properties on different areas causing considerable deviations in the measurement results. The moisture content also influences the resistance against indentation. The resin content of wood may also be important depending on its non-volatile portion (di- and triterpene), especially in close grained timbers. It is practically impossible to account for all influencing factors and, therefore, strong analytical relationships are not known.

The use of hardness measurements began more than one hundred years ago. At the turn of last century, Janka did pioneering work in this field (Janka 1915). He introduced the use of a hemispherical indenter and carried out a vast amount of measurements on nearly 300 wood species. The measurement results were plotted as a function of volume density of the samples. The compression strength of the wood was also examined as a possible independent variable.

Due to the wide range of volume density (0.3–1.25 g/cm^3), Janka found a definite correlation between hardness and density but with considerable scattering of the measurements. Figure 2.86 shows the trend line of Janka's results and new measurements which follow the old trend line (Soerianegara and Lemmens 1994). The scattering of measurement results is also quite similar to those of hundred years ago.

The use of a spherical indenter allows an approximate mathematical treatment using the appropriate relationships of contact mechanics and mechanics of the elastic half space (Timoshenko and Goodier 1951). The loading of the surface by a spherical indenter may be static or dynamic, Fig. 2.87.

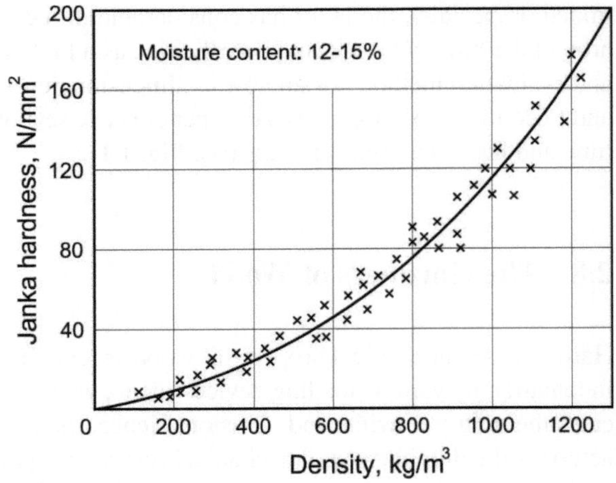

Fig. 2.86 Janka hardness as a function of wood density. Solid line: Janka's average for 286 wood species; points: new measurements mainly for SE-Asian species

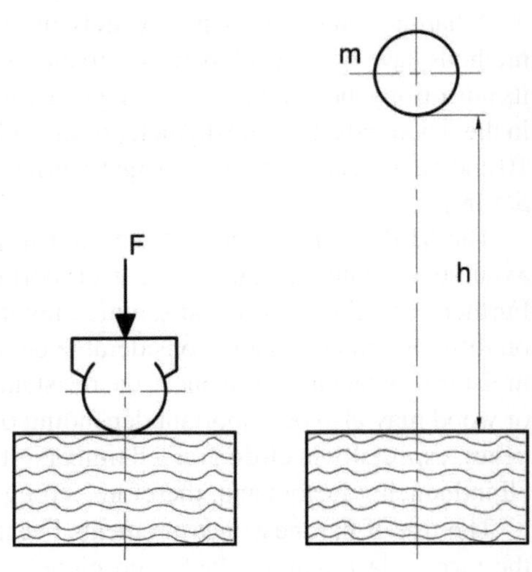

Fig. 2.87 Static and dynamic measurement method for hardness

Using the static measurement method, a spherical indenter with a radius R is pressed into the wood surface and the force F is calculated with the following equation:

$$F = \frac{4}{3} \cdot \frac{E}{1 - \nu^2} \sqrt{R} z^{3/2} = B \cdot z^{3/2} \tag{2.53}$$

where

E is the modulus of elasticity,
ν is the Poisson's ratio,
z is the deformation under the indenter.

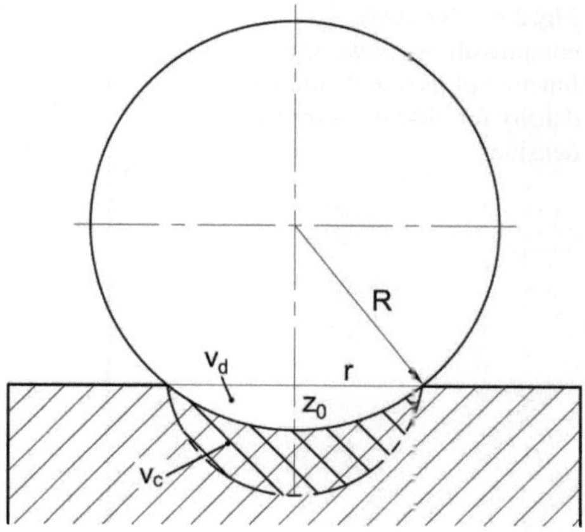

Fig. 2.88 Displaced and compacted volume under an indenter

The hardness now can be characterized by the material property $E/(1-\nu^2)$, and we may write

$$H = \frac{E}{1-\nu^2} = \frac{3F}{4\sqrt{R}z^{3/2}} \tag{2.54}$$

The result seems to equal the modulus of elasticity of the material but it is not the case. Due to the large and three dimensional deformations and the combined elastic and plastic deformations, that is a *modulus of deformation* characterizing the resistance of material against the indenter penetration under the given conditions. This is why different measurement methods supply different hardness values for the same material.

Janka has used a steel ball 11.28 mm in dia. and pressed half of it into the wood surface with a deformed cross-section of 100 mm^2. The hardness was defined as the required force divided by the deformed cross-section:

$$H_\mathrm{J} = \frac{F}{100} \quad \mathrm{N/mm^2} \tag{2.55}$$

The measurement of Janka's hardness is associated with large permanent deformation with a displaced volume of 0.376 cm^3. This displaced volume will be compacted and accommodated under the steel ball in a given volume which shows permanent deformation. The energy of compaction is absorbed by this volume, Fig. 2.88.

In order to calculate the compacted volume, its volume density should be known. Sadly, such experiments have not been done. Some reference may be obtained from uniaxial compaction measurements.

The compressed volume under the indenter is given by

$$V_\mathrm{c} = \frac{\rho_\mathrm{v}}{\rho_0 - \rho_\mathrm{v}} V_\mathrm{d}$$

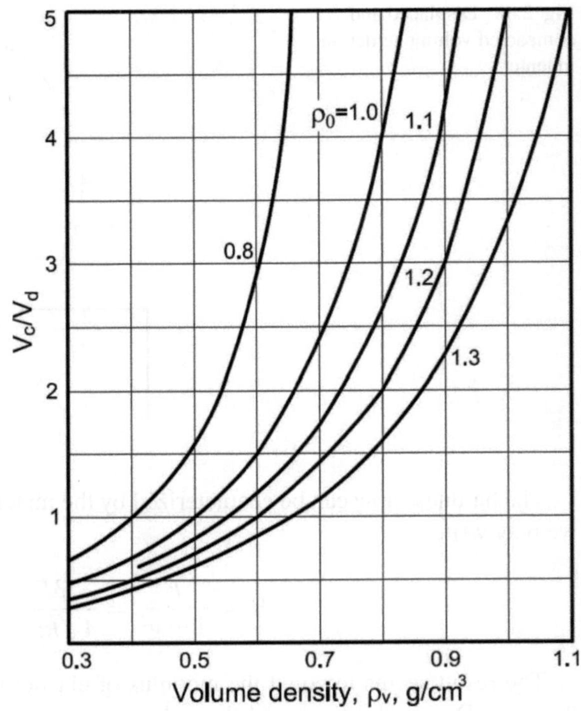

Fig. 2.89 Relative compressibility of wood as a function of its initial volume density for different terminal densities

where

V_d is the displaced volume,
ρ_v is the volume density of the wood,
ρ_0 is the density of the compacted volume.

The variation of relative compacted volume V_c/V_d as a function of timber density is shown in Fig. 2.89.

With higher densities, the compacted volume of wood rapidly increases.

Researchers in the 1930s already observed that Janka's measurement method supplies inflated values for high-density wood (Pallay 1951). The reason can be found in the increasing compacted volume under the indenter. This compacted volume increases the virtual diameter of the indenter which requires higher forces and gives higher hardness values.

In order to eliminate these shortcomings, a wider contact area and lesser displaced volume were suggested [method of Krippel and Pallay (Pallay 1951)] which would be less sensitive to the modifying effect of the compacted volume under the indenter. Using a 5/4″ dia. steel ball with a constant 2 mm indentation ($V_d = 0.191$ cm^3) and 200-mm^2 contact surface, the hardness values as a function of timber density are given in Fig. 2.90 (Pallay 1951).

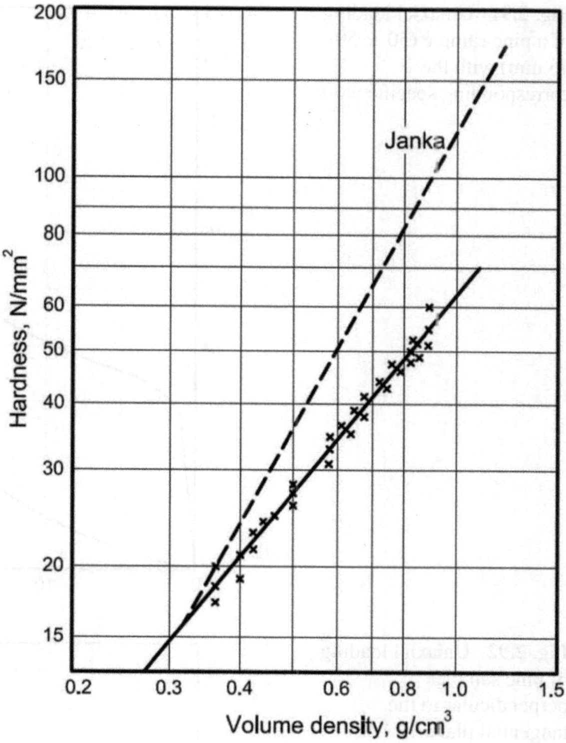

Fig. 2.90 Krippel—Pallay hardness as a function of wood density. The dotted line indicates the corresponding Janka hardness

The correlation function has the following form:

$$H_{\mathrm{KP}} = 65 \cdot \rho_{\mathrm{v}}^{1.25} \tag{2.56}$$

while Janka's hardness according to Fig. 2.86 is described by a quadratic function

$$H_{\mathrm{J}} = 118 \cdot \rho_{\mathrm{v}}^{2} \tag{55a}$$

where the wood density must be substituted in g/cm^3. For lower densities around 0.4 g/cm^3, both methods supply the same hardness values.

The work done by the indentation will be absorbed by the compacted volume, and it can be expressed in J/cm^3 unit. In order to get some information about its magnitude, a uniaxial compression test was performed on light pine samples ($\rho_{\mathrm{v}} = 0.38$ g/cm^3), Fig. 2.91.

After a short steep course of the force, the skeleton of the material loses its stability and the force (or stress) slowly increases with considerable deformations. As the bigger voids are used up by the compaction, the force rapidly increases again. The work done by the compaction increases continuously according to the equation

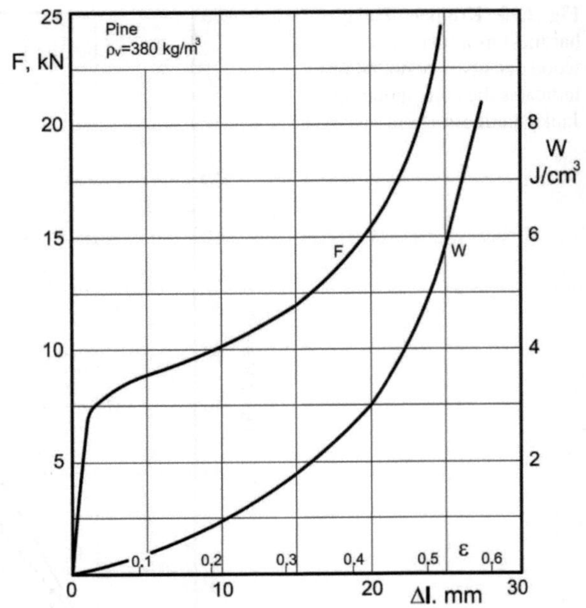

Fig. 2.91 Uniaxial loading of a pine sample (50 × 50 × 46 mm) with the corresponding specific work

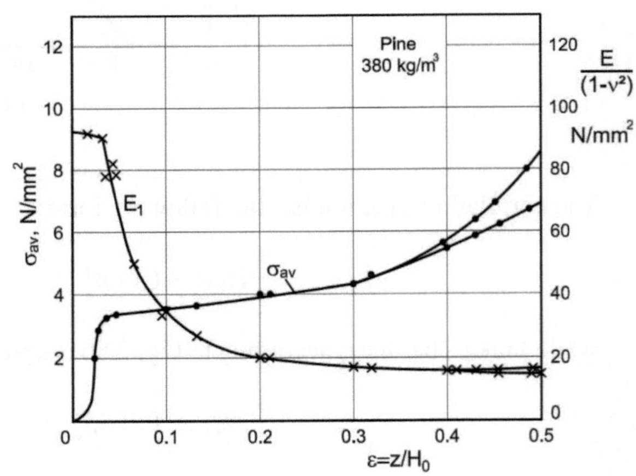

Fig. 2.92 Uniaxial loading of pine samples perpendicular to the tangential plane

$$W = 2.1 \cdot \frac{\varepsilon}{\varepsilon_{max} - \varepsilon} \quad \text{with} \quad \varepsilon_{max} = 1 - \frac{\rho_v}{\rho_{max}}$$

where $\rho_{max} \cong 1.5\ \text{g/cm}^3$ is the density of solid wood.

Figure 2.92 shows the stress–strain relationship and the calculated modulus of elasticity which rapidly decreases with increasing deformations.

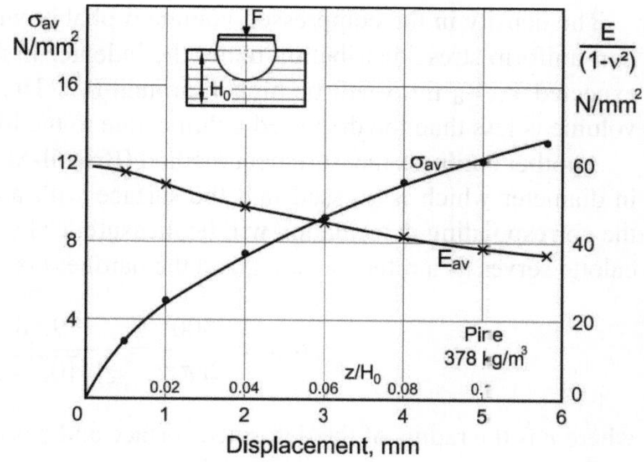

Fig. 2.93 Loading of pine samples with a steel ball 11.8 mm in diameter. Average of four measurements

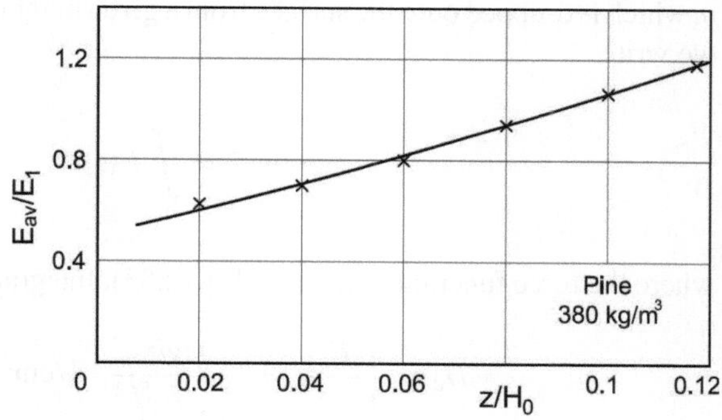

Fig. 2.94 Relationship between the E_{av}/E_1 ratio and the relative deformation

In the following, the same pine wood was used to perform hardness measurements with a steel ball 11.8 mm in dia. (it is very close to Janka's steel ball 11.3 mm in dia.). The average of four measurements is depicted in Fig. 2.93.

The apparent modulus of elasticity was determined from the average stress and deformation using Hertz's theory. The modulus of elasticity moderately decreases with increasing deformations. It is interesting to compare the moduli of elasticity given in Figs. 2.92 and 2.93 which is presented in Fig. 2.94 as a function of relative deformation.

The E_{av}/E_1 ratio increases almost linearly with the relative deformation and indicates that between uniaxial loadings, and loading with a spherical indenter, a definite correlation may exist despite the nonlinear behaviour of the material. Cushioning materials show a quite similar behaviour (Fig. 2.74).

Using Figs. 2.92 and 2.93, the density of the compacted volume under the indenter can be approximately estimated. In Fig. 2.93, the average stress at maximum indentation was 13 N/mm^2. In the uniaxial test (Fig. 2.92), this average stress required $\varepsilon \cong 0.65$ strain. The corresponding compressed density is $380/0.35 = 1086$ kg/m^3. Using this value, $V_c/V_d = 0.54$.

The density in the compressed volume probably will not be constant, due to the non-uniform stress distribution under the indenter in the half space. Therefore, the expected V_c/V_d ratio will be higher, around 0.7. That means that the compressed volume is less than the displaced volume, due to the low density of the wood.

Another hardness measurement method (Brinell-Mörath) uses a steel ball 10 mm in diameter which is pressed into the surface with a constant force of 500 N and the corresponding deformation will be measured. The contact surface as a spherical calotte serves as a reference area, and the hardness is calculated as

$$H_{BM} = \frac{500}{2r\pi z} = \frac{79.58}{z\sqrt{10z - z^2}} \qquad (2.57)$$

where r is the radius of the deformed surface and z is the vertical deformation.

The dynamic measurement method generally uses a metal sphere with the mass m which is dropped onto the surface from a given height h. Using the energy equation, we write

$$mgh = \int_0^{z_0} F(z)\mathrm{d}z$$

where the force function is given by Eq. (2.53). Integrating the above equation yields

$$H_d = \frac{E}{1 - \nu^2} = \frac{mgh}{\frac{8}{15}\sqrt{R}z_0^{5/2}} \quad \mathrm{J/cm^3 \ or \ N/cm^2} \qquad (2.58)$$

where z_0 means the maximum deformation. This value is obtained from the measured radius of the imprint by the simple relation:

$$z_0 = R - \sqrt{R^2 - r^2}$$

The maximum deformation can also be expressed from Eq. (2.58) as follows

$$z_0 = \left(\frac{2.5\,mgh}{B}\right)^{0.4}$$

Keeping in mind Eq. (2.53), the maximum force is calculated as

$$F_{max} = B^{0.4}(2.5\,mgh)^{0.6}$$

The use of the dynamic method is demonstrated with the following example, using a steel ball 7/4 in. in dia. ($R = 22.2$ mm), its mass is 0.375 kg and the drop height is 0.5 m. The potential energy is mgh = 1.751 Nm. Measurements were performed on samples of ash, spruce and jarrah. The deformations and calculated results are summarized in Table 2.6.

Table 2.6 Evaluation of dynamic hardness measurements

Species	r (mm)	z_0 (mm)	V_d (cm^3)	H_d (N/mm^2)	T (s)	v (m/s)
Spruce	7.5	1.305	0.1164	358	3.3×10^{-4}	2.1
Ash	5.5	0.692	0.033	1750	6.19×10^{-4}	2.1
Jarrah	6.2	0.883	0.041	950	4.2×10^{-4}	2.1

Table 2.7 Different hardness values, N/mm^2

Wood species	H_J	H_{BM}	H_d
Spruce	25	14	360
Linden	25	15	360
Pine	33	18	600
Larch	38	20	850
Beech	60	32	1400
Oak	70	34	–
Ash	70	34	1750
Hornbeam	75	36	1900
Sugar maple	80	40	–
Blue gum	100	48	2600
Black locust	80	40	–
African ironwood	130	70	–

The wood is a viscoelastic material with a time effect. Therefore, it is interesting to calculate the average loading velocity which is determined by the deformation z_0 and the impact duration T:

$$v = z_0/T \quad \text{and} \quad T = 0.946\frac{\pi}{2}\frac{z_0}{v_0}$$

where v_0 is the falling velocity $\left(v_0 = \sqrt{2gh} = 3.132 \text{ m/s}\right)$.

The obtained hardness values are high and near the real modulus of elasticity of wood. The average loading velocity is always the same, $v = 2.1$ m/s, which is high enough to neglect the viscoelastic effect.

This method is not free of systematic errors. Due to the elastic rebound, the actual deformation is higher than the measured permanent deformation. Furthermore, the energy assigned to the permanent deformation is less than the full potential energy of the steel ball.

In summary, the measurement and evaluation methods of hardness never supply exact results. Nevertheless, they are suitable to compare the hardness of different wood species. In order to interpret the different hardness values properly, an approximate correspondence between the methods is useful. Table 2.7 shows comparative measurement results for selected wood species.

From this table, it follows that Janka's hardness is roughly twice as much as Brinell-Mörath hardness and the dynamic hardness has the following correlation equation with Janka hardness:

$$H_{\mathrm{d}} = 3.6 \cdot H_{\mathrm{j}}^{1.43}$$

These established hardness measurement methods require a more detailed analysis and revision to develop a single and more accurate hardness evaluation.

2.9 Abrasion Resistance Property

There are utility surfaces (kitchen table, flooring) subjected to different, mostly frictional loading, causing wear. The wear process causes material loss, defects in coating, change in colour and scratches. These changes generally lower the utility value of the product considerably, especially its aesthetic value.

The loading conditions on these surfaces may be quite different in their kinds and intensity, and almost always have a probabilistic nature. Most of the loading is friction, and the ability of the surface to resist wear is characterized by the abrasion resistance of the given material.

The abrasion resistance of wood materials is also a secondary and complicated mechanical property without an exact definition. There are few experimental methods and measurement units which allow a comparison of abrasion resistance for different wood materials.

The most commonly used measurement unit is the Taber abraser and its standardized evaluation method. The abraser consists of two rollers coated with sandpaper and mounted on a balanced axle. The grade of sandpaper is specified for solid wood, coated and laminated surfaces. The axle is loaded with a given vertical force, and the sample is rotated under the rollers. The axis of the wheels does not coincide with the centre of rotation giving an additional abrasion effect similar to the "drifting wheel" action (Csanády et al. 2015).

The abraded surface is a circular path about 12 mm wide. The measure of abrasion resistance is the material loss after 100 revolutions.

The main shortcomings of the Taber abraser are:

- in the first minutes of running, an unsteady "running-in" process takes place which does not correspond to the subsequent average wear rate,
- the removal rate of abrasive material (sandpaper) considerably varies in time due to its wear process depending on the wood species,
- the effect of wheel load is strongly nonlinear,
- using fine-grade abrasive materials, the abrasive surface may be clogged to a certain extent but only the path on the wood surface is cleaned by suction.

The very simple standardized evaluation method of the Taber abraser can be improved by analysing the measurement results (Csanády et al. 2015). Figure 2.95

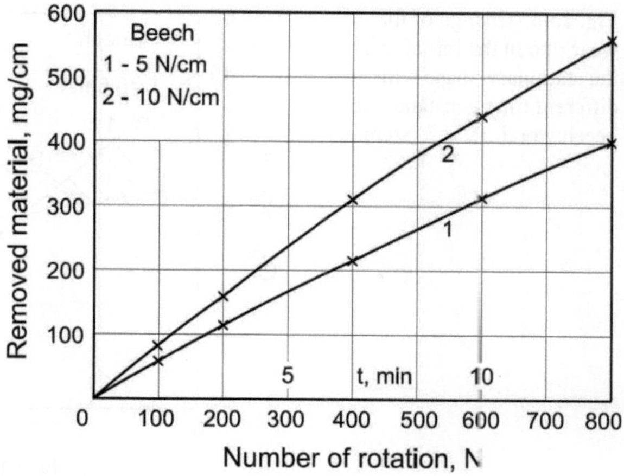

Fig. 2.95 Material loss of beech samples as a function of number of rotations with two wheel loads

shows experimental results on beech samples with two different wheel loads as a function of rotation number. The elapsed time (t) is given by the number of rotations as $t = N/60$. In contrast with common sanding, the removed material here is not proportional to the surface pressure or wheel load.

The performance of sandpapers generally decreases as a function of time due to the wear process. The decrease of wear rate, similarly to the common sanding, can be described with an exponential function in the following form:

$$\frac{\mathrm{d}G}{\mathrm{d}t} = G_0 e^{-\beta \cdot t} \quad \text{mg/cm} \cdot \text{min} \tag{2.59}$$

where G_0 is the initial wear rate and β characterizes the rate of decrease as a function of time. The reciprocal value of β is often referred to as the time constant T. The integral of the above equation gives the removed material as a function of time

$$G = \frac{G_0}{\beta}\left(1 - e^{-\beta \cdot t}\right) \quad \text{mg/cm} \tag{2.60}$$

The experimental results fairly follow Eq. (2.60) except the first hundred rotations during which the rate of material removal is systematically higher than the extrapolated average value, Fig. 2.96.

This deviation is probably due to the influence of the initial value of surface roughness on the removal rate. Indeed, measurements have shown that the initial surface is much smoother than the abraded surface. Using rough abrasive materials (P-32, P-40) with average grit diameters of 400 μm, in the initial stage each grit can bite a full chip from the relatively smooth material. As the surface becomes rough with deep furrows, the bite of some grits may be reduced resulting in less material removal. That means that a "running-in" period takes place to produce the steady-

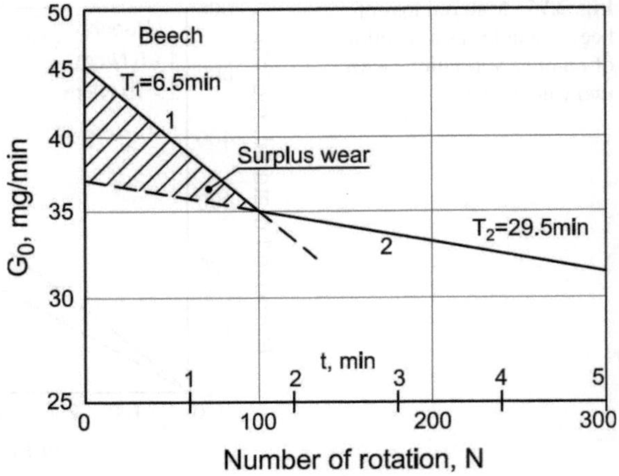

Fig. 2.96 Change of the wear rate in the initial and the stationary phase with different time constants for beech wood. $P' = 5$ N/cm

Table 2.8 Abrasion resistance classification

mg/100 rotation	G_0, (mg/cm · min)	Classes
<80	<50	Very good
80–150	50–85	Good
150–220	85–120	Satisfactory

state material removal which then slightly decreases due to the continuous wear of the abrading material.

The proposed new evaluation method is to construct the steady-state line of material removal according to Fig. 2.96 without the use of the first measured point after 100 rotations. The extrapolation of the steady-state line to $N = 0$ gives the material removal rate the each case (37 mg/cm · min in Fig. 2.96 for beech wood). The time constant T calculated for the steady-state line may supply some further information. Its value is the result of interactions between the wood and sandpaper. Since a coarse grade of sandpaper was used, one should not reckon with clogging of the sandpaper which would lower the time constant. A hardwood can accelerate the wear of sandpaper and decrease the time constant. The time constant T_1 for the initial section in Fig. 2.96 is to be determined with the following condition: the initial curve (1) intersects the stationery curve at $N = 100$ rotations, and its integral between zero and $N = 100$ rotations gives the experimentally measured value after $N = 100$ rotations.

A finer abrasive sandpaper *P*-150 is used to test coated utility surfaces (flooring). Figure 2.97 shows processed experimental results for clear coated Jatoba flooring (Hymenaea courbaril).

The main difference to Fig. 2.96 is the much smaller time constant T which may be the result of a light clogging on the abrasive surface due to the fine and sticky fragments of coating material.

The abrasion resistance of wood surfaces can be classified based on the measurement results. One of the possible classifications is given below in Table 2.8.

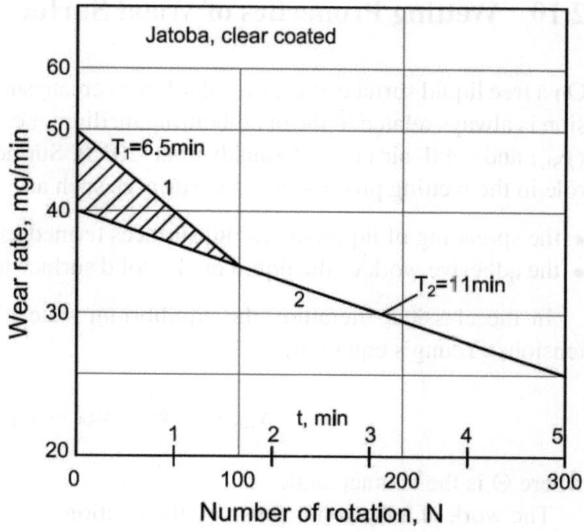

Fig. 2.97 Wear rate as a function of number of rotations for clear coated parquet

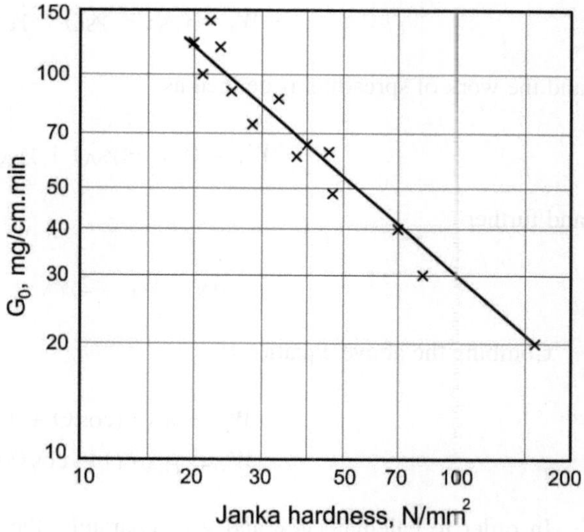

Fig. 2.98 Specific abrasion resistance as a function of Janka hardness

The abrasion resistance of wood materials may correlate with other mechanical properties such as hardness or density. Such correlations are associated with considerable scattering due to many disturbing factors. Figure 2.98 represents a correlation function between the specific wear rate and Janka hardness for selected wood species. Using the correlation among different hardness values, the wear resistance can be related to other properties.

Among the European timber species hornbeam and black locust have excellent abrasion resistance. The resistance of coated surfaces depends mainly on the abrasion resistance of the coated layer but it also requires a firm support. Therefore, the top layer of coated flooring is made of wood with proper hardness.

2.10 Wetting Properties of Wood Surfaces

On a free liquid surface, the molecular forces create surface tension. The surface tension is always related to the neighbouring medium, i.e. liquid–air (γ_{LA}), solid–liquid (γ_{SL}) and solid–air (γ_{SA}) (Csanády et al. 2015). Surface tension plays an important role in the wetting process of wood surfaces such as

- the spreading of liquid on a solid surfaces termed as "spreading wetting",
- the adhesive work of the liquid on the solid surface termed as "adhesion wetting".

In the classical literature, the equilibrium state is calculated with the surface tensions (Young's equation):

$$\gamma_{LA} \cdot \cos\Theta = \gamma_{SA} - \gamma_{SL}$$

where Θ is the contact angle.

The work of adhesion is given by the relation

$$-W_a = (\gamma_{SA} - \gamma_{SL}) + \gamma_{LA}$$

and the work of spreading is defined as

$$W_S = (\gamma_{SL} - \gamma_{SA}) + \gamma_{LA}$$

and further

$$W_S - W_a = 2\gamma_{LA}$$

Combing the above equations

$$\begin{aligned} -W_a &= \gamma_{LA} \cdot (\cos\Theta + 1) \\ W_S &= \gamma_{LA} \cdot (1 - \cos\Theta) \end{aligned} \qquad (2.61)$$

In order to calculate g_{SA} and g_{SL} separately, the validity of the Antonov rule should be accepted. This is based on the assumption that the interfacial tension is the difference between the surface tensions of the contacting solid and liquid:

$$\gamma_{SL} = \gamma_{LA} - \gamma_{SA}$$

With this assumption

$$\gamma_{SA} = \frac{\gamma_{LA}}{2} \cdot (\cos\Theta + 1)$$

If $\Theta = 90°$, then $\gamma_{SA} = \gamma_{SL}$.

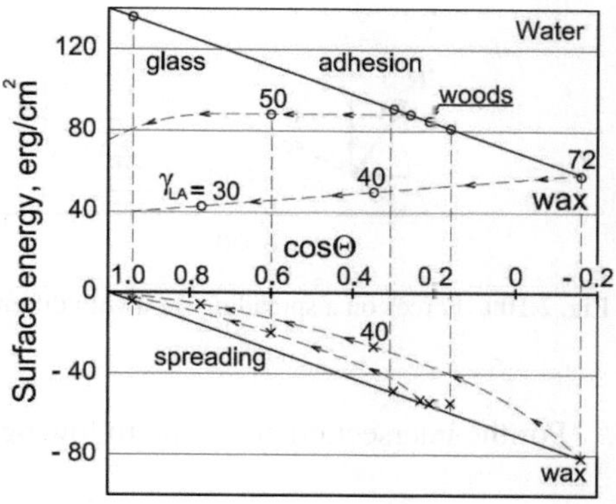

Fig. 2.99 Surface energies as a function of contact angle and surface tension

Wood surfaces have an average contact angle of 75° and for water $\gamma_{LA} = 72$ dyn/cm. Using these data, $\gamma_{SA} = 45, 26$ dyn/cm the energy for spreading $\gamma_{SL} = 53$ is and energy for adhesion is 91 erg/cm^2. Various measurement methods give similar values.

A general overview is given in Fig. 2.99 (Csanády et al. 2015). The contact angles for wood species vary between 70° and 80°. Using an oil or wax treatment, the contact angle increases. The glass surface shows the other extreme of possible variations. The substitution of water with a fluid having lower surface tension (petroleum, $\gamma_{LA} = 24$ dyn/cm) lowers the contact angle and also the energy required for spreading but the energy of adhesion remains almost constant.

Dropping fluid onto the surface, the contact angle continuously decreases due to spreading and infiltration in the first seconds and later infiltration. Because these phenomena are exponential, the plot of the contact angle as a function of time on a half-logarithmic scale yields two straight lines intersecting of which indicates the end of spreading. In the initial section, when both spreading and infiltration occur, the relative decrease of the contact angle as a function of time may be given as

$$\frac{\theta}{\theta_i} = e^{-k_1 t}$$

In the second phase, only infiltration occurs and its equation reads

$$\frac{\theta}{\theta_i} = B \cdot e^{-k_2 t}$$

where

θ_i is the initial contact angle,
B is the intersection with the ordinate of the straight lines of the second phase,
k_1 and k_2 are constants estimated from the straight lines in the half-logarithmic plot (somewhat similar to Fig. 2.96).

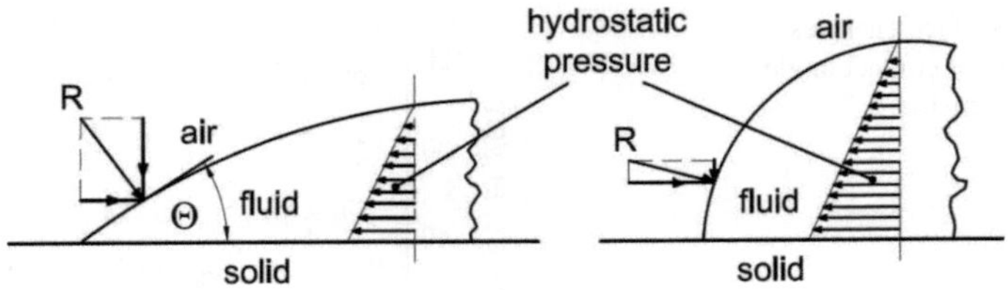

Fig. 2.100 Forces on a spreading fluid with different contact angles

For the intersection point, the following equality holds

$$e^{-k_1 t} = B \cdot e^{-k_2 t}$$

and the time t_0 to the intersection point, which indicates the end of spreading, is given by

$$t_0 = \frac{\ln B}{-k_1 + k_2}$$

As an example, a pine wood with 670 kg/m^3 density was sanded with P-80 grit size and the following constants were found: $k_1 = 0.55$, $k_2 = 0.188$, $B = 0.41$, $t_0 =$ 2.46 s and the initial contact angle was 70°. With planing, it was: $k_1 = 0.325$, $k_2 =$ 0.029, $B = 0.47$, $t_0 = 2.55$ s and $\theta_i = 73°$.

Evaluation of experimental measurements on different wood species, with and without heat treatment, shows that the average time of spreading fluctuates around 2.5 s using water or low-viscosity stains. Sanded surfaces require slightly less spreading time, due to their better conduction of fluids in the crushed upper layer in horizontal directions. Paints with higher viscosity require a longer time for spreading, around 10 s.

Due to the unreliability of measurement methods, no reliable relationships as a function of influencing parameters are known. The effect of viscosity on spreading is also almost unknown.

Due to these shortcomings of the existing methods, it may be useful to turn to engineering methods. The force relations of a spreading fluid are shown in Fig. 2.100. Inside the fluid, a hydrostatic pressure develops depending on the fluid thickness. There is an external pressure in the opposite direction due to surface tension ($\Delta p = \gamma_{LA}/R$, where R is the radius of fluid curvature). The solid surface tends to increase the contact surface until equilibrium is arrived. It is obvious that a fluid which spreads more easily will be thinner on a given surface. Measurements have shown that there is a definite correlation between fluid thickness and the contact angle, Fig. 2.101 (Csanády et al. 2015). The wettability of a surface can be estimated by fluid thickness measurements.

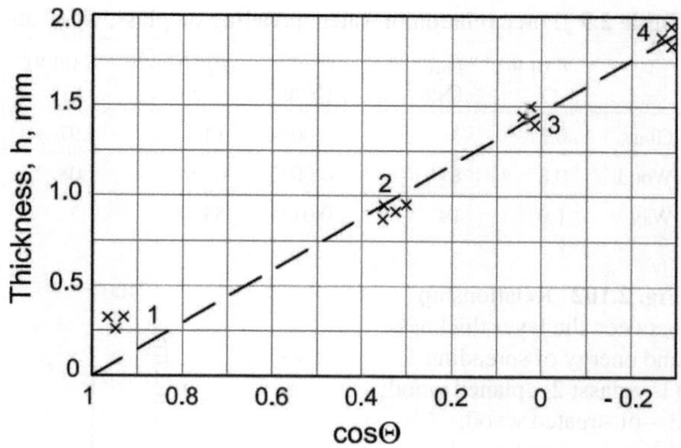

Fig. 2.101 Relationship between the layer thickness and contact angle (1—glass; 2—planed wood; 3—oil-treated wood; 4—wax-treated wood)

There is a complicated method to measure the energy of spreading. This energy can be approximated by a far simpler calculation method. Using Fig. 2.100, the difference of internal and external forces related to the unit length (N/m) should originate from the contact surface in order to hold an equilibrium.

The maximum hydrostatic pressure is

$$p_{\max} = h \cdot \gamma \quad \mathrm{N/m^2}$$

where h is the fluid thickness,
γ is the specific weight of water.

This pressure gives a force related to the unit length as follows

$$P' = p_{\max} \cdot \frac{h}{2} \quad \mathrm{N/m}$$

The external pressure due to surface tension is

$$p_\gamma = \frac{\gamma_{LA}}{R}$$

where the radius R is calculated as

$$R = \frac{h}{1 - \sin(90 - \Theta)} \quad \Theta \leq 90°$$

$$R = \frac{h}{1 + \sin(\Theta - 90)} \quad \Theta > 90°$$

The force related to the unit length reads

$$P'' = p_\gamma \cdot h \quad \mathrm{N/m}$$

Table 2.9 Force relation in water spreading on glass, wood and wax

	h (mm)	p_{max} (N/m^2)	p' (N/m)	p_g (N/m^2)	R (mm)	P'' (N/m)	$P'' - P'$ (N/m)	W_S (J/m^2)
Glass	0.3	3	0.00045	14.5	4.97	0.0044	0.004	0.0043
Wood	0.8	8	0.0032	66.67	1.08	0.0533	0.05	0.053
Wax	1.9	19	0.0181	55.4	1.3	0.1052	0.087	0.093

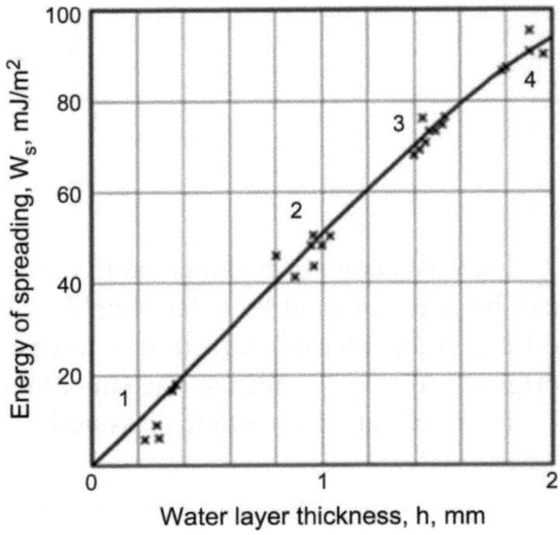

Fig. 2.102 Relationship between the layer thickness and energy of spreading (1—glass; 2—planed wood; 3—oil-treated wood; 4—wax-treated wood)

The difference of forces P'' and P' gives the force originated from the contact surface and governs the spreading. That is

$$W_S = \gamma_{LA}\frac{h}{R} - \gamma\frac{h^2}{2} \tag{2.61a}$$

Calculations for glass, wood and wax are summarized in the following Table 2.9.

Calculations show that in all cases, the calculated values agree surprisingly well with the theoretically calculated energy values for spreading. This finding demonstrates the successful use of engineering methods for solving physical problems.

Using the above calculation method, the energy of spreading was calculated based on water thickness measurements on glass and wood surfaces including treated surfaces. The results are depicted in Fig. 2.102 which gives a practical relationship between water layer thickness and energy of spreading for the natural spreading process. The energy of spreading is roughly proportional with the layer thickness. Using other fluids with different physical properties (viscosity, polar, non-polar), the above relationship will be modified in a certain extent.

The approximation of the energy of adhesion with an engineering method is more difficult but some estimation may be done. It is based on the calculation of the heat of sorption for bound water in the monomolecular layer. This layer is bound to the surface with the highest molecular forces. The diameter of water molecule is about 2.1 Angstroms and, therefore, the mass of the monomolecular layer is 2.1×10^{-4}

g/m^2. Monomolecular absorption is typical up to about 5% moisture content and 16–18% relative humidity. The corresponding radius of micro-capillary is around 0.6 nm. In this case, the heat of sorption of bound water can be calculated (Siau 1995) as

$$E_S = \frac{4.4 \times 10^{-6}}{r} \quad \text{J/mol}$$

where the radius r must be substituted in m. This equation gives $E = 7.333$ J/mol = 408 J/g. Multiplied by the specific layer mass, 2.1×10^{-4} g/m^2, we get 0.0856 J/m^2 surface energy which agrees well with the above calculated value of 0.091 J/m^2. It is obvious that the energy of adhesion originates from molecular forces.

Returning to Fig. 2.99, the energy of adhesion remains more or less constant if water is substituted for a fluid with lower surface tension. It may be supposed that the molecular forces on the wood surface remain the same in contact with other fluids. The energy of spreading decreases with lower of surface tension.

There are many disturbing factors influencing the wettability of a given wood surface. Due to its cellular structure, the presence of early and latewood of different porosity, the effect of machining on the surface integrity, all have a definite effect on wettability. As a hygroscopic material, the moisture content of wood may also have some effect on its wettability because the diffusion coefficient of wood depends on its moisture content. The viscosity of resins and paints determines their penetration into a wood surface.

The machining of wood surfaces has a definite influence on its wetting and adhesion. A fine knife-planed surface retains its anatomical characteristics while sanding homogenizes the surface due to its more crushing effect and the anatomical structure (early and latewood, vessels) is much less expressed. This more homogeneous surface may contribute to improved spreading and penetration of coatings (de Moura and Hernandez 2006). The time of spreading was shorter on a sanded surface compared to a planed surface, indicating that the fine-crushed surface has more hygroscopic water conduction parallel to the surface in all directions. Consequently, the same wood species may have different liquid spreading behaviour depending on the machining process.

The spreading property of a liquid on wood surfaces can simply be compared by measuring the spreading radius R of a droplet with a known mass (Csanády et al. 2015). The relationship between mass, height and radius of the spreading droplet can be approximated by an equation

$$m = V \cdot \rho = 3 \cdot \rho \cdot h^{0.88} \cdot R^2$$

or

$$h = \left(\frac{m}{3 \cdot \rho \cdot R^2} \right)^{1.13}$$

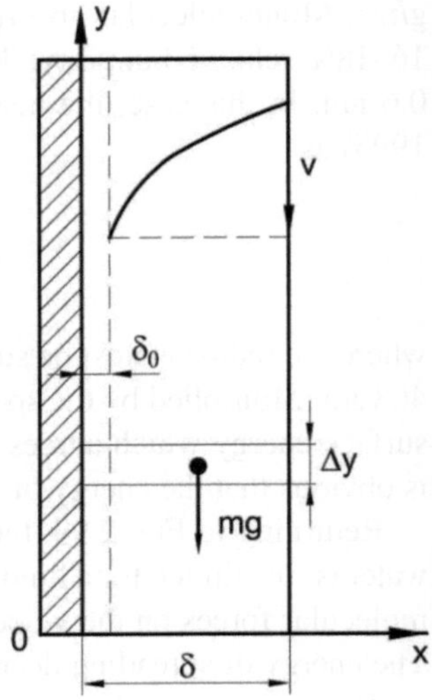

Fig. 2.103 Equilibrium of liquid layer on a vertical plate

where

m is the mass of the droplet, mg,
ρ is the density of liquid, mg/mm^3,
R is the radius of spreading, mm.

The efficiency of spreading may be characterized by the ratio of mass to the spreading area ($m/R^2\pi$) which is proportional to the spreading thickness. For example, the spreading of water on a wood surface is around 0.08 g/cm^2 or 0.8 kg/m^2 with a 0.8 mm thick layer. The viscosity of liquid determines the amount and time of spreading. The above spreading value for water should correctly be interpreted which is related to natural spreading without any external forces. In practice, we always employ forced spreading using a brush or a rotating cylinder providing a much higher coverage of between 10 and 15 m^2/litre lacquer.

Some oily wood species, such as teak require a freshly machined or sanded surface for successful gluing and staining. The oily species have contact angles between 90° and 110°.

A practical question in coating with lacquers and varnishes is that what is the minimum layer thickness which does not flow down on a vertical surface, Fig. 2.103.

The flow of Newtonian liquids on a vertical surface has been treated long ago, but this does not give any information about the thickness of the layer remaining on the surface. To get information about this minimum layer thickness, non-Newtonian liquids should be supposed. The characteristic flow curve for an ideal plastic flow is described by the equation

$$\tau - \tau_0 = \mu \frac{dv}{dx}$$

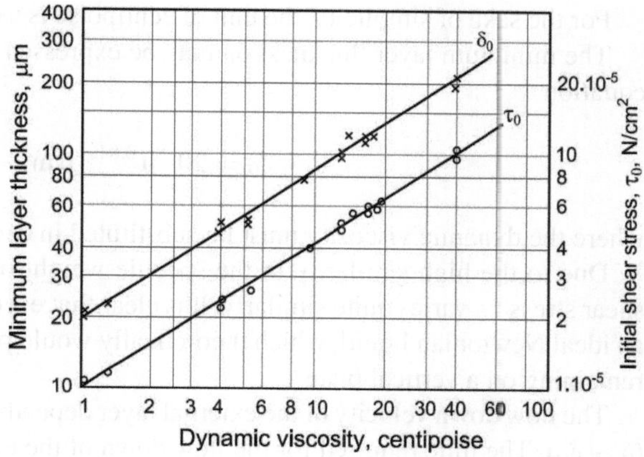

Fig. 2.104 Minimum layer thickness and initial apparent shear stress as a function of dynamic viscosity on a vertical plate

where a minimum shear stress τ_0 is required to initiate flow, but the gradient of flow velocity is inversely proportional to the dynamic viscosity μ similarly to Newtonian liquids.

The liquid film flows downward by gravity, and its thickness continuously decreases to a minimum value. The film velocity down the plate can be calculated from the equilibrium condition for a finite element Δy of the film under consideration:

$$\gamma \cdot \delta \cdot \mathrm{d}y = \tau \cdot \mathrm{d}v$$

or

$$\gamma \cdot \delta = \mu \frac{\mathrm{d}v}{\mathrm{d}y} + \tau_0$$

which is integrated for the flow velocity to the following relation

$$v = \frac{\gamma}{2\mu}\delta^2 - \frac{\tau_0}{\mu} \cdot \delta$$

where γ means the specific weight of the liquid.

If $v = 0$ and $\delta = \delta_0$, then we get the minimum layer thickness

$$\delta_0 = \frac{2\tau_0}{\gamma}$$

Values for the initial shear stress τ_0 can be obtained by viscosity measurements but the above simple relationship enables researchers to use thickness measurements on vertical surfaces.

Using different liquids with known viscosity, including water, the minimum layer thickness was experimentally determined. The relationship between layer thickness and dynamic viscosity is shown in Fig. 2.104.

For the sake of simplicity, the unit of centipoise is used which is 1×10^{-3} Ns/m^2.

The minimum layer thickness δ_0 may be expressed with the following empirical equation

$$\delta_0 = 20 \cdot \mu^{0.64} \quad \mu\text{m}$$

where the dynamic viscosity must be substituted in centipoise.

Due to the high similarity in the specific weight of the liquids used, the initial shear stress τ_0 varies quite similarly. It is clear that even water cannot be regarded as an ideal Newtonian liquid, which theoretically would produce a zero layer thickness remaining on a vertical plate.

The flow down velocity of the external layer depends on the excess layer thickness $(\delta - \delta_0)$. The time required for the flow down of the external layer is given by

$$t_0 = \frac{2\mu \cdot L}{\gamma(\delta - \delta_0)^2}$$

where L means the length of the vertical plate. For example, taking $L = 1$ m, $\mu = 16 \times 10^{-3}$ Ns/m^2, $\gamma = 10{,}000$ N/m^3, the adhering layer thickness is $\delta - \delta_0 = 50$ µm, then the time for flow down is 21 min. A 200 µm excess layer thickness will flow down in 1.33 min. That means that a small excess layer does not disturb the handling of a freshly coated plate. Using a smaller viscosity, the flow down time decreases in direct proportion to the viscosity.

The flow down process is asymptotic, due to the parabolic velocity distribution in a laminar layer. That means that the velocity of last, most internal layer tends to zero and requires an infinitely long time to flow down. In practice, the short-time processes are more interesting. The results shown in Fig. 2.104 refer to short-time measurements.

The role of surface roughness in the wetting process is much debated. It is probable that the roughness increases the spreading as it increases the hygroscopic conduction parallel to the surface. It is also probable that liquids with different viscosity require different surface structures to enhance spreading.

The adhesion of coatings correlates with the wetting properties (de Moura and Hernandez 2005). The maximum adhesion strength at rupture may vary in a wide range; typical values fluctuate between 400 and 800 N/cm^2.

Spreading glue on joints (gluing), requires less surface tension γ_{LA}, while joint strength needs a higher surface tension. The solid surface tension γ_{SA} should be as high as possible, and surface liquid tension γ_{SL} should be as low as possible. The optimum choice is always limited by various factors, and practical observations for a given wood species are important.

2.11 Colour Properties of Woods

The aesthetic value of wood products, beside form design, will be determined by its colour hues and its gloss. Especially, the inhomogeneity of natural colour and with given gloss may be extremely beautiful. At the same time, the evaluation of colour and gloss is not easy and for a long time it was much more an artistic subject without numerical representation.

Today colour can be measured with acceptable accuracy, for instance, in the $L* - a* - b*$ evaluation system. At the same time, the standard does not give a correspondence between colour coordinates and colour hue. Therefore, knowing only the colour coordinates gives no indication of the colour hue. This greatly hinders the wide-ranging use of colour description and numerical evaluation. Further efforts are needed to develop practical methods and a correspondence principle between colour coordinates and colour hue.

2.11.1 *Properties of the Reflection Curve, the Colour Hue*

The colour of a material is most accurately represented by its reflection curve. Thus, it is important to make a detailed analysis of the reflection curve.

A vast amount of measured reflection curves uniquely show that they never have a local maximum within the visible range between 400 and 700 nm but continuously increase towards 700 nm, which means that wood always has a *mixed colour*. A true blue, green or red colour always has a distinct maximum in its reflection curve.

A selection of typical reflection curves is illustrated in Fig. 2.105 (Sitkei 2013). They can be described with acceptable accuracy in the following general form:

$$R = a + b \cdot \lambda^n \quad (\%) \tag{2.62}$$

where λ is the wavelength and the constants a, b and n should be determined from the reflection curve.

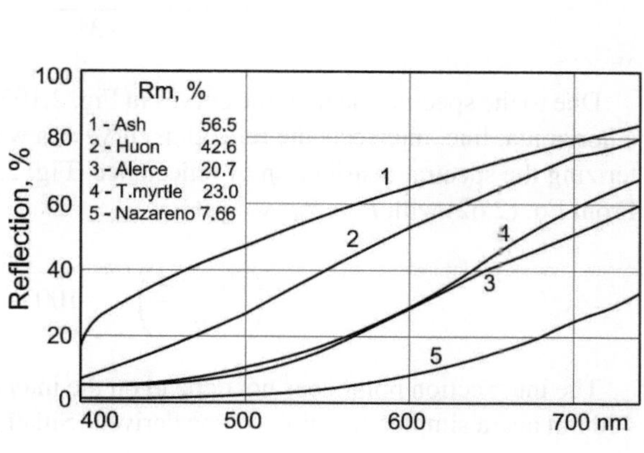

Fig. 2.105 Reflection spectra for different wood species

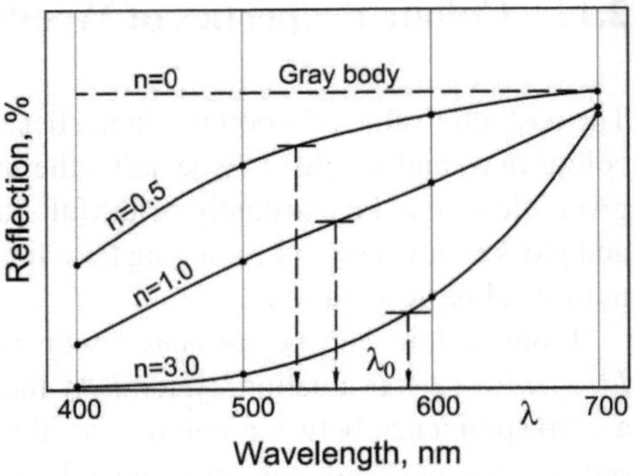

Fig. 2.106 Intersection points (λ_0) for different spectral distributions

This equation allows a detailed examination of the reflection curve. The exponent n is an important variable uniquely characterizing the colour hue of a given wood. The exponent n generally varies between 0.4 and 4, while the colour hue varies from grey to red. $n = 0$ means a horizontal line with a constant reflection. A grey body is defined by Stefan (1879) as having the same absorptivity or reflectivity for all wave lengths ($n = 0$). Indeed, a weathered pine wood with grey colour has almost constant reflection in the visible range (Sitkei 2013).

The area under the reflection curve gives its theoretical lightness termed as *physical lightness*. The physical lightness R_m is calculated by integration in the visible range from 400 to 700 nm as follows:

$$R_m = \frac{3a + \frac{b}{n+1} \cdot 3^{n+1}}{3} = a + \frac{b}{n+1} \cdot 3^n \quad (2.63)$$

with $x = \lambda/100$ and $0 \leq x \leq 3$ (Csanády et al. 2015; Sitkei 2013).

The area under the reflection curve can be related to the rectangular area (see Fig. 2.106), and this relative area is very characteristic for a given colour hue. Using Eq. (2.63), the relative area under the reflection curve is expressed by the equation

$$\overline{R} = \frac{\frac{a}{b} + \frac{1}{n+1} \cdot 3^n}{\frac{a}{b} + 3^n} \quad (2.64)$$

Due to the specific shape of the curves in Fig. 2.105, the average reflection R_m, as a horizontal line, intersects the reflection curve at a wavelength λ_0 uniquely characterizing the spectral distribution of reflectance (Fig. 2.106) and also the colour hue. From Eq. (2.62) with $R = R_m$, we get

$$\lambda_0 = \left(\frac{R_m - a}{b} \right)^{1/n} \cdot 100 + 400 \quad (2.65)$$

The intersection point does not depend on the intersection of a reflection curve at 400 nm and a simpler equation can be derived (Sitkei 2013).

Fig. 2.107 Characteristic wavelength λ_0 as a function of exponent n and the corresponding colour hue

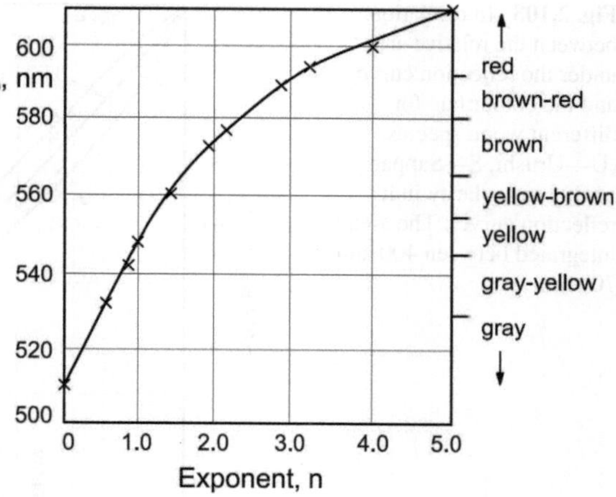

$$\lambda_0 = \frac{300}{(n+1)^{1/n}} + 400 \tag{2.65a}$$

which is the only function of exponent n. For $n = 0$, the denominator tends to the limit value of $e = 2.71$, and therefore, the lowest value of λ_0 is 510 nm. Equation (2.65a) is illustrated in Fig. 2.107 with the corresponding colour hues.

Equations (2.64) and (2.65a) can now be connected through exponent n to yield the relationship

$$\overline{R} = f\left(\frac{a}{b}, \lambda_0\right)$$

which is plotted in Fig. 2.108. Solid lines denote the theoretical relationship for different *a*/*b* ratios while the points indicate the measurement results for different wood species. The points follow well the solid lines. The range of different colour hues is also given in Fig. 2.107. These results prove the assumption that the course of a reflection curve uniquely defines the colour hue. If some irregularity occurs in the reflection curve, it may cause slight deviations. As an example, Fig. 2.109 shows reflection spectrums for urushi (Rhus verniciflua), Sappan (Caesalpinia sappan) with the rare orange colour. Especially, the latter does not fit in well with Eq. (2.62). Tasmanian blackwood (Acacia melanoxylon) and Kashmir walnut (Juglans regia) with the characteristic greyish brown colour have also irregular reflection.

It is striking that for the quite different colours of blackwood and Kashmir walnut, the standard method gives the same colour hue ($h^* = 1.82$ and 1.85) but our new estimation methods put both species to the more right and different places ($\lambda_0 =$ 589 nm, $K_2 = 0.27$ and $\lambda_0 = 564$ nm, $K_2 = 0.64$).

The accurate estimation of the area under experimentally determined reflection curves is time-consuming. There are simpler methods to characterize the shape of

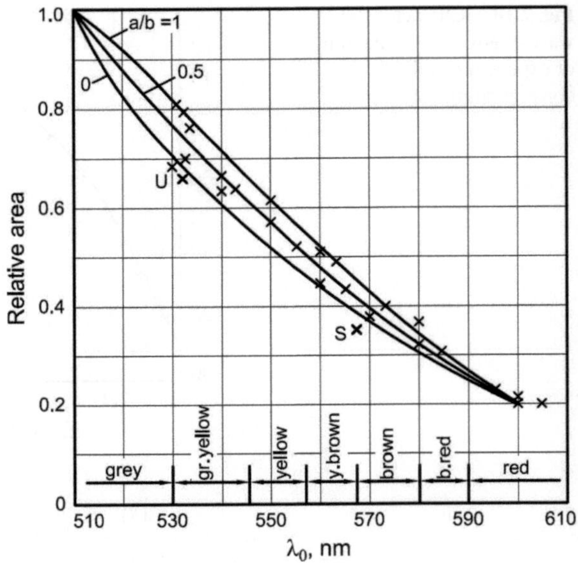

Fig. 2.108 Interrelation between the relative area under the reflection curve and the colour hue for different wood species. (U—Urushi, S—Sappan, having irregularity in its reflection curve). The area is integrated between 400 and 700 nm

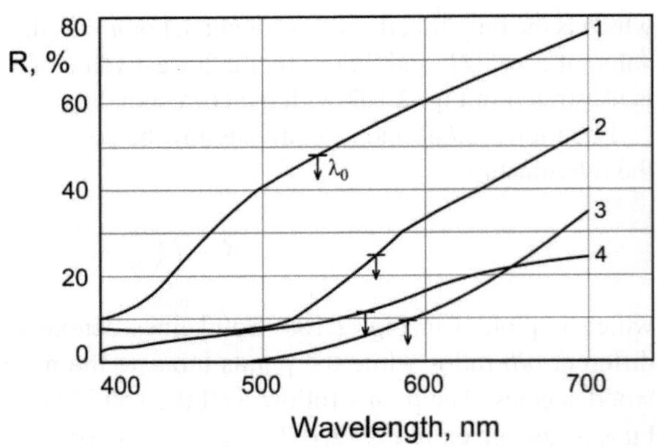

Fig. 2.109 Reflection curves for Urushi and Sappan wood species. 1—Urushi, $R_m = 47.6$, $\lambda_0 = 530$ nm, $h^* = 15$, $K_2 = 1.50$. 2—Sappan, $R_m = 23.66$, $\lambda_0 = 571$ nm, $h^* = 1.57$, $K_2 = 0.47$. 3—Blackwood, $R_m = 9.346$, $\lambda_0 = 589$ nm, $h^* = 1.82$, $K_2 = 0.27$. 4—Kashmir walnut, $R_m = 11.25$, $\lambda_0 = 564$ nm, $h^* = 1.85$, $K_2 = 0.64$

the reflection spectrum. The simplest way to take the R_{450}, R_{550} and R_{650} ordinates and make the following ratio (Fig. 2.110):

$$K_1 = \frac{R_{550} - R_{450}}{R_{650} - R_{550}} \tag{2.66}$$

If Eq. (2.62) holds, the following expression is equivalent

$$K_1 = \frac{1.5^n - 0.5^n}{2.5^n - 1.5^n} \tag{2.66a}$$

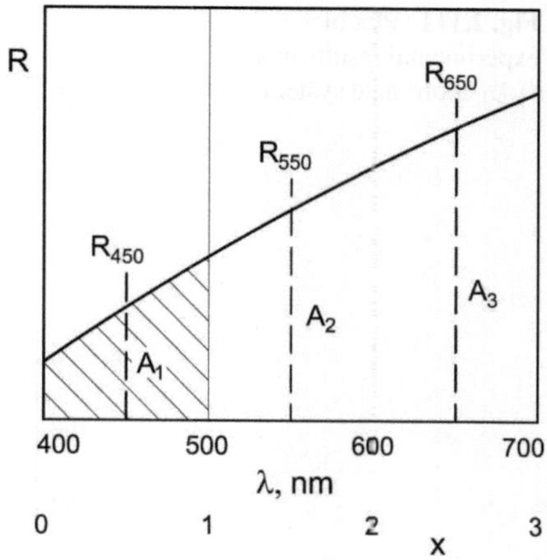

Fig. 2.110 Calculation of partial areas under the reflection curve

Equations (2.66) and (2.66a) are exact for $n = 1$ and good approximations for exponents different from one.

A similar but more accurate method is to take the partial areas A_1, A_2 and A_3 and make the same ratio as in Eq. (2.66):

$$K_2 = \frac{A_2 - A_1}{A_3 - A_2} \tag{2.67}$$

Using Eq. (2.62), each partial area can be integrated and the area ratio becomes the form

$$K_2 = \frac{2^{n+1} - 2}{3^{n+1} - 2^{n+2} + 1} \tag{2.67a}$$

which depends only on the exponent n.

Equations (2.65a) and (2.66a) or (2.67a) can be related which gives a theoretical curve in the $K - \lambda_0$ coordinate system and the measured values can be plotted in the same coordinate system, Fig. 2.111. The measured colour hue values follow well the theoretical curve from grey to red for a range of wood species. Figure 2.111 determines the colour hue in a simple way. Using Eq. (2.66) or (2.66a), when the K value is calculated and from Fig. 2.111, the corresponding λ_0 and the colour hue are simple to read.

If the reflection curve less accurately follows the theoretical Eq. (2.62), then the use of Eq. (2.67) calculated directly from the reflection curves gives better results.

Finally, we can examine the increasing cumulative area under the reflection curve related to the entire area, which is

$$\frac{R_{mx}}{R_m} = \frac{\frac{a}{b} \cdot x + \frac{1}{n+1} \cdot x^{n+1}}{\frac{a}{b} \cdot 3 + \frac{1}{n+1} \cdot 3^{n+1}} \qquad 0 \le x \le 3 \tag{2.68}$$

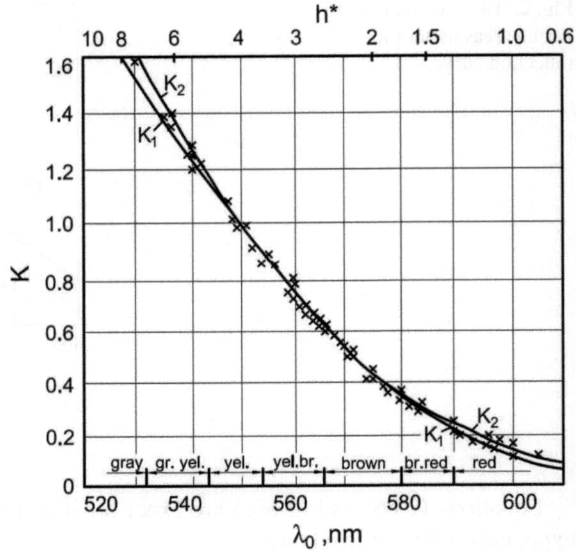

Fig. 2.111 Plot of experimental results in a $K - \lambda_0$ coordinate system

A grey body ($n = 0$) has a straight line giving a linear increasing area. In reality, the curves have an almost parabolic shape depending on the reflection spectrum and the colour hue. Figure 2.112 illustrates data for different wood species and colour hues. The ash has a greyish colour, and its cumulative area curve is closest to that of a true grey body. On the contrary, the red ivory, with its vivid red colour, has the lowest curve.

In Fig. 2.112, the area under a given curve is obtained by the integration of Eq. (2.68):

$$\int_0^3 \frac{R_{mx}}{R_m} \mathrm{d}x = \frac{\frac{a}{b} \cdot 1.5 + \frac{1}{(n+1)\cdot(n+2)} \cdot 3^{n+1}}{\frac{a}{b} + \frac{1}{n+1} \cdot 3^n} \tag{2.69}$$

The evaluation of this integral as a function of exponent n gave an interesting result. Using the x-unit between 0 and 3, the relative area of a triangle for the grey body is $3 \times (1/2) = 1.5$. For increasing exponents, the relative area decreases according to an exponential function with the following form

$$\int_0^3 \frac{R_{mx}}{R_m} \mathrm{d}x = A \cdot e^{-\beta n} = 1.5 \cdot e^{-0.24n} \tag{2.69a}$$

which is illustrated in Fig. 2.113.

Equations (2.65a) and (2.69) can be connected to constitute a cumulative relative area—colour hue (λ_0) function which is plotted in Fig. 2.114. In the same figure, data

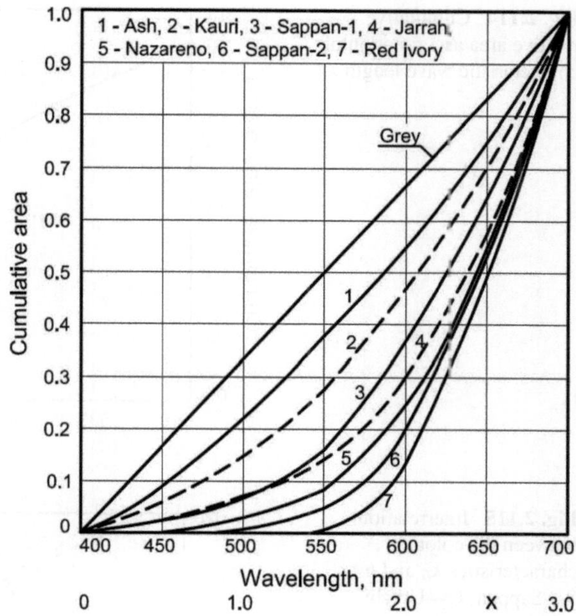

Fig. 2.112 Cumulated area under the reflection curve as a function of wavelength for species with different colour hues

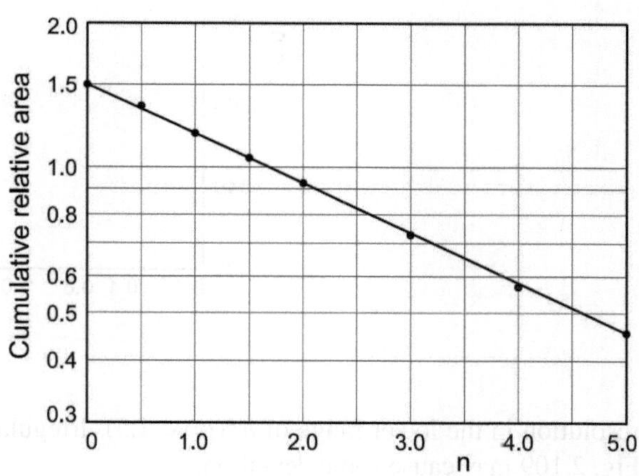

Fig. 2.113 Cumulative relative area under the reflection curve as a function of exponent n

from Fig. 2.112 are also drawn in for comparison. The measured values follow the theoretical curve with acceptable accuracy. The reflection spectrum uniquely defines the colour hue of a given wood species.

There are different methods to describe the reflection spectrum to predict the colour hue. The simplest method is to use the ordinate or area ratio according to Eq. (2.66) or (2.67), and the colour hue can be read from Fig. 2.111.

The colour hue characteristic λ_0 correlates with the standardized colour hue $h^* = b^*/a^*$ using rectangular coordinates. This correlation is depicted in Fig. 2.115 for wide range of wood species. The new colour hue characteristic λ_0 has a better

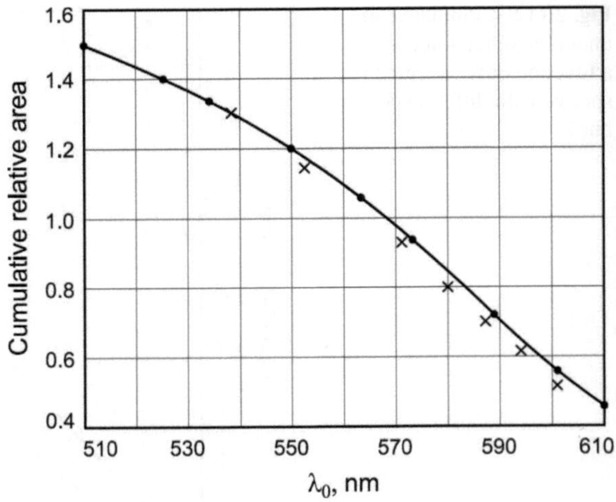

Fig. 2.114 Cumulative relative area as a function of characteristic wave length λ_0

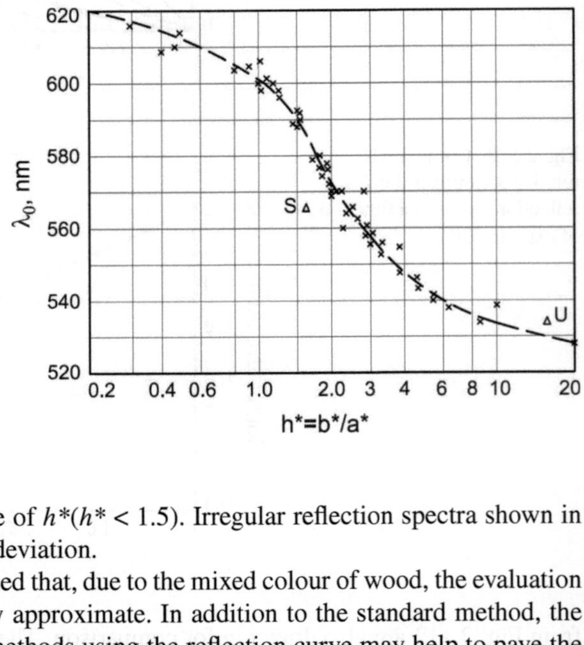

Fig. 2.115 Interrelation between the colour characteristics λ_0 and h^*. S—Sappan; U—Urushi

resolution in the lower range of $h^*(h^* < 1.5)$. Irregular reflection spectra shown in Fig. 2.109 may cause some deviation.

Finally, it may be concluded that, due to the mixed colour of wood, the evaluation of a colour hue may be very approximate. In addition to the standard method, the introduced new evaluation methods using the reflection curve may help to pave the road towards a better new standard method.

2.11.2 Lightness of Colour

Beside colour hue, the other main colour characteristic is lightness which is calculated in the standard CIE Lab system only from the green Y tristimulus. The green colour

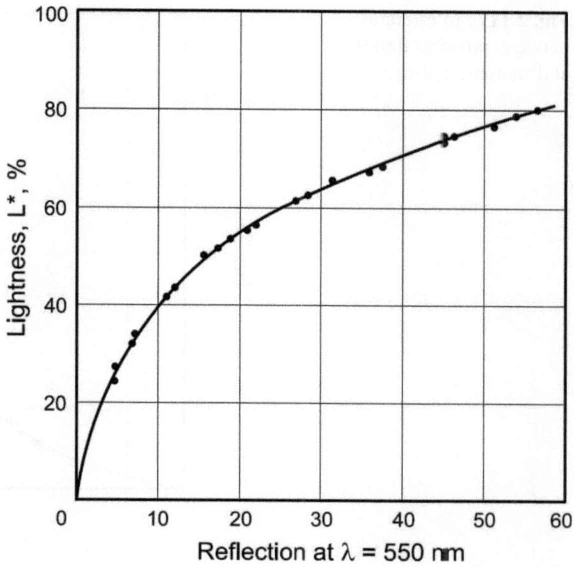

Fig. 2.116 Correlation between lightness and reflection at the 550 nm wavelength

has its maximum reflection wavelength at 550 nm, and it is obvious to seek a direct relationship between the standard lightness L^* and the reflection R_{550} at 550 nm wavelength. This relationship is seen in Fig. 2.116. The correlation is very good which is quite understandable from the definition of lightness. The correlation equation assumes the form

$$L^* = 16 \cdot R_{550}^{0.4} \tag{2.70}$$

The exponent of this equation slightly differs from that of the standard equation having the value of 1/3. But the standard equation is a shifted one with a constant term. Equation (2.70) accurately determines the lightness directly from the reflection curve.

Physical lightness R_m is given by integration of the area under the reflection curve [see Eq. (2.63)] and correlates with the standardized lightness L^* which is depicted in Fig. 2.117.

The correlation is fairly good and described with the following simple equation

$$R_m = 0.0086 \cdot L^{*2} \tag{2.71}$$

Substituting Eq. (2.70) into Eq. (2.71), we get

$$R_m = 2.2 \cdot R_{550}^{0.8} \tag{2.71a}$$

which is demonstrated in Fig. 2.118. The correlation is also fairly good using the data of European, South African and Southeast Asian wood species.

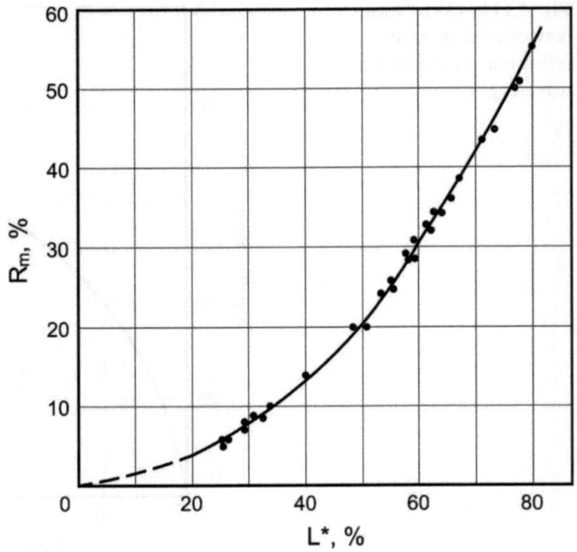

Fig. 2.117 Interrelation between physical lightness and standard lightness

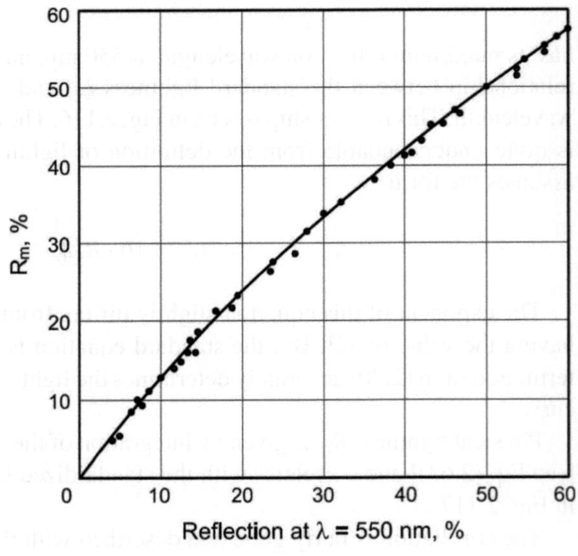

Fig. 2.118 Physical lightness as a function of the reflection at the 550 nm wavelength

These Figures clearly show that the main colour characteristics can also be determined direct, simply and accurately from the reflection curves.

A less used characteristic is ***colour saturation***. It is defined in the CIE Lab system, using the $a^* - b^*$ plane, as the distance of a given colour point from the origo of the coordinate system:

$$C = \sqrt{a^{*2} + b^{*2}}$$

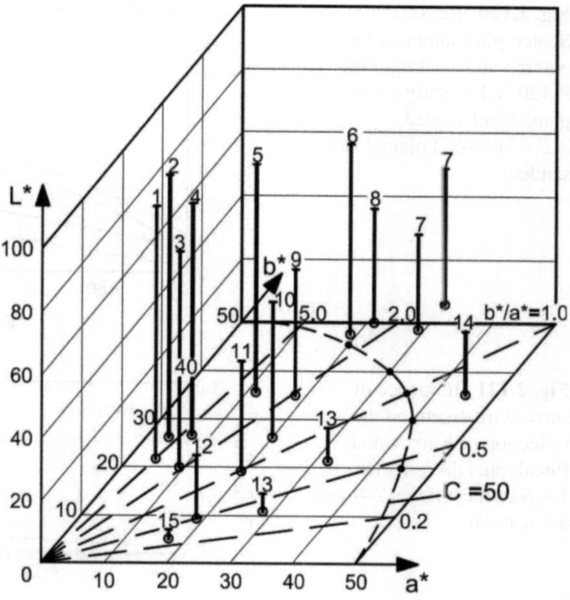

Fig. 2.119 Wood species with different colour hues and saturation on the $b^* - a^*$ plane. 1—*Fraxinus excelsior*, 2—*Tilia tomentosa*, 3—*Fagus sylvatica*, 4—*Pinus sylvestris*, 5—*Lagarostrobos franklinii*, 6—*Pinus sylvestris*, latewood, 7—*Caesalpinia sappan*, 8—*Xanthocyparis vietnamensis*, 9—*Dacrydium cupressinum*, 10—*Fitzroya cupressoides*, 11—*Eucalyptus diversicolor*, 12—*Acacia donaldsonii*, 13—*Dalbergia maritima*, 14—*Berchemia zeyheri*, 15—*Xanthorrhoea preissii*

There may be an infinitely high number of differently saturated colour points along the colour line for a given colour hue (b^*/a^* = constant) (Fig. 2.119). Therefore, colour saturation is independent of other colour characteristics. Nevertheless, some tendencies may well be observed which are valid rather within species and not between species.

In Fig. 2.119, we see some wood species with an exceptionally high saturation. The latewood of Scotch pine may have high saturation giving an enhanced colour. A mature Huon pine, despite its yellow colour, may have also a high saturation, endowing it with a golden yellow shade. The highest saturation was measured on a rainbow wood sample (Caesalpinia sappan) with an orange-red colour.

Colour saturation is important in the interpretation of ***colour enhancement***. It is widely known that after woodworking operations, especially after sanding, the surface of wood has a greyish veil, which is the result of the reflection of broken cell walls. This can be recognized by the increased reflection in the 400–500 nm range. Even between fine-planed and sanded surfaces, a perceptible difference may occur. Figure 2.120 shows the reflection of a Scotch pine sample after fine planing and sanding with P120. The highest relative change in the reflection occurs in the 400–500 nm wavelength range which is responsible for its greyish veil.

Different colour enhancing methods can be used to produce a more distinct and vivid colour hue. The most common enhancing procedure is surface treatment with vegetable resins, gums, oils, beeswax and lacquers, or their mixture with shellac. Shellac may be combined with Kauri resin and coloured by adding pigments. It is a spirit varnish, yielding a tough film, resulting in a high polish with $GU_{60} = 80–90\%$. Kauri resin mixed with shellac gives the coating good water resistance and elasticity.

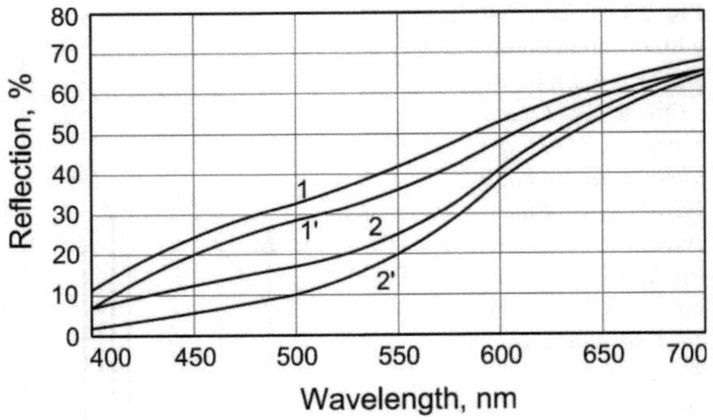

Fig. 2.120 Reflection of a Scotch pine sample after planing and sanding with P-120. 1.1′—earlywood planed and sanded, 2.2′—latewood planed and sanded

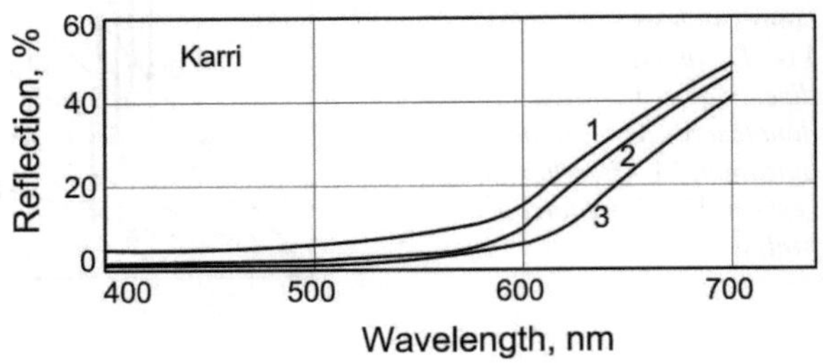

Fig. 2.121 Influence of surface treatment on the reflection of Karri wood, Eucalyptus diversicolor. 1—Natural surface, 2—wax, 3—teak oil

The latter is important to avoid fine cracks and fissures from by the flight of time. It is no wonder that makers of fine furniture have widely used this surface finish over centuries. Today, several leading wood turners use a mixture of linseed oil and shellac, claiming that shellac will ensure a gloss for many years.

Experimental results on Karri wood sample (Eucalyptus diversicolor) are shown in Fig. 2.121 in their natural condition and treated with beeswax or teak oil. After treatment, the reflection in the 400–500 nm range decreased considerably. The colour hue remained much the same but the lightness decreased by 30%, while the saturation of colour C increased about 40%. This fundamental increase in saturation will be perceived by an observer as a colour enhancement. Physically, the oil or wax treatment almost removes the reflection in the lower wavelength range, and the colour appears more vivid.

Polishing a wood surface has similar effect. The broken cell structure will be transformed into a smooth surface reflecting much less in the lower wavelength range. This effect is demonstrated in Fig. 2.122 with Kauri samples. The machined surface has a considerable reflection in the 400–500 nm range which fully disappeared after treatment. At the same time, the lightness was reduced 30% while the saturation increased by nearly 60%.

There was no definite correlation between the reduction of lightness and increase in saturation. This finding is explained by the different dependence of these colour characteristics on the reflection distribution. The dependence of lightness L^* is given by Eq. (2.70). The redness a^* and the yellowness b^* are approximated by the following relations

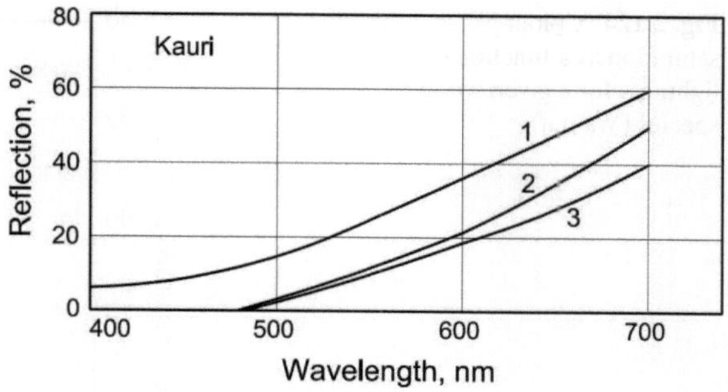

Fig. 2.122 Reflection for Kauri wood (Agathis australis) 1—machined, 2—polished, 3—swamp Kauri polished

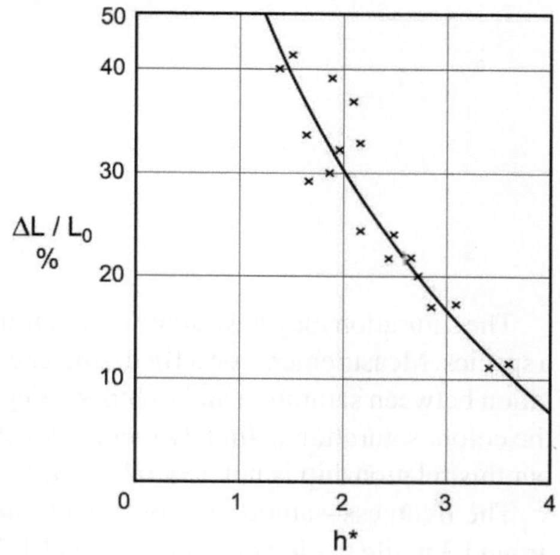

Fig. 2.123 Lightness reduction ratios as a function of colour hue after treatment with rice bran oil (Malaysian wood species)

$$a^* = 100 \cdot \left[\left(\frac{R_{600}}{R_{550}} \right)^{1/3} - 1 \right] \quad b^* = 51 \cdot \left(1 + 0.44 \cdot R_{550}^{0.4}\right) \cdot \left[1 - \left(\frac{R_{450}}{R_{550}} \right)^{1/3} \right] \tag{2.72}$$

The colour saturation is determined by both the redness and yellowness in the following fashion

$$C = 20 \cdot \left(1 + R_{550}^{0.4}\right) \cdot \left[\left(\frac{R_{600}}{R_{550}} \right)^{1/3} - \left(\frac{R_{450}}{R_{550}} \right)^{1/3} \right] \tag{2.73}$$

These equations show that both the yellowness b^* and the saturation C are slightly dependent on the lightness.

Some correlation has been found between lightness reduction and colour hue for different wood species due to treatment with rice bran oil. This correlation is given in Fig. 2.123 for Malaysian wood species. The scattering is higher for reddish-coloured species for which the colour hue characteristic h^* is not very accurate.

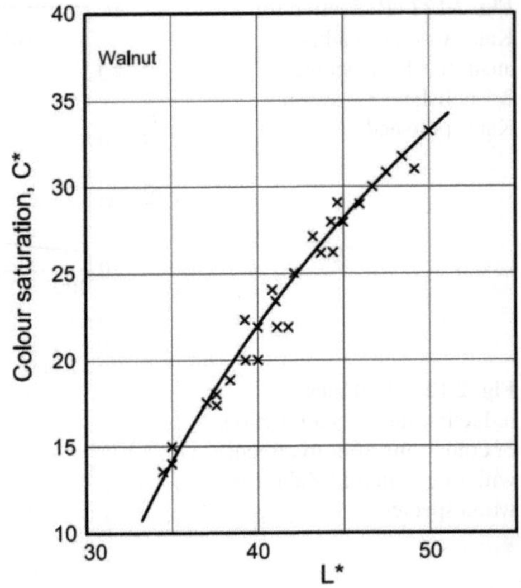

Fig. 2.124 Colour saturation as a function of lightness for a given wood species (Walnut)

The saturation may have some correlation with other colour characteristics within a species. Measurements on a Biedermeier cabinet made of walnut show a good correlation between saturation and lightness, Fig. 2.124 (see also in Appendix A53). Here the colour saturation definitely increases with the lightness with moderate scattering but this relationship is not generally valid.

The lightness–saturation ratio usually has values around two; the upper limit is around 3 while the lower limit is around 1.2. Surface treatment lowers the lightness and enlarges the saturation making the colour more vivid.

The most important application of colour characteristics are the

- colour selection and matching,
- colour enhancement,
- colour homogenization,
- colour change.

They are important value-adding methods, contributing to the increase of aesthetic value of a given product. Some types of wood require colour homogenization before they can produce a saleable product (Csanády et al. 2015).

2.11.3 Practical Application of Colour Properties

Colour is a unique property of wood giving it aesthetic beauty and value which was always highly respected in the history of mankind. A practising engineer is mainly interested in the identification of lightness and colour hue. One of the most important tasks is the connection of measured colour characteristics to the colour hue and shade of a given wood sample. Standards do not give any correspondence between measured values and colour hue, which may be explained by the mixed colour of wood.

In Sect. 2.11.1, several colour hue estimation methods are presented based on extensive measurements and new processing techniques (see Figs. 2.107, 2.108, 2.111 and 2.115). Wood materials never have a uniform and continuous colour distribution over their surface due to their cellular structure. The colouring materials, (extractives), are never evenly distributed but packed first of all in the cellular cavities of hardwoods. On the surface, the cut cell walls show their own colour which is generally greyish yellow. Due to the small size of cell walls and cavities, our eyes see a more or less uniform colour. Some wood species, such as the jarrah eucalypt or teak wood, possess an extremely uniform colour in beautiful shades suitable to make high-quality furniture.

Most of the wood materials show various colour inhomogeneities due to

- differently coloured grains and rays,
- local colour differences in the trunk, especially in the bottom part near the roots (walnut),
- irregular growth with vivid colour differences of various pattern,
- infection with staining fungi (Tasmanian Sassafras),
- soil constituents and the age of the tree,
- the direction of cut in relation to the longitudinal axis given various figure patterns.

Colours can be modified by heat treatments, including steaming, heated dry or humid air. The purpose of a heat treatment may be to achieve a more characteristic brownish colour, equalize and modify the colour of sapwood and heartwood, and stabilize the colour. A long-term colour modification may occur as a consequence of sun irradiation (fading).

Outdoor wooden structures are exposed to the combined effects of temperature and humidity, sun irradiation, wind and pollution over a long time.

Due to artificial treatments or environmental actions, the colour may change towards grey or brown. It is important to know that a heat treatment always changes the colour towards brown. On the contrary, light irradiation containing rays in the UV range changes the colour towards grey. On hot summer days, due to sun irradiation can warm a wood surface up to temperatures of 60–70 °C and then the colour changes towards brown. Under outdoor conditions over time, "cool" irradiation dominates and the resultant colour change shifts towards grey.

These questions summarized above are treated in detail in the special literature (Csanády et al. 2015), and the interested reader is referred to the references for additional information.

2.12 Gloss Properties of Wood Surfaces

The gloss of wood surfaces plays and important role and determines its aesthetic properties. The gloss of a surface is the result of reflection due to the incident light shining from different directions. Due to the highly anisotropic texture of wood, the reflection from a wood surface is a very complicated phenomenon depending on many factors. The direction of cut in relation to the main axis of bole can result very many different surface structures with different reflection properties.

The general reflection behaviour of machined surfaces was investigated 80 years ago (Schmidt and Eckert 1935; Eckert 1936). The light flow impinging per unit time on a unit area of the interface is the *incident radiation* or *irradiation.* Part of the radiation will be reflected back into the air, and its ratio to the incident radiation is called *reflectivity.* The rest of the radiation penetrates into the wood and will mostly be absorbed; the ratio of absorbed to the incident radiation is called *absorptivity.*

We can distinguish two different extreme cases of reflection: fully diffuse reflection, which means a constant reflection in all directions, and specular reflection (perfect mirror) which means reflection only under the mirrored angle, Fig. 2.125. Neither of these extremes occurs naturally.

The emissivity and reflectivity of real materials depend on their surface properties, especially on their roughness (Schmidt and Eckert 1935; Eckert 1936). Therefore, the emissivity of all materials partly depends on the viewing angle. Due to this directional reflectivity, the *brightness* (or gloss) appearing to the eye depends on the light flux originating from the radiating surface as it appears to the eye. That means the apparent brightness is proportional to the directional reflectivity. A black body surface has the same brightness regardless of the viewing angle. Wood materials may have different reflectivity depending on their surface topography, especially viewing angles over 60°. This circumstance is very important when using different incident and measuring angles.

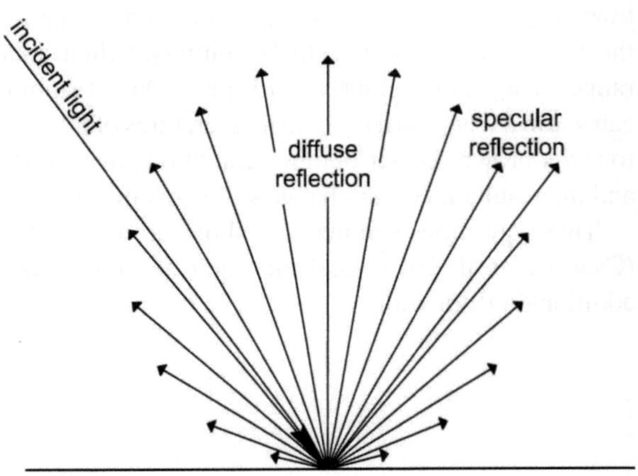

Fig. 2.125 Diffuse and directional radiation of a surface

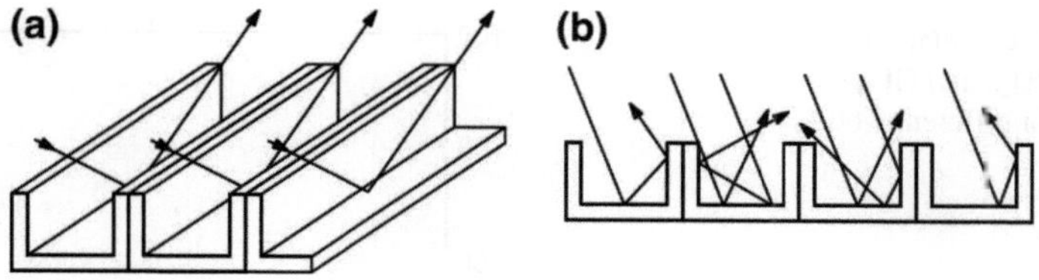

Fig. 2.126 Schematic diagram of the reflection of light on a wood surface, parallel to the grain (**a**) and perpendicular to the grain (**b**)

In wood, the slender cells are arranged in layers in one direction. These cells are cut in a given plane in relation to the grain. Depending on the direction of cut, a different micro-relief will be achieved having different reflection as a function of the direction of incident light. Figure 2.126 shows a schematic diagram of light reflection depending on the irradiating direction (Masuda 2001). Cutting of vessels and lumens, their bottom surface reflects incident light similarly to bundles of small mirrors. The light is much more reflected parallel to the grain giving a higher gloss. On the contrary, the incident light will be more scattered (diffuse reflection) perpendicular to the grain and from the broken edges of cell walls.

The higher directional reflectivity parallel to the grain can be modified considerably by breaking the continuity of the reflecting mirrors using a proper surface finish, for instance, sanding perpendicular to the grain (Martin 2015; Sitkei 2014; Csanády and Magoss 2016a, b). Sanding is always associated with bristle fracture of cell walls causing more diffuse reflectance.

For practical uses, commercial gloss-measuring units are available with different measuring angles and measuring area. The most common measuring angle is the 60°, but 20° for high gloss and 85° for low gloss are also recommended. It should be stressed that the reading of a gloss-meter is not the percentage of the reflected light, which may cause some trouble with interpretation from using different incident angles. A black polished glass standard is used with a given refractive index. The measured value is set to 100 gloss unit (*GU*) for this standard.

In practice, there are different gloss grades:

high gloss	70–100 GU
gloss	40–70 GU
silky gloss	25–40 GU
silky mat	15–25 GU
mat	less than 15 GU.

Establishing gloss grades is always conditional and the number and boundary values of gloss grades may vary considerably. New standards tend to classify only three grades, such as gloss, semi-gloss and low gloss. The gloss grade "silky gloss" (Seidenglanz) is popular and has been the preferred glossiness for a long time.

Earlier experiments observed that the reflection of a given surface is more directed for inclined incident light (Eckert 1936). This circumstance alone may cause difficulties in the correct interpretation of reading for different measuring angles. A

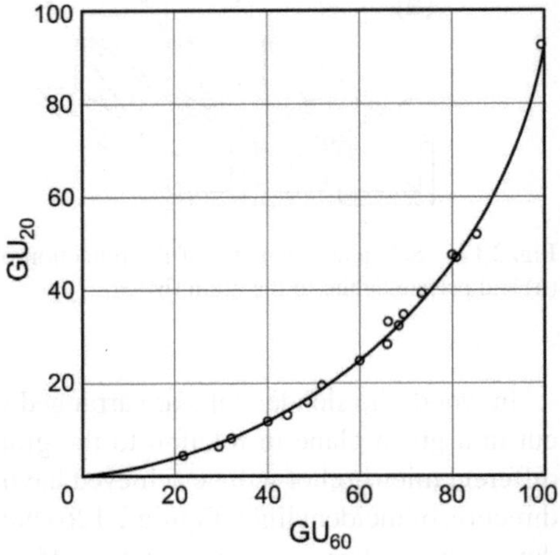

Fig. 2.127 Correlation between GU_{20} and GU_{60} readings for different wood surfaces

wide-ranging investigation on surfaces of different glosses showed a strong correlation between GU_{20} and GU_{60} readings which is plotted in Fig. 2.127 (Sitkei 2014). The correlation function is nearly a quadratic one in the simple form:

$$GU_{20} = 0.0078 \cdot GU_{60}^2$$

This fairly good correlation may probably be explained by the fact that the incident angles 20° and 60° are large enough to be relieved of excessive influence on surface roughness. An 85° incident angle is much more sensitive to surface roughness, grain angle and irregular growth.

Natural wood surfaces are generally good light absorbers and, therefore, their GU_{60} readings are rarely higher than 10. Polished and treated surfaces have a higher gloss depending on the material (wax, oil, lacquer, shellac) used for treatment and on the density (hardness) of the wood which definitely influence the attainable gloss. Due to the micro- and macro-inhomogeneity of wood surfaces, the gloss values on a surface area vary within a given range.

Preliminary experiments have shown that the gloss ratios measured with 85° and 60° incident light are distinct for a given surface and their ratios are almost constant. Therefore, we illustrated the experimental results in a $GU_{85} - GU_{60}$ coordinate system. The points indicate the range of gloss on a given surface.

Figure 2.128 shows the correlation between GU_{85} and GU_{60} values for a planed Sugi heartwood (Cryptomeria japonica) parallel and perpendicular to its grain. Coherent values measured on the same surface have a good correlation, and the directional tangent goes through the origin of coordinate system. This means that the GU_{85}/GU_{60} ratio is approximately constant for a given wood species and surface condition. The surface condition is mostly determined by the surface roughness R_z or $(R_{pk} + R_k)$. It is remarkable that the ratios are distinctly different in the direction of the fibres and perpendicular to the fibres. The ratios are 1.2, 0.89, 0.76 and 0.55.

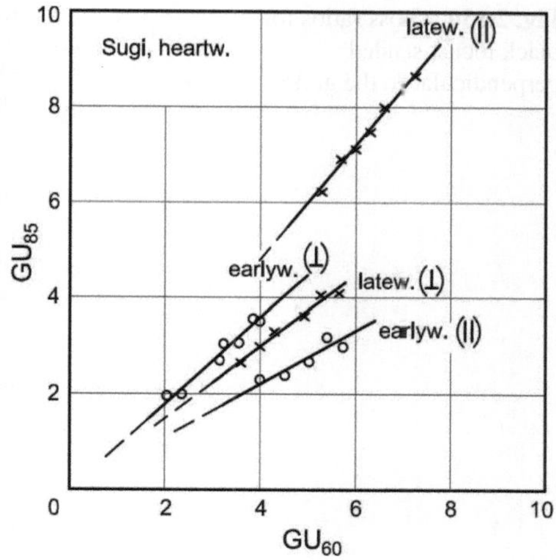

Fig. 2.128 Gloss ratios for Sugi heartwood (Cryptomeria japonica), $\rho_V = 0.43$ g/cm^3, $R_z =$ 23–25 μm

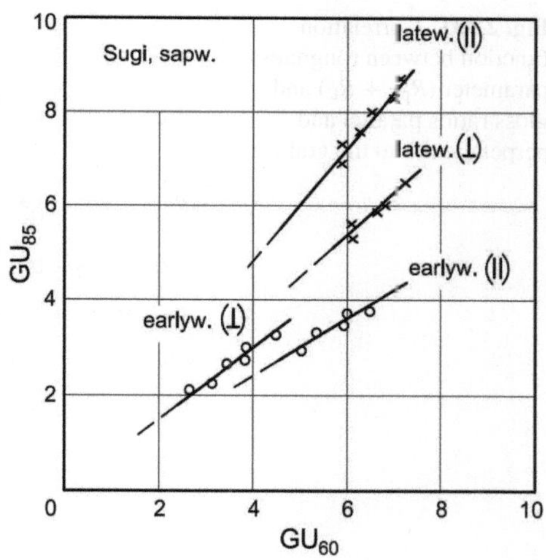

Fig. 2.129 Gloss ratios for Sugi sapwood (Cryptomeria japonica), $\rho_V = 0.43$ g/cm^3, $R_z = 22$–25 μm

Similar results for Sugi sapwood are presented in Fig. 2.129 (Sitkei 2014). The range of gloss readings is quite similar but the earlywood ratios are narrower.

The picture is changed if the surface was sanded perpendicular to the grain. The fine grits break the continuity of reflecting mirrors and considerably alter the directional reflection of the surface. Figure 2.130 shows results on black locust samples using different sandpaper grades. The fine scratches made of grits behaves similarly to cut lumens and vessels having directional reflection parallel to the grain. Therefore, we get here higher GU_{85}/GU_{60} ratio perpendicular to the grains. Furthermore, selecting a proper sandpaper grade, both ratios parallel and perpendicular to the grain are fairly close together giving a homogeneous mat surface (Martin 2015; Sitkei 2014) which may sometimes desirable. Quite similar results were obtained on ash and Sugi samples (Csanády et al. 2015).

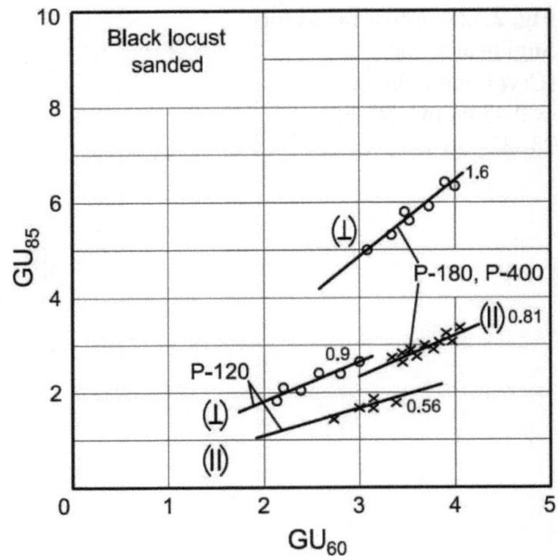

Fig. 2.130 Gloss ratios for black locust sanded perpendicular to the grain

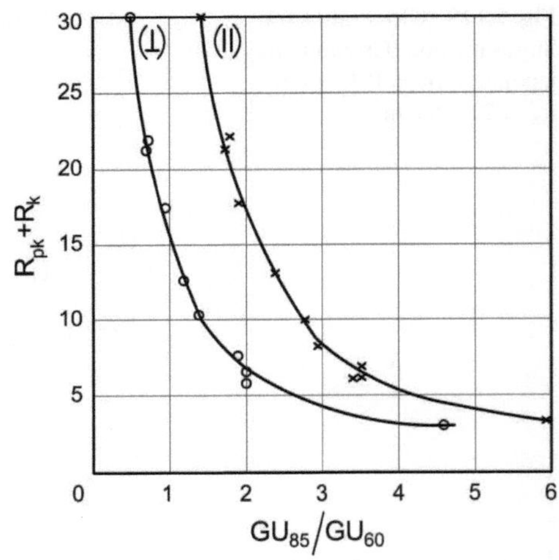

Fig. 2.131 Correlation function between roughness parameter ($R_{pk} + R_k$) and gloss ratios parallel and perpendicular to the grain

Many measurements revealed that a correlation may exist between gloss ratios and the ($R_{pk} + R_k$) roughness parameter. It is explained by the fact that a smoother surface reflects light from a larger viewing angle. Figure 2.131 shows the expected correlation functions between the roughness parameter ($R_{pk} + R_k$) and the gloss ratios parallel and perpendicular to the grain. This correlation is valid when the dominant effect influencing the reflection originates from the surface structure of wood. This constraint is fulfilled by using natural wood surfaces without treatment or treated with only a thing layer of wax, oil, stain or shellac. Resin clear coatings depending on their thickness may considerably change the reflection properties. Irregular growth structures may diminish or even eliminate the difference between gloss ratios parallel and perpendicular to the grain. At the same time, many veneered surfaces treated with thin stain or clear lacquer closely correspond to the above correlation function.

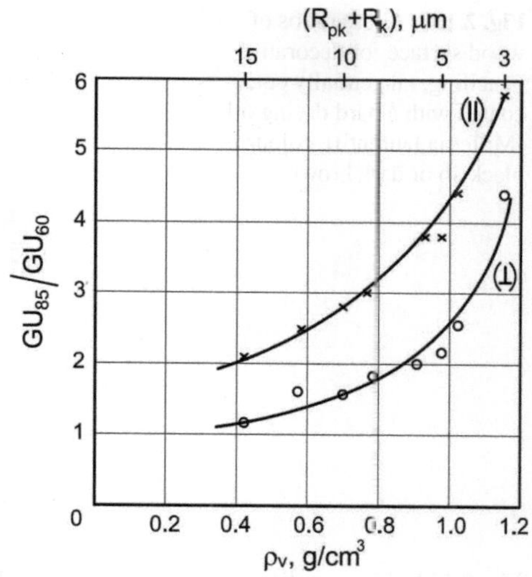

Fig. 2.132 Gloss ratios for polished samples as a function of volume density

Table 2.10 Gloss properties of different strip flooring

Wood species	Maximum gloss, 60°		Gloss ratio, 85°/60°	
	(II)	(⊥)	(II)	(⊥)
Ash (*F. excelsior*)	15	14	2.3	1.15
Guapinol (*Hymenaea courbaril*)	20	13	1.5	0.7
Merbau (*Intsia palembanica*)	20	13	1.45	0.55
Doussie (*Afzelia rhomboides*)	20	13	1.65	0.62
Oak (*Quercus robur*)	52	37	1.33	1.1

The anatomical roughness of a wood sample is considerably determined by its void ratio or density. High-density woods (ironwoods) always take a better finish and therefore, there is a correlation between the density and gloss ratios. This correlation is depicted in Fig. 2.132 showing a fairly good correlation. There will surely be differences in the gloss ratios of soft and hardwoods with the same density. Nevertheless, the gloss ratios definitely increase as a function of density, confirming the overwhelming effect of density on gloss properties.

Furniture and strip flooring are coated with resins of different glossiness fulfilling consumer requirements. Generally, gloss ratios decrease and come nearer to each other as GU_{60} values increase. The following Table 2.10 summarizes different measurement values. In this case, the resin used for clear coating determines the gloss grade of the surface and the wood species has little influence on the gloss.

The behaviour of hard drying oils used on high-quality decorative panelling is interesting. Due to the high aesthetic requirement, the top layer sheets are carefully cut to give the maximum achievable figure pattern beauty which will be further enhanced by coating with a hard drying oil. These results are depicted in Fig. 2.133 using the very distinct coloured Wenge (Millettia laurentii). In contrast with other

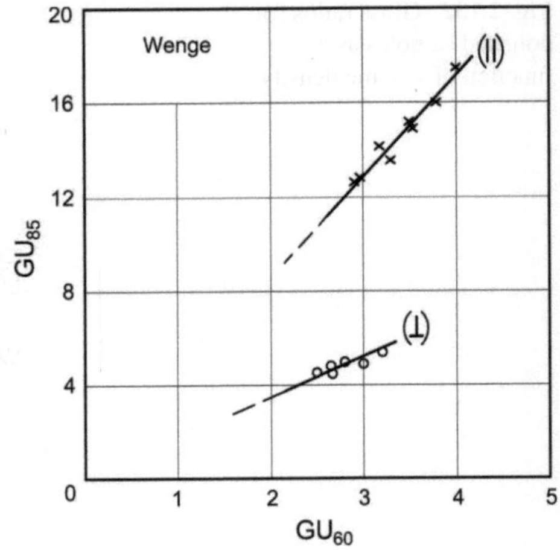

Fig. 2.133 Gloss ratios of wood surface for decorative panelling, tangentially cut, coated with a hard drying oil (Millettia laurentii), colour is blackish or dark brown

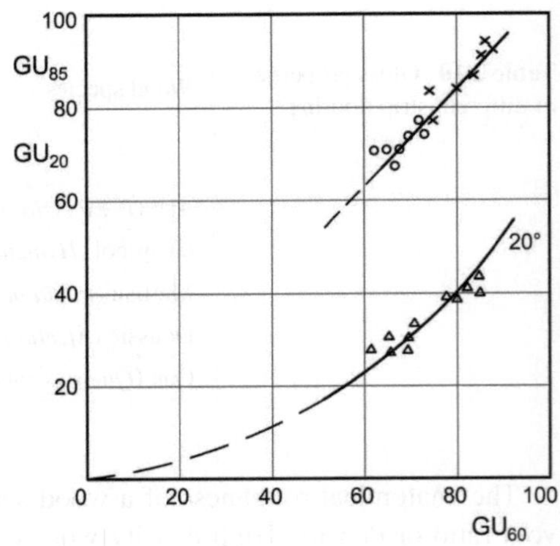

Fig. 2.134 Gloss readings on an old cabinet using all three measuring angles. (x)—parallel to the grain, (o)—perpendicular to the grain, (Δ)—20°, mixed grain angles

results, the gloss ratios are very different parallel and perpendicular to the grain. As a consequence, the grain is clearly accentuated and the gentle figures of the surface appear in their full beauty. The latewood is pitch-black which has considerably higher gloss values in both directions compared to those of the earlywood, similarly to Figs. 2.128 and 2.129.

The highest gloss values, around $GU_{60} = 90$, have been measured on an old Biedermeier cabinet with shellac (politur) coating Fig. 2.134 (Sitkei 2014). GU_{20} readings are also given which correspond to Fig. 2.127. The comparatively thick shellac layer hides the fine details of the surface texture and only the strikingly marked figures and the differences in colour hue will be effective. Practically, there are no differences in gloss values measured in different directions and the gloss ratio is very close to one.

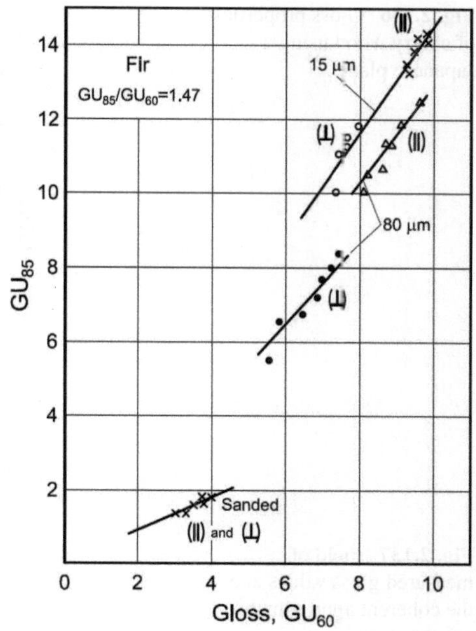

Fig. 2.135 Gloss properties of fir planed with Japanese plane

The gloss properties of surfaces made of the Japanese planer provide interesting results. This distinct cutting method produces an extremely clean surface (see Fig. 2.23) with excellent gloss properties giving a gentle lustre, vivid colour and beauty to the surface. Comparative colour measurements have shown that the reflection in the 400–500 nm range decreases considerably, due the clean cutting, and removes the common grey veil.

Figure 2.135 shows gloss measurements on fir samples (Abies alba) planed with 15 and 80 μm chip thicknesses. For comparison, gloss values for fine-sanded surfaces are also included. The GU_{85} readings are at least 50% higher compared to conventional planing. Taking a finer chip thickness, the gloss further increases.

Measurement results on cherry wood (Prunus avium) are given in Fig. 2.136 showing the effect of a chip breaker used on the knife. On cherry wood, the chip breaker improved the gloss values, which are quite remarkable. Combining the beautiful reddish colour of the cherry with its high natural gloss and coating with a transparent drying oil, an excellent ornamental piece can be achieved.

Based on the large amount of experimental results, a general tendency in the variation of gloss ratios can be formulated. The following general behaviour can be observed from the results measured parallel to the grain.

The maximum gloss ratios vary between 5 and 6, providing a very smooth surface and gloss readings are GU_{60} less than 20. As the readings increase, the expected gloss ratios will decrease quickly and they tend to one as the gloss readings approach 100. With low gloss ratios, the surface roughness plays a decisive role. Using high gloss coatings, the gloss ratios follow the general rule.

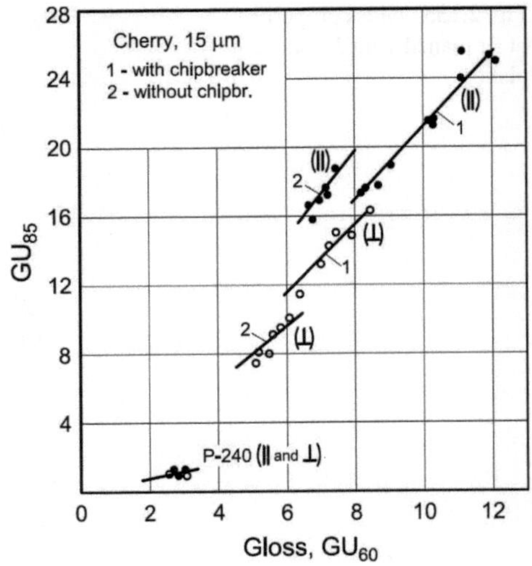

Fig. 2.136 Gloss properties of cherry wood using a Japanese plane

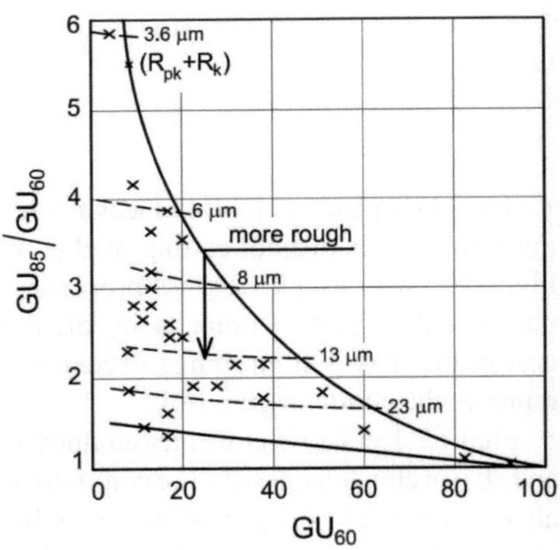

Fig. 2.137 Field of measured gloss values and the coherent approximate roughness values

The experimental results are given in Fig. 2.137 with the coherent roughness values (Sitkei 2014; Csanády and Magoss 2016a). A natural wood surface has a high absorption and, therefore, high glosses are always produced from surface treatments.

The following main conclusions may be drawn from the experimental results:

- The gloss property of a wood surface depends on both the incident and the viewing angles. In general, a 60° incident and viewing angle is suitable but it gives limited information.
- The use of the GU_{85}/GU_{60} ratio gives further information due to the difference in the ratio parallel and perpendicular to the grain.
- The gloss ratio parallel and perpendicular to the grain can be modified using a proper surface finish.

- The gloss properties of wood surfaces are closely correlated with its surface roughness.
- The density of wood indirectly influences its gloss properties considerably.

2.13 Objective Functions

General remarks

Objective functions formulate the general relationships among the system (process) parameters as influencing variables, including equality and inequality constraints. Any optimization method deals with the selection of design or manufacturing variables to optimize the given objective function subject to the satisfaction of stated constraints. That means that the objective function $f(x)$ attains an extreme value which can be expressed in the following abbreviated form

$$\min\{f(x); b(x) = 0; c(x) \leq 0\}$$

where $b(x)$ and $c(x)$ are the equality and inequality constraints, and $\underline{x}$ means a vector of design variables. Furthermore, the feasible interval of parameters needs to be specified (i.e. cutting speed, tooth bite, feed speed, etc.). The objective function and the constraints are commonly nonlinear functions of the design variables which may cause problems in the optimization procedure. A solution may be obtained from a sequence of unconstrained minimizations that use the Davidon–Fletcher–Powell algorithm with the constraints in a quadratic penalty function (Davidon 1959; Fletcher and Powell 1963). A more general philosophy of design and planning is the use of multi-criterion analysis. In this case, the designer or planner can provide a set of good alternative solutions to decision-makers and clients (Stadler 1988). Because the pure analytical solutions are limited due to the lack of reliable functional relationships, the use of a knowledge-based expert system may help produce solutions to problems (Bertling 1991). The working principle in an expert system is similar to the reasoning of a human expert to arrive at conclusions to make decision to solve a problem.

The objective or criterion functions may be formulated in different ways, according to the overall production philosophy:

- minimization of total production time, (maximization of production rate),
- minimization of total costs,
- maximization of product value or total profit.

The first two strategies are most commonly used in optimization procedures. The maximization of total profit as a governing rule may be used in cases when high-quality or precious wood is available to make furniture of special value, decorative panelling, wooden artefacts and ornaments. In this case, the aesthetic values, the eternal beauty and everlasting value are the most important decision-making factor.

The processing operations in the woodworking industry may be classified into the following main groups:

1. Primary woodworking operations
   - debarking logs,
   - sawing logs into boards,
   - cutting and splitting small logs into firewood,
   - chipping wood for the paper industry,
   - chipping wood for composites (particle and fibre board),
   - chipping wood for oil extraction by steaming.
2. Processes used to enhance or change the properties of work materials
   - steaming to enhance the dimensional stability of wood,
   - steaming to change or homogenize the colour of wood,
   - temporary softening by heat treatment (hot water or steam) for bending, peeling and knifing,
   - treatment to change the sorption properties.
3. Processes used for machining parts to fixed shape and dimensions such as turning, planing, shaping, milling, drilling, sanding, and water jet machining.
4. Processes used for surface treatments including polishing, staining, painting, coating with veneer, lacquer, paper foil, laminate or low-pressure melamine. Sometimes high-pressure water jet is used for cleaning.
5. Processes used for joining and assembly the parts: nailing, screw fastening and adhesive joining (dowel, mortise and tenon, dovetail, spline, lap, bridle, or finger joint).

Processes listed above, which are relevant to manufacturing a given product, should be considered in the objective function.

2.13.1 Optimization of Machining Processes

Machining operations play the governing role in manufacturing wood products. Most of the parts of a wood product will be shaped and finished by woodworking operations. Machining is a process of removing some material using woodworking machines with cutting tools. The manufacturer strives to achieve an optimum combination of minimum cost and maximum production rate with good product quality.

The machining process is influenced by an array of input and output variables. The input variables are the process-independent variables which are given below

- woodworking machine, its capacity, accuracy and rigidity,
- cutting tool, its material, edge coating, tool geometry,
- cutting conditions: rotation speed, feed speed, tooth bite, depth of cut, cutting angle, cutting direction to the grain,

- work material properties: strength, hardness, density, moisture content, growth structure and colour, machining properties.

The output variables are process-dependent variables and include the following:

- tool lifetime, wear rate,
- cutting forces,
- cutting power requirement,
- surface roughness, finish, material defects coming to light,
- dimensional accuracy,
- material removal rate,
- vibration and noise,
- cutting edge surface temperature, chip characteristics, especially in target chip production,
- grain structure, figures, colours left on decorative surfaces.

Most of the process-dependent variables can be predicted with variable accuracy using

- collected knowledge and experience in practice,
- functional relationships based on the kinematics of different cutting tools (Csanády and Magoss 2013),
- functional relationships based on experimental results (force and energy requirements, roughness, tool lifetime and vibration (Csanády and Magoss 2013; Sitkei 2013).

Some input variables are well-defined, such as the cutting tool characteristics and the cutting conditions. On the other hand, some machine properties and generally not known, such as its rigidity, damping properties, natural frequency of the spindle-tool system. Due to the variability of wood, some material properties cannot be predicted without proper measurements. Internal defects can hardly be forecast in advance.

Striving after an economical production, the machining time should be as short as possible which can be achieved by high-speed woodworking operations (high cutting speed, feed speed and depth of cut) resulting in high material removal rate. The implementation of such production line requires additional investment and a well-organized plant management. High-speed cutting requires higher machine rigidity, accurately balanced tools in order to avoid harmful vibration. Due to the high thermal load of the cutting edge, it is necessary to use tungsten carbide, ceramic or diamond tools. The same is true for narrow kerf sawing which would save valuable timber material (e.g. incense cedar for pencils). A thin blade is sensitive to vibration and loss of stability (Csanády and Magoss 2013), and therefore it requires a careful preparation after each sharpening.

A series of woodworking operations may strongly interact. A larger tooth bite in planing or milling, which in itself would be more economical, would then require more stock removal in sanding due to the increased waviness of the surface. If fine sanding is required, then more than one grit size should be used.

Inaccurate sawing producing a rough surface requires higher target size to allow the planer to produce a smooth surface with the prescribed size (Csanády and Magoss

2013). The higher target size means a loss in wood material and also more energy consumption for wood removal. The price of raw timber material is a considerable cost factor and, therefore the utilization of raw material in the process of conversion into a final product is of great importance. The starting material is generally in boards or cants of given size in length, width and thickness. This raw material will be cut into pieces required for further woodworking operations as parts for the product.

The utilization of raw material is influenced by the oversizing of pieces for further operations and by the proportion of waste material from the sizing of pieces. As an example, in manufacturing strip flooring, the utilization of raw material may fluctuate between 60 and 70% but in machining a window frame it may be only 50%.

2.13.2 *Maximization of Production Rate*

In this case, the objective function contains the time elements of processes needed to produce a given product which should be minimized for the entire production with the given constraints. Here we treat the machining process but other processes such as gluing, pressing and surface treatment have similarities in setting up the time elements. The total production time of a machining process is composed of the following time elements:

Machine preparation time related to the total number of components machined

$$T_p = T_s/N$$

where T_s means the set-up time and N is the number of components.

Loading and unloading time (T_L),
Process adjusting and return time (T_a),
Machining time T_m which is given as

$$T_m = L_f/e \quad \text{and} \quad e = e_z \cdot z \cdot n$$

where L_f is the feed distance,
e is the feed speed,
e_z is the tooth bite,
z is the number of teeth,
n is the spindle speed.

Tool changing time per component related to the tool lifetime T

$$T_c = T_{ch} \cdot \frac{T_m}{T}$$

where T_{ch} means the time for changing a dull tool.

For a given tool and work material, the tool lifetime mainly depends on the cutting speed and the allowable maximum edge radius. If the surface quality is crucial, then wear limit and tool lifetime must be decreased (see Sect. 2.5).

The Taylor tool lifetime equation should be used to determine the proper tool lifetime T (see in Sect. 2.5). It requires the knowledge of edge reduction as a function of cutting velocity and true cutting length working with the given wood material. This measurement is time-consuming, and therefore, such relationships are rarely known. If the cutting velocity varies in a narrower range, it is more useful to determine the variation of roughness as a function of working time and from these results to establish the tool lifetime. Nevertheless, the general regularities presented in Sect. 2.5 may assist in predicting the expected effect of process variations.

Constraints may be formulated from the following:

Machine capabilities

- maximum power on spindle,
- minimum and maximum spindle speed,
- range of feed speed,
- availability of tools in different diameters and number of teeth,
- natural frequency of spindle with a given tool.

Quality requirements

- waviness,
- roughness parameters,
- dimensional accuracy.

Work material properties

- defects, knots and irregular growth,
- direction of cut of boards,
- target size and end size ratio.

The process parameters are connected in the following manner

$$e_{z\min} = \frac{e_{\min}}{z \cdot n_{\max}}$$
$$e_{z\max} = \frac{e_{\max}}{z \cdot n_{\min}}$$
$$v_{\min} = \frac{D \cdot \pi \cdot n_{\min}}{60}$$
$$v_{\max} = \frac{D \cdot \pi \cdot n_{\max}}{60}$$

The material removal rate (MRR), which is a key factor in production rate optimization, is given by

$$\mathrm{MRR} = eH \cdot b \quad \mathrm{cm}^3/\mathrm{min}$$

which is theoretically independent of rotation speed or cutting speed. Naturally, in order to maintain the tooth bite in a feasible range, feed speed and rotation speed must be proper selected.

The cutting depth H linearly increases the material removal rate but its maximum value may be limited by the oversize of the workpiece. The cutting depth decreases the tool life and the feed distance (see Figs. 2.19 and 2.20), but the volume of material removal surpasses the decreasing effect of tool life reduction. As a rule of thumb, the removal increases approximately with the square root of cutting depth, that is

$$V_T = eH \cdot b \cdot T = L_{fT} \cdot H \cdot b$$

Substitution of expression (2.20a) for L_{fT} yields

$$V_T = \frac{e_z \cdot z \cdot b}{1.42 \cdot v^{n/m}} \cdot \sqrt{\frac{H}{R}} \cdot \left(\frac{y - y_0}{A}\right)^{1/m} \quad \mathrm{m}^3 \qquad (2.74)$$

which is an important relationship for maximization of production rate.

To optimize sanding, the main governing factors are the amount of material to be removed, the cutting speed, feed speed and surface pressure. The constraints are the required surface roughness and the service life of the sandpaper. There are differences between aluminium oxide and silicon carbide abrasive grains. Silicon carbide abrasives are better for sanding composite materials containing adhesives due to their higher heat conduction coefficient. Generally, aluminium oxide abrasives perform better for sanding solid wood.

Using the same nominal grit size, the grit geometry of aluminium oxide and silicon carbide may slightly differ. Grits of aluminium oxide are more rounded with a slightly higher equivalent radius. Aluminium oxide abrasives gave somewhat higher average roughness R_a compared to silicon carbide (de Moura and Hernandez 2006). Furthermore, the grits of aluminium oxide produced deeper surface damage, similarly to a dull knife (de Mourea and Hernandez 2006; Ratnasingam and Scholz 2004).

2.13.3 Minimization of Machining Costs

Machining costs and other machine-related operation costs are considerable part of the total manufacturing cost. Due to the high capital investment, these machines need to operate as efficiently as possible. The costs of machining are the following:

- machine cost per hour,
- labour cost per hour,
- tool cost, either for sharpening or changing knives, tool change cost,
- set-up cost.

The machine and labour costs depend on the time required for machining.

In general, the constraints are the following:

- maximum cutting force depending on the maximum chip thickness must be limited to a given value.
- maximum cutting power is limited to the maximum power available on the spindle.
- minimum and maximum spindle speed,
- range of feed speeds,
- allowable surface roughness including waviness. Different values may be established for roughing and finishing operations,
- tool life which depends on the optimum cutting speed, depth of cut and also on the allowable roughness.

The optimum number of passes should be determined when a higher depth of cut is required. Often in practice, a deep cut can be made in one pass. For example, window frames, using special profile cutters, are machined with cutting depths up to 40–45 mm. At the same time, only 50% of the raw material is used. Using high-quality and expensive wood, the minimum utilization of raw material also must be put on the list of constraints which require a change in the structural design.

Chapter 3
Principles of Optimization

3.1 Introduction

Optimization is a powerful tool in the hand of design and manufacturing engineers for finding the best possible solution from alternatives. The classical function of an engineer is to design new products with better performance, functional and aesthetic values and improved operating characteristics for less manufacturing costs.

In this chapter, we present guidelines and alternative approaches to formulating and solving problems to produce optimum results. Any optimization method consists of a performance measure (objective function) with independent variables, a set of equality and inequality constraints that constitutes the model of the system. The most important step in optimization is the formulation of a model representing the real system. In formulating this model, it is necessary to select the appropriate functional relationships describing the system behaviour as a function of independent variables. In order to facilitate this approach, an extensive and separate chapter explains the most important functional relationships and guidelines.

This chapter discusses the close relationship between science and natural laws, the mathematical and engineering methods of optimization with an array of practical examples.

3.2 General Remarks

The classical function of a product designer is to design new, better, more aesthetic products for improved functionality for customers using raw materials optimally. Similarly, the manufacturer wants to carry out the necessary manufacturing operations (machining, surface treatments, and assembly etc.) to minimize the total manufacturing time and cost. This requires careful decision-making based on a properly selected optimization technique. The optimization should determine the best value of

E. Csanády et al., *Optimum Design and Manufacture of Wood Products*,
https://doi.org/10.1007/978-3-030-16688-5_3

influencing factors resulting in a highest possible product utility value at a minimum total manufacturing cost.

As a first step, it is necessary to clearly delineate the problem boundaries of the design or manufacturing system to be optimized. A quantitative criterion, an objective function, should be selected and defined which determines the best selection of system variables. To accomplish the objective function within the necessary constraints, the functional relationships are needed which express the relation of the system variables (see Chap. 2). Actually, this activity formulates the model for the solution of the given optimization problem.

The success of an optimization process fundamentally depends on the available functional relationships and their reliability. If functional relationships are available, then finding the optimum generally does not mean mathematical problems, even in the nonlinear cases. Therefore, we attached great importance to develop the widest circle of functional relationships in Chap. 2.

The implemented model may consist of the whole system under consideration or the whole system may be broken down into smaller subsystems that can be dealt with individually. The model of a whole system may be quite complex and its solution may be more difficult. It has the advantage of dealing with the possible interaction among the subsystems. A smaller subsystem can be solved individually with fewer difficulties, and it may supply a reliable optimum or near-optimum solution by taking the possible interactions among subsystems into account.

The objective function is formulated in accordance with the selected criterion on the basis of which the performance of design or manufacture of the product can be evaluated. In many cases, an economic criterion is selected, e.g. the best value for money, high value-added design, minimum material use, highest product value from high-quality (precious) wood or a multi-functional product, etc.

In manufacturing, technological factors are usually considered such as the minimum production time, maximum production rate, minimum energy consumption, optimum use of machine capacity and maximum use of raw materials.

These different criteria may be in conflict with each other. This means that it is not possible to simultaneously maximize or minimize several criteria. Generally, one main criterion is primary and others as secondary. The latter are treated as constraints having prescribed maximum or minimum values. Therefore, decision-making should always make compromises in the selection of primary criterion and the importance of secondary criteria.

Mankind did not invent the optimization method. Nature has used this method for an infinitely long time in evolution and always uses its resources optimally. In the natural laws, we can very often recognize an inherent striving after optimality. Therefore in the following short section, we try to give a glimpse into nature's methods, demonstrated by examples.

3.3 Science and Natural Laws. Optimum "Design" of a Growing Tree

Natural processes are governed by natural laws which have existed from time immemorial. A tree growing in the forest is also subjected to natural laws provided that artificial actions are excluded. An inherent property of nature is to use its resources optimally. Mankind long ago recognized this principle of economy and tried to use it to solve everyday problems and also in science. (Science is the totality of our knowledge based on natural laws.) For instance, engineering mechanics extensively uses the principle of minimum potential energy to solve a wide range of problems.

Due to natural laws, a growing tree strives to develop its trunk and branch structure in such a way that their diameters in a given place optimally correspond to the loads originated from its own weight and external loads (wind, snow), to provide stability. We look at how nature designs the height and trunk diameter of trees. We take characteristic examples from the plant world illustrated in Table 3.1.

The wheat culm has a fantastic hight/diameter ratio of 250, whereas the Douglas fir has a ratio of only 25. An empirical equation describes the relationship in the form

$$\frac{H}{d} = \frac{250}{\sqrt{H}} \tag{3.1}$$

where H is the height,
d is the trunk diameter at ground level.

From this equation, the minimum or optimum diameter is

$$d_{\text{opt}} = \frac{H^{1.5}}{250}$$

And now, look at the result derived from engineering mechanics. For the sake of simplicity, we consider a trunk with a constant diameter and in this case, its weight is

$$G = \frac{d^2\pi}{4} H \cdot \gamma$$

Table 3.1 Corresponding plant sizes

Name	Height, m	Diameter, m	Height/diameter
Wheat culm	1.0	0.004	250
Giant bamboo	10	0.12	83
Douglas fir[a]	100	4.0	25

[a]Dörner tree in Oregon

where γ is the specific weight, N/m^3. The maximum compression stress is calculated as

$$\sigma_c = \frac{G}{A} = H \cdot \gamma \tag{3.2}$$

where A means the cross-section of the trunk. It is interesting to note that a tree has a compressive strength that would allow it to grow to a height of 1000 m!

Slender objects are sensitive to losing their stability. The critical force to fracture is given by the Euler equation

$$F_{cr} = \text{const}\frac{E \cdot I}{H^2} \quad \text{with } I = \frac{d^4\pi}{64}$$

and E means the modulus of elasticity. The critical stress to fracture is simple

$$\sigma_{cr} = \frac{F_{cr}}{A} = \text{const} \cdot E \cdot (d/H)^2$$

Rearranging this equation gives

$$\frac{H}{d} = \text{const}\sqrt{E/\sigma_{cr}}$$

and making use of Eq. (3.2), we have

$$\frac{H}{d} = \text{const}\sqrt{E/H} \tag{3.3}$$

For given tree species, the modulus of elasticity E may be taken as constant which yields

$$\frac{H}{d} = \frac{\text{const}}{\sqrt{H}} \tag{3.3a}$$

which is exactly the same as we first learned from nature. That means that natural laws and sciences do not contradict each other.

Further analysis has shown that similar to engineering structures, the trees have a coefficient of safety to withstand temporary external loads such as wind and snow (Sitkei 2007). That means that a natural optimum does not correspond to a mathematical optimum according to a constant maximum stress design or minimum weight design. Therefore, stability or instability conditions are equally important in comparison with optimality.

This last problem can best be demonstrated concerning the equilibrium of a side branch of a tree (Fig. 3.1). Its substitute model is a cantilever with a varying cross-section and equally distributed load. The latter assumption takes the weight of the

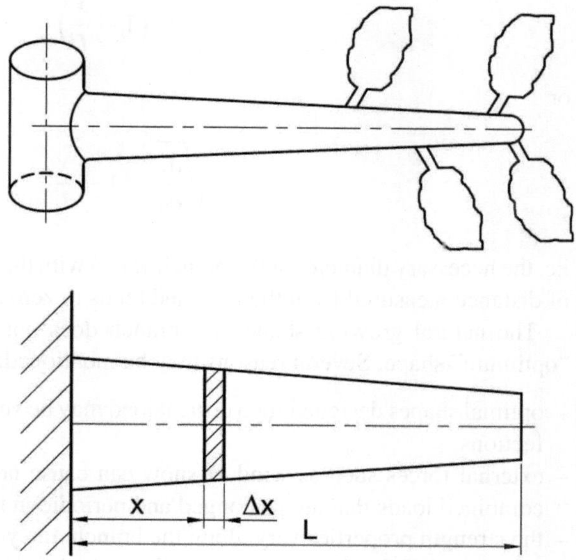

Fig. 3.1 Branch of a tree and its substitute model

secondary branches and foliage into account which is placed to the outer end of the main branch.

The bending moment at an arbitrary distance x is given by

$$M(x) = \frac{p}{2}(L - x)^2$$

with the distributed load on the unit length

$$p.= \frac{d_0^2 \pi}{4} \gamma$$

where γ means the specific weight of the branches, N/m^3. The resistance moment W at an arbitrary distance x is calculated as

$$W = \frac{d_x^3 \pi}{32}$$

The equilibrium equation reads

$$\sigma(x) = \frac{M(x)}{W(x)} = 8\gamma \frac{(L - x)^2}{d_x} \tag{3.4}$$

$$\text{and}\, \sigma_0 = 8\gamma \frac{L^2}{d_0} \quad \text{if}\, x = 0$$

If we wish to have a constant, maximum allowable stress along the branch, then

$$\frac{d_x}{d_0} = \left(1 - \frac{x}{L}\right)^2 \tag{3.5}$$

or

$$\sqrt{\frac{d_x}{d_0}} = 1 - \frac{x}{L}$$

i.e. the necessary diameter of the branch varies with the square root of d_0 as a function of distance measured from the base and tends to zero at the outer end, Fig. 3.2.

The natural growing shape of a branch does not follow the above theoretical "optimum" shape. Several reasons may be mentioned:

- optimal shapes designed for a critical load may be very sensitive to material imperfections,
- external forces such as wind or snow can cause not only pure bending but also combined loads that are prolonged and periodic in time,
- the strength properties vary along the branch, the younger parts are weaker,
- smaller side-branches further complicate the local stress state,
- the fluid flow along the branch to the foliage requires a minimum cross-section (functionality criterion),
- due to the internal structure of wood (vessels, tracheids), an average strength or other property is valid only over a limited size which cannot be too small,
- Nature surely sets some aesthetic requirements to its creations.

The above discussions teach us about nature's design principles which are always a multi-criteria optimum design corresponding to the best choice from an array of possible partial optimums regarding different criteria. If one criterion prevails (e.g. stability), nature's choice generally agrees with our best finding based on scientific considerations. It is obvious that our creative design should follow a multi-criteria optimization procedure taking into account criteria for rational material use, requirements for strength, stability, functionality (ergonomic requirements) and aesthetic value, requirements for rational manufacturing, storage, transport and maintenance.

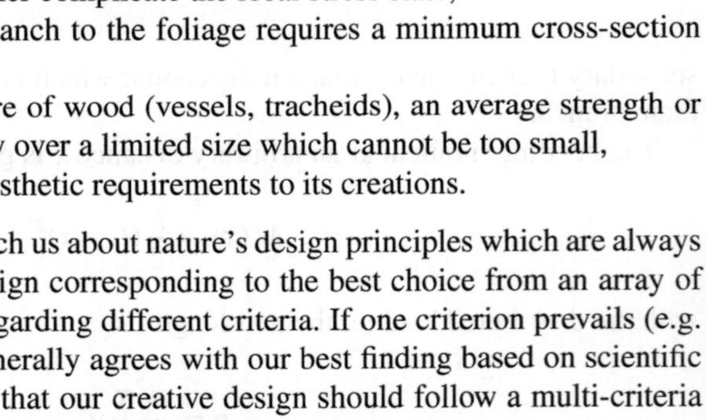

Fig. 3.2 Natural shape (1) and shape for constant maximum stress design (2)

3.4 Scientific Methods of Optimization

The goal of an optimization is the solution of the objective function subjected to constraints in equality and inequality form to find its maximum or minimum. The single-objective optimization problem with m variables and equality and inequality constraints can be written as

$$\begin{aligned}
&\text{maximize(minimize)} F(x_1, x_2, \ldots x_m) \\
&\text{subjected to} \\
&g_1(x_1, x_2, \ldots x_m) = 0 \qquad h_1(x_1, x_2, \ldots x_m) \le 0 \\
&g_2(x_1, x_2, \ldots x_m) = 0 \qquad h_2(x_1, x_2, \ldots x_m) \le 0 \\
&. \\
&g_m(x_1, x_2, \ldots x_m) = 0 \qquad h_m(x_1, x_2, \ldots x_m) \le 0
\end{aligned}$$

In the real world, most of the processes are nonlinear and, therefore, generally, we should select an appropriate nonlinear solution technique such as the Fletcher–Powell algorithm which is widely used with success (Davidon 1959; Fletcher and Powell 1963). The solution of the above problems is obtained with a transformation method in which the original constrained problem is transformed into a sequence of unconstrained problems using the penalty concept. The penalty term forms a barrier along the boundary of the feasible region. One part of the constraints limits the range of variables. For instance, the stress may not be higher than the allowable stress, the cutting speed or feed speed cannot be higher than the capability of the machine. The acceptable waviness or roughness of a surface may also be prescribed in the form of a constraint.

It should be stressed that the constraints should carefully be selected to avoid misleading solutions. Relatively simple problems may result in a quite affordable computational problem. The following example will be discussed to show the essence of this difficulty and a possible way to avoid it.

Let us consider a simple clamped cantilever made of wood and loaded on the free end by a force 220 N. The length of the beam is 50 cm (Fig. 3.3).

The cross-section of the beam is to be selected so that the volume of the beam is minimized. The variables are the width b and height h of the cross-section and the objective function to be minimized will be

$$V \approx F(x_1, x_2) \approx L \cdot b \cdot h = L \cdot x_1 \cdot x_2 = \min!\left(\text{cm}^3\right) \tag{3.6}$$

where $b = x_1$ and $h = x_2$

Only distinct combinations of x_1 and x_2 are allowed for the structure to support the required load. The feasible region of the variables is delimited by functional relationships expressed in the form of inequalities. The first one is the allowable bending stress σ_b on the beam

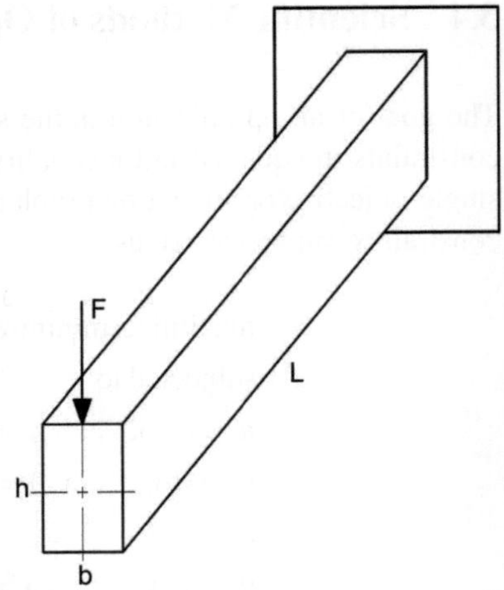

Fig. 3.3 Cross-section of the beam

$$h_1(x) = \sigma_b - \sigma(x) \geq 0 \text{ and } \sigma(x) = \frac{6F \cdot L}{b \cdot h^2} \tag{3.7}$$

$$\sigma_b = 2000\,\text{N/cm}^2$$

If the ratio $h/b = x_2/x_1$ becomes large, it involves the risk of the beam to buckling. Therefore, those combinations of x_2 and x_1 that will cause buckling must be disallowed. The critical buckling load can be calculated as:

$$F_{cr} = \frac{4}{L^2}\frac{b^3 h}{6}\sqrt{E \cdot G}\left[1 - \frac{h}{4L}\sqrt{\frac{E}{G}}\right] \tag{3.8}$$

where the modulus of elasticity E and the shearing modulus G must be taken in the h–b plane ($E = 150{,}000$ N/cm^2, $G = 40{,}000$ N/cm^2). Using the given material property values, the critical load may be well approximated by the equation

$$F_{cr} \cong 49{,}316\frac{b^3 h}{L^2} \tag{3.8a}$$

and the constraint is given by

$$h_2(x) = F_{cr}(x) - F \geq 0$$

A further constraint may be the allowable deflection of the beam under the given load:

$$f = \frac{4F \cdot L^3}{Ebh^3} \quad (E = 1{,}500{,}000\,\text{N/cm}^2) \tag{3.9}$$

and

$$h_3(x) = f\text{max} - f(x) \geq 0 \quad (f\text{max} = 0.45\ \text{cm})$$

A non-negativity restriction on h gives the last constraint

$$h_4(x) = x_2 \geq 0$$

The formulation of this solution to the problem in an unconstrained form yields

$$F(x) = 50x_1x_2 + R\langle h_1(x)\rangle^2 + R\langle h_2(x)\rangle^2 + R\langle h_3(x)\rangle^2 + R\langle h_4(x)\rangle^2 \qquad (3.10)$$

where R is a set of penalty parameters and $\langle\rangle$ is the bracket operator with the following meaning

$$\langle a\rangle = a \quad \text{if}\, a \leq 0$$
$$\langle a\rangle = 0 \quad \text{if}\, a > 0$$

Using the Fletcher–Powell algorithm, the following solution will be obtained using increasing penalty parameter R. (More about the algorithm is given in the next Section). The optimum solution yields $b = 1.317$ cm, $h = 5.009$ cm and the minimum volume is 330 cm^3.

In the following, we show an analytical approach, supplying practically the same results with the advantage that it can simply be extended to arbitrary conditions. It is important to recognize that the main constraint is the critical buckling load. Therefore, the following basic equations will be used.

From Eq. (3.7) with $\sigma_b = 2000\ \text{N/cm}^2$, we get

$$F = 333.3 \cdot \frac{bh^2}{L} \qquad (3.7\text{a})$$

If we take the condition $F_{cr} = F$, then, using Eq. (3.8a), the optimum beam height has the value

$$h = 148 \cdot \frac{b^2}{L} \text{ or } b = \frac{\sqrt{h \cdot L}}{12.16} \qquad (3.11)$$

Multiplying Eqs. (3.7a) and (3.8a) yields

$$F_{cr} \cdot F = 1.6433 \cdot 10^7 \frac{b^4h^3}{L^3}$$

Extracting a root from the above equation gives

$$F = 4054 \frac{b^2 h^{3/2}}{L^{3/2}} \text{ or } h = \frac{1}{254} \left(\frac{F \cdot L^{1.5}}{b^2} \right)^{2/3} \tag{3.12}$$

From Eqs. (3.11) and (3.12), the optimum value of b and h can uniquely be determined. Combining these two equations, we obtain an implicit equation for h:

$$h = 0.11 \left(\frac{F \cdot L^{0.5}}{h} \right)^{2/3} \tag{3.12a}$$

The previous numerically calculated example is demonstrated in Fig. 3.4.

Starting from a quadratic cross-section, $b = h = 3.2$ cm, the cross-section is around 10 cm^2 and the volume of the beam is 510 cm^3. The critical buckling load is high enough and therefore b may be decreased and h increased. As the critical buckling load nears the service load F, the smallest cross-section is achieved which agrees well with that of the numerical method. Selecting a higher beam height h, the sufficient cross-section increases.

The above calculation method always supplies the optimum values for b and h, and this method is suitable to compile a chart which gives the optimum solution for an arbitrary service load F and beam length L. This is shown in Fig. 3.5 using the material properties for F, G and σ_b given above.

Another interesting problem is the *optimal column* which was already considered by Lagrange in the 1770s. The question is the following: what is the shape of an axially loaded column which has the minimum volume for a given material, length and load. It is obvious that the main problem is the buckling of the column and therefore, the optimal shape should be similar to a barrel (Fig. 3.6).

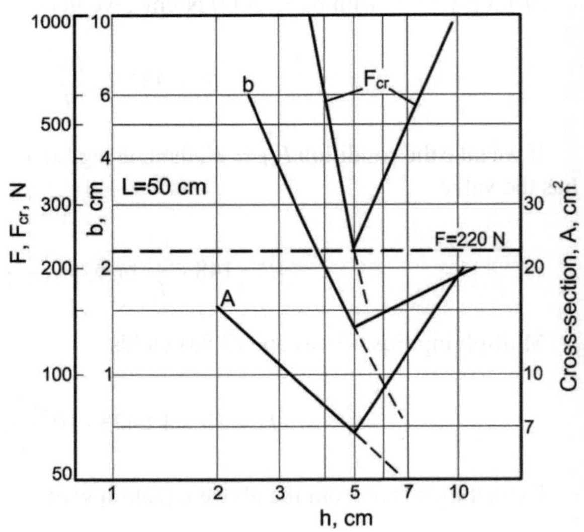

Fig. 3.4 Variation of system variable to the clamped cantilever problem

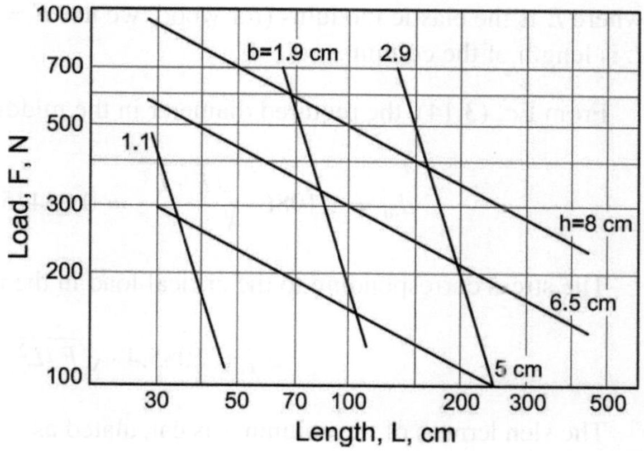

Fig. 3.5 Chart for optimum solution of the clamped cantilever

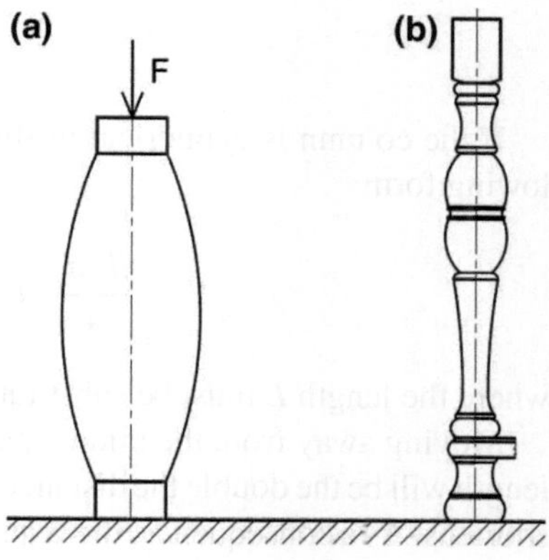

Fig. 3.6 Barrel-shaped column to minimize volume (*A*) and a turned table leg (*B*)

Although this problem has been rigorously treated and solved for a given case by the numerical method (Carter and Ragsdell 1974), no generalized engineering solution exists.

The diameter at the ends of the column is determined by the allowable compressive strength of the material

$$d_0 = \sqrt{\frac{4F}{\pi \sigma_c}} \tag{3.13}$$

where F is axial load,
σ_c is the allowable compression strength (for woods it may be taken as 3500 N/cm^2).

The critical load for buckling is given by the Euler equation

$$F_{cr} = \frac{\pi^2 EI}{L^2} \text{ and } I = \frac{d^4 \pi}{64} \tag{3.14}$$

where E is the elastic modulus (for wood, we use $E = 1{,}500{,}000\ \mathrm{N/cm^2}$),
L is length of the column.

From Eq. (3.14), the required diameter in the middle of the column is given by

$$d_{\mathrm{m}} = 1.1986 \cdot \sqrt[4]{\frac{F \cdot L^2}{E}} = 0.03425 \cdot \sqrt[4]{F \cdot L^2} \tag{3.15}$$

The stress corresponding to the critical load in the middle of the column is

$$\sigma_{\mathrm{cr}} = 1085.4 \cdot \sqrt{F/L^2} \tag{3.16}$$

The slenderness of the column λ is calculated as

$$\lambda = \frac{4L}{d_{\mathrm{m}}} \tag{3.17}$$

If the column is cylindrical in shape, then its volume is simply given in the following form

$$V = \frac{d_{\mathrm{m}}^2 \pi}{4} \cdot L = 9.21 \cdot 10^{-4} L^2 \sqrt{F}\ \mathrm{cm}^3 \tag{3.18}$$

where the length L must be substituted in cm.

Moving away from the middle part of the column towards the ends, the effective length will be the double the distance to the end. So, the apparent length of the column shortens. As a consequence, the required diameter decreases towards the ends. Using Eq. (3.15), the diameter can be calculated stepwise. As an example, Fig. 3.7 shows the variation of diameter for a wood column with $L = 100$ cm and $F = 4000$ N. It is surprising that the minimum volume, 302.5 cm^3, almost only the half (52%) of the cylindrical shape, 582.7 cm^3.

Except for the short cylindrical parts at both ends of the column, the required diameter varies according to a square root function along the half-length, that is

$$d_x = 1.1986 \cdot \sqrt[4]{\frac{F}{E}} \cdot \sqrt{L_x}$$

or

$$d_x^2 = 1.4367 \cdot \sqrt{\frac{F}{E}} \cdot L_x$$

The volume of the half barrel shape can be integrated, and doubled, as

$$V' = \frac{\pi}{4} \int_{L_{\mathrm{cr}}/2}^{L/2} d_x^2 \cdot \mathrm{d}L = 0.5642 \cdot \sqrt{\frac{F}{E}} \cdot \left[L^2 - L_{\mathrm{cr}}^2\right]$$

where the critical length L_{cr} is determined at the point when the critical buckling force equals the load on the column, $F_{cr} = F$. This condition yields the following expression:

$$L_{cr} = 0.7855 \cdot d_0 \cdot \sqrt{\frac{E}{\sigma_c}} = 1085.4 \cdot \frac{\sqrt{F}}{\sigma_c}$$

The corresponding volume is simply calculated as

$$V'' = \frac{d_0^2 \cdot \pi}{4} \cdot L_{cr} = \frac{F}{\sigma_c} \cdot L_{cr}$$

The total volume of the optimal column is given by the relation

$$V_{min} = 0.5642 \cdot \sqrt{\frac{F}{E}} \cdot \left[L^2 - L_{cr}^2\right] + \frac{F}{\sigma_c} \cdot L_{cr}\ \mathrm{cm}^3 \tag{3.19}$$

The minimum volume may be divided by the cylindrical volume according to Eq. (3.18), and this ratio as a function of slenderness is also given in Fig. 3.7. Under the critical slenderness ratio of 65, the volume ratio is always one. With increasing slenderness, the optimum volume ratio quickly decreases and tends asymptotically to the 0.5 value. This means that an optimal column can support the same load with 50% of the volume compared to a cylindrical column.

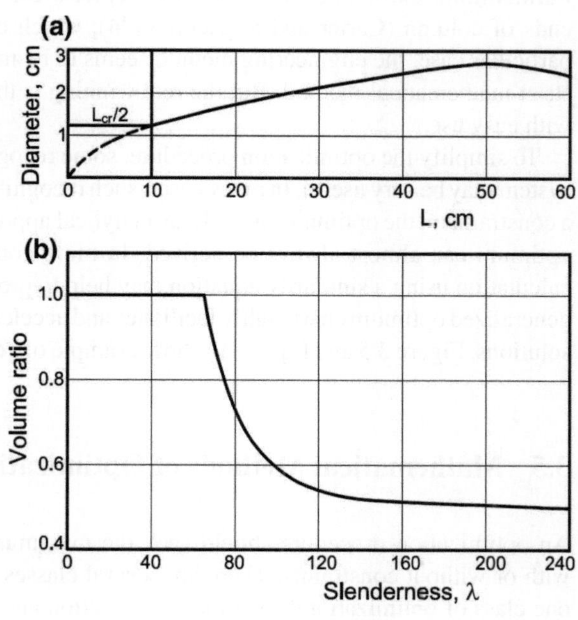

Fig. 3.7 Shape of an optimal wood column of 100 cm long, loaded with 4000 N (**a**) and the optimum volume ratio compared to the cylindrical shape as a function of slenderness (**b**)

The minimum volume can also be estimated, similarly to Eq. (3.18), by the following equation

$$V_{min} = 4.78 \cdot 10^{-4} L^2 \sqrt{F}\ \mathrm{cm}^3 \tag{3.19a}$$

corresponding to a V_{min}/V ratio of 0.519 as an average in the slenderness range from 120 to 240.

Equation (3.19a) differs from Eq. (3.18) only in its constant, and this means that the similarity holds for both cases. The accuracy of Eq. (3.19a) slightly decreases for slenderness ratios under 100. It is clear that a short and thick column does not tend to buckle and the allowable compressive strength determines the necessary diameter.

No buckling occurs when the buckling stress equals the allowable compressive stress. It is always the case if the equality $L = 0.31 \cdot \sqrt{F}$ holds. The corresponding slenderness is $\lambda = 65$.

The constants in the above equations refer to the specified material constant E and σ_c. If needed, the material constants may be altered slightly, modifying the constants of the equations.

It is interesting to note that table legs turned from solid wood often have a thicker middle part with some ornamental profiles which, willy–nilly, increases the stability of the leg.

Different optimization problems can also be handled by engineering methods which supply generally valid solutions with acceptable accuracy. The calculations are far simpler than by using numerical methods which are also not exact. The selection of approximation functions is always associated with some inherent errors. Furthermore, the mathematical method neglects the two cylindrical parts at both ends of column (Carter and Ragsdell 1974), which causes a certain error. In this particular case, the engineering method seems to be more accurate compared to the strict mathematical method. But the real winning is the general functional solution with easy use.

To simplify the optimization procedure, some recognition of the behaviour of the system may be very useful. In many cases, such recognition is the decisive influence of a constraint on the optimum value. If an analytical approach is possible, a generalized optimum can almost always be derived. In more complicated cases, a systematic calculation using a similarity equation may help to produce an optimum solution. A generalized optimum chart highly facilitates and accelerates the selection of optimum solutions. Figure 3.5 and Eq. (3.19) show example of generalized optimum solutions.

3.5 Mathematical Methods of Optimization

An optimization procedure should seek the maximum or minimum of a function with or without constraints. There are several classes of optimization problems. In one class of optimizations, the objective function and the constraints of the system can be expressed as linear equations as a function of design variables. This kind of optimization problem can be solved by linear programming, which is widely used for solving economic and industrial problems. The availability of commercial softwares makes it easy to use this method.

To demonstrate the linear optimization, let us examine a smaller economic problem which may be encountered in a factory. A company manufacturing strip flooring uses visual inspection with two groups of inspectors. Members of the first group can inspect 60 finished parquet strips per hour with 98% accuracy while the members of the second group examine only 50 pieces with 96% accuracy. The wage for the first group is \$10/h and for the second group is \$8/h. An error made by an inspector causes a loss of a \$10/piece to the company because it must be sold at half price. The daily production is 3200 pieces which should be inspected. The number of available inspectors is limited to four and five persons in each group. The economic question is how many inspectors should be assigned from each group?

The objective function is the total cost of the wages and losses due to erroneous inspection. When denote the number of inspectors with x_1 and x_2 in each group, then the cost function reads

$$F = 8.10 \cdot x_1 + 8.8 \cdot x_2 + 8.60 \cdot 0.02 \cdot 10 \cdot x_1 + 8.50 \cdot 0.04 \cdot 10 \cdot x_2$$

or

$$F = 176x_1 + 224x_2$$

Constraints are given by the limited number of inspectors

$$x_1 \leq 4 \text{ and } x_2 \leq 5$$

Further, the daily production of 3200 pieces must be inspected, that is

$$8.60 \cdot x_1 + 8.50 \cdot x_2 \geq 3200$$

or

$$6 \cdot x_1 + 5 \cdot x_2 \geq 40$$

The graphical representation of this example is given in Fig. 3.8.

The feasible region is given by the constraints. The last inequality constraint requires that any feasible solution should be on one side of the straight line. Inside the feasible region ABC, there are many feasible points but with different values of the objective function F. The optimum solution is given by the feasible point giving the lowest value for F. The objective function F is also a straight line with a given function value. Changing the value of F, the straight line moves parallel to itself. Translating the straight line towards the origin, the value of function decreases but

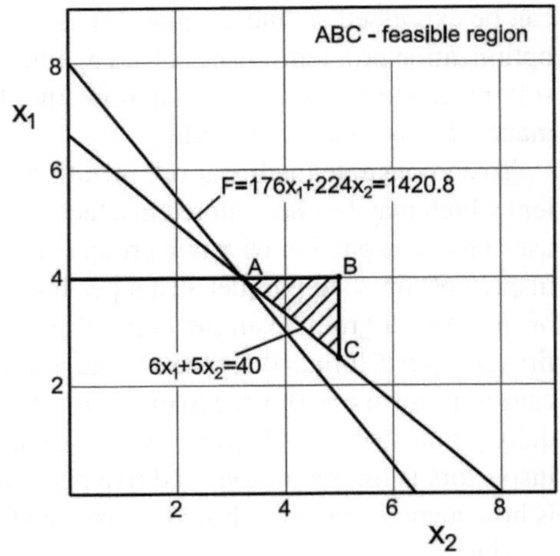

Fig. 3.8 Feasible region and optimum point

the straight line must contain at least one point in the feasible region. This point is the upper-left corner of the triangle ABC, and the objective function has the minimum value of 1420.8. The optimum number is of inspectors $x_1 = 4$ and $x_2 = 3.2$. The fractional value for x_2 means $x_2 = 3$ with some overtime work.

The result is a unique optimal solution because no other feasible point exists which would give a lower function value.

The objective function shows that the multiplier of x_1 is considerably smaller than that of x_2. This makes the preference of inspectors in group 1 very likely which is verified by the calculations. The feasible region, in Fig. 3.8, is small enough to select the optimum process variables without any mathematical calculation (see the problem in Sect. 3.6.10).

Natural laws are seldom linear and therefore, in both design and manufacturing engineering, the objective functions and the constraints are generally nonlinear.

The general formulation of the problem reads

$$\begin{aligned} &\text{Min/Max}\, F(x_i)\ i = 1, 2, 3, \ldots \text{n} \\ &\text{subject to} \qquad g_j(x_i) \geq 0\ \ j = 1, 2, 3, \ldots \\ &\qquad\qquad\qquad h_k = 0\ \ k = 1, 2, 3, \ldots \end{aligned}$$

The original constrained problem is transformed into a sequence of unconstrained problems using penalty functions. The penalty function contains the constraint function and the set of penalty parameters R in the form of a multiplier. The unconstrained objective functions can be solved with successive approximation using gradient-based methods. The gradient method uses the first derivatives to seek the steepest gradient towards the minimum or maximum point. There are several modifications of the gradient method and one of the widely used algorithms is the Davidon–Fletscher–Powell method (Davidon 1959; Fletcher and Powell 1963).

Various penalty terms may be used but one of the most preferred forms is the parabolic penalty term used for equality constraints.

For inequality constraints, the parabolic penalty function can be used with the bracket operator. The bracket operator is defined as

$$\langle a \rangle = a \quad \text{if}\, a \leq 0$$
$$\langle a \rangle = 0 \quad \text{if}\, a > 0$$

The bracket operator produces an exterior penalty function (barrier), and no penalty is assigned to the boundary and feasible points.

The penalty concept is demonstrated with the following simple examples which can be handled analytically.

Let us consider the following objective function

$$F(x_1, x_2) = x_1^2 - 6 \cdot x_1 + x_2^2 - 4 \cdot x_2 + 15$$

subjected to $x_1 + x_2 = 4$.

The derivatives of the unconstrained function are

$$\frac{\partial F}{\partial x_1} = 2 \cdot x_1 - 6 = 0 \quad \text{and } x_1 = 3$$
$$\frac{\partial F}{\partial x_2} = 2x_2 - 4 = 0 \quad \text{and } x_2 = 2$$

The equality constraint gives the penalty term in the following form

$$F(x, R) = x_1^2 - 6 \cdot x_1 + x_2 - 4 \cdot x_2 + 15 + R \cdot (x_1 + x_2 - 4)^2$$

The derivatives read

$$\frac{\partial F}{\partial x_1} = 2 \cdot x_1 - 6 + 2R(x_1 + x_2 - 4) = 0$$
$$\frac{\partial F}{\partial x_2} = 2 \cdot x_2 - 4 + 2R(x_1 + x_2 - 4) = 0$$

Solving the above two equations, we obtain

$$x_1 = \frac{5 \cdot R + 3}{2 \cdot R + 1} \text{ and } x_2 = \frac{3 \cdot R + 2}{2 \cdot R + 1}$$

If $R \to \infty$ and using L'Hospital's rule, the following limit values are produced

$$x_1 = 2.5 \text{ and } x_2 = 1.5$$

which satisfy the equality constraint. For $R = 0$, we get the extremum points $x_1 = 3$ and $x_2 = 2$ of the unconstrained function. Using $R = 100$, we get $x_1 = 2.5025$ and $x_2 = 1.5025$ which are quite near the exact values. This finding is important for numerical calculations.

The above equality constraint can also be given in an inequality form

$$g(x_i) = 4 - x_1 - x_2 \geq 0$$

In this case, we employ the bracket operator

$$F(x, R) = x_1^2 - 6 \cdot x_1 + x_2^2 - 4 \cdot x_2 + 15 + R\langle 4 - x_1 - x_2 \rangle^2$$

We take the derivatives as formerly

$$\frac{\partial F}{\partial x_1} = 2 \cdot x_1 - 6 - 2R\langle 4 - x_1 - x_2 \rangle$$
$$\frac{\partial F}{\partial x_2} = 2 \cdot x_2 - 4 - 2R\langle 4 - x_1 - x_2 \rangle$$

We assume that $(x_1 + x_2) \leq 4$ and therefore the penalty term may be written as $+2R(x_1 + x_2 - 4)$ which equals that of the equality with the same solution.

Mixed constraints may also be expressed as follows

$$h(x) = x_1 - x_2 = 0$$

$$g(x) = 8 - x_1 - x_2 \geq 0$$

In this case, the transformed unconstrained function reads

$$F(x, R) = F(x_1, x_2) + R(x_1 - x_2)^2 + R\langle 8 - x_1 - x_2 \rangle^2$$

Taking the derivatives equal to zero, yields

$$x_1 = \frac{8R + 3}{2R + 1} \text{ and } x_2 = \frac{8R + 2}{2R + 1}$$

and the limit values for $R = \infty$ will be

$$x_1 = 4, x_2 = 4 \text{ and } F(x_1, x_2) = 7.$$

Numerical methods must be used for more complicated objective and constraint functions. Let us examine our $F(x_1, x_2)$ function with the following constraints

$$h(x) = x_1 - 2x_2 = 0$$

Table 3.2 Subsequent approximation values

	x_1	x_2
$R = 1$	4.556	2.448
$R = 10$	4.945	2.496
$R = 100$	4.994	2.4996
$R = 1000$	4.9994	2.49996

and

$$g(x) = \left(62 \cdot 5 - x_1^2 \cdot x_2\right) \geq 0$$

The transformed unconstrained function is given by

$$F(x, R) = F\left(x_{1,}x_2\right) + R(x_1 - 2x_2)^2 + R\left(62.5 - x_1^2 \cdot x_2\right)^2$$

The partial derivatives are as follows:

$$\frac{\partial F}{\partial x_1} = 2x_1 - 6 + 2R(x_1 - 2x_2) + 2R\left(62.5 - x_1^2 \cdot x_2\right) \cdot (-2x_1 \cdot x_2)$$

$$\frac{\partial F}{\partial x_2} = 2x_2 - 4 + 2R(x_1 - 2x_2)(-2) + 2R\left(62.5 - x_1^2 \cdot x_2\right)\left(-x_1^2\right)$$

Using increasing R values, the subsequent approximation yields the following x_1 and x_2 values: Table 3.2.

The optimum values are $x_1 = 5$ and $x_2 = 2.5$ and the value of the objective function is $F(x_1, x_2) = 6.25$ (Fig. 3.9).

The optimum is always determined by the constraints which is important for using engineering optimization methods.

If the partial derivatives are not given in an analytical form, the program calculates the derivatives with a numerical method which is less accurate and has a slower convergence.

3.6 Application of Engineering Optimization Methods

A subsystem of the whole manufacturing system can often separately be handled and optimized. If there are interactions among the subsystems, they can also be taken into account with acceptable accuracy. In the following subsystem, optimizations will be discussed. Identifying the decisive factors determining optimum will simplify the solution procedure. Furthermore, a generalized solution allows decision-making with a beneficial compromise between optimal cost and maximum production rate.

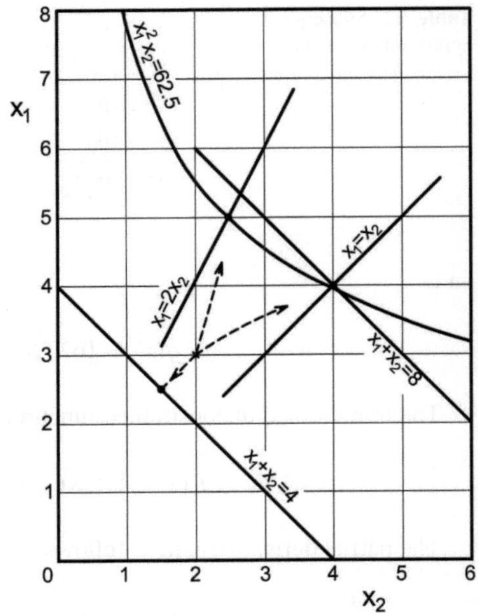

Fig. 3.9 Displacement of unconstrained stationary points ($x_1 = 3, x_2 = 2$) due to different constraints

3.6.1 Dust Collecting System

Woodworking operations produce chips and dust which must be collected by an appropriate dust collecting system. The dust collecting system consists of a fan, pipes and suction hoods. It is important to minimize the energy consumption of the system. The energy consumption depends on the volume of air flow in the unit time and the pressure drop. Therefore, the objective function has the following general form

$$\sum V_i \cdot \Delta p_i = \text{Min!} \tag{3.20}$$

with $V_i = \frac{d_i^2 \pi}{4} v_i \text{m}^3/\text{s}$

$$\Delta p_i = \lambda \frac{l_i}{d_i} \cdot \frac{\rho}{2} \cdot v_i^2 \text{ (pipe lines)}$$
$$\Delta p_i = \xi_i \cdot \frac{\rho}{2} \cdot v_i^2 \text{ (hoods,connecting moulds)} \tag{3.21}$$

where λ is the air resistance coefficient,
l_i is a given pipe length,
d_i is a given pipe diameter,
ρ is the air density,

v_i is the air velocity,
ξ_i is the resistance coefficient of suction hoods and connecting elements.

The transport of chips and dust modifies the pressure compared to the air motion. The concentration of transported material is generally low, and therefore it may be assumed that if a system is designed calculating only air motion, which satisfies the optimum conditions, then it will operate just as well with material transport.

Here is an example using the classical optimization method. The layout of the system is given in Fig. 3.10.

The pipe length, the coefficients and the constraints are as follows: $l_1 = 5$ m, $l_2 = 5.5$ m, $l_3 = 10$ m, $\lambda = 0.024$, $\xi_1 = 3$, $\xi_2 = 2.8$, $\xi_{13} = 0.2$, $\xi_{23} = 0.4$. The constraints $V_1 \geq 1000$ m^3/h, $V_2 \geq 1440$ m^3/h, $v_1 \geq 16$ m/s, $v_2 \geq 16$ m/s, $v_3 \geq 16.5$ m/s.

The air volumes are prescribed by the machine manufacturers while the minimum air velocities must ensure the system will not clog. Using these figures, Eq. (3.20) has the following form

$$\sum V_i \cdot \Delta p_i = U = \frac{\rho\pi}{8}\left[\begin{array}{r} d_1^2 v_1^3\left(\lambda\dfrac{l_1}{d_1} + \xi_1 + \xi_{12}\right) \\ +d_2 v_2\left(\lambda\dfrac{l_2}{d_2} + \xi_2 + \xi_{23}\right) + \lambda \cdot l_3 \cdot d_3 \cdot v_3^3 \end{array}\right]$$

or

$$U = d_1^2 \cdot v_1^3\left(\frac{0.0584}{d_1} + 1.5582\right) + d_2^2 \cdot v_2^3 \cdot \left(\frac{0.0643}{d_2} + 1.5582\right) + 0.1076 \cdot v_3^{2.5}$$

where d_3 was substituted using the known air volume $V_3 = 2400$ m^3/h $= 0.667$ m^3/s to relate d_3 to v_3, that is, $d_3 = 0.9213/\sqrt{v_3}$. In the common point of pipes 1 and 2, the pressure must be the same and it is expressed by the following equality relation

$$A = \left(\lambda \cdot \frac{l_1}{d_1} + \xi_1 + \xi_{12}\right)v_1^2 - \left(\lambda \cdot \frac{l_2}{d_2} + \xi_2 + \xi_{23}\right)v_2^2 = 0$$

or

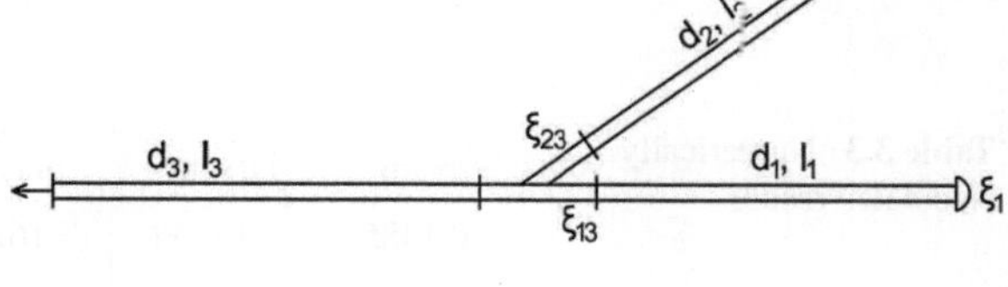

Fig. 3.10 Layout of a dust collector for two workplaces

$$A = \left(\frac{0.12}{d_1} + 3.2\right)v_1^2 - \left(\frac{0.132}{d_2} + 3.2\right)v_2^2 = 0$$

The five constraints given above will be expressed by inequality relations. The minimum suction volume

$$V_1 = \frac{d_1^2\pi}{4}v_1 \geq \frac{1000}{3600}\,\text{m}^3/\text{s}$$

from which

$$X = d_1^2 - 0.3537/v_1 \geq 0$$
$$Y = d_2^2 - 0.495/v_2 \geq 0$$
$$Z = v_1 - 16 \geq 0$$
$$B = v_2 - 16 \geq 0$$
$$C = v_3 - 16.5 \geq 0.$$

The objective function U is completed with one equality and five inequality relations. The unconstrained objective function is obtained by adding the equality and inequality functions as quadratic penalty functions in the form

$$F = U + R \cdot A^2 + R \cdot \langle X\rangle^2 + R\langle Y\rangle^2 + R\langle Z\rangle^2 + R\langle B\rangle^2 + R\langle C\rangle^2 \tag{3.22}$$

where the penalty parameters R varies from 1 to ∞. This means that the original constrained problem is transformed into a sequence of unconstrained problems using the penalty functions.

Using the Fletcher–Powell algorithm, the following results were obtained in m and m/s dimensions: Table 3.3.

The function U has the minimum value of 538 Nm/s and the calculated value of d_3 is 0.2268 m. It is obvious that for the pipe diameters must be the nearest standard size.

Now let us look behind the obtained results. Equations (3.20) and (3.21) clearly show that the most important influencing factor is the velocity of air flow which determines the energy consumption. The obtained velocities are within the allowable range. The small difference between v_1 and v_2 is due to the equality condition.

If we use the minimum allowable velocities as definite values, then a direct calculation supplies the same results in a much shorter time. One may encounter more problems in the proper design of a dust collecting system for varying operational conditions when the number of machines in operation is not constant. The clearest design principle is to connect each machine to the dust collector with an individual pipe.

Table 3.3 Numerically calculated results

d_1, m	d_2, m	v_1, m/s	v_2, m/s	v_3, m/s
0.1485	0.1754	16.01	16.12	16.51

3.6.2 *Wood Machining Process*

The economy of manufacturing considerably influenced by the woodworking operations including knife machining and sanding. Different criteria (objective functions) may be posed to achieve maximum production rate, minimum production costs, but in some cases, the maximum tool life may also be important. An array of constraints must always be taken into account. Constraints of the machining parameters are

- the available feed rates,
- the available rotation speeds and cutting speeds,
- the maximum power or torque available,
- stable cutting area excluding certain combinations of cutting speed and feed speed causing vibrations.

Technological requirements are another class of constraints which may be regarded as *goal constraints*:

- the allowable waviness of the surface,
- the allowable maximum surface roughness specified by a single or several roughness parameters,
- the material removal rate (MRR) should be specified or constrained to a minimum rate,
- the tool life may be specified to achieve optimum time between tool changes or constrained to a minimum tool lifetime.

The maximum production rate requires a maximum feed rate. The feed rate is in a simple relation with the main operational parameters as

$$e = e_z \cdot n \cdot z \text{ m/min}$$

where

e_z is the tooth bite,
n is the rotation speed,
z is the number of teeth.

If we use the waviness t_1 as a prescribed constraint which is given by

$$e_z = 2\sqrt{D \cdot t_1}$$

where D means the tool diameter, then the feed speed is calculated as

$$e = 2 \cdot \sqrt{D \cdot t_1}(n \cdot z) \tag{3.23}$$

which is plotted in Fig. 3.11 for different tool diameters and at a waviness of 2 μm.

The production rate depends on the material removal rate (MRR), which is the material volume removed in the unit of time

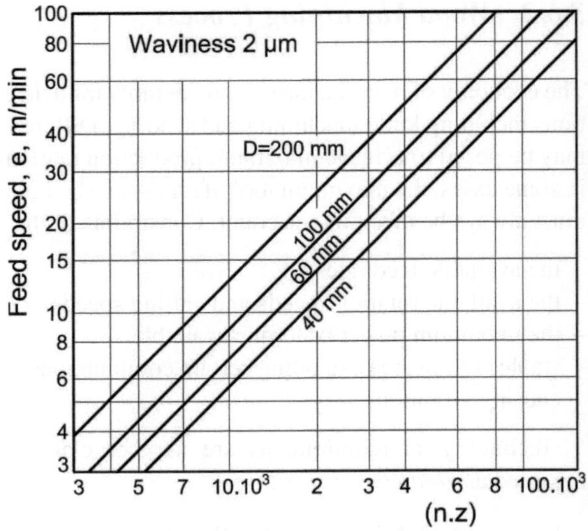

Fig. 3.11 Feed speed as a function of rotation speed x number of edges for different tool diameters

$$V = eH \cdot b = 2\sqrt{D \cdot t_1} \cdot H \cdot b \cdot (n \cdot z)\,\text{cm}^3/\text{min} \tag{3.24}$$

where H is the depth of cut,
b is the width of cut.

Because the width of cut may often vary, it is more convenient to relate the material volume to the unit width

$$\frac{V}{b} = 2 \cdot \sqrt{D \cdot t_1} \cdot H \cdot (n \cdot z)\,\text{cm}^3/\text{min.cm} \tag{3.24a}$$

Taking $D = 10$ cm and $t_1 = 2\,\mu\text{m}$, Eq. (3.24a) is plotted in Fig. 3.12 which allows a quick estimate for different conditions.

The next important constraint may be a prescribed upper limit for roughness parameters. As explained in Sect. 2.6.5, the average roughness parameter R_z depends on the feed per tooth e_z in the following fashion (see Eq. 2.38 and Fig. 2.47):

$$R_z = A + B \cdot e_z^n\,\mu\text{m}$$

where the constants B and n may generally be taken as $B = 43$ and $n = 0.6$, but the constant A is determined by the anatomy of the given wood species. It may be express either by the first term of Eq. (2.33)

$$A = 123 \cdot \Delta F^{0.75}$$

or with the Abbott-ratio (see in Fig. 2.29)

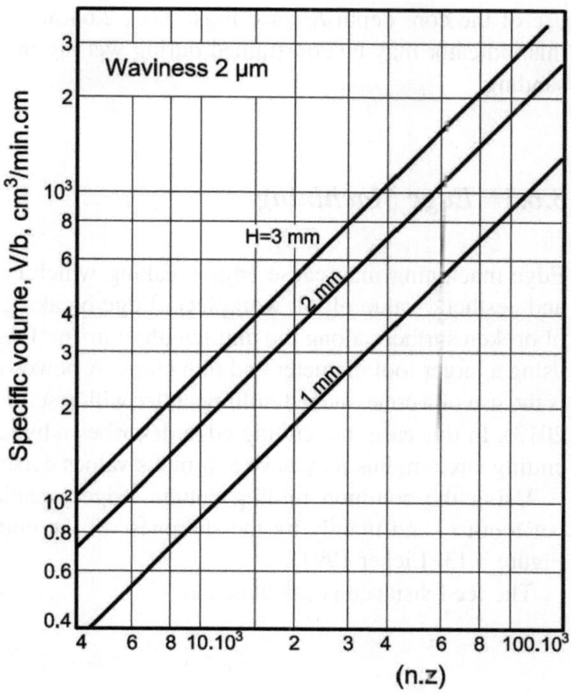

Fig. 3.12 Removal rate as a function of rotation speed x number of tool edges for different depth of cut. $D = 100$ mm, $e_{zmax} = 0.9$ mm

$$A = \left(\frac{36}{\frac{R_{pk} + R_k}{R_{vk}}} + 0.4 \right)^{1.05}$$

For example, when machining Scotch pine, we limit the roughness to $R_z = 40\ \mu$m, the feed per tooth must not be higher than $e_z = 0.5$ mm. The constraint for waviness of 2 μm allowed feed per tooth values up to 0.9 mm and, therefore, the roughness constraint will reduce the material removal rate in a ratio of $0.5/0.9 = 0.55$ compared to Fig. 3.12.

In many cases, the core depth R_k is an important roughness parameter and an indicator for tool sharpness which correlates with other roughness parameters such as R_a and R_z (see Figs. 2.39 and 2.40). The interrelations among the roughness parameters given in Chap. 2 highly facilitate the use of various roughness parameters as a constraint. For example, Fig. 2.40 clearly shows that a small feed per tooth considerably reduces the relative core depth which is the essence of a smoothing pass and has been known for a long time.

The use of water-based paints is spreading which is associated with a **wetting process** and may raise questions about the optimum machining conditions for this purpose (see Sect. 2.6.6). It turned out that the initial roughness properties of a surface may dramatically be changed after wetting highly depending on the kind of machining. To characterize surface stability during wetting, a good indicator is the use of the core depth R_k (see Figs. 2.63, 2.64 and 2.66). The relative increase of this indicator may be constrained during wetting in order to avoid an intermediate sanding.

3.6.3 Edge Machining

Edge machining may cause edge breaking which considerably lowers the quality and aesthetic value of the workpiece. Edge breaking is measured by a summation of broken surfaces along the unit length, mm²/m. Edge breaking can be reduced by using a larger tool diameter and thin chips. A powerful help to avoid edge breaking is the use of a cone-shaped milling cutter with a setting angle (Csanády and Magoss 2013). In this case, the cutting edge describes a hyperbolic pathway and its virtual cutting circle radius may have enormous values depending on the setting angle.

Using the common milling cutters, edge breaking is influenced by the feed per tooth e_z and mainly by the sharpness of the cutting edge or feed distance L_f. Figure 3.13 (Licher 1991).

The feed distance is calculated as

$$L_f = e \cdot T = e_z n \cdot z \cdot T$$

while the true cutting distance is given by

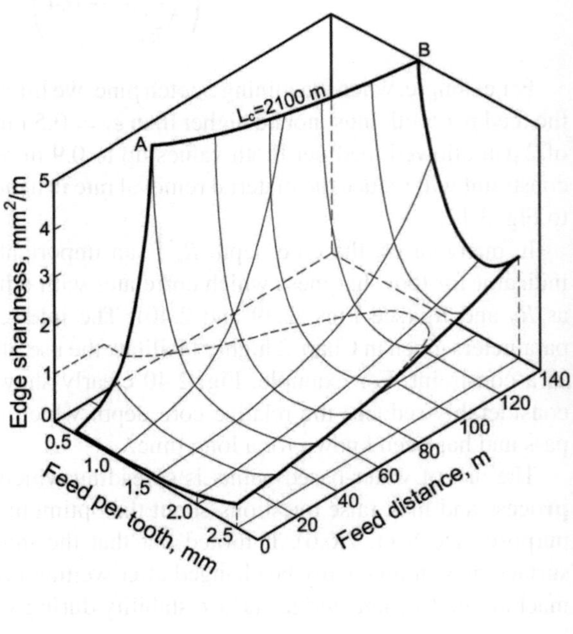

Fig. 3.13 Edge sharpness as a function of feed per tooth and feed distance. $D =$ 180 mm, $H = 2$ mm, $z = 1$, $v_c = 60$ m/s. Particle board and hard metal edge

$$L_c = L_f \frac{R \cdot \varphi}{e_z} \quad \text{with } R \cdot \varphi = 1.42 \cdot \sqrt{H \cdot R} \tag{3.25}$$

where

T is the tool lifetime,
R is the tool radius,
φ is the angle of cutting,
H is the depth of cut.

It is obvious that the wear of the edge is directly influenced by the cutting distance and not by the feed distance. Evaluation of Fig. 3.13 revealed that along the AB-line $L_c = 2100$ m = constant, which means, constant shardness corresponds to a given constant cutting distance. This recognition facilitates the correct description of the experimental results. From Eq. (2.31), we get

$$\frac{L_f}{e_z} = \frac{L_c}{R\varphi} = 110.230$$

which is constant along the AB line. The edge sharpness due to the wear is governed by the cutting distance L_c or its equivalent $R \cdot \varphi(L_f/e_z)$.

The smaller component of shardness as a function of tooth bite e_z, for $L_f = 0$, may be approximated in the following form:

$$S(1) = A \cdot e_z^2$$

while the major component depending on the cutting distance has an exponential character

$$S(2) = e^m - 1 \quad \text{with } m = K\left(\sqrt{R \cdot H}\frac{L_f}{e_z}\right)^n$$

The resultant edge shardness is the sum of the above two equations

$$S = A \cdot e_z^2 + e^m - 1 \text{ mm}^2/\text{m} \tag{3.26}$$

where A, K and n are constants and they can be determined from experimental results. In Fig. 3.13, the constants have the following values: $A = 8 \times 10^4$. $K = 2.755 \times 10^{-4}$ and the exponent $n = 1.2$, if metric dimensions are used.

In practical cases, the allowable shardness is prescribed and the feed distance L_f is the dependent variable as a function of tooth bite e_z, the cutting depth H and the tool radius R. From Eq. (2.32), we get

$$L_f = \frac{e_z}{\sqrt{R \cdot H}}\left(\frac{\ln\left(S + 1 - A \cdot e_z^2\right)}{K}\right)^{1/n} \tag{3.26a}$$

Using the results given in Fig. 3.13, the maximum feed distance L_f can easily be calculated for different conditions. For example, if the allowable shardness is 1 mm^2/m, then the maximum feed distance per edge is $L_f = 82.1$ m at a tooth bite of $e_z = 2.31$ mm which corresponds to 677 m cutting distance per edge and a feed speed of $e = 14.7$ m/min. If the allowable shardness is doubled ($S = 2$ mm^2/m), then the corresponding values are: $L_f = 179.8$ m, $L_c = 1007.4$ m, $e_z = 3.4$ mm and $e =$ 21.6 m/min per edge. Furthermore, the optimum values of L_f and e_z to the chosen shardness S have the following simple relationship:

$$\frac{(L_f/e_z)_{opt}}{S^{0.59}} = \text{const.}$$

or for our particular case given in Fig. 3.13.

$$S = 1.57.10^{-5} \cdot L_c^{1.7}\ \text{mm}^2/\text{m}$$

where L_c must be substituted in m. This last equation also indicates that the shardness is primarily determined by the edge wear. It is interesting to note that tool radius and cutting depth considerably influence the optimum feed distance but not the optimum tooth bite which depends only on the allowable shardness.

Using tools with multiple edges, the feed distance increases according to the number of edges. In the above example, due to the high cutting speed of 60 m/s, the pure cutting time between two sharpenings is relatively short. Calculating for different allowable shardness values, the cutting time T may be expressed by a simple power equation:

$$T = \frac{L_f}{e} = \frac{82.1 \cdot S^{1.13}}{14.7 \cdot S^{0.55}} = 5.58 \cdot S^{0.58}\ \text{min}$$

where the shardness S must be substituted in mm^2/m.

Cutting speed affects the feed distance and cutting time T. Lowering the cutting speed, the feed speed obviously decreases and cutting time increases. The feed distance may be increased in a given extent. Look at Fig. 2.19 for similar case (particle board and a hard metal edge), we may write

$$T \cdot v^{2.2} = \text{const.}$$

or

$$T_1 v_1^{2.2} = T_2 v_2^{2.2}$$

If we lower the cutting speed from 60 to 40 m/s, then the cutting time increases as follows:

$$T_2 = \left(\frac{v_1}{v_2}\right)^{2.2} \cdot T_1 = 2.44\, T_1$$

$$e_2 = 0.667 e_1$$

$$L_{f2} = e_2 T_2 = 1.627 e_1 T_1$$

where e_1 and T_1 are the known values for feed speed and cutting time at 60 m/s cutting speed. The cutting time T is more than doubled, the feed speed decreases by one-third and the feed distance increases some 60%. This may profitably aid the organization of tool change and tool costs.

Sadly, Fig. 3.13 is not generally valid for other wood materials. The form of Eq. (2.32) may be used to process arbitrary experimental results.

3.6.4 Lifetime of the Tool

The lifetime of the cutting tool is also an important constraint depending on many factors (see Sects. 2.5 and 2.6.). Different approaches may be used depending on objective function and machining conditions as follows:

- high cutting speed depth of cut and feed speed are selected for a maximum production rate. The consequence is a relatively short tool life with frequent tool change,
- a high requirement for surface quality also shortens the tool life,
- the edge lifetime is bound to one shift in order to allow workers to change the tool during a shift change. This limits the cutting speed to a certain extent,
- the minimum cost requirement generally limits the cutting speed and depth of cut. This strategy is always in conflict with the maximum production rate. A compromise may balance the difference between the two strategies,
- optimum selection of edge materials to ensure a desired production rate and cost-effectiveness for a given machining process. Depending on the priority, either the production rate or cost-effectiveness will be taken as an objective function while the other is taken as constraint.

Section 2.5 explains that the material removal is proportional to the feed distance but the wear process is proportional to the true cutting distance. The ratio of these distances is given by

$$\frac{L_c}{L_f} = 1.42 \frac{\sqrt{H \cdot R}}{e_z} \tag{3.27}$$

or

$$\frac{L_c}{L_f} = 1.42 \cdot \frac{n \cdot z}{e} \sqrt{H \cdot R} \tag{3.27a}$$

where

H is the depth of cut,
R is the radius of tool,
e is the feed speed.

Using a small tooth bite or low feed speed, the tool life will be reduced in respect to the material removal. Using multi-pass machining, there may be economic to use several rough passes with a bigger tooth bite and the final pass with small tooth bite. A deep depth of cut and larger tool diameter also increases the relative value of a cutting length. The effect of depth of cut can be seen in Figs. 2.19 and 2.20 in Sect. 2.5.

The material removal is directly proportional with the depth of cut. Therefore, the material removal during the lifetime of the tool increases with the square root of cutting depth (see in Sect. 2.13)

$$\frac{V_T}{b} = \frac{e_z \cdot z}{1.42 \cdot v^{n/m}} \sqrt{\frac{H}{R}} \left(\frac{y - y_0}{A} \right)^{1/m} \tag{3.28}$$

The exponent of cutting speed is generally $n/m = 1.2$, so the material removal slightly decreases with higher cutting speeds due to the more intensive wear. The following example shows the use of the above equation.

Beechwood was used to conduct experiments as a function of feed distance (see Figs. 2.58, 2.59 and 2.60). The following tool and operational parameters were used, tool diameter $D = 120$ mm, number of teeth $z = 4$, rotation speed $n = 6000$ rpm, cutting speed $v = 37.7$ m/s, feed speed $e = 12$ m/min, tooth bite $e_z = 0.5$ mm and depth of cut $H = 2$ mm. The cutting speed was examined in the 3000–8000 rpm range. The hard metal edge had a sharpening angle of $\beta = 50°$, and therefore the relation between the edge radius and theoretical edge reduction is $\rho = 0.73y$. Using Eq. (2.16), the edge wear after 1800 m feed distance has the following expression

$$\rho = \rho_0 + 0.0297 \cdot v_{\mathrm{c}}^{0.65} \cdot L_{\mathrm{c}}^{0.52} \tag{3.29}$$

or

$$y = y_0 + 0.034 \cdot v_{\mathrm{c}}^{0.65} \cdot L_{\mathrm{c}}^{0.52}$$

The initial values of ρ_0 and y_0 were 15 and 20.5 μm, respectively, while after 1800 m feed distance $\rho = 60$ μm and $y = 82.2$ μm. The total feed distance refers to four teeth and therefore for one individual tooth the feed distance is 450 m and the cutting distance is 14,000 m. Similarly to Eq. (2.18), the Taylor tool life equation has the form

$$v^{2.25} \cdot T = \left(\frac{\rho - \rho_0}{A} \right)^{1.92} \cdot \frac{0.0737}{\sqrt{H/R}} \tag{3.30}$$

or substituting the corresponding values from Eq. (2.37), we get

$$v^{2.25} \cdot T = \frac{96.365}{\sqrt{H/R}}$$

or

$$v \cdot T^{0.44} = \frac{164}{(H/R)^{0.22}} \quad \rho = 60\,\mu\text{m} \tag{3.31}$$

From Eq. (2.38), it is clear that the constant of the Taylor equations is determined by the wear limit which is here $\rho = 60\,\mu\text{m}$. If a roughness constraint allows only a smaller tool edge radius, then the constants decrease and tool life will be shorter. Using a deeper cut, the cutting length increases and the tool life decreases. The roughness parameters R_z of beechwood in Fig. 2.58 are related to the edge radius with the empirical equation

$$R_z = 40 + 0.45(\rho - \rho_0)\,\mu\text{m} \tag{3.32}$$

where the edge radius ρ must be substituted in μm. Selecting a maximum allowable roughness value and using Eq. (2.38), the lifetime of the tool can easily be calculated.

Figure 3.14 shows the relationship between tool life and allowable edge radius for two cutting speeds which is near a quadratic function. Using different depths of cut, the tool life may be longer or shorter depending on the depth of cut compared to $H = 2$ mm used in Fig. 3.14. For example, in the case of $H = 1$ mm, the tool life increased by $\sqrt{2} = 1.414$ compared to Fig. 3.14.

The cutting speed greatly shortens the tool life, but a higher cutting speed also means a higher material removal rate. Therefore, it is interesting to examine the true effect of cutting speed. The specific material removal between two sharpenings is expressed for our beechwood according to Eq. (2.36) as

$$\frac{V_T}{b} = \frac{e_z \cdot z}{1.42 \cdot v^{1.25}} \sqrt{\frac{H}{R}} \left(\frac{\rho - \rho_0}{A}\right)^{1.92} \text{m}^3/\text{m} \tag{3.33}$$

which is plotted in Fig. 3.15.

Using higher cutting speed, the material removal during the lifetime of the tool is decreasing with a factor of $(v_2/v_1)^{1.25} = 1.42$. Increasing the depth of cut, the material removal increases only with the square root of the *H/R* ratio.

Selecting the allowable surface roughness and the corresponding edge radius fundamentally influences the material removal during the lifetime of the tool. A tool edge radius that is too low substantially decreases the production rate. With a high roughness requirement, it is worthwhile to make a final smoothing pass with a small depth of cut. For example, comparing a smoothing depth of cut $H_1 = 0.2$ mm to the common value $H_2 = 2$ mm, then the feed distance increases in a ratio of $\sqrt{H_2/H_1} = 3.16$ which is quite considerable. Due to the small depth of cut, the

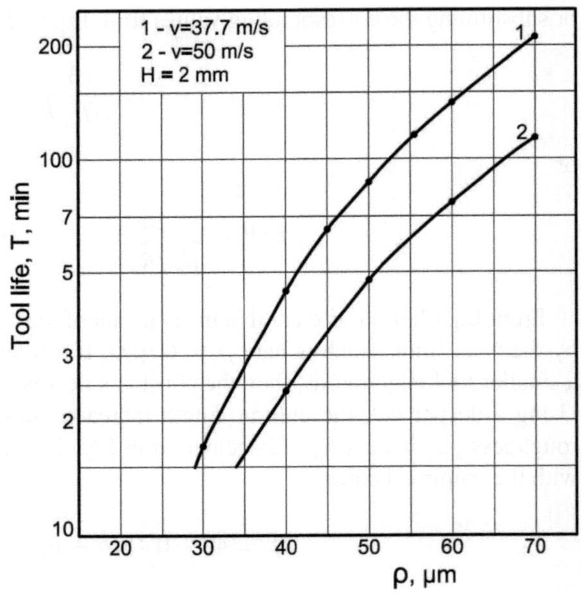

Fig. 3.14 Tool life versus allowable edge radius for different cutting speeds on beechwood

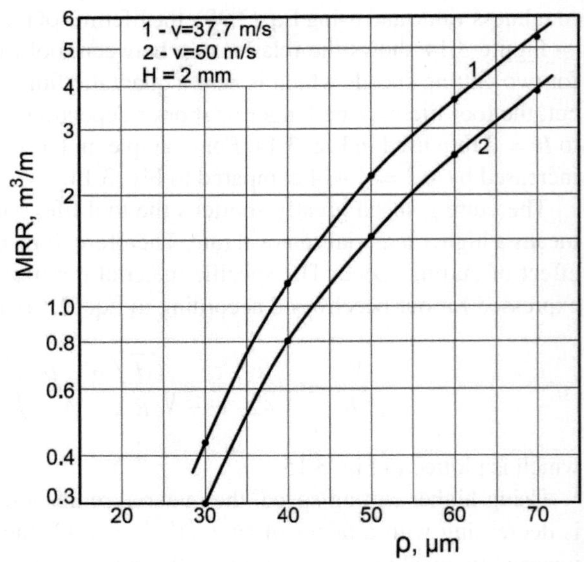

Fig. 3.15 Specific material removal rate as a function of allowable edge radius for beechwood

machined surface related to the lifetime of the tool may be more useful. The area of the machined surface is proportional to the feed distance and the width of cut

$$A_{\mathrm{fT}} = L_{\mathrm{fT}} \cdot b$$

and

$$\frac{A_{\mathrm{fT}}}{b} = L_{\mathrm{fT}} = \frac{0.704 \cdot e_z \cdot z}{v^{1.25} \cdot \sqrt{H \cdot R}} \left(\frac{\rho - \rho_0}{A} \right)^{1.92} \ \mathrm{m^2/m} \tag{3.34}$$

which is plotted in Fig. 3.16 ($e_z = 0.5$ mm and $z = 4$).

Increasing cutting speed and depth of cut decreases the area that can be machined during the life of the tool. The allowable maximum edge radius is also a major factor determining the time a machine can be used between two edge sharpenings.

Using different edge materials, the general form of Eq. (2.37) remains unchanged, but the constants vary according to the material properties of both the tool edge and wood. Furthermore, the sharpening angle of the tool has also some influence on the exponent of the cutting distance. A higher sharpening angle creates less heat load therefore causes less wear resulting in a somewhat lower exponent. Composite materials containing adhesives increase the wear of a tool and shorten its life.

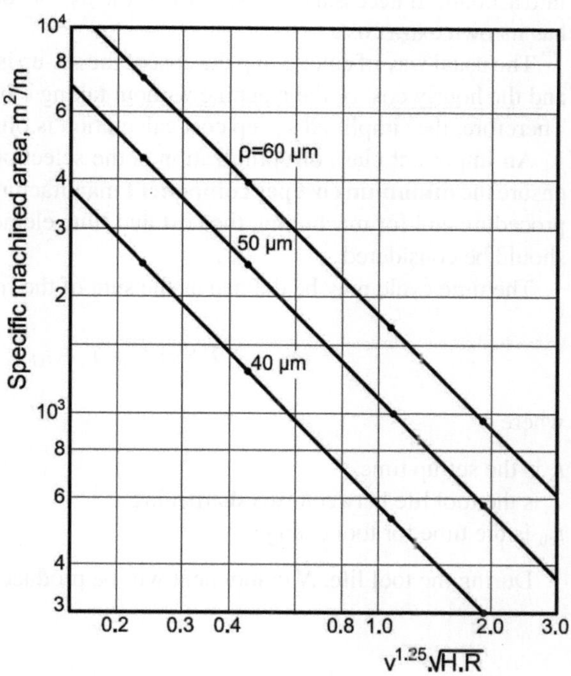

Fig. 3.16 Specific machined area as a function of cutting speed, depth of cut and edge radius on beechwood, $e_z = 0.5$ mm, $z = 4$

3.6.5 Manufacturing Costs

So far, we discussed pure machining without any additional time elements required in an actual manufacturing process (see Sect. 2.13). The *main time components* are as follow:

- machine preparation (set-up) time,
- loading–unloading time,
- process adjusting and return time,
- machining time,
- tool-changing time.

Using CNC machining centres, the set-up time may be reduced dramatically with the concept of an external set-up. This means that part of the set-up operation which can be performed while the machine is operating on the previous series. Another way to reduce set-up time is the clever design of tools and fittings enabling a quick change of tools and fittings.

The estimation of the loss of time for set-up may be quite different. If the machine is not working at its full capacity, then a longer set-up time will cost not too much. On the contrary, if the machine is a "bottleneck" constraining the total production, then the lost production due to set-up time is a loss in production for the whole factory. Accordingly, the set-up costs should be determined by taking the real consequences into account. If necessary, one solution is the use of additional overtime work, which has its own extra cost.

The usual way of calculating the cost of the set-up is to take the normal labour cost and the hourly cost of the machine without taking other cost elements into account. Therefore, the simplified set-up cost calculation is often inappropriate.

An important class of optimization is the selection of operational conditions to ensure the **minimum cost per component** manufactured. That is, a cost optimization procedure and for machining, the cost and time elements between two sharpenings should be considered.

The time cycle may be defined as the sum of the following time elements

$$\Sigma t_{\mathrm{i}} = t_{\mathrm{s}} + T + t_{\mathrm{ch}} \tag{3.35}$$

where

t_{s} is the set-up time,
T is the tool life between two sharpenings,
t_{ch} is the time for tool change.

During the tool life, N component will be produced

$$N = \frac{T}{t_{\mathrm{m}} + t_{\mathrm{n}}} \tag{3.36}$$

where

t_m is the machining time for one component,
t_n is the loading and unloading time.

The calculation of machining time is based on the feed distance per component L_f^c and feed speed e in the following form

$$t_m = K \cdot \frac{L_f^c}{e} = K \cdot \frac{L_f^c}{e_z n \cdot z} = K \cdot \frac{2 \cdot \pi}{60} \frac{L_f^c R}{e_z v \cdot z} \tag{3.37}$$

where K is a correction factor taking the necessary approaching and return time into account ($K = 1.05$–1.1). The total cost related to a time cycle is given by

$$C = x[t_s + T + t_{ch}] + y \tag{3.38}$$

where
x is the labour and machine cost in the unit time,
y is the cost of sharpening or replacing the edge of the tool.

The machine cost in the unit time depends on the cost of amortization, interest rate, maintenance cost, and space and energy costs. The tool cost includes the original cost of the tool and sharpening cost. The cost of a tool depends on its total lifetime while the sharpening cost is determined by the number of workpieces machined between two sharpenings. The number of sharpenings is given by the total lifetime of the tool and the tool life between two sharpenings.

The cost of sanding depends on the cost of the sanding belt and the time needed to replace it.

Using Eq. (2.38), the tool life T reads

$$T = \left(\frac{\rho - \rho_0}{A}\right)^{1/m} \cdot \frac{0.0737}{v^{(n/m)+1} \cdot \sqrt{H/R}} = \frac{B}{v^{(n/m)+1} \cdot \sqrt{H/R}} \tag{3.39}$$

The unit cost of components is simply given by

$$C_u = \frac{C}{N} \tag{3.40}$$

The following example shows the use of these relationships (see also Eqs. 2.37 and 2.40). Tool diameter $D = 120$ mm, $z = 4$, surface quality constraint $\rho_{max} = 50\ \mu$m, $e_z = 0.5$ mm, $H = 1$ mm, $n = 6000$ rpm, $e = 12$ m/min, $L_f^c = 10$ m, the set-up time $t_s = 10$ min, $t_{ch} = 5$ min, $t_n = 1$ min, specific machining cost $x = 50$ \$/h = \$0.833/min, tool-sharpening cost $y = 10$ \$. Using Eq. (2.47) and with $1/m = 1.92$ and $(n/m) + 1 = 2.25$, the tool life is 127.8 min. The machining time is calculated from Eq. (2.45) with $K = 1.1$, and it results $t_m = 0.916$ min. The number of components produced during the life of the tool is $N = 127.8/(0.916 + 1) = 66.7$. The total cost per time cycle is \$129, and the unit cost is $C_u = 129/66.7 = \$1.935$/piece. This result may not be optimum. We can easily find the optimum by varying the cutting speed and the tooth bite. These calculation results are illustrated in Fig. 3.17.

The optimum cutting speed and its corresponding specific cost depend on the tooth bite. The course of the curves on the right side of the minimum line (dotted line) is flat allowing a good compromise to increase the production rate with a slightly higher specific cost. For example, using $e_z = 0.5$ mm and $H = 1$ mm, the minimum cost is \$1.93/component and the feed speed is 13 m/min. Selecting a 60 m/s cutting speed, the unit cost will be \$2.10 (9% higher) but the feed increases to 19.1 m/min which means a 46% increase in the production rate.

The depth of cut modifies the effect of cutting speed on the unit cost. Increasing the depth of cut decreases the tool life and increases the unit cost. The material removal rate (or feed speed) always surpasses the rate of cost increase when using a higher cutting speed, Fig. 3.18.

For cut 3 mm deep, the optimum cutting speed is 34 m/s but increasing the cutting speed to 60 m/s, the unit cost will increase 24% while the feed speed increases 76%.

Manufacturing high-quality furniture may require the production of *curved surfaces.* Curved surfaces may be two- or three-dimensional. A typical example is the curved door of a cabinet which is curved horizontally but straight vertically (two-dimensional). The curved surface may be convex or concave or both.

Curved surfaces can be manufactured by bending or pressing technology, and by machining. Sitting surfaces of seats are often made by bending or pressing plywood sheets to a given profile. If only a few parts are required or the bending or pressing technologies cannot be used, machining is commonly used to produce curved surfaces

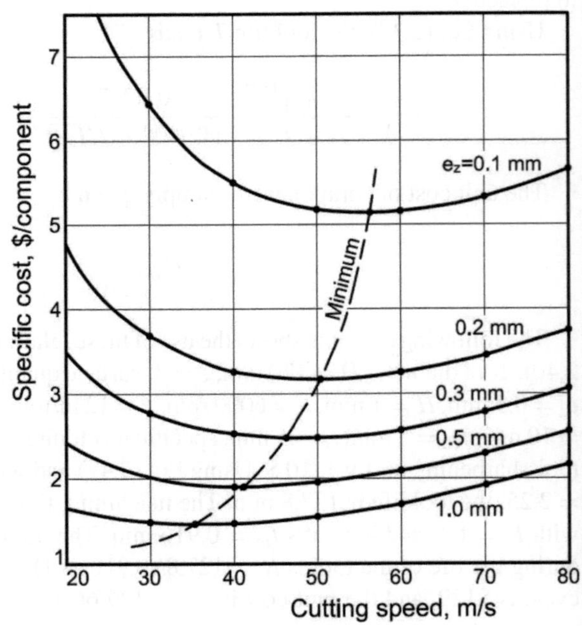

Fig. 3.17 Influence of cutting speed and tooth bite on the specific cost of machining. Beechwood $H = 1$ mm, $\rho_{max} = 50\,\mu m$

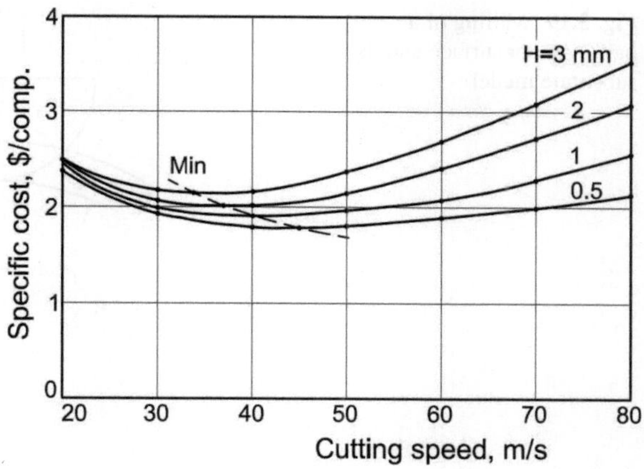

Fig. 3.18 Specific cost as a function of cutting speed for different depths of cut. Beechwood, $e_z = 0.5$ mm, $\rho\text{max} = 50\,\mu\text{m}$

with arbitrary profiles. CNC machining centres are suitable; they allow tools to be changed according to the requirements to achieve maximum efficiency in the production rate or cost.

The main problem in the production of curved surfaces is the inherent geometric roughness which should be kept as small as possible. Geometric roughness is the result of interaction of the tool profile with the surface. Generally, a ball-end milling is used and the main process variables are the tool-profile selection, the tool path, tool-surface inclination and the cutting conditions (cutting speed, feed rate, tooth bite, depth of cut and number of passes). A multi-tool milling operation is used when more than one operation will be done, such as face milling, corner milling, pocket milling or slot milling. Multi-pass milling first does rough machining followed by finish machining.

To machine-curved surfaces, it is important to select the correct tool profile. Figure 3.19 shows an example of machining a half-cylinder surface with ball-end milling. If the radius of the workpiece is much greater than the radius of tool, then a plain substitute model may be used with simpler relations.

The height of geometric roughness z can be obtained from the following equation:

$$r^2 = \left(\frac{b}{2}\right)^2 + (r - z)^2$$

where r is the tool radius and b is the distance between two adjoining passes. The above equation yields

$$z = r - \sqrt{r^2 - \frac{b^2}{4}}$$

or in dimensionless form

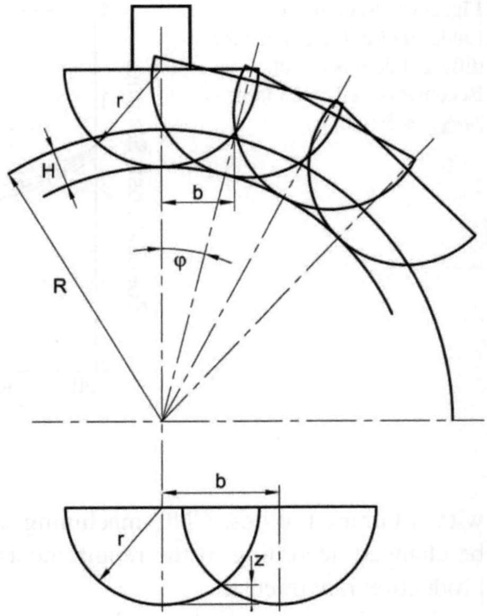

Fig. 3.19 Milling of a half-cylinder surface and its substitute model

$$\frac{z}{b} = \frac{r}{b} - \sqrt{\left(\frac{r}{b}\right)^2 - \frac{1}{4}}$$

This equation can be accurately converted into the simpler equation

$$\frac{z}{b} = \frac{0.125}{\left(\frac{r}{b}\right)}$$

or

$$z = 0.125 \cdot \frac{b^2}{r} \tag{3.41}$$

This equation clearly shows that the cutting width b and the tool radius r fundamentally influence the geometric roughness z. The plain substitute model does not give a correct result if the tool radius tends to infinity, i.e. if the tool has a flat face. In the latter case, the geometric roughness is determined by the local radius of the machined surface and by the cutting width b:

$$z_0 = R \cdot \frac{1 - \cos\varphi/2}{\cos\varphi/2} \quad \text{and} \quad \varphi = \frac{b}{R} \text{ radian} \tag{3.41a}$$

which is equivalent to

$$z_0 = 0.125 \cdot \frac{b^2}{R} \tag{3.41b}$$

It is obvious that the geometric roughness cannot exceed the depth of cut H, therefore

$$z \leq H \text{ or } z_0 \leq H$$

and that is always fulfilled if the maximum width of cut is

$$b_{\max} \leq \sqrt{\frac{H \cdot r}{0.125}} \text{ or } b_{\max} \leq \sqrt{\frac{H \cdot R}{0.125}}$$

for ball-end milling and flat-face milling. Convex, axisymmetric tool profiles, such as conical or spherical, have the inherent drawback that their cutting velocity in the centre line is zero. As a consequence, they need a higher feed force, produce more surface roughness in the vicinity of the centre line with low cutting speeds, especially when the feed direction is perpendicular to the grain. Softwoods are more sensitive to low cutting speeds than hardwoods and using a sharp cutting edge and higher rake angles (around 15°) may help to overcome the problem.

Another possibility is the selection of a proper tool-surface inclination angle, Fig. 3.20. In this case, the centre line and its vicinity do not take part in the cutting of final surface. If the allowable geometric roughness is prescribed, then the width of cut b may not be higher than

$$b_{\max} = \sqrt{\frac{z \cdot r}{0.125}}$$

For example, if $z = 0.1$ mm and $r = 30$ mm, then $b_{\max} = 4.9$ mm.

To avoid the cutting action of the axisymmetric point, the minimum inclination angle should be

Fig. 3.20 Using inclination angle in ball-end milling

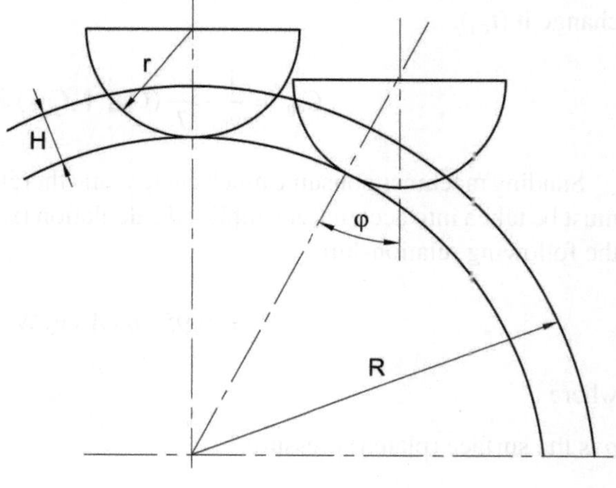

$$\varphi_{\min} = \arccos\left(1 - \frac{z}{r}\right)$$

or its accurate substitute function

$$\varphi_{\min} = 81.7 \cdot \sqrt{\frac{z}{r}}$$

In practice, it is selecting a somewhat higher inclination angle will exclude the low-speed part of the tool from the cutting process.

The general rules of wood machining are also valid here. For instance, increasing feed speed increases roughness, especially when the feed direction is perpendicular to the grain. Climb cutting (down milling) is more demanding to edge sharpness than counter cutting (up milling).

The cost of *sanding* is related to the feed length or the sanded surface in the unit time, and occasionally to the material removal rate. The cost elements are similar to those of knife machining. The machine and labour costs can be given by

$$C_{\mathrm{m}} = \frac{1}{v_{\mathrm{f}}}(C_{\mathrm{ma}} + C_{\mathrm{lab}}) \cdot \left(1 + \frac{t_{\mathrm{n}}}{t_{\mathrm{pr}}}\right) \ \$/\mathrm{m} \tag{3.42}$$

where

v_{f} is the feed speed,
C_{ma}, C_{lab} are the machine and labour costs in the unit time,
t_{n}, t_{pr} are the non-productive and productive time.

Cost of the sanding belt is related to its service life T:

$$C_{\mathrm{b}} = \frac{1}{v_{\mathrm{f}}} \cdot \frac{C_{\mathrm{belt}}}{T} \ \$/\mathrm{m} \tag{3.43}$$

The cost of changing the sanding belt depends on the tool life and time needed to change it (t_{ch}):

$$C_{\mathrm{ch}} = \frac{1}{v_{\mathrm{f}}} \cdot \frac{t_{\mathrm{ch}}}{T}(C_{\mathrm{ma}} + C_{\mathrm{lab}}) \ \$/\mathrm{m} \tag{3.44}$$

Sanding machines consume much energy, and therefore, their energy consumption must be taken into account. *A* simplified calculation (see Eq. 2.11a in Sect. 2.4) gives the following relationship

$$P = 0.95 \cdot p \cdot A \cdot v_{\mathrm{c}} \ \mathrm{W} \tag{3.45}$$

where

p is the surface (platen) pressure,

A is the sanded surface, $A = b \cdot L_c$,
v_c is the cutting speed.

Taking one hour productive time and the unit price of electricity C_{el} (\$/kWh), the specific cost of energy is

$$C_e = \frac{P \cdot C_{el}}{60 \cdot v_f}\left(1 + \frac{t_n}{t_{pr}}\right) \text{\$/m} \tag{3.46}$$

where P must be substituted in kW. The main problem is the correct determination of the belt life. The most reliable method is to determine the economic lifetime based on the material removal rate. The following example will demonstrate this evaluation method. Taking a belt sander 1 m wide, 30 cm long, surface pressure $p = 0.5$ N/cm^2, belt speed 20 m/s, $v_f = 8$ m/min, to sand beechwood (see Figs. 2.11 and 2.14). The sanded surface is 0.3 m^2, the power requirement is approximately 23.5 kW. The maximum removal rate is 0.45 cm^3/cm^2 min, which means 1350 cm^3/min for the entire sanded surface. This maximum value continuously decreases due to its wear. Using the experimentally obtained removal rate in Fig. 2.11, for the next ten-minute intervals the material removal can be calculated as shown in Table 3.4.

Taking the machine cost \$20/h, labour cost \$10/h, sanding belt \$20 and set-up time of 20 min, time for belt change 30 min, the operational time is variable. If the calculated cost at the end of each interval is divided by the volume of removed material, then we obtain a minimum value at 50 min (Fig. 3.21). At the same time, the cost related to the sanded surface area is continuously decreases. The belt should be used as long as the surface quality requirements are met. Sadly, no well-established quality criterion has been worked out yet.

The continuous decrease of specific stock removal due to wear may cause some problems in the sanding operation. If the thickness of material removal is important, then it may be corrected by varying the surface pressure or feed speed. Fortunately, a tenth of a mm deviation is generally not a problem in evaluating the surface quality. If a given thickness of the workpiece is required, a contact wheel sander will be used (see Chap. 5).

Table 3.4 Material removal in the subsequent 10 min intervals and the corresponding specific costs. ($v_f = 8$ m/min)

Intervals min	0–10	10–20	20–30	30–40	40–50	50–60	60–70	70–80
$Q \times 10^{-3}$ cm^3/min	12	7.26	5.05	3.47	2.526	1.89	1.42	1.1
\$/m^3	4166	2855	2468	2340	2310	2330	2360	2448
\$/m^2	2.08	1.146	0.83	0.677	0.58	0.52	0.476	0.44

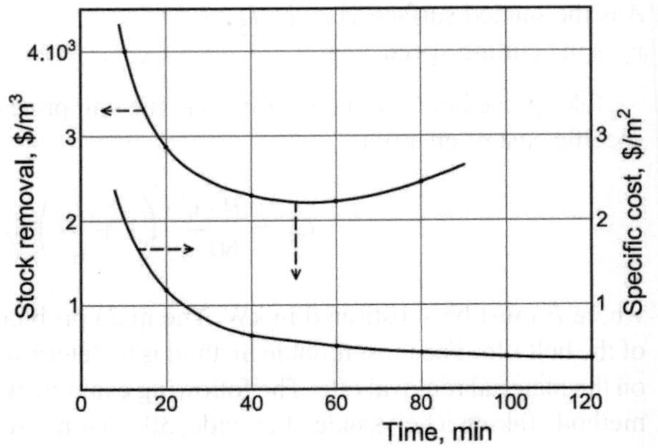

Fig. 3.21 Optimum lifetime of a sanding belt related to stock removal and the specific cost of sanding a surface area

3.6.6 *Primary Wood Processing*

The primary work of the *wood industry* is to cut the raw material into lumber with a given cross-section. The requirement for surface quality is not too stringent, but the cutting accuracy is very important to reduce the necessary tolerance for subsequent woodworking operations. There are many reasons for surface irregularities: blade instability (washboarding), uneven tooth setting, low lateral stiffness and high feed speed (Csanády and Magoss 2013). The feed speed is an especially important factor because the sawing performance primarily depends on the feed speed. Hence, it is important to correctly tension the saw blades.

In sawing, the performance and energy consumption are determined by the cut cross-section in the unit time, which is calculated as the feed speed multiplied by the height of the sawn cross-section *eH*. The maximum allowable feed speed is constrained by the gullet feed index which is the ratio of the loose chip volume and gullet feed volume (gullet area × kerf width). The chip produced has to fit in the gullet until its exit from the wood. The maximum allowable feed speed is theoretically equal to

$$e_{\max} = \frac{V_g \cdot v}{\varepsilon \cdot V_w} \quad \text{with } V_w = H \cdot t \cdot b \tag{3.47}$$

where

V_g, V_w are the gullet and solid wood volume,
ε is the volume ratio for chip and solid wood which is generally around 3,
t is the tooth pitch,
b is the kerf width.

From the above equation, it is obvious that there is hyperbolic relation between feed speed and cross-section height: as the sawn height increases the allowable feed speed decreases. The loose chip can be slightly compacted with a small pressure.

Therefore, the theoretical allowable feed sped may be increased by 20% without the risk of developing higher lateral pressure in the gullet causing additional friction. Figure 3.22 shows the relationship between feed speed and sawn height for a bandsaw cutting oak logs with different gullet feed indexes and also measurement results in a sawmill. An experienced worker could accurately select the appropriate feed speed according to the diameter of the sawn log (Déry 1985). The optimum performance of a sawmill mainly depends on the feed speed and minimization of non-productive time. Due to the limited feed speed, the optimum cost of the mill is obtained at the maximum allowable feed speed and maximum share of productive time. The latter requires skill and good organization of individual operations. Smaller sawn boards have a greater surface compared to their volume. Cutting small logs into boards or small pieces considerably increases production costs. The specific cutting surface of logs from 30 to 45 cm in diameter is approximately 23 or 37 m^2/m^3 if sawing 1″ or 2″ boards.

Experiments were conducted on a smaller bandsaw with wheel diameter of 1100 mm, a gullet area of 218 mm^2, a 30-mm tooth pitch and with a cutting speed of 30 m/s. The theoretical feedspeed was $e = 4.36/H$ where the cutting height H must be substituted in m. The bandsaw cuts in one direction and the backward motion occurred at a speed of around 40 m/min. The coefficient of time use was around 80%. Oak logs were sawn into 1″ (26 mm) boards using logs from 20 to 45 cm in diameter and of 4 m long.

Figure 3.23 shows the specific energy consumption, without edge cutting, as a function of log diameter.

The board/log volume ratio fluctuates between 70 and 75%. The sawn volume moderately increases with the log diameter. Accordingly, the specific cost decreases to the same extent, Fig. 3.24. The following cost components are taken into account: machine cost \$80/h, cost of three workers is \$30/h and the tool cost with change is \$15/h.

Figure 3.24 clearly shows that, doubling the log diameter, the sawn volume in the unit time increases only some 30% and the specific cost decreases to the same extent.

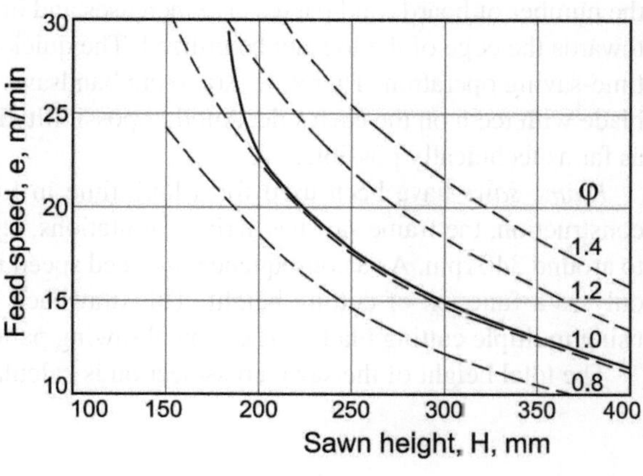

Fig. 3.22 Relationship between feed speed and sawn height for different gullet feed indexes φ. Solid line shows measurement results in a sawmill for oak, $v = 30$ m/s, $t = 30$ mm, $V_g = 500$ mm^3, $\varepsilon = 3$

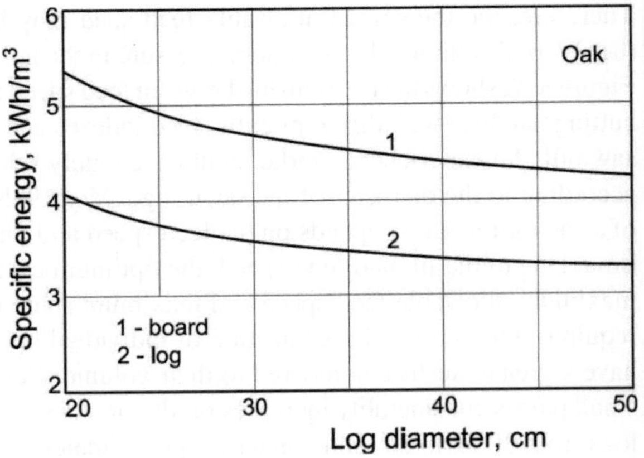

Fig. 3.23 Specific energy consumption of a bandsaw as a function of log diameter cutting 1″ boards

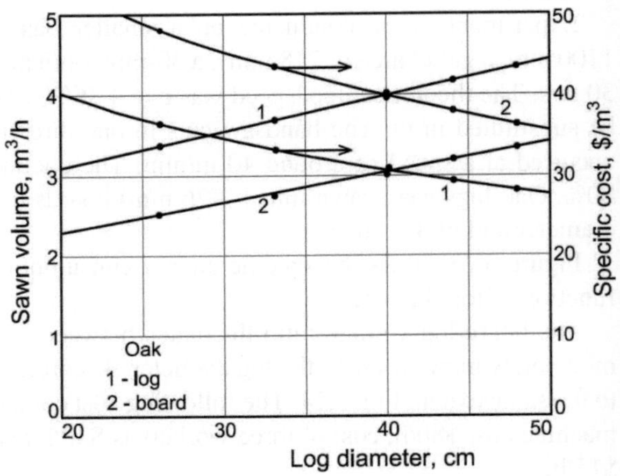

Fig. 3.24 Sawn volume and specific cost for a bandsaw cutting 1″ boards

This is due to the fact that in each pass only one board is cut and, at the same time, with an increasing diameter, the feed speed decreases. With increasing diameter, the number of boards and passes also increases and only the smaller cutting heights towards the edge of the log can be utilized. The quick return motion is an important time-saving operation. Therefore, some big bandsaws cut forward and back using a blade with teeth on the each side. Another possibility is to use higher cutting speeds as far as technically possible.

Frame saws have been used for a long time in European sawmills. Due to its construction, the frame saw has serious limitations, e.g. its rotation speed is limited to around 340 rpm. As a consequence, the feed speed is also limited and it may vary only as a function of cutting height. This drawback is, however, compensated by using multiple cutting blades. The typical sawing pattern is shown in Fig. 3.25.

The total height of the sawn cross-section is calculated as

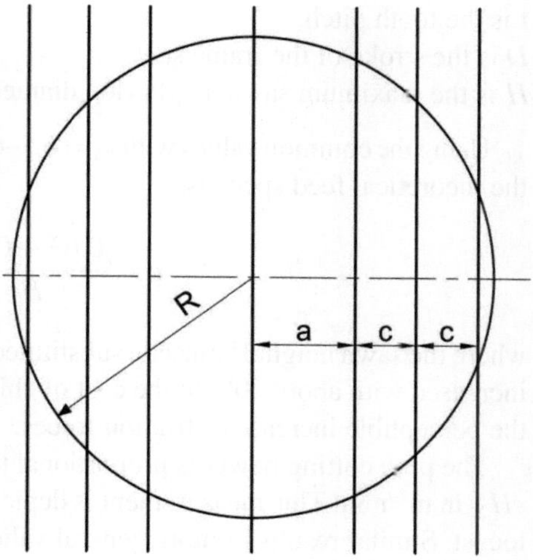

Fig. 3.25 Sawing pattern with a frame saw

$$H_{\Sigma} = 4 \cdot \left[\frac{R}{2} + \sum_{n}^{i=0} \sqrt{R^2 - (a + i \cdot c)^2} \right] \tag{3.48}$$

where $R > a + i.c$. The total sawn height approximately depends on the log diameter with an exponent of 2.2. For example, sawing board thicknesses of $a = 7.5$ cm and $c = 4$ cm wide, the total sawn height is

$$H_{\Sigma} = 0.07\, d^{2.2} \text{ cm}$$

while for $a = 5.2$ cm, and $c = 2.6$ cm

$$H_{\Sigma} = 0.112\, d^{2.2} \text{ cm}$$

where the log diameter d must be substituted in cm.

The theoretical feed speed (under the assumption that the chip has to fit in the gullet) can be calculated as

$$e = \frac{A_g \cdot v_a \cdot 60}{2 \cdot \varepsilon \cdot H \cdot t} \text{ m/min} \tag{3.49}$$

with

$$v_a = \frac{D \cdot n}{30} \text{ m/s}$$

where

A_g is the gullet area,

t is the tooth pitch,
D is the stroke of the frame saw,
H is the maximum sawn height (log diameter).

Using the common values with $v_a = 6.5–6.8$ m/s, $t = 25–30$ mm, $A_g = 2.4–2.8$ cm^2, the theoretical feed speed is

$$e = \frac{0.63 - 0.76}{H}\ \text{m/min} \tag{3.49a}$$

where the sawn height H must be substituted in m. In practice, this feed speed may be increased with about 20% to the cost of chip compaction in the gullet space without the perceptible increase of friction forces.

The pure cutting power is proportional to the sawn cross-section in the unit time, eH_Σ in m^2/min. This measurement is depicted in Fig. 3.26 for Scotch pine and black locust. Similar results of more general validity are given in Fig. 2.9.

Keeping in mind that Eq. (2.60a) may increase, the volume of sawn logs in the unit time is proportional with the log diameter and the efficiency of time utilization η_T

$$V \cong 45 \cdot d \cdot \eta_T\ \text{m}^3/\text{h} \tag{3.50}$$

where d must be substituted in m.

The specific energy consumption (kWh/m^3 log) depends on the sawing pattern, wood species, log diameter and the power consumption of the frame saw when it is idling.

The idling power consumption of frame saws with a hydraulic drive is relatively high, around 20 kW. Figure 3.27 shows measurement results in a sawmill cutting

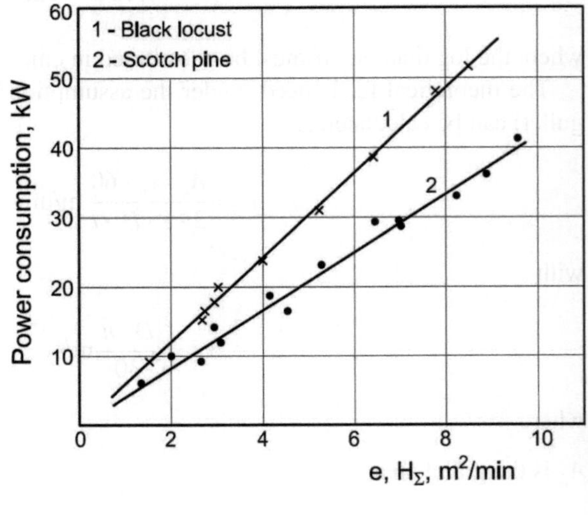

Fig. 3.26 Cutting power consumption for black locust and Scotch pine

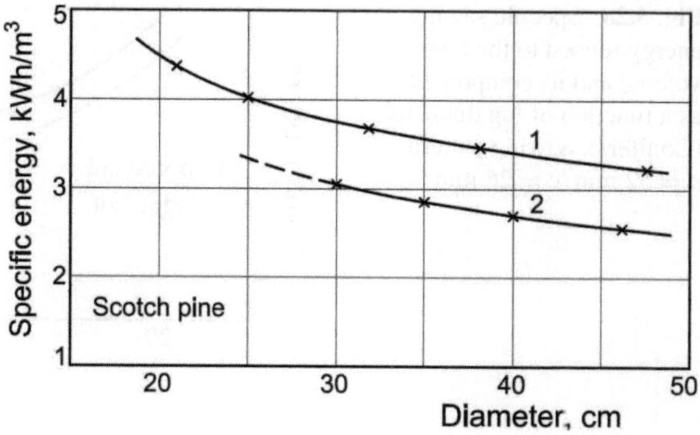

Fig. 3.27 Specific energy consumption of a frame saw cutting Scotch pine, using two sawing patterns as a function of log diameter. 1. $a = 5.2$ cm, $c = 2.6$ cm 2. $a = 7.5$ cm, $c = 4$ cm

Scotch pine and using two sawing patterns (Sitkei et al. 1988). The time utilization was 80%, and the energy consumption is only from the main cutting without edge cutting. The latter was made by a double-circular saw with an adjustable cutting distance.

The specific energy consumption was calculated for Scotch pine as

$$\frac{P}{V} = \frac{(1 - \eta_T) \cdot P_0 + \eta_r \cdot (P_0 + 4.1 \cdot eH_\Sigma)}{\eta_T \cdot 45 \cdot d} \text{ kWh/m}^3 \tag{3.51}$$

with $P_0 = 20$ kW idling power and $\eta_T = 0.8$.

The production of boards from the raw material may be as high as 70% if the logs are properly sorted according to their diameter. That means that the specific energy consumption related to the volume of boards will be 43% higher compared to Fig. 3.27.

Using double-circular saws for edge cutting, each saw requires a maximum driving power of 10 kW to use feed speeds up to 70–75 m/min for boards one or two inches thick. Cutting three-inch boards, the feed speed should be reduced to approximately 50 m/min. The specific energy requirement for edge cutting fluctuates between 0.45 and 0.3 kWh/m^3 board as a function of log diameter (conifers). Hardwoods (oak, black locust, beech) require 40–50% more energy for sawing (see Fig. 3.26.). The energy requirement and its components as a function of log diameter are demonstrated in Fig. 3.28 which is related here to the unit board volume. 70% of the log volume is turned into boards.

The specific sawn surface area and its components vary as a function of log diameter, using a given sawing pattern, Fig. 3.29. The sawn surface area of boards cut slightly increases while that of the edge cut slightly decreases with increasing log diameters. As a result, the total sawn surface does not vary much and its average value is roughly 45 m^2/m^3 board for each sawing pattern.

Due to the big difference in the feed speeds for frame saws (2–4.5 m/min) and circular saws (50–70 m/min), both sawing operations can be synchronized in time and the edge cut does not influence the capacity of the frame saw. The organization

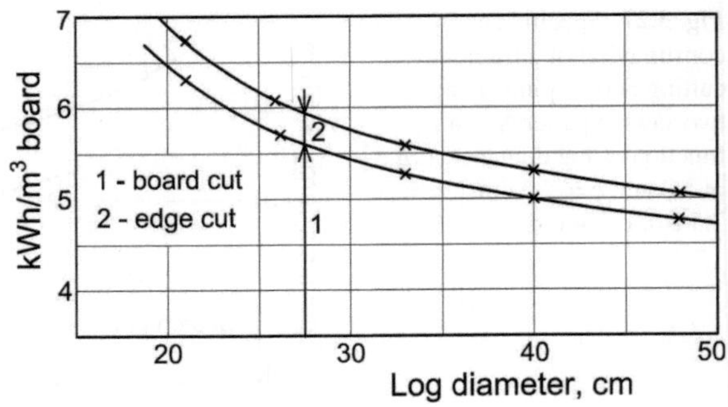

Fig. 3.28 Specific sawing energy related to the board volume and its components as a function of log diameter (Conifers). Sawing pattern: $a = 52$ mm, $c = 26$ mm

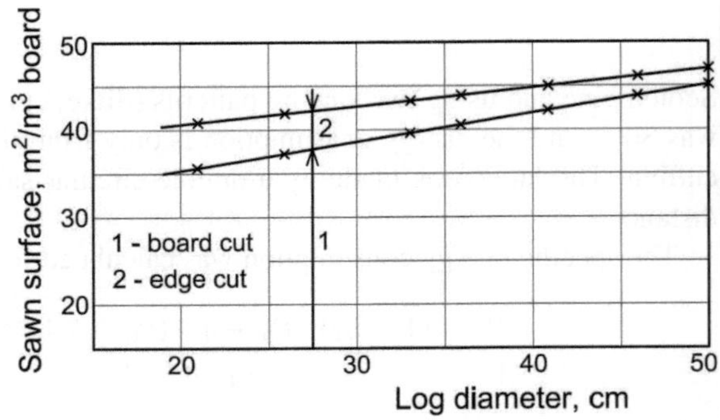

Fig. 3.29 Specific sawn surface area related to the board volume and its components as a function of log diameter. Sawing pattern: $a = 52$ mm, $c = 26$ m

of material handling is very important because the volume of sawn logs in the unit time directly depends on the efficiency of time use, as shown in Eq. (2.61).

The specific cost of the boards ($/m^3) depends on the various cost elements such as the machine, labour, energy and overhead costs, tool and sharpening costs, and further the setting, tool-changing and non-productive times. Taking the prime cost of all the machines as $900,000, the machine costs $160/h, the labour for three workers costs $30/h and the tool and tool-changing cost 30 $/h, the total cost in the unit time is $220/h. Using the above example given in Figs. 3.28 and 3.29, the sawn volume and its specific cost as a function of log diameter are shown in Fig. 3.30. Clearly, the log diameter influences the unit cost considerably.

The best way to reduce costs is good organization of subsequent operations, decreasing the non-productive time. It is to ensure the highest possible board outcome using log sorting and an appropriate sawing pattern.

The energy cost of sawing is not very significant compared to other costs. The cost of 1 kWh energy is approximately $0.10, and therefore the net energy cost is around $0.50–$0.60/m^3 board (see Fig. 3.28).

Due to the strong limitation of system variables (cutting and feed speed), optimization is much more a striving after the ideal case where there is no non-productive time and there are no waste products. At the same time, the use of multiple blades linearly increases the cutting capacity, according to Eq. (2.61), as a function of log

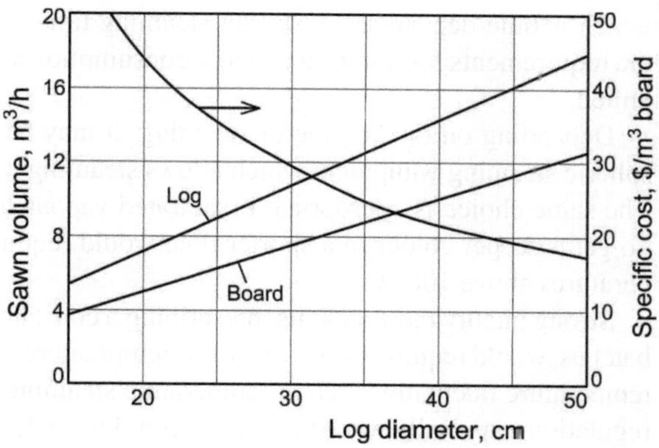

Fig. 3.30 Sawn volume and specific cost of board production as a function of log diameter. Sawing pattern: $a = 52$ mm, $c = 26$ mm

diameter. With increasing log diameters up to its maximum allowable, it is fully competitive with a bandsaw.

Cutting precious wood (e.g. incense cedar for pencils) or thin lamellae, thin kerf sawing is necessary. For this purpose, there are special frames and circular saws with a kerf width down to 1.2 mm.

3.6.7 Steaming of Wood

Steaming is a common value-adding process suitable for

- form stabilization,
- colour homogenization,
- colour modification.

There are wood species with good mechanical properties (black locust, rubberwood), but their original colour is not very attractive to use for quality wood products. The colour of heart and sapwood may differ in a manner which is aesthetically not acceptable. In these cases, steaming of wood may be a successful method of adding value. The optimization of steaming is a typical multi-criteria decision with the following possible objective functions:

- minimum energy consumption,
- minimum steaming time,
- minimum investment,
- minimum specific cost,
- maximum quality requirements.

Much energy is to warm the boards up to steaming temperature. The amount of heat mainly depends on the density of the wood and its moisture content. Additional heat introduction is needed during steaming to cover heat losses into the environment, which is time-dependent. A shorter steaming time reduces these losses. Therefore, the requirements for minimum energy consumption and minimum steaming time are united.

Depending on the volume of steaming, it may be rational to use simpler atmospheric steaming equipment which allows steaming temperatures up to only 100 °C. The same choice is appropriate if saturated vapour is available for other purposes. To get a deeper colour in a shorter time would require overheated steam with temperatures above 100 °C.

Strong quality requirements, prescribing a constant colour hue for the subsequent batches, would require a lower steaming temperature which is less sensitive to smaller temperature fluctuations. High-temperature steaming requires accurate temperature regulation, especially when the required colour hue is in the mid-range of the possible colour change. In the lower range of possible colour change (dark colour hues), the rate of colour change for every steaming temperatures decreases and also the risk to colour deviation.

Colour homogenization always requires overheated steam around 110 °C or somewhat higher.

The colour change must penetrate the entire volume of the wood, so its entire volume must be warmed. A board 1″ thick needs around 8–10 min to warm which will be accelerated when water vapour condenses on the board surface. The result of condensation may be a liquid film or as a droplet. The formation of droplets is easier on rough surfaces such as sawn boards. The condensate highly increases the heat transfer coefficient on the surface decreasing the time needed for the warming-up period.

For planning and optimization of a colour modification, the corresponding functional relationships are needed. In the following, the steaming of black locust, an important wood species in Central and Eastern Europa, will be discussed in detail.

Figure 3.31 shows the lightness variation as a function of time at different steaming temperatures. A logarithmic scale must be used to read lower time values accurately. It is clearly seen that higher temperatures dramatically accelerate the browning process, decreasing the necessary time for treatment. Furthermore, lightness reduction proceeds to lower limit values around $L^* = 35\%$.

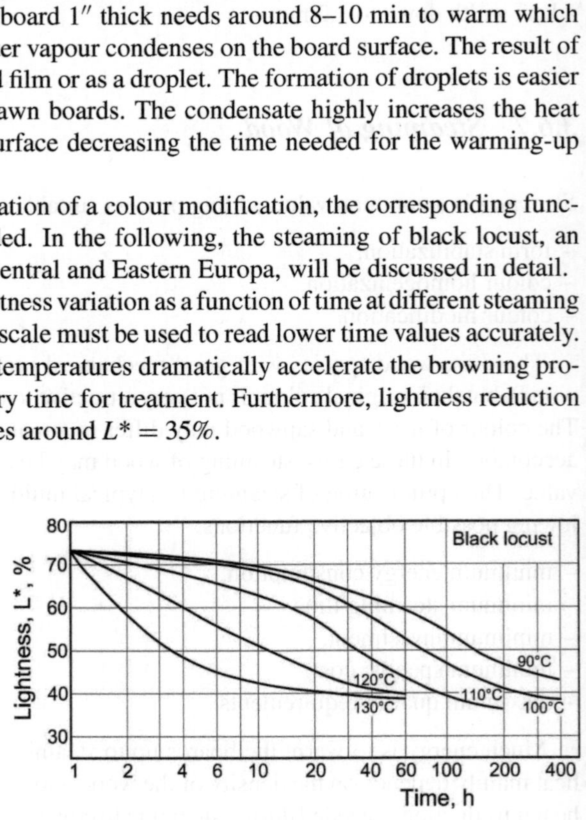

Fig. 3.31 Lightness reduction by steaming as a function of time at different steaming temperatures

The next important question is the variation of colour hue as a function of steaming temperature and time. Fortunately, the colour hue has a uniquely defined relationship as a function of lightness reduction as depicted in Fig. 3.32.

It means that the colour hue is determined alone by the lightness and it does not depend on the steam temperature. This connection between lightness and colour hue means that we cannot produce arbitrary lightness and colour hue independently of each other.

Another plot is given in Fig. 3.33 where the relationship between steaming temperature and time is represented for a constant lightness and colour hue. A considerably longer steaming time is needed to obtain a deep red-brown colour compared to a mid-brown colour. Both colour hues may be used for quality strip flooring.

If the decision-making for a given colour hue is done, the next question is the necessary steam quantity related to the unit board volume (m^3). Heating occurs with

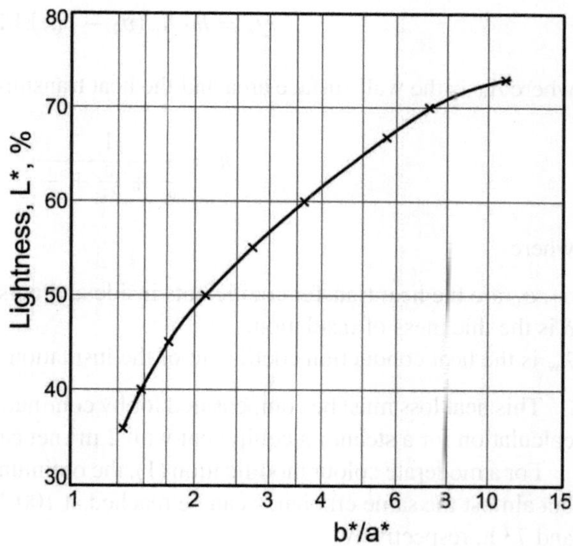

Fig. 3.32 Relationship between lightness and colour hue in black locust

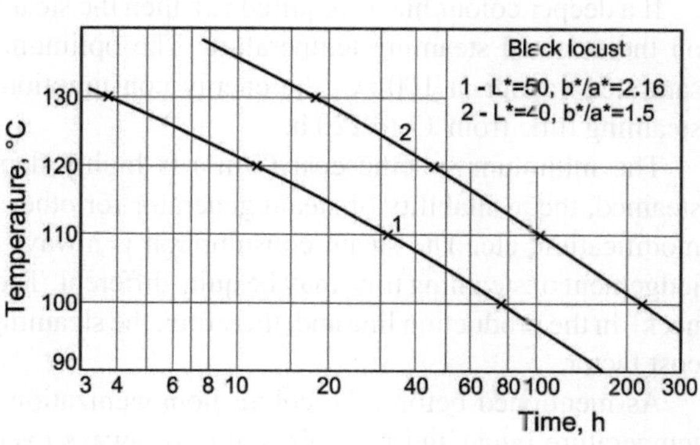

Fig. 3.33 Relationship between steaming temperature and time for constant lightness and colour hue. Black locust

the heat of evaporation through condensation. The heat necessary for warming up a unit board volume is calculated as

$$Q = c_p \cdot \rho \cdot (\vartheta_1 - \vartheta_0)\ \mathrm{kJ/m^3}$$

where

c_p—is the specific heat of wood, kJ/kg °C,
ρ—is the wood density, kg/m^3,
ϑ_1—is the temperature of steaming,
ϑ_0—is the temperature of the environment.

During the steaming process, heat will be transferred onto the external wall surface of the steaming equipment into the environment. This heat loss is given by the equation

$$Q_2 = h \cdot A_w(\vartheta_1 - \vartheta_0)\ \mathrm{kJ/h}$$

where A_w is the wall surface area and the heat transmission coefficient h is given by

$$h = \frac{1}{\frac{1}{\alpha_1} + \frac{\delta}{\lambda_w} + \frac{1}{\alpha_0}}$$

where

α_1, α_0 are the heat transfer coefficients inside and outside,
δ is the thickness of insulation,
λ_w is the heat conduction coefficient of the insulation.

This heat loss must be compensated for by continuous heat introduction. A model calculation for a steaming equipment with 2 m^3 net capacity is shown in Fig. 3.34.

For a moderate colour modification (1), the optimum steam temperature is 115 °C but almost the same efficiency can be reached at 100 °C. The steaming times are 19 and 73 h, respectively.

If a deeper colour hue is required (2), then the steam consumption strong depends on the selected steaming temperature. The optimum is nearly 130 °C, and using saturated vapour at 100 °C, the energy consumption increases with 50% and the steaming time from 17 to 220 h.

The minimum specific cost ($/m^3) is highly dependent on the volume to be steamed, the availability of steam generator for other purposes, the required colour modification, etc. The steam consumption is always a definite cost factor, but the judgement of steaming time may be quite different. The steaming is seldom a "bottle-neck" in the production line and, therefore, the steaming time may be a less important cost factor.

As mentioned before, the colour homogenization is more bound to a narrower temperature range and it requires almost always overheated steam in the range of

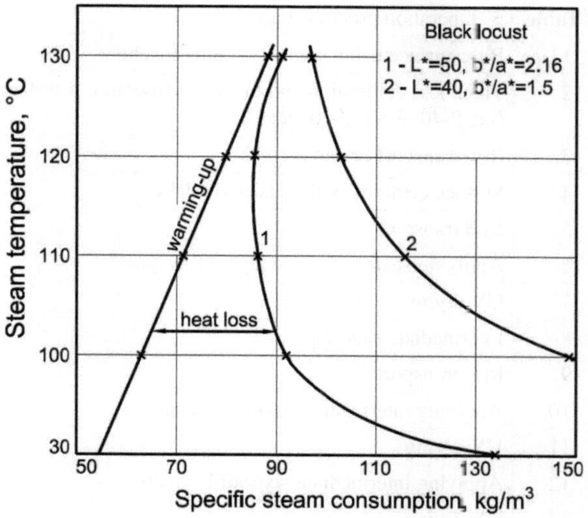

Fig. 3.34 Steam consumption of one m^3 timber as a function of steam temperature for two different colours. The wood density is 700 kg/m^3

110 and 115 °C (Csanády et al. 2015). In this case, the quality requirement, as a homogeneous colour, surpasses all other desires.

3.6.8 Surface Finish of Wood

Coating of wood surfaces with lacquer or hard-drying oils is a frequent manufacturing operation. In the following, we examine the finishing process of strip-flooring parquet requiring a set of subsequent operations such as calibrating to ensure uniform thickness, surface finishing with sanding grit sequencing to achieve the desired surface quality, application of layers of sealer, mid- and top layers of lacquer with intermediate sanding and cleaning.

Generally, the complete manufacturing process is accomplished in a sequence of single operations and the layout of equipment follows the sequence of operations in a line. The sequence of operations is drawn up in an *operation process chart*, which is in use for many years to display the operations, material transportation and inspection in a compact form (Table 3.5). The movement of parts within a plant does not add any value and functional utility to the product and, for this reason, the single criterion for optimum equipment layout is the minimum material handling cost with the shortest movements between equipment. In this case, the operations are purposefully linked together in a line: when one operation is finished, the product moves directly to the next operation.

It is obvious that the whole line should possibly operate with constant or nearly constant feed speed. The throughput is determined by the slowest operation (bottleneck) and, therefore, the main task of optimization is the synchronization of feed speed for the single operations to ensure the highest possible overall feed speed.

Table 3.5 Operation process chart

1.	Base surface using contact sanding machine
2.	Thickness calibration and making a smooth top surface using sanding grit sequencing (i.e. P-40, P-80, P-100 and P-150)
3.	Roll transport of parts
4.	Surface cleaning with brush cylinders
5.	Roll transport
6.	Applying sealer and base layer of lacquer
7.	UV drying
8.	Intermediate sanding
9.	Roll transport
10.	Applying intermediate layer of lacquer
11.	UV drying
12.	Applying intermediate layer of lacquer
13.	UV drying
14.	Intermediate sanding
15.	Surface cleaning
16.	Roll transport
17.	Applying layer of lacquer
18.	UV drying
19.	Applying top layer of lacquer
20.	UV drying
21.	Applying final top layer of lacquer
22.	UV drying
23.	Roll transport
24.	Inspection
25.	Transport to workstation making the connecting profiles
26.	Packing for distribution
27.	Storage

The feed speed of sanding may be around 10 m/min or even more. The feed speed in the UV-drying channel requires some considerations. The time needed for hardening of the coating film depends on the type of coating film and the UV irradiation intensity (kW/m^2). Selecting a high feed speed, the drying channel will be longer which increases the investment costs.

The relationship between hardening time T and the UV irradiation intensity I has the following form

$$T = A + \frac{B}{I} \text{ min}$$

where A and B are constants and I means the irradiation intensity, kW/m^2. The required drying channel length L is determined by the feed speed and hardening time

$$L = e \cdot T = e \cdot \left(A + \frac{B}{I}\right) \text{m}$$

The cost per hour of a drying channel depends on the length of the channel, electricity and labour costs. The drying channel has a minimum length because space is needed for the UV lamp. Therefore, the cost of drying channel is not a simple linear function of its length.

There are different cost-calculating methods taking into account the prime cost of equipment, interest on capital, depreciation, energy consumption, maintenance and space requirement. A simplified method takes 40% of equipment prime cost and divides that by the utilization in hour, per years, to give the operating cost per hour. The energy and labour costs should be added.

Following the simplified method, the operational cost of a drying channel with a given length L can be given as

$$C_{\text{e}} = D + E \cdot L + C_{\text{lab}} \,\$/h$$

where D and E are constants, and C_{lab} means the labour cost per hour.

The surface dried in the unit of time will be

$$Q = b' \cdot e \cdot 60 \,\text{m}^2/\text{h}$$

where b' is the effective width of the drying material and feed speed e must be substituted in m/min.

The total power consumption is calculated as the channel area multiplied by the irradiation intensity.

$$I_{\text{t}} = b \cdot L \cdot I = b \cdot I \cdot e \cdot \left(A + \frac{B}{I}\right) \text{kW}$$

where b means the width of the drying channel.

The total operational cost (equipment + energy + labour) divided by the dried surface gives the specific cost in the following:

$$\frac{C_{\text{t}}}{Q} = \frac{D + E \cdot e \cdot (A + B/I) + C_{\text{lab}} + c_{\text{e}} \cdot b \cdot I \cdot e \cdot (A + B/I)}{60 \cdot e \cdot b'} \,\$/\text{m}^2 \qquad (3.52)$$

where c_{e} is the specific cost of electricity in \$/kWh. (Generally, \$0.10/kWh may be used). From the above equation, the optimum irradiation intensity and the channel length can be derived. Taking the partial derivative with respect to I and equating it to zero, it yields

$$I_{\text{opt}} = \sqrt{\frac{E \cdot B}{c_e \cdot b \cdot A}}\ \text{kW/m}^2 \tag{3.53}$$

and the optimum channel length is calculated as

$$L = e \cdot \left(A + \frac{B}{I_{\text{opt}}}\right)\ \text{m}$$

The optimum irradiation does not depend on the feed speed, but the optimum channel length does and it linearly increases with the feed speed.

The following example shows the use of the derived equations. The relationship between hardening time and irradiation intensity is given in Fig. 3.35.

The irradiation intensity is related to the area of the drying channel. The UV-irradiation lamp is 1.3 m long, and its line intensity varies between 60 and 120 W/cm. The constants in the corresponding equation have values of $A = 0.05$ and $B = 1.35$ for hard-drying acrylic lacquer. Different lacquers may have slightly different constants and the approximate field of variation is also given in Fig. 3.35. The constants in the operational cost equation are $D = 3.5$ and $E = 0.5$, while the labour cost is \$10/h. The width of the drying channel is 1.3 m, and the effective material width is 1.152 m (6 × 192 mm).

The calculated specific cost as a function of irradiation intensity and for different feed speeds is shown in Fig. 3.36. The specific cost hardly varies with the irradiation intensity, except for small intensities.

Increasing feed speeds considerably decreases the specific cost. The course of the curve in Fig. 3.35 will change the optimum irradiation intensity to a certain

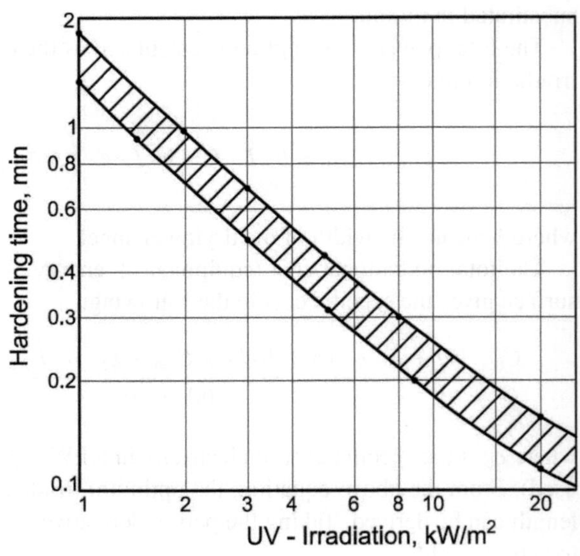

Fig. 3.35 Correlation between irradiation and hardening time

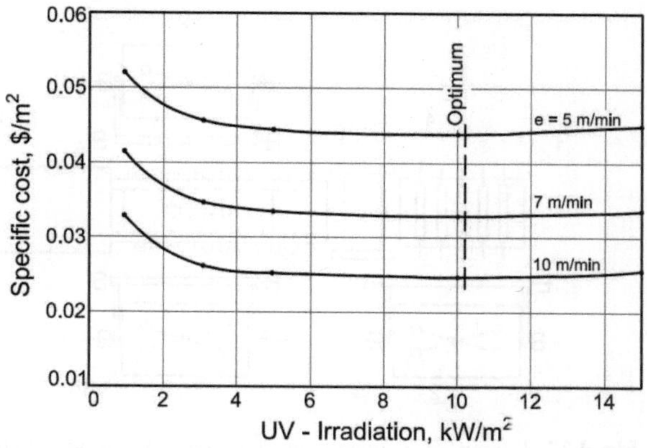

Fig. 3.36 Specific cost of the irradiation

extent. For our example, the optimum irradiation intensity is 10.19 kW/m^2 and the corresponding optimum channel length is 1.82 m.

The UV-irradiation lamps produce considerable heat which will heat the air in the channel up to 60–70 °C depending on the ventilation rate of the drying channel. The ventilation rate for a channel 2.5 m long is generally 1600 m^3/h.

Due to the thin parts used in producing strip-flooring parquet production, 55–60% of the raw board material is used. Therefore, it is important to use thin kerf sawing, the minimum planing thickness and the correct sizing of raw material.

The above example shows that the most important variable is the feed speed, which should be as high as possible. Using modern technology, the average feed speed is today around 10 m/min.

3.6.9 *Production of Veneered Panels*

Panels are frequently veneered on both sides, generally using particle board. The production is done on a specially assembled line consisting of manipulators, conveyors, adhesive applier, loading and unloading devices and a hot press, Fig. 3.37.

Individual operations require different times and the most important time elements are the following:

t_1 is the preparation time needed to assemble the lower and upper veneer sheets with the glued mid-part (particle board). Typical values are between 30 and 45 s. It is often the "bottleneck" using fast-hardening adhesives.
t_2 is loading (and unloading) time, including the time for opening and closing of the press. Values are around 20 and 25 s.
t_3 is the pressing time which may vary in wide range depending on the adhesive used.
K is the time utilization factor which may vary between 0.9 and 0.95 s.

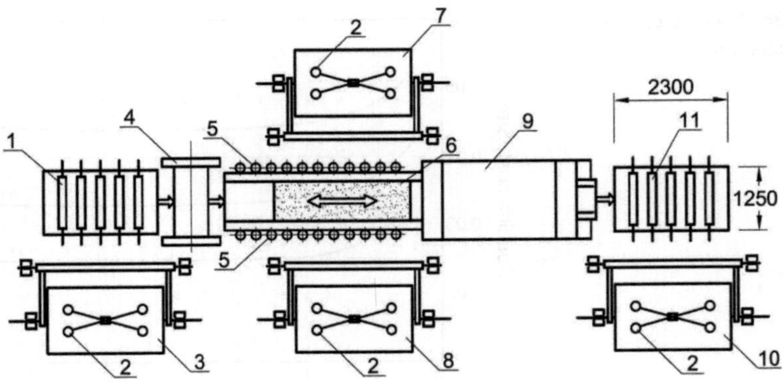

Fig. 3.37 Layout of a manufacturing line making veneered boards. 1—feeding roller table, 2—stacker with suction discs, 3—panel to be veneered, 4—glue application by roller, 5—clamping stacker, 6—band-tray conveyor, 7—bottom veneer, 8—top veneer, 9—hot press, 10—veneered panel, 11—gravity roller conveyor

The pressing time highly depends on the hardening time of the adhesives as a function of temperature. As an example, Fig. 3.38 shows hardening times for different kinds of carbamide-formaldehyde adhesives.

The experimentally determined curves are described with the following simple equation

Fig. 3.38 Time required for hardening of different carbamide-formaldehyde adhesives as a function of platen temperature

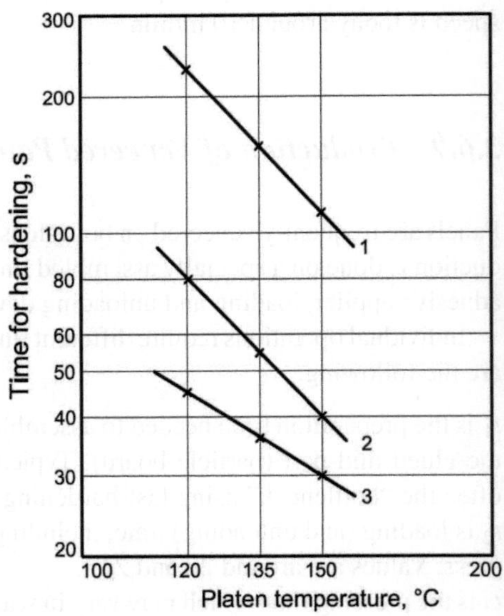

$$t_3 = \text{const}\left(\frac{\vartheta}{100}\right)^{-n}$$

where the exponent n has values of $n_1 = 3.3$, $n_2 = 3.1$ and $n_3 = 1.82$. For fast-hardening adhesives, the rate of change with temperature decreases compared to slow-hardening ones.

The warming-up of the adhesive layer is an unsteady process and therefore the hardening occurs at varying temperature within a given range.

Figure 3.39 shows warming-up curves for the adhesive layer as a function of heating time. Dotted lines intersect the hardening time for adhesives 2 and 3 shown in Fig. 3.38.

The number of pressing cycles per hour is calculated as

$$N = \frac{K \cdot 3600}{t_2 + t_3} \quad \text{if } t_1 \leq t_3$$

or

$$N = \frac{K \cdot 3600}{t_1 + t_2} = \text{const} \quad \text{if } t_1 > t_3$$

The specific cost (\$/panel) is given by the following relation

$$C = \frac{C_\text{m} + C_\text{l} + C_\text{e} + N \cdot C_\text{a}}{N} \text{ \$/panel} \tag{3.54}$$

Fig. 3.39 Warming up of the glue layer as a function of heating time at different platen temperatures. 1-mm-thick veneer on 17-mm particle board. Platen pressure is 20 bar

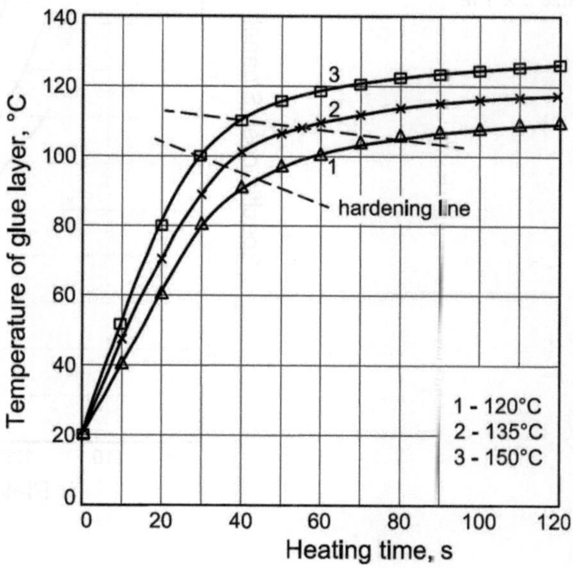

where N means the number of pressing cycles per hour.

C_m is the machine cost, \$/h, and it depends on the level of automation. Typical values may fluctuate between 60 and \$110/h,
C_l is the labour cost, typically \$10/h,
C_e is the cost of electricity for heating. In our case, the averaged power requirement is 3.5 kW and its price is \$0.35/h,
C_a is the cost of adhesive per panel. Using 160 g/m^2 adhesive, its cost for 4 m^2 (2 × 1 m panel) is \$1.92/panel (cost of adhesive is 3 \$/kg).

Calculations were performed using the three different adhesives specified in Fig. 3.38. The results are depicted in Fig. 3.40.

In our case, $t_1 = 28$ s and it is smaller than the shortest pressing time which is 30 s. Therefore, the pressing time can be shortened. If the preparation time t_1 nears the pressing time t_3, then pressing time cannot be shortened.

Figure 3.40 clearly shows that it is impractical to increase the platen temperature over 150 °C because around 150 °C, the difference among the adhesives decreases to a minimum. Moreover, the differences in prices at 150 °C may fully vanish if the price of faster hardening adhesives is higher. The price of veneer and particle board is not included because they are constant in a given case and can be added to the specific cost.

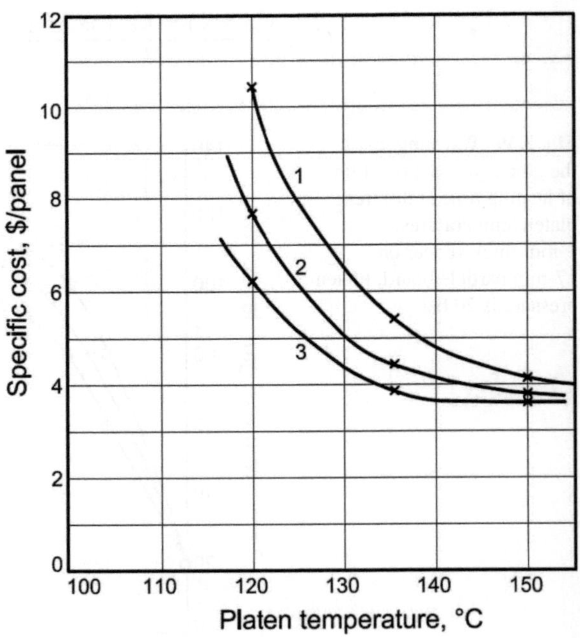

Fig. 3.40 Specific cost of veneered panel as a function of platen temperature for different adhesives. Panel size 2 × 1 m

3.6.10 Bending of Solid Wood

Bending of solid wood in forms is a technique which has been used since ancient times. First, the bow was developed and used for hunting and battle. Bent parts were generally used to decrease the water resistance of rafts and boats. For the same reason, the front end of sleigh runners was also bent upwards. Wooden carriages also had many bent parts, i.e. wheel felloe. The technique of solid wood bending for furniture was first developed by M. Thonet around 1830 and later the Thonet-furniture became popular worldwide (Andres 1923). The theory of wood bending is more than 100 years old (Exner 1887). Basic research was later conducted by Prodehl (1931). Thonet used a steel strap to shift the neutral axis to the tension side.

Carpenters also knew how to make bent furniture correctly. Much later, theory explained why the carpenter made the bending in that way he did.

Some Basic Rules of Wood Bending

The bending of wood for shaping parts has the following advantages:

- there is less material waste than by woodworking,
- shaping is simple and quick,
- bending machines and equipment are comparatively cheap,
- energy consumption for bending is low,
- mechanical properties (strength and stiffness) after bending are better than after shaping with woodworking operations.

Disadvantages of bending process:

- many wood species are not suitable for bending,
- quality requirements are relatively high,
- set the moisture content into a range between 20 and 25%,
- softening wood by steaming is time-consuming,
- drying and stabilizing of the bent parts also require time.

The main problem of wood bending is the low strain to failure of wood at room temperature and dry condition. The strain to rupture is around 1.0%, and the compressive strain to failure is 2–3%. The bending of wood stick to a round shape is only possible if the following condition holds

$$\varepsilon_t \geq \frac{h}{2R}$$

where

ε_t is the strain to rupture at tension,
h is the thickness of bent part,
R is the radius of bending measured to the neutral axis.

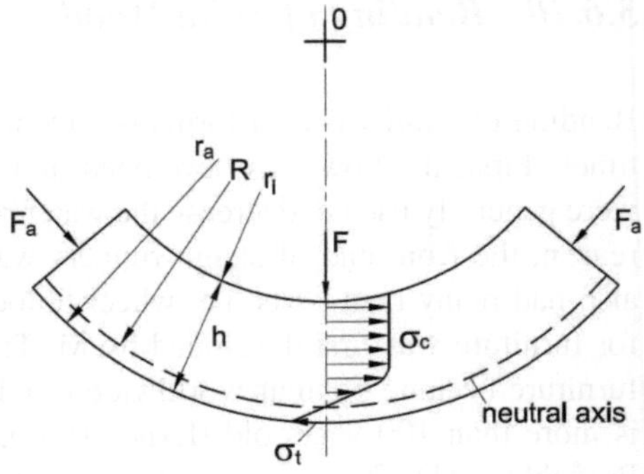

Fig. 3.41 Compressive (σ_c) and tensile stress (σ_t) distribution for wood bending with end pressure

Thonet observed that by softening the strain to rupture cannot be increased, but the compressive strain to failure increases enormously. The load-bearing capacity of the compressed side may only be utilized if the neutral axis is quite near the tension side. It can be achieved by axial pressure on both ends of the stick which can be accomplished by using a metal strap on the tension side (Fig. 3.41).

The achievable minimum bending radius related to the thickness is estimated by the following equation (Vorreiter 1958).

$$\frac{h}{r_i} = \frac{\varepsilon_t + \varepsilon_c}{1 - \varepsilon_c}$$

where the strains ε_t and ε_c must be in the feasible range causing no breakage. Because, the strains are seldom available with acceptable accuracy, the above equation gives approximate values. A selection of experimentally determined values is given in Table 3.6 (Vorreiter1958; Peck 1957).

Experimental results clearly show that hardwoods generally have much better bending properties than conifers.

Practical Recommendations

The first important step in the manufacturing process is to select wood species suited for bending In general, hardwoods are better suited for large plastic deformation than softwoods. Within hardwoods, ring-porous species are also generally better than diffuse porous ones. Hardwood with a higher density has better bending properties because there is less difference in its mechanical properties between early and latewood.

Only a few species of wood are used for making bent parts. Beech is the most common, followed by oak, birch and black locust. For special purposes, ash and elm are also used.

The boards used for bending should come from straight trunks with a circular cross-section. The raw material must be straight-grained and free from any defect. The juvenile wood near the pith is not suitable.

Table 3.6 Bending properties of different wood species, $h = 2.54$ cm, saturated steam for 40 min

Wood species	ε_c	Without		With	
		Metal strap			
		r_i, cm	h/r_i	r_i, cm	h/r_i
Beech (Fagus silv.)	0.3	35	0.073	5.9	0.43
Black locust (Robinia)	0.35	30	0.085	4.6	0.55
Birch (Betula)	0.25	43	0.059	7.6	0.33
Ash (Fraxinus)	0.29	30.5	0.084	6.3	0.40
Oak (Quercus)	0.33	33	0.077	5.1	0.50
Mahagony (Swietenia)	0.078	71	0.036	30.5	0.084
Horse-chesnut (Aesculus)	0.195	20.5	0.125	10.5	0.24
Elm (Ulmus)	0.375	24.5	0.10	4.2	0.6
Karri (Eucalyptus)	0.11	32	0.079	20	0.127
Sitka spruce (Picea)	0.033	140	0.015	75	0.034

The next important question is the initial moisture content prior to softening treatment. Increasing moisture content generally makes bending easier, but high moisture content above the fibre saturation point has the risk of bursting internal cells due to hydraulic liquid pressure. Therefore, the best initial moisture range is between 20 and 25% moisture content.

Prior to bending the parts, it is desirable to finish the wood parts as much as possible. A rough surface may have areas with stress concentration causing fractures. Furthermore, straight samples are cheaper to process than bent pieces. In certain cases, additional processing is required if the final product must have a special cross-section profile.

Since dry wood breaks easily, it must be softened (plastification). The best practice is to use saturated steam at a temperature close to 100 °C. The equilibrium moisture content of wood in saturated steam at ambient pressure is around 19% which avoids the desiccation of the sample with 20–25% moisture content. Although overheated steam would decrease the necessary heating time to a certain extent, it would lead to desiccation of the surface due to the strong decrease of the equilibrium moisture content, Fig. 3.42 (Shubin 1973).

Furthermore, the use of saturated steam at ambient pressure causes some condensation which will enhance the surface heat transfer and prevent any desiccation on the surface.

The necessary heating time in saturated steam is about one and a half minute for each mm of thickness. That means, for a piece 30 mm thick requires 40–50 min in the steaming unit.

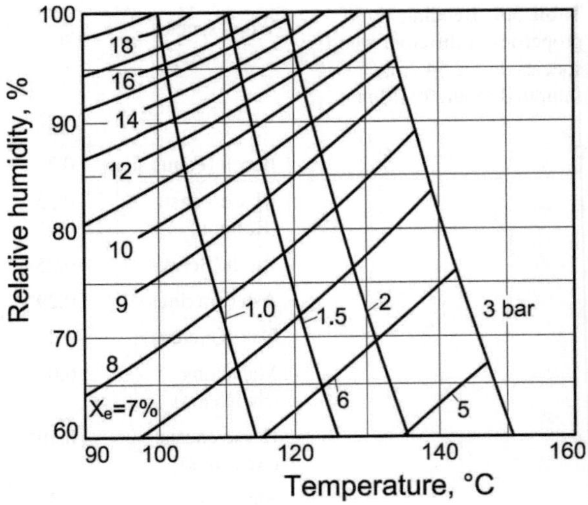

Fig. 3.42 Equilibrium moisture content of timber X_e in overheated steam

Since both heat and moisture affect the stress–strain relation, the sample should be bent with as little delay as possible after softening.

The bending process can be accomplished by hand or machine and made without end pressure (free bending) and with end pressure (restrained bending).

Free bending can be used to make a slight curvature. For example, chair-back rails and similar pieces are often bent in hot-plate press without end pressure. Free bends are not very permanent since the deformations during bending are relatively small. In some cases, it may be necessary to overbend the piece slightly. In most cases, an axial end pressure is necessary to shift the neutral axis to the tension side and prevent tensile failure. End pressure is applied by means of a metal strap with end fittings. The metal strap is placed along the convex side of the specimen to support tensile forces which otherwise would be supported by the wood. The smallest possible bending radius is achieved if we utilize both the allowable tensile and compressive strains. That creates an optimum load distribution between the tension and compression sides. Sadly, it is practically impossible to accurately place the neutral axis..

In practice, one of the main problems is to ensure possible constant end pressure during the entire bending. Figure 3.43 shows a simple device that excerpts pressure on the convex face of a bent specimen and prevents the end block from overturning, assuring a near-constant end pressure (Peck 1957).

Bending technology depends on the intricacy of bent pieces. The easiest task is a simple bend in a single plane which can be made by hand, in a hot-plate press or in a bending machine. The workpieces may be bent singly, in group of several pieces, or in multiple widths which will later be sawn into final pieces. Using a hot-plate press, the bent workpieces are held to shape and dried between the heated plates. A more complicated task is to make S-type bends in a single plane or compound bends in more than one plane. They are mostly made by hand and require highly skilled

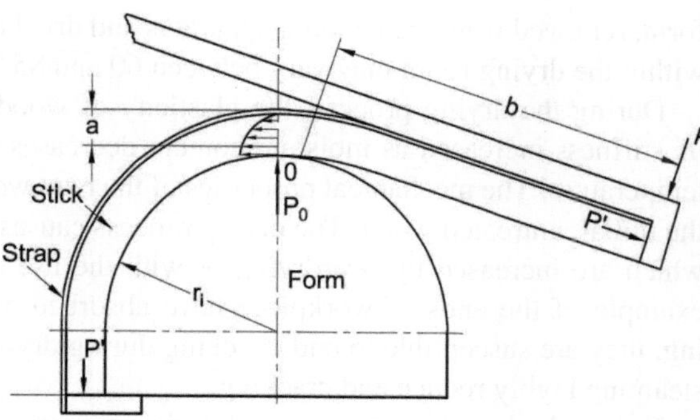

Fig. 3.43 Regulation of end pressure using an end block equipped with a reversed lever

workmen. S-type bends consist of several simple bends but reversing in direction, Fig. 3.44.

The convex and concave sides of the workpiece are interchanged and, therefore, two straps are needed to be fastened to the form. The ends of both straps are equipped with end blocks.

After the bending process and while the wood still hot and wet, it tends to spring back after it is removed from the bending apparatus. This phenomenon is due to the remaining tensile and compressive stresses in the wood. To prevent the spring back, the workpiece must be held in its bent shape until it cools and dries. In order to dry and fix (set) the bend, it either remains in the bending apparatus or it is clamped to a

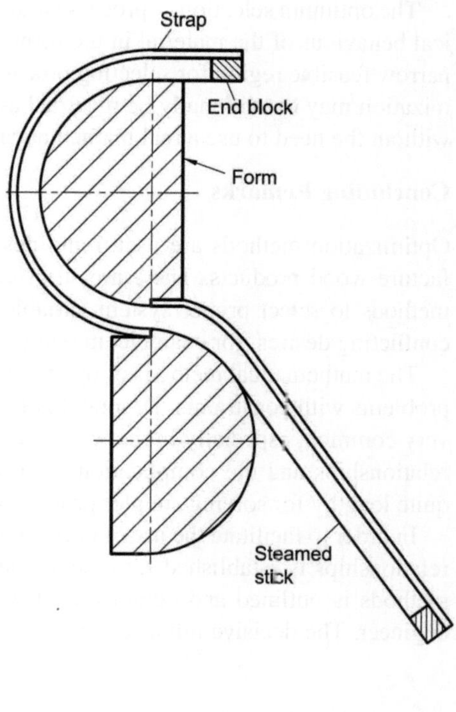

Fig. 3.44 Bending apparatus for making an S-type bend

form, removed from the bending apparatus and dried in a heated room. Temperature within the drying room may vary between 60 and 85 °C.

During the drying process, the plasticity of wood is considerably reduced and its stiffness increased as moisture content decreases to the equilibrium at ambient temperature. The mechanical properties of the bent wood become more like those of the initial, untreated wood. The drying process causes shrinkage and drying stresses which are increased by overdrying or with the use of high moisture content. For example, if the ends of workpiece have absorbed excess moisture during steaming, they are susceptible to end checking during drying. Using end coatings during steaming highly reduce end cracking.

Due to the lengthy softening and drying process in conventional wood bending, an attempt was made to reduce the production time by using high-frequency heating and drying (Sandberg and Johansson 2005). This method uses wood with moisture contents of 20–25% and the workpiece is heated, bent and dried in single sequence, and reducing the cycle time to about 10 min. The use of this method is strongly limited to a simple bend in a single plane with a moderate bending radius.

Practice of one and a half centuries shows that the optimum process parameters of wood bending fall into a narrow feasible range. The most important parameters are the moisture content, the temperature and the limited value of tension strain.

The moisture content has also a limited feasible range between 20 and 25%. In principle, the heating temperature might be 100–130 °C. The use of steam for heating limits the temperature to 100 °C, because overheated steam would quickly dry out the surface of the workpiece making a successful bending impossible. The allowable tension strain is strongly limited which is ensured by using the metal strap.

The optimum selection of process parameters is uniquely determined by the physical behaviour of the material in the form of constraints. These constraints assign a narrow feasible region for selecting process parameters Fig. 3.45. This type of optimization may conditionally be regarded as a pure physically constrained optimum, without the need to use a real mathematical optimization procedure.

Concluding Remarks

Optimization methods are useful and desirable to successfully design and manufacture wood products. There are strict mathematical procedures and engineering methods to select proper system variables to ensure an optimum combination of conflicting desires, for example, maximum production rate and minimum cost.

The mathematical methods of optimization are well-established for the nonlinear problems with constraints. Despite this fact, the use of optimization methods is not very common, especially in the wood industry. The reason is the lack of functional relationships and the complication of strict mathematical methods which may be quite lengthy for solving simpler problems.

In order to facilitate the use of optimization methods, a wide variety of functional relationships is established for practical uses. The use of engineering optimization methods is outlined and demonstrated, which are more familiar to the practising engineer. The decisive influence of a given constraint must be recognized, and this

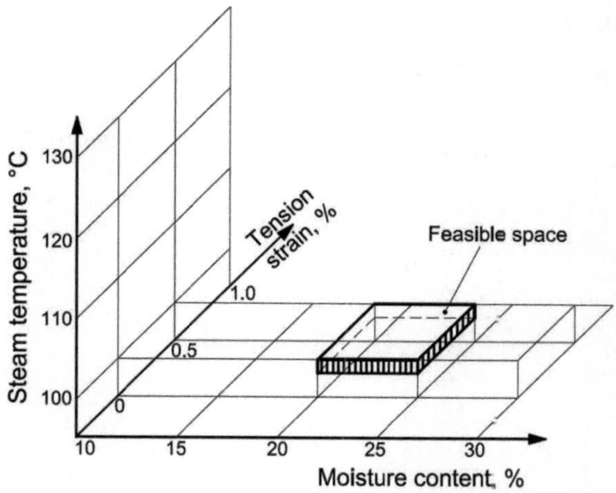

Fig. 3.45 Feasible space for selecting optimum parameters for wood bending

influencing variable will be used in the following calculations with its constrained value. Engineering optimization methods may provide generalized optimum solutions which are generally valid for arbitrary conditions. A generalized optimum solution has the advantage that it enables quick decision-making when conflicting criteria are handled. Optimization problems in the wood industry may be divided into subsystems which may be optimized with an easier implementation. Subsystems may have interactions with each other which should properly be handled. Generalized optimum solutions highly facilitate making the necessary corrections. A further special feature of the design and manufacture of wood products is the use of value-adding methods to enhance quality. This activity is an important part of the system optimization, and it includes special surface treatments, colour and gloss enhancement, colour homogenization and colour modification. The additional cost of these treatments should be in proportion with the increase in system value which can be sold.

Chapter 4
Design Principles

4.1 Introduction

This chapter deals with general questions of product design and development in the wood industry. The current methodological aspects of other branches of industry are compared, and the specialities related to wood products are highlighted. Special attention is paid to the need and possibilities of engineering design of wood products, focusing on reliability-based strength design with a tentative application of orthotropic strength criterion for solid wood parts. A number of examples aid the understanding of each demonstrated concept. Load-carrying capacity of furniture joints is analysed using the method of dimensional analysis leading to similarity relationships that allow assessing joint resistance values in a dimensionally correct way, as opposed to equations arrived at through simple regression analysis and using numerical constants that depend on the used unit system.

Questions related to the design of capable products are treated in 4.6 in conjunction with machining accuracy. The relationship of tolerance assignment and machine capability is connected to the manufacture and assembly of wood parts on the basis of laboratory and in-production measurements. Worked-out examples demonstrate the proposed way to optimize fit and tolerances.

4.2 Overview of Current Practice in Design of Furniture and Other End-User Wood Products

4.2.1 Market-Pull and Technology-Push Product Development

Wood products, such as products of many other branches of industry, can be either market-pull products or technology-push products.

E. Csanády et al., *Optimum Design and Manufacture of Wood Products*,
https://doi.org/10.1007/978-3-030-16688-5_4

In a *market-pull* situation, a company begins product development with a market opportunity and uses available technologies to satisfy the market needs. In most sub-branches of woodworking industry, this situation prevails.

Developing *technology-push* products means that the company bases its product development on a new proprietary or licensed technology, looking for an appropriate market in which to apply this technology. Thonet's new technology to produce bent wood pieces triggered the development of a vast variety of sitting furniture with novel structure and appearance at relatively low prices. The technologies developed for producing wood-based composites in sheet form soon found their place in the market as case furniture.

A further driving force to new product development may arise from available raw materials with unique aesthetic properties. This *raw-material-driven* product development is fed by the existence of precious wood materials in smaller size and quantity which is characteristic of some subtropical and tropical areas in SE Asia, Africa and South America. These, mainly handicraft products, carvings and turnery pieces, are high value-added products with unique quality, beauty and imperishable value.

4.2.2 Predominant Issues in New Product Development in Woodworking Industry and the Role of Intuition

Today, one of the key issues in developing wood products is the competition of other materials in the manufacture of many products traditionally made of wood. This affects manufacturers of wooden doors and windows first of all, and to a less extent flooring and panelling, as well as the furniture industry. The superiority of the replacing materials—plastics, metals, ceramics and composites—with regard to the performance of the product is questionable in most of the applications; however, the price advantage is often indisputable. In addition, campaigns claiming a negative ecological impact of the use of wood may influence customers. Therefore, in developing new products made of wood, competition with products made of other materials has to be accounted for, besides similar wood product competitors in the marketplace. Misperception of the adverse ecological inferences sometimes needs to be combated. There are three ways to cope with the price advantage of replacement materials. The first is to reduce manufacturing costs while keeping product performance constant. Relatively little potential is left in this respect, unless a technological breakthrough emerges. The second is to counterbalance higher production costs by the value added by the use of wood. The third opportunity is imparting added real value to the product through enhanced usability and functional capabilities. Designers need to give these two latter options special consideration.

Another issue relates mostly to furniture. This is a category of wood products where aesthetics generally dominate and the properties related to usability very often remain secondary. In other words, the ratio of aesthetic to real value is higher than in the case of, e.g., household appliances. Therefore, form design, or industrial design in general, is decisive in the success of the product in the market. The decision-maker in the design process may opt to follow a given trend (current or retrospective), or to come out with a novel solution of form. This latter requires strong intuitive qualities of the designer.

Besides the increased need for form design, the market requires a variety of appearance for the same functional capabilities, whereas the company's interest is not to diversify the manufacturing process.

A next issue has emerged in these last decades in the market of some consumer goods including furniture: the market is growing for both low priced mass products regardless of the loss in quality and durability, and for high-level exclusive furniture. That puts manufacturers of middle-range price furniture in a hard situation.

Utilization of wood after the end of product life is an issue of growing importance. Designers have the responsibility to think about the ways of treating the products after their use: reusing, recycling or disposal. While wood itself is an environment-friendly material, using it in a product may involve polluting substances that affect the after-life requirements.

4.2.3 Methodological Problems in Comparison with Other Branches of Industry

Woodworking has traditionally been regarded as a specialty having little in common with other branches of manufacturing. The main reason is the particular behaviour that wood as a raw material exhibits, making experience and skill, consequently tradition a major ingredient of competence. That may explain why the developments in design methodology of other engineering areas have not penetrated into this speciality.

Product development processes in woodworking depend much on the production volume of the company as in other areas, and a formal procedure generally exists for batch or mass production. However, beyond the formalities of checking and decision-making, little if any detail is specified in how a product proposal is developed to the detailed documentation of an optimal realization of the proposal.

Some typical weaknesses will be mentioned in what follows. To start with, in Fig. 4.1, we demonstrate a generalized process flow of designing new furniture and compare it to a desirable process flow based on the practice in some other branches of industry.

Comparing the two process flows, the concept of a new furniture product is defined through form variants rather than by function. After a product concept is selected for further development, construction will be designed with shape, dimensions and material assigned to the parts. Taking into account the manufacturing possibilities of the company, the final working up of parts will be decided and prototypes will be manufactured for evaluation by sense perception and physical tests. Evaluation of

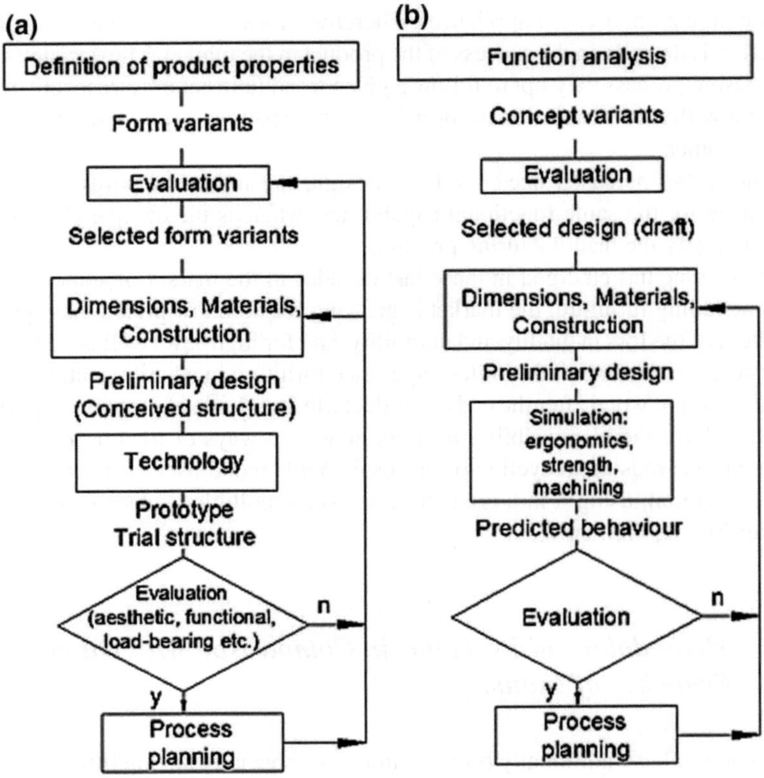

Fig. 4.1 Process flows of designing new furniture; current practice (**a**); desirable process flow (**b**)

the design using visual presentation by computer or mock-ups in the earlier stages of layout development may be part of the process; however, these are limited to visual assessment. This means that the early stages of design are not used efficiently to assess the need for corrections. Verification of the rightness of the whole design process is charged onto the prototype, and when it fails, a whole process will have to be repeated with delay and costs incurred.

The desirable process flow includes the evaluation of the design in the different engineering areas by simulation at each stages of development; this makes timely corrections possible at lower costs and in a shorter time.

Another flaw of the design methods of wood products other than building structural members is that load-carrying capacity is not regarded as an explicit design requirement. For the same reason, there are no stipulations, recommended methods or tools for strength design of parts or construction of wood products like furniture, doors and windows, although they do carry loads under use.

Finally, for wood products, no tools or measures directly supporting the quality are generally used in the course of design unless there is a design review at the end of the various design phases.

4.2.4 *Need for Systematic Product Development and Engineering Design*

In order to succeed in the market, companies must be efficient in product development in terms of customers' expectation, economic feasibility of the product and a timely introduction of the new product into the market. This needs a well-defined procedure that guides the design activities in an expedient way to find optimum solutions in a vast field of possible solutions.

A design procedure has to be a plan of action linking working steps. It also has to include strategies and principles to achieve goals, and methods to solve individual design problems. It does not mean that intuition is not important; systematic procedures cannot replace it. They can only increase the designer's inventiveness.

In all phases of the product development process, whether systematic or not, the cycle of setting of objectives (task), search for solution variants, evaluation and decision-making repeats where a key question is solution finding. The same cycle applies to the whole of the product development process (or on the system level); see Fig. 4.2. The process of designing has two indispensable accompaniments; one is the iteration of the cycles stepwise to higher levels of development. The second is divergence, then convergence with solutions, including variants and selection.

Solution finding remains a methodological challenge. The methods to be used may be those related to either intuitive thinking or discursive thinking. Both have their role, depending on the character of the design task and the type of product to be developed.

A model of the designer's problem-solving is shown in Fig. 4.3. Intuitive and discursive steps alternate, complementing each other. Besides the intuitive steps, there is a need for systematic processes, because a real idea is unpredictable and cannot be forced out. The result always depends to a great extent on the experiences and attitude of designer. The two types of approaches may appear in different ratios depending on the type of the design problem at hand. The intuitive approach is also important when interpreting the "soft" requirements and wishes. Such "soft" requirements related to aesthetics, comfort and novelty in appearance may dominate in a broad category of wood-based products, furniture and items of interior design.

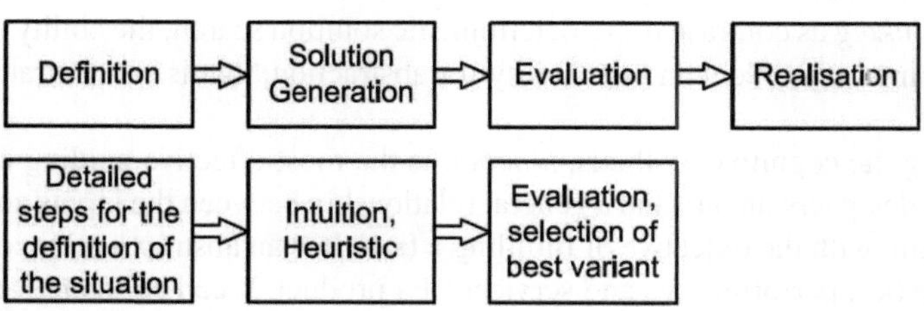

Fig. 4.2 Working cycle of the design process

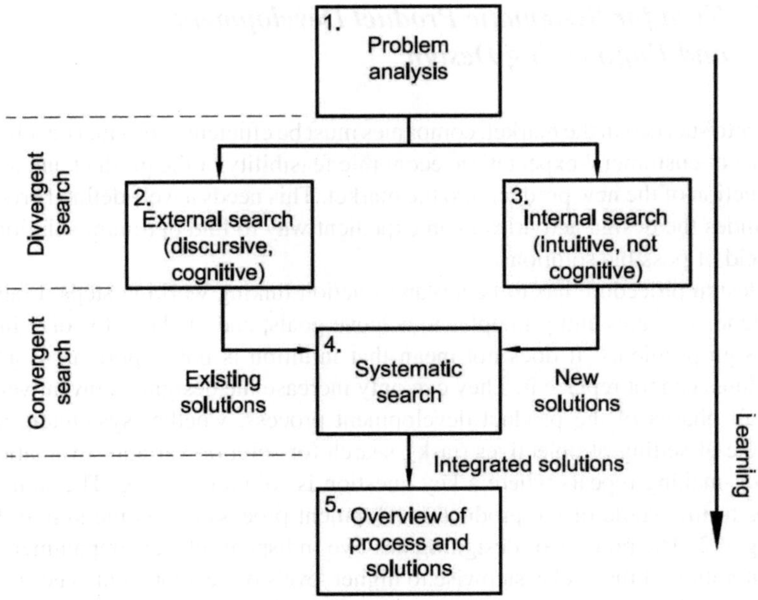

Fig. 4.3 Model of designer's problem-solving

Design problems themselves are inherently problematic because:

- they are underdefined,
- they are not deterministic,
- they are complex.

Very often the designer has the task to design a complex artefact, meeting requirements that have not yet been formulated. The more these problems present, the higher level of intuition seems to be required. Systematic design helps effectively rationalize the design. Structuring the problem facilitates recognition of either new applications from similar design tasks, or new solutions. The critical step in the design process is the conceptual phase. The solution lies in clarifying the need (task, problem), breaking it down to subproblems and providing useful solution alternatives to the important problems.

Designers need the ability to develop the right alternative via original thinking, *lateral thinking* as contrasted to a deterministic solution search, the ability to decompose a complex problem and the ability for abstraction, that is concentration on the essentials.

Among the cognitive methods, *function* is the most effective guiding aspect. By function, designers understand a general relationship between the inputs and outputs of a system, with the objective of fulfilling a task. It is an abstract notion, describing the properties, performances and services of a product. It can be treated as a causal relationship for which it holds:

$$\text{Input} \rightarrow \text{System} \rightarrow \text{Output}$$

where System = functional relationships.

One of the main strengths of systematic design, the approach of functionality means:

- Customer needs are converted into functions, and then, we search for alternatives to perform those functions.
- Design problem decomposition—breaking the problem into subfunctions.
- Thinking in functions is abstraction, abstracting from the concrete appearance and physical realization of the product.

4.3 Application of Principles of Systematic Product Development

4.3.1 General Flow of the Development Process, Possible Process Models

In the broad sense of the word, product development denotes a process starting with the perception of a market need and ending with the new product introduced into the market. Current design methodologies, integrating the different schools of design, emphasize one or another aspect of the activities making part of the whole process, depending on the nature of products to be developed, on the production volume, or on the preferences of the authors. The model based on L. Miles' value method (Miles 1961; Hegedűs 2001) lays emphasis on the economic side. The systematic approach by Pahl et al. puts the engineering aspect forward (Pahl et al. 2007), spending more on conceptualizing, embodying, detailing, along with the necessary computation and information processing; others pay more attention to clarifying objectives (Cross 1996) or treat system-level or embodiment design and detail design in a particular way (Ulrich and Eppinger 2014). In spite of these differences, they all cover the design activities needed for conceiving, developing, producing and introducing the new product into the market.

In general, this whole process from market to market also called *integrated product development* can be divided into the three major stages of product planning (task setting), strict development and product realization as shown by the flow chart after Roosenburg (Roosenburg and Eekels 1995) in Fig. 4.4.

The methodologies are intended to be flexible enough to apply to any type of product, yet one cannot find examples of their application for wood products in the literature. The process model to be followed is in principle the same as for other products and can be outlined as shown in Fig. 4.5. The individual phases of the development process will be next described in some detail explaining the particular-

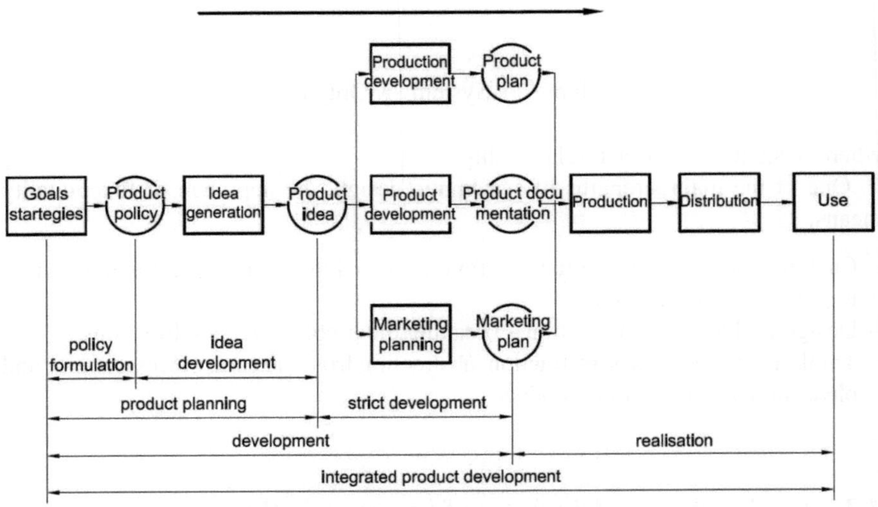

Fig. 4.4 Process flow of integrated product development

ities of wood products such as furniture and more technical products like building components. Distinctions for one-off products made of wood as compared to the design process requirements of batch-produced wood products will be outlined. It is important to realize that with the organization of design activities into the phases shown below does not mean that the phases and the steps within them strictly follow after the completion of a preceding one. Whenever possible, designers may work on a downstream step in advance, parallel with a phase or step actually dealt with, shortening the total time of development process. This is not the only advantage; efforts for product and process optimization will also benefit. In fact, the whole process of product development should consist of overlapping periods of optimization: product concept optimization, solution principle optimization, embodiment optimization and production optimization (Pahl et al. 2007). Application of the concept of an integrated product model allows the completion of phases and working steps of the development process in a parallel way. This will be dealt with for wood products in Sect. 4.3.3.

Planning phase

Working steps required in this phase are:

- analysis of the situation; determination of search fields for product ideas,
- generation of product ideas,
- evaluation and selection,
- elaboration of selected ideas in more detail,
- definition of product proposal.

It is not always necessary to go through all these steps explicitly, but it is recommended for the designer or to the team of designers to mentally follow the logic of the sequence of these working steps.

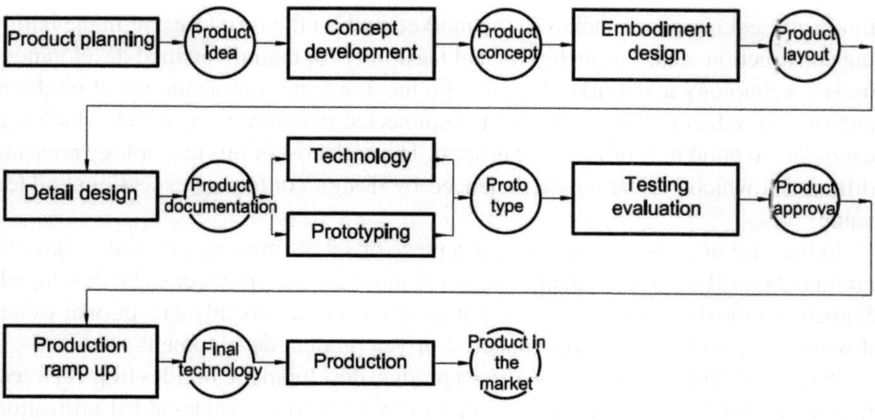

Fig. 4.5 Process model for the development of wood products

In batch-produced furniture, the search field is most often determined in terms of appearance or style. This will lead to the choice of material (wood species), general product architecture, overall shape, lightness or robustness and form details that are an important part of the product idea. In other instances, users' expectations for comfort may be the main focus.

The search for ideas may be directed to function structures when the company adopts a strategy of innovative products aiming at furniture having added functions, or at "smart" furniture.

This latter may relate to the planning phase of one-off products; however, appearance remains a decisive factor in the search for ideas.

As outlined in Sect. 4.2.2, companies with middle-range price furniture will be unable to keep their profitable market share against the price advantage of mass produced products unless they offer particular quality and/or performance features. That is, the focus should be directed to added value.

For the batch production of more technical wood products such as fenestration products, parquet, and panelling, a thorough analysis of the situation is indispensable to success. Especially in the case of wooden doors and windows, competitiveness may be achieved through enhanced functional capabilities. This, however, will be based on an often innovative development of components including glazing, hardware, weather strips, ventilating and shading systems made of materials other than wood. Similarly, novel solutions for the operation of the product may define the search field for ideas. Improvement of the performance of the wood parts—frame and sash members—by combining solid wood or veneer and other materials (plastics, metals, wood composites) is another area of new ideas resulting in added value. This applies to parquet and panelling where, in addition to performance, appearance that wood may impart is of special interest.

Currently, there seem to be two areas of *technology-push* product development. One is wood composite elements of any shape, with planned physical properties, continuously seeking new matches in the market, both in the furniture and in the building construction area. The technological feasibility for their diversified development makes technology and market progress go hand in hand, promising novel products and/or cost reduction. The other area is connected to compressed wood, which is a candidate to build new product ideas upon. The high cost of this technology presents difficulties which can be counterbalanced by design conferring exceptional added value.

In the case of *one-off products* (e.g. a piece or set of furniture), designers have to confine themselves to the client's imagination on the new product to be developed. Sometimes the designer's work is about an artefact of artistically exceptional value; this situation is close to the raw-material-driven product development.

In the context of *raw-material-driven* product development, besides tropical trees, to a lesser extent temperate wood species can be used for value-added utilization due to their appearance, which also should be a challenge for designers. As for the design activities in the course of the planning phase, analysis of the market situation may need less emphasis. At the same time, idea generation requires more in-depth knowledge of wood as a material, than in a generic process of product development.

Conceptual phase

The conceptual phase includes the following steps:

- identifying customer needs,
- establishing target specifications,
- generating product concepts,
- selecting product concept(s),
- testing product concept(s),
- setting final specifications.

The first two steps (identifying customer needs and establishing target specifications) are regarded in some process models as a distinct phase called "task clarification" (Pahl 2007), the output of which is a requirement list.

We suggest using the term "product specifications" in the development process of wood products, meaning the precise description of the function that the product has to perform. Target specifications derived from customer and company needs are the input for concept generation and will later be refined to establish final specifications.

For concept generation, further clarification needs abstraction and is facilitated by interpreting a product idea as an overall function to be realized. Developing a general understanding is most often followed by decomposing the problem into subproblems. When generating concepts, or solution principles to the problem, one has to focus on critical subproblems first.

For the development of technical products, this is best done by establishing a function structure and identifying the critical subfunctions. This is the case with fenestration products and other building components. This will lead to a general definition of the profile system to be used for a window or door. For furniture, unless the product idea is about new or added functions, decomposition by functions may not work the best.

A focus on key customer needs (Ulrich and Eppinger 2014) is more suitable, when form, or appearance in general, rather than working principles are of primary importance. This holds for batch-produced and one-off products alike. For mass produced furniture, functional decomposition can be suggested, whereby variation of appearance (and function too) is best achieved through platform design and/or applying modular system.

Problem decomposition according to the sequence of user actions may be the best choice for kitchen furniture and may work with office furniture.

Decomposition of the design problem may not be useful for simple products. Sometimes, excellent concepts for such products may emerge in a heuristic way which cannot be forced. Therefore, a design project cannot be based exclusively on such an eventuality. Ideas arrived at by heuristics are more typical when form is of primary concern. They may be essential when form is a candidate feature to win a market segment. In these cases, generation of product ideas and concept development cannot be separated or at least overlap to a great extent.

Manufacturers of furniture or fenestration products sometimes design hardware and accessories either for improved aesthetic features or for added functions. These items need distinct conceptual, embodiment and detail design phases.

System-level (embodiment) design

In the course of embodiment design, the designers develop function carriers, that is, functional units for the solution principles of subproblems of the selected product concept. The next step is to develop a preliminary layout and form design variants. For complex systems, this is done in several cycles, working first with the main functional units that determine layout and overall form. After selecting the suitable layout and form, they are further developed to accommodate the remaining main functional units and then the auxiliary function carriers. In each cycle, a number of layouts are generally developed and evaluated against technical and economic criteria for selection and decision-making. The form design of the finally selected preliminary layout variant may be completed and finalized at this stage, and a preliminary part list prepared.

Embodiment design of furniture most often may be done parallel with concept development. This is because the materials to be used have already been decided in the conceptual phase, or even with product idea selection, since appearance is an important, if not the most important ingredient of the product idea. In this way, the main dimensions, arrangement of functional units and their build-up (e.g. veneer-faced and edge-banded MDF panels) as well as the way of assembling them are mostly decided in the concept development phase. Likewise, the use of fittings, hinges and other functional hardware components has mainly been decided with the finalized concept. Therefore, the conceptual phase includes a larger part of the development activities than with more technical products. The rest of embodiment design—completing and optimizing form and the other appearance-related features, the final choice of the type and brand of fittings and other hardware and appliances—can be done along with detail design.

This leaves the team of concept development with increased responsibility and a greater need for competences both in form design and in technological possibilities.

For furniture with simpler function, or with an overall function that is best determined by customer needs (e.g. sitting furniture), it is better to separate concept development and embodiment design, even if some of the raw materials (species of wood) used for visible parts have been decided beforehand. Overall dimensions, shape and relative position of structural parts, as well as deformability, thermal behaviour and vapour permeability of surfaces supporting the human body influence sitting comfort, giving ground to embodiment design. What is more, the requirement of resisting service loads needs special engineering considerations.

It must be noted that due to the inevitable iteration of the clarification–creation–evaluation–decision cycle in the conceptual phase may result in the embodiment design of a new chair with adequate technological knowledge of the designer(s).

Embodiment design of windows means the definition of material or material combination (layered systems) of the frame and sash/casement members, dimensioning of the main units, definition of the type of glazing/panelling, specification and positioning of hardware, sealing, weather stripping, shading and ventilating components.

The basic rules of embodiment design are stated earlier (Pahl et al. 2007); that is, clarity, simplicity and safety have to be taken into consideration. By clarity, we mean here the avoidance of misuse, which relates mainly to fenestration products but may apply to furniture as well. Simplicity or functional efficiency affects not only economic feasibility but reliability and duration of useful product life as well. Safety requires the consideration of all aspects of accident/injury prevention, including strength.

Another aspect of embodiment design is design for assembly (DFA), which aims to reduce cost of assembly in the first place. Some basic rules developed in mechanical and precision engineering apply to wood-based products as well. Among those rules, the benefits of part integration could be better exploited in storing and kitchen furniture, and definitely in fenestration products. Maximizing the ease of assembly is important for knockdown or flat-pack furniture for customers' satisfaction. For the same reason, it is important for home maintenance such as hardware replacement or adjustment.

Detail design

The working steps in this phase are as follows (Pahl et al. 2007):

- finalize details, complete detail drawings,
- integrate into overall layout drawings, assembly drawings and parts' lists,
- complete production documents with manufacturing, assembly, transport and operation instructions,
- check all documents for standards, completeness and correctness.

As a result, the embodiment is completed with a final definition of shapes, dimensions and surface properties, tolerances and fits of all individual components, and a final scrutiny of the production methods, operating procedures and costs.

The above steps of detail design, with the exception of the third one, overlap with steps of embodiment design, especially in wood-based panel constructions, where the check for standard parts and accessories has to be done earlier as they influence the final working details and assembly of panel parts. This is true for fenestration products; however, the final shape of parts for assembly can be specified during detail design.

In designing chairs, generally much is left for detail design. In order to avoid major changes in design at this stage, thorough technological knowledge in the embodiment phase is indispensable.

Production volume may affect the content of documentation produced at the end of this phase. For one-off and small batch products, more detailed manufacturing, assembly and operation instructions are needed.

Embodiment and detail design of case furniture, fenestration products and other technical products using wood as the main construction material comprises the proper selection of non-wood parts and accessories that are supplied by other manufacturers and may be very complex. Typically, these are either catalogue or standard parts; catalogues themselves or technical handbooks may give guidance for their use. It is advisable to avoid custom tailored parts except for one-off products.

While in the field of mechanical engineering, design codes and standards deal with the design of machine elements, there are virtually no such documents for wooden parts. Over time, rules of thumb have been developed by experience. Often we need to depart from them on account of the design intent or because of new materials. That would need testing or simulating performance of the new solution before a decision is made on the final working up of the details. The main concern in the case of wood and wood composites is generally strength and dimensional stability.

Two other aspects have to be addressed in the course of detail design. These are production costs and product quality, which predominantly determine the success of a product in the market.

Production costs can be controlled through the use of the method of design for manufacturing (DFM). Indeed, DFM has to be performed throughout the development starting with the conceptual phase, when the final product specification is developed. Cost estimates and ensuing decisions made at these earlier stages of the development process contribute to the avoidance of process iteration at the detail phase because of major changes needed in parts and/or assembly on cost considerations.

Since detail design is critical for a part's technical function, production processes and elimination of production errors, it has a major influence on product quality. Quality, however, cannot be assured in this final design phase only. Rather, it has to be designed into the product and production processes from the very early phases of the product development process. Section 4.3.2 deals with the adaptation of this concept in the woodworking industry.

We emphasize here the role of the appropriate working up of parts in designing a capable and reliable product; more will be demonstrated on this subject in Sect. 4.7.2.

Prototyping

Prototyping seems to be an inevitable step in the development process of furniture, except for one-off products. There are two main reasons for making prototypes of a piece or set of furniture. One is testing and checking against standard or contractual requirements. The other reason is testing acceptance in the marketplace or by the client.

While a prototype should preferably be manufactured using the same type of machining operations and tools as planned in the production ramp-up, the documentation for the latter will be produced after the approval of the prototype documentation.

Physical testing of prototype pieces of furniture may reveal flaws of the design. Depending on the seriousness of flaws, either refinements in design or iteration of part of the development process may become necessary in order to remedy the flaws. Changes could affect detail or even embodiment design, in the worst-case concept development as well. In order to eliminate such hazards, it is advisable to apply proper methods for the evaluation of the design with respect to the requirements throughout the development process. Sects. 4.4 and 4.5 deal with the requirements of design for strength. If we use these methods at the appropriate stages of product development, this reduces the need for frequent changes in the design. Then, we have a greater chance that the prototype will satisfy all the requirements.

In fenestration products, prototypes may be needed for the standard tests by government bodies, for the compulsory certification of key performance characteristics or for enhancing its attractiveness if the product is in the marketplace. There are procedures in the literature that may facilitate achieving the desired performance levels by design.

Making prototypes of fenestration products is typical when novel solutions in operating hardware or other auxiliary appliances are the motor of product development and mainly serve for arousing interest in the marketplace.

Production ramp-up

In this phase, details of the intended production system are worked out and the product is manufactured in pilot lot using this system. The aim is to work out any remaining problems in the production process. It is an uneconomic practice to rely on pilot lots to explore possible remaining alterations of product or part properties due to the differences of manufacturing details as related to prototype making. Intensive use of the methods described in Sect. 4.7 may prevent these problems from occurring.

4.3.2 Designing Quality into the Products and Manufacturing Processes—Methods and Tools to Be Used

As developments of design methodology in general, tools and methods of quality management used in other engineering areas have not yet become standard practice in the woodworking industry for several reasons. The particularities in the behaviour of wood have already been mentioned. What is more, properties of each piece of wood are unique because they all come from different parts of a once living tree with varying growing patterns. That makes machining properties, visual characteristics and response to changes in ambient conditions vary from piece to piece. Machining properties vary in different anatomical directions. Wood is sensitive to humidity. Its colour, shine and pattern have a decisive effect on the aesthetic effect of the product. While a key issue in quality is repeatability of product properties, this cannot be strictly kept in a system where the input raw material is not homogeneous and the adjustment of processing parameters can only be approximatively optimal. Lastly, most of the processing defects cannot be repaired; the reparable ones require high manual input and skill.

Most industrial engineers know that a very high proportion (up to 80%) of defects can be traced back to the product development process. In addition, flaws in design manifest themselves much later. At the same time, the cost of making corrective changes in the design increases according to a power rule as the course of development progresses. It follows that quality can only be achieved if it is built into the product and the production process from the very beginning of product development. The early phases of the development process need special attention; however, with wood products one hardly meets any examples of using proper methods on an industrial level.

The optimum matching of the customer needs and the product properties can be assured early in the development process by the proper interpretation of the former as engineering characteristics. Quality function deployment (QFD) is a method broadly used in this context. For those, who are unfamiliar with this method, the website of the American Society for Quality (ASQ) offers more information, besides a number of textbooks on design and quality. The QFD procedure is conducted in teamwork. Structured customer requirements serve as input, and technical properties of the product are associated with them. These properties may have different levels of concreteness, depending on the stage of the development process, and thus, it may be applied stepwise; see in ASQ. The core of the QFD scheme is the correlation matrix whose elements reflect the strength of the relationship between a customer requirement and a technical property, along with the relative importance of the requirement in question. Values in the matrix summed up in a column express the relative importance of a technical property. The final output is a set of target values for the technical properties.

Some methods of complex fault analysis, especially ABC or Pareto analysis, are built in the quality system in a number of woodworking companies. Typically, they are used as tools for evaluating the situation for follow-up decisions rather than for identifying potential causes of occurrence of faults. At the same time, both Ishikawa's method of cause and effect analysis and Pareto analysis already used in the development process may largely contribute to achieving a fault-free design and manufacturing processes. Likewise, fault tree analysis complemented with hazard assessment may enable early detection of costly consequences.

The use of QFD is well known for the functional requirements; however, it is not practised in furniture design where many of the user needs are related to ergonomics rather than functioning. It has been shown recently (Horváth et al. 2012) that QFD works with requirements for comfort. In that study, the optimum geometry and some other characteristics of a seat for general use were aimed with the use of QFD, combined with the *application of design of experiments (DOE)*. Table 4.1 lists the customer needs as surveyed by the authors. In Table 4.2, engineering characteristics of the sitting furniture considered to be influential on the fulfilment of customer needs can be read. These have to be measurable, rateable or binary quantities.

From the 22 engineering parameters shown here, five were ranked the most important as follows: sitting height, width of the sitting surface, its inclination angle, height of back support, and depth of sitting. It means that these parameters are of key importance in achieving customer satisfaction. The expected output of the procedure was their target values. Because of the need for a number of decisions based on the more or less subjective QFD methodology, the authors used DOE to optimize the performance of the product. A two-level factorial design with five factors was planned. In the individual runs, trial chairs were assembled setting the chair's engineering parameters of interest at their preselected levels. Sixteen individuals tried sitting and evaluated the level of the fulfilment of each of the customer needs on a scale of 1–10

Table 4.1 Customer needs for a seat of general use

WHATS (What customers are asking for)
Stability
Load-bearing capacity
Ease of standing up
Comfortable sustained sitting
No risk of injury
Prevention of unhealthy position
Easy to move without exertion
Pleasant to touch
Easy to clean surfaces
Provision of relaxing posture
Relaxation of the upper body
Relaxation of legs
Durable
Fits to the table

Table 4.2 Engineering characteristics influencing the satisfaction of customer needs

	HOWS
How to deliver **WHATS** (Measurable Quantities)	Structural stiffness
	Strength of structural joints
	Resistance to abrasion of the surfaces
	Width of seat
	Depth of seat
	Height of seat
	Slope of seat
	Distance of armrests
	Height of armrest
	Inclination angle of back
	Width of armrests
	Height of back
	Length of armrests
	Curvature of back
	Radius of file on frame members
	Thickness of upholstering
	Hardness of upholstering
	Weight of the chair
	Surface quality
	Air permeability of the cover fabric
	Thermal conductivity of upholstery
	Vapour resistance of the cover fabric

for every chair setup from the factorial design. Factor main effects, interaction effects and their significance level were quantified by analysis of variance, leading to the optimal setting of the engineering parameters.

The method of failure mode and effect analysis (FMEA) is a quality tool used successfully in mechanical engineering and many other areas to minimize the risk of failure or imperfect functioning of the product in use conditions because of deficiencies either in design or in manufacturing. Those who are not familiar with this method may refer to the website of the American Society for Quality (ASQ) besides a number of textbooks on design and quality. An FMEA conducted by a cross-functional team is useful to identify potential failure modes of a system and to prioritize the risks associated with these failure modes. The process includes decomposition of the system according to the functions to be fulfilled, identification of the consequences and possible causes of the failures, as well as the tools currently used to detect them. Weights are given to severity of consequences (S), probability of occurrence (O) and lack of detectability (D) in scales of 1–10. The product of the three weights is the Risk Priority Number (RPN); it is assigned to each failure mode for all causes of failure. The two versions, Design FMEA and Process FMEA, may be used either for improving a product and manufacturing process, or in the design of a new product and process. In principle, there would be no differences in applying them to wood products.

Features related to aesthetics of furniture are difficult to treat in a systematic way. However, the FMEA approach may be useful in such design problems too. Design of FMEA starts with the decomposition of the product to structural units with assignable functions. Features defining the aesthetic values of a piece of furniture may be treated as functions. Failure in the fulfilment of these functions can be associated with specific causes, which may occur with assessable probabilities. In addition, the degree of detectability of these occurrences can be assessed. The establishment of the relative severity of the failure modes is the most demanding problem, requiring a sophisticated sense of style, form, rhythm, symmetry, harmony and competence in technology.

Design FMEA analysis of a cupboard was conducted in the frame of a research (Antal 2007). Unlike technical functions, it proved to be expedient to consider larger structural units consisting of several parts (e.g. body of the cupboard), when investigating the fulfilment of aesthetic functions. The results reveal those possible flaws of design which may significantly endanger the aesthetic quality of the product. It is interesting to note that some of the causes of failures of high risk in this study were related to manufacturing and assembly ensuing from the design. At the same time, one has to understand that Design FMEA cannot replace genuineness and a sense of style when exercised in the furniture design process. Rather, it provides possibilities to avoid design mistakes originating from inadequate knowledge of style and form on one hand, and from the ignorance of achievable quality of manufacturing and assembly on the other, including possible changes in precision under service conditions.

When defining failure modes for aesthetic functions, the question of how the function may not be realized is not as straightforward to answer, as with technical functions. In order to facilitate the analysis, the sequence of steps to be followed should be purposefully expanded as shown in Fig. 4.6. Describing the aesthetic properties before the definition of aesthetic functions is helpful in establishing function structure and defining functions. In addition, specifying failure modes is facilitated this way; therefore, they are defined on the basis of properties described, rather than on the basis of the aesthetic functions. As intermediary means, features corresponding to the critical functions are defined and suited to the individual aesthetic properties as can be seen in the figure.

Table 4.3 shows the explored FMEA chain of failure events for a door of a cupboard along with the corresponding scores of failure mode severity (S_i), probability of occurrence of cause (O_{ij}), detectability offered by the applied control system (D_{ij}) as well as Risk Priority Number (RPN_{ij}). Items with an *RPN* value higher than 125 need special attentions during design, and remedial measures need to be formulated.

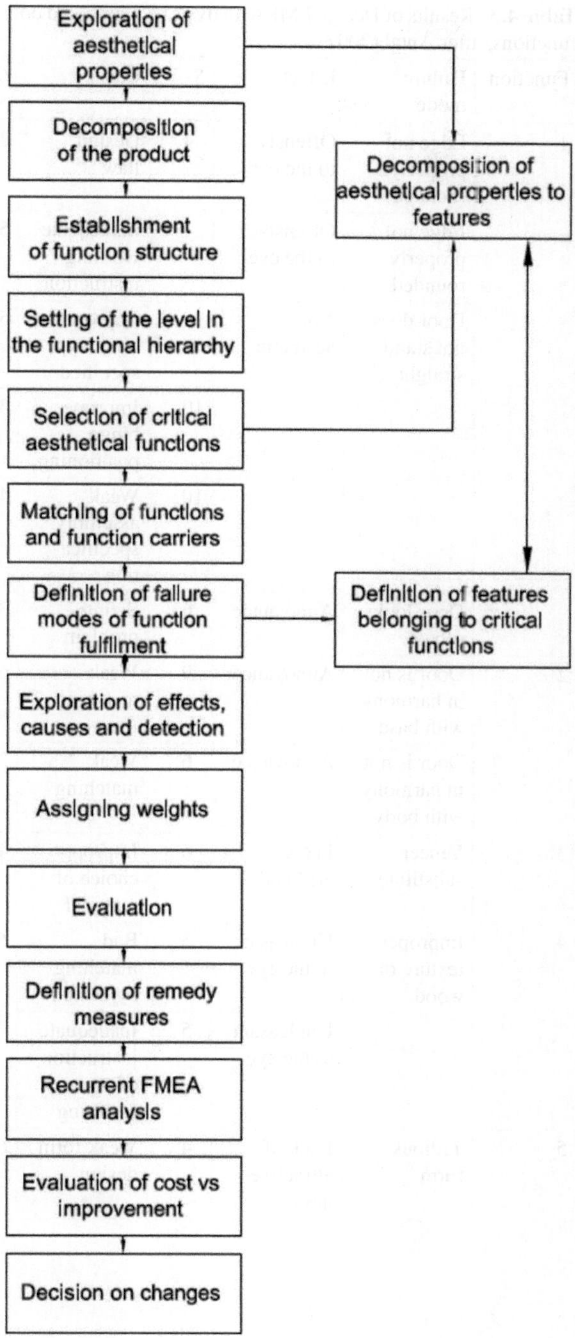

Fig. 4.6 Flow chart of the FMEA process as applied to the design of a piece of furniture

Table 4.3 Results of Design FMEA analysis of a cupboard door from the point of view of aesthetic functions, after Antal (2007)

Function	Failure mode	Effect	*S*	Cause	*O*	Detection	*D*	*RPN*
1	Edge not properly rounded	Offensive to the eye	4	Design flaw	2	Design review	3	24
	Edge not properly rounded	Offensive to the eye	4	Inadequate working instruction	5	Design review	3	60
	Door does not stand straight	Not aesthetic	10	Improper hinge type specified	5	Standards, catalogue inform	4	**200**
			10	Improper hinge positioning	3	Design review	6	**180**
			10	Weak assembly specification	4	Design review	7	**280**
	Door looks robust	Annoyance	6	Sizing problem	3	Design review	2	**270**
2	Door is not in harmony with base	Annoyance	9	Weak matching	6	Form jury	5	72
	Door is not in harmony with body	Annoyance	6	Weak matching	3	Form jury	2	36
3	Veneer substitute	Feels artificial	6	Improper choice of material	4	Design review	3	72
4	Improper texture of wood	Unpleasant to the eye	5	Bad matching by design	6	Form jury	3	90
		Unpleasant to the eye	5	Inadequate instruction of manufacturing	7	Form jury	4	**140**
5	Tedious form	Lack of attractiveness	4	Weak form design	6	Form jury	6	**144**

(continued)

Table 4.3 (continued)

Function	Failure mode	Effect	*S*	Cause	*O*	Detection	*D*	*RPN*
6	Improper colour and shine	Not aesthetic	4	Improper choice of material	4	Design review	8	**128**
			6	Improper choice of colour	5	Design review	5	**150**
7	No characteristic form	Visually neutral	5	Weak form design	5	Form jury	3	75
8	No sign of style	Lack of style	4	Weak form design	6	Form jury	4	96
9	Outdated	Lack of attractiveness	7	Form design based on short-term trend	7	Form jury	3	**147**

The aesthetic functions numbered from 1 to 9 in the first column of the table are as follows:

1—carries beauty,
2—conveys harmony,
3—influences taste,
4—helps fixing visual experience,
5—creates individual form,
6—visualizes a unique range of tones,
7—shows originality,
8—signs of a given style,
9—resists changes of fashion.

As shown in Table 4.3, six among the thirteen failure modes proved to be critical; one on account of weak consideration of manufacture and assembly. another on account of several reasons, including manufacturing and assembly issues.

We always want *a process or product to be robust*, i.e. insensitive to disturbing effects. It can be achieved by a choice of the design parameters influencing the output parameters of interest. Disturbing effects include the unwanted variation of the actual value of design parameters with respect to their specified value, the inevitable variation of raw material properties, inaccuracies in machined sizes and assembly within tolerance limits, as well as the changing environmental factors beyond our control. Full knowledge of the functional relationships governing a given product property or process output parameter allows us to optimize them to attain the best level and the least variability in a deterministic way. However, this is seldom the case, and often, it would need enormous effort to explore the relationships in questions.

Other techniques like the design of experiments (DOE) may be helpful in such problems. It has to be recognized however that the use of factorial experiments does not explore the causal relationships; rather it helps describe the dependence of the output parameters on the input parameters by means of a polynomial-type relationship. Since finding the optimum means attaining the minimum or maximum, or a target value of this function over the meaningful range of design parameters with the least variation, this latter issue should be addressed too.

In the next example, we consider a bathroom commode with wide drawers. Figure 4.7 shows the front view and side section of the upper part. This piece of furniture is made of solid beech wood, except for the inside of the drawers and the back wall. The sidewalls of the drawers are hung on the inside face of the commode sides by means of commercial drawer slides. The nominal height of the opening around the drawer front panel is $H = 300$ mm. The dimensions shown in the figure are those specified for the manufacture and assembly of parts and hardware and are subject to inaccuracy controlled by tolerances.

A common problem in case of commodes is the irregular position of front panels of drawers with respect to the surrounding parts or to each other, especially with wide drawers. The inaccuracy in position may occur to an extent that makes pushing-in difficult or simply disturbs the appearance of a piece of furniture. The objective of robust design is to avoid interference of the front panel with the framing parts at the bottom and top alike. On the other hand, we want the bottom and top gap sizes as

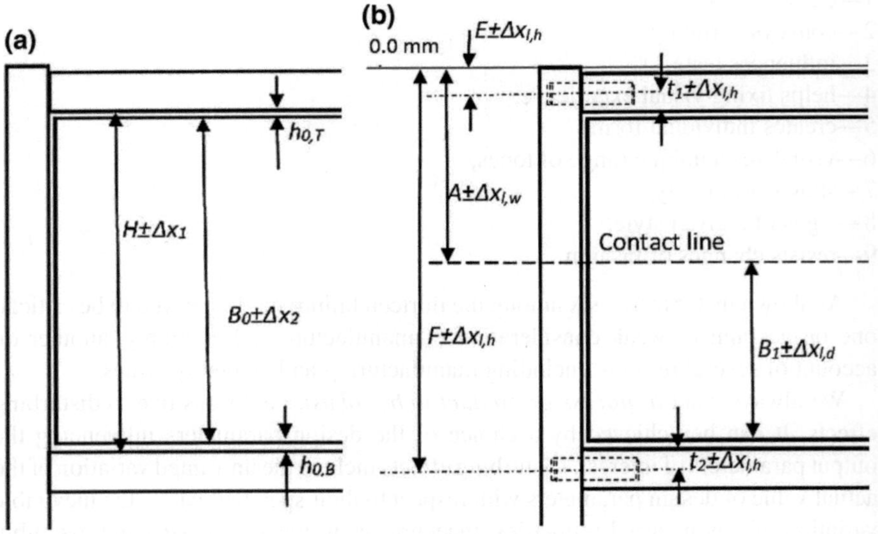

Fig. 4.7 Construction and dimensions of the commode under study, top drawer left corner; drawer front panel and opening height, top and bottom gap (**a**); components of the manufacturing inaccuracies expressed as location tolerances of dimensions given in technical drawings (**b**), where E is design data and $F = E + t_1 + H + t_2$

uniform as possible within changing circumstances of manufacturing inaccuracies and conditions of use.

In order to demonstrate the fulfilling of robust design objectives, the dimensions shown in Fig. 4.7 will be used, where

A (mm)—distance of the contact surface of the slide from the baseline,
H (mm)—opening height for the drawer's front panel, design value,
B_0 (mm)—front panel height, design value,
B_1 (mm)—optional design value,
t_1 and t_2 (mm)—half thickness of the horizontal members.

Inaccuracies that apply in the assembly of drawers and body of commode are as follows:

$\pm\Delta x_1$—manufacturing inaccuracy of the drawer opening height,
$\pm\Delta x_2$—manufacturing inaccuracy of the front panel height,
$\pm\Delta x_3$—assembly inaccuracy of the slide parts.

Further factors causing the gap size to vary are the weight of drawer content causing vertical displacement Δx_w, as well as across-the-grain hygroexpansion $\pm\Delta x_{\text{higr}}$ (swelling and shrinking) of the drawer front panel. Swelling/shrinking of sidewall in the grain direction for a length of H is negligible. The inaccuracy $\pm\Delta x_1$ has to be split between the bottom and top edge when the drawer slide is fixed at a height between the top and bottom of the drawer front. The same applies to the across-the-grain hygroexpansion of the front panel.

As restraints, we may set lower limits of the gap sizes at the top and bottom:
$h_{xT} \geq 0.1$ mm
$h_{xB} \geq 0.1$ mm.

Taking a full load of the drawer and swelling of the front panel to represent the worst-case situation, these restraints can be written for the gap sizes at the top and bottom, respectively, as

$$h_{xT} = h_{0T} \pm \frac{\Delta x_1}{2} \mp \frac{\Delta x_2}{2} \pm \Delta x_3 + \Delta x_w - \frac{\Delta x_{\text{higr}}}{2} \geq 0.1\ \text{mm}$$

and

$$h_{xB} = h_{0B} \pm \frac{\Delta x_1}{2} \mp \frac{\Delta x_2}{2} \mp \Delta x_3 - \Delta x_w - \frac{\Delta x_{\text{higr}}}{2} \geq 0.1\ \text{mm}$$

From these equations, the worst-case minimum requirement of gap sizes would be

$$h_{0T} = h_{xT} - \frac{\Delta x_1}{2} + \frac{\Delta x_2}{2} - \Delta x_3 - \Delta x_w + \frac{\Delta x_{\text{higr}}}{2}$$

and

$$h_{0B} = h_{xB} - \frac{\Delta x_1}{2} + \frac{\Delta x_2}{2} + \Delta x_3 + \Delta x_w + \frac{\Delta x_{\text{higr}}}{2}$$

It has to be noted that the worst-case situation for the top gap is different because an unloaded rather than a loaded drawer creates more severe conditions; therefore, the term $-\Delta x_w$ is to be omitted:

$$h_{0T} = h_{xT} - \frac{\Delta x_1}{2} + \frac{\Delta x_2}{2} - \Delta x_3 + \frac{\Delta x_{\text{higr}}}{2}$$

The numerical values of the inaccuracies used in the above equations are: $\pm\Delta x_2$ is the size tolerance typically of ± 0.4 mm when machining the front panel to an exact height; $\pm\Delta x_1$ is the resultant of four location tolerances ($\Delta x_{l,h}$) of ± 0.2 mm each, namely hole positioning in the sidewall and in the edge of the horizontal member both above and below the drawer. Likewise, $\pm\Delta x_3$ is the resultant of two location tolerances of ± 0.2 mm each ($\Delta x_{l,d}$ and $\Delta x_{l,w}$). We have to measure the distances from the same baseline. This is the upper edge of the sidewall which is used as the baseline for the specifications for machining; see Fig. 4.7b. In the symbols of inaccuracies, subscripts l and s denote location and side tolerance, respectively. As the second letter in the subscript h refers to dowel hole, d to drawer side and w to cabinet sidewall.

Swelling of the drawer front across the grain direction is taken into account on the basis of the two ambient conditions of 60% relative humidity (RH) at 20 °C and 70% RH at 26 °C. Eventual shrinking due to 45% RH and 20 °C may occur. Actual changes in ambient conditions may be considerably higher for short periods; however, there is a delay in attaining equilibrium moisture content in wood behind changes in the environment (Siau 1995; Scaar 1988). The calculated maximum swelling and shrinkage values are 1.3 mm and 0.65 mm, respectively, using average value of the swelling coefficients of beech wood in the radial and tangential direction (Molnár 1999), and taking the restraining effect of surface treatment into account by halving the actual values. It has to be noted that these high-dimensional changes expected under normal conditions could be substantially diminished by drying wood to reduce shrinking and swelling. Expansion coefficients and the slope of sorption isotherms can be reduced to less than half of the static ones in this way (Scaar 1988). The traditional cabinet making industry carefully seasoned the raw material over long periods of time, resulting in shrinking and swelling of furniture parts not exceeding the manufacturing inaccuracies.

The front panel of a fully loaded drawer displaces downwards depending on the spring constant of the drawer hanging system (slide and its fixing/embedment). For our drawer 800 mm wide, there will be a vertical displacement of 0.3 mm with the use of commercial cabinet hardware. (This displacement does not occur when a bottom-mount slide is supported by a structural member in addition to the sidewall.)

When the drawer slide is fixed at a height other than middle, the inaccuracy $\pm\Delta x_2$ and the shrinking/swelling $\pm\Delta x_{\text{higr}}$ are to be split between the bottom and top edges according to the quotient $a = B_1/B_0$ shown in Fig. 4.7.

For design purposes, minimum requirements have to be met at the bottom and top alike; gap sizes assigned accordingly are treated as nominal sizes.

In order to achieve the design objectives of safe operation and uniformity of gap sizes, the optimum of the nominal gap sizes and the position of slide measured from the bottom of the front panel of the drawer has to be defined.

A maximum bottom gap size is produced by an unloaded drawer and an upward shift of the drawer front due to assembly inaccuracies, while a fully loaded drawer and downwards shift of drawer front causes the maximum top gap, according to the formulas below, taking the absolute value of the inaccuracy components.

Bottom gap:

$$\begin{aligned} h_{\max,B1} &= h_{0,B} + \Delta x_{l,d} + \Delta x_{l,w} + \Delta x_s \cdot a + 0.65 \cdot a + 2 \cdot \Delta x_{l,h} \\ &= 2 \cdot \Delta x_{l,d} + 2 \cdot \Delta x_{l,w} + 2 \cdot \Delta x_s \cdot a + 1.5 \cdot 1.3a + w + 4 \quad \Delta x_{l,h} \end{aligned}$$

Top gap in the same circumstances (bottom-gap-based situation):

$$\begin{aligned} h_{\max,T1} &= h_{0,T} - \Delta x_{l,d} - \Delta x_{l,w} + \Delta x_s \cdot (1-a) + 0.65 \cdot (1-a) + 2 \cdot \Delta x_{l,h} \\ &= 2 \cdot \Delta x_s \cdot (1-a) + 1.5 \cdot 1.3 \cdot (1-a) + 4 \cdot \Delta x_{l,h} \end{aligned}$$

Top gap:

$$\begin{aligned} h_{\max,T2} &= h_{0,T} + \Delta x_{l,d} + \Delta x_{l,w} + \Delta x_s \cdot (1-a) + 0.65 \cdot (1-a) + w + 2 \cdot \Delta x_{l,h} \\ &= 2 \cdot \Delta x_{l,d} + 2 \cdot \Delta x_{l,w} + 2 \cdot \Delta x_s \cdot (1-a) + 1.5 \cdot 1.3 \cdot (1-a) \\ &\quad + w + 4 \cdot \Delta x_{l,h} \end{aligned}$$

Bottom gap, in the same circumstances (top-gap-based situation):

$$\begin{aligned} h_{\max,B2} &= h_{0,B} - \Delta x_{l,d} - \Delta x_{l,w} + \Delta x_s \cdot a + 0.65 \cdot a - w + 2 \cdot \Delta x_{l,h} \\ &= 2 \cdot \Delta x_s \cdot a + 1.5 \cdot 1.3 \cdot a + 4 \cdot \Delta x_{l,h} \end{aligned}$$

Table 4.4 contains the minimum and maximum gap sizes calculated for the sample commode using three selected values of the ratio a of the fixing height of drawer slide, namely $a = 0$ (bottom mount or undermount), $a = 0.5$ (side mount, mid-height of drawer) and $a = 1$ (top of drawer).

The last row in the table shows that the ratio of bottom-to-top gap is under 1 for bottom-mount drawers and exceeds 1 for top-hung drawers, while positioning of the slider at mid-height is much favourable. It is possible to find the optimal value of the ratio a on condition of the equality of the bottom and top gap size. Table 4.5 shows the optimum ratios for several circumstances that may be relevant as worst cases.

It can be seen that optimal values of the ratio a do not vary much, being below 0.5 when criteria for the bottom gap are used and above 0.5 for the top gap. It might be the designer's discretion to choose the most relevant circumstances to evaluate extreme gaps, knowing that height above the floor level has an influence on the visual appearance of gaps.

In practice, nominal gap sizes larger than the minimum requirements are often decided either for the appearance's sake or to stay on the safe side. As minimum gap

Table 4.4 Minimum gap requirements and maximum gap sizes at the bottom and top of drawer front for three distinct positions of the slider; $h_{max,1}$ refers to bottom-gap-based situation, $h_{max,2}$ refers to top-gap-based situation

$a = B_1/B_0$	0			0.5			1.0		
	h_{min} mm	$h_{max,1}$ mm	$h_{max,2}$ mm	h_{min} mm	$h_{max,1}$ mm	$h_{max,2}$ mm	h_{min} mm	$h_{max,1}$ mm	$h_{max,2}$ mm
Bottom	1.1	1.90	0.80	1.95	3.28	2.18	2.8	4.65	3.55
Top	2.5	3.55	4.65	1.65	2.18	3.28	0.8	0.80	1.90
Ratio of bottom-to-top gap		0.54	0.17		1.50	0.66		5.8	1.87

Table 4.5 Optimized slider positioning in different circumstances of gap increase

Circumstances	Bottom-gap-based, drawer fully loaded	Bottom-gap-based, drawer empty	Top-gap-based, drawer fully loaded	Top-gap-based, drawer empty
$a = B_1/B_0$	0.41	0.30	0.70	0.59
gap size mm	2.725	2.725	2.725	2.725

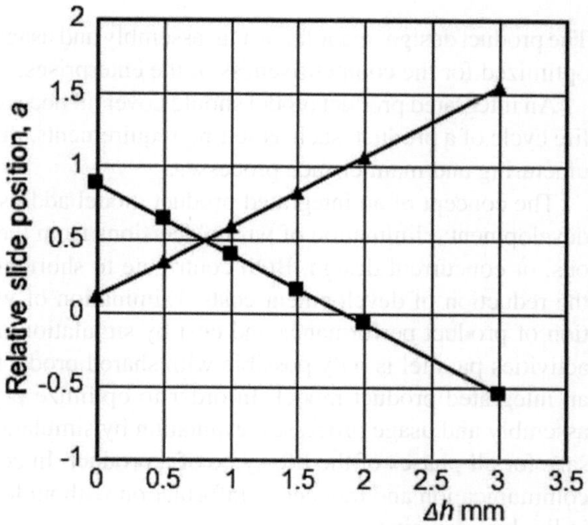

Fig. 4.8 Relationship between the difference of bottom and top nominal gap sizes Δh and the relative height a of the slide position in the case of top-gap-based (square data points) and bottom-gap-based design (triangle data points), obtained by using worst-case situations

requirements are met, the ratio of $a = B_1/B_0$ may be optimized in a similar way as before, using the values of design gap sizes in place of $h_{0,B}$ and $h_{0,T}$, respectively. Numerical results of optimum values of the ratio a are plotted against the difference between bottom and top nominal gap sizes in Fig. 4.8. The top-gap-based optimization leads to preferring a close to top positioning of the slide with identical nominal sizes of the bottom and top gaps. However, a bottom gap 1 mm larger forms nearly mid-height of the side of the drawer which is optimum. In contrast, bottom-gap-based optimization sets forth close to bottom positioning of the slide with uniform nominal gap sizes, and a nominal top gap 1 mm larger shifts the optimum towards mid-height. Larger differences lead to either negative a-values (meaning a slide position below the drawer) or to values larger than 1 (position above the drawer top). Figure 4.8 shows that this approach of optimization for similar top and bottom gap sizes also allows the choice of nominal gap sizes on the basis of slide positioning preference. For example, when an undermount slide is chosen ($a = 0$), a top-gap-based design (use conditions include fully loaded drawer and dryer than average climate) requires a nominal gap 1.85 mm larger at the bottom.

Finally, it has to be emphasized again that the calculations in this section were based on the worst-case approach, i.e. on the simultaneous occurring of unfavourable extreme values of inaccuracies and other deviations, which is not likely in real life. Therefore, this problem will be revisited in Sect. 4.7 in connection with the statistical tolerance design.

4.3.3 Use of the Concept of Integrated Product Model

The product design, manufacturing, assembly and usage processes have to be globally optimized for the competitiveness of the enterprises.

An integrated product model should cover all necessary information for the whole life cycle of a product, such as setting requirements, different design activities, manufacturing and maintenance processes.

The concept of an integrated product model addresses two key issues in product development: elimination of wrong decisions from the early phases, and simultaneous, or concurrent design. Both contribute to shortening the time to market and to the reduction of development costs. Elimination of wrong decisions needs prediction of product performance and cost by simulation on models; conducting design activities parallel is only possible with shared product information. The solution is an integrated product model. In order to optimize product design, manufacturing, assembly and usage processes, evaluation by simulation and improvement is necessary for all phases of the life cycle of a product. In concurrent design, this requires communication and transfer of information without losses between the actors of the individual domains.

The up-to-date product model is a complex, integrated model with multiple functions; it is a computer-based representation, describing an artificial object in the different stages of its life, such as design, manufacture, assembly, usage and disposal by means of technical, economic and other data. Figure 4.9 demonstrates the life cycle phases of product modelling.

A product model is a structured conglomerate of data, functioning as a tool for creating, modifying, storing and visualizing all product-related data. It is not monolithic, rather complex, made up of a number of aspect models. Aspect models are components depending on the phase of product life and method of handling this phase in design. Due to its integrated nature, the IT relations between the aspect models can be unequivocally described, so that sharing of information between the different domains of design is assured.

In general, the domains of product modelling, that is, groups of aspect models, are as below:

- concept models,
- geometric models,
- assembly models,
- visualization models,
- simulation models,
- application models.

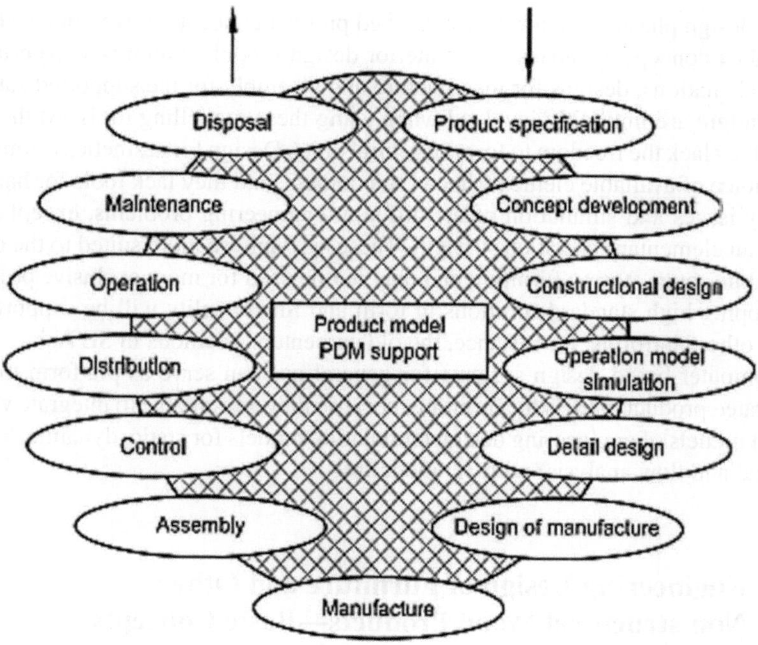

Fig. 4.9 Conceptual build-up of the integrated product model, with computerized product data management (PDM) for the support of information transfer

During the last decades, various software-based applications have appeared in different branches of industry, parallel with the developments in information technology. The original concept of grouping the individual modules around a kernel based on the geometrical model of the product further developed to the so-called shared project models, first in the building industry (Kiviniemi et al. 2005). One of the main problems with their implementation into practical applications has been the internal structure of the different software products that does not support the information needed for the whole process. These difficulties have led to different developments. There are solutions (e.g. model with interoperable solutions) that limit losses of information using incremental data flow that is not file based. FunStep initiative in the furniture industry allows the interoperability of software solutions wherever they are used within the company. An important achievement of this initiative is the formalized structure of product information defined now in the ISO 103030-236 standard for furniture. In spite of these developments, the integrated use of product modelling applications has not been established in the woodworking industry. Instead, a growing variety of software solutions supporting individual design tasks is in use, taking one's mind off the overall opportunities for creating efficiency. Product information transfer from CAD platform to CAM, CNC and production planning environment is nowadays in practice like the 2020 CAD. Such modelling systems are useful in minimizing design efforts in the embodiment and detail design phases with a well-established product concept, or even in developing a product concept by means of an interior design module. Often raw material and cost calculations, designs for manufacture and assembly for the supported category of furniture are implicitly involved when using these modelling tools. At the same time, they lack the freedom to form structural units. Design for aesthetics is limited to the choice of available element shapes and colours, and they lack tools for handling quality issues and simulation of the different engineering problems, except sometimes an elementary check on strength. These systems are more suited to the design of middle-range storing furniture and may not be used for more exclusive products; developing high standard solutions in form and functionality will be supported by using other platforms, for instance, the old carpenters' methods in SE Asia.

Computer-based design systems for general use can serve as platform to build integrated product models for furniture, because they are suitable to integrate various aspect models of engineering design (simulation models for static, dynamic, kinetic, thermal and flow analysis) with 3D modelling.

4.4 Engineering Design of Furniture and Other Non-structural Wood Products—Basic Concepts of Design

This section focuses on engineering tasks where a technical need is to be satisfied with optimal safety. The probability of failing to satisfy the needs at any point of time during the planned working life is to be controlled by means of reliability-based approach. One has to assume a risk of failure determined on economic criteria. A tentative proposal is demonstrated in the next section to quantify acceptable risk levels for the load-carrying capacity of furniture and other wood products, primarily not load-carrying. Two basic models of controlling the probability of failure will be discussed:

- stress–strength interference (SSI) method,
- stress–strength–time (SST) method.

The SSI method applies for cases when neither needs nor capacity varies with time. The SST model is useful for treating problems when either the technical need or capacity, or both change in time. These models are generally valid for any engineering problems including load-carrying capacity, functional properties and production requirements.

4.4.1 Design for Performance—Optimum Risk of Failure

Design for performance (dimensioning) is that part of the engineering design whose task is to define all the properties of a structure that result in the desired capacity with optimal probability in the course of the planned service life referred to in the area of structural design as design working life.

The basic relationship of probability-based dimensioning can be written in a general form as below (Mistéth 1976):

$$P\{[R(t) - S(t)] \geq 0\} \geq 1 - \frac{1}{r} \quad 0 < t \leq T \tag{4.1}$$

where

S(t)—technical demand that may change in time,
R(t)—time-dependent capacity of the planned object for the satisfaction of the demand,
T—planned service life,
1/r—acceptable risk of the capacity being inferior to the demand at any time within the service life.

For strength design $S(t)$ means the loads that the structure is required to carry without failing.

This relationship implies that the structure should be dimensioned so that the capacity during its service life will be higher, at a predetermined probability than the technical requirement. In principle, the optimal risk should be determined on the basis of minimal cost. The total cost of failing to satisfy the demands consists of the rth part of damage costs (D) caused by the failure, including profit loss if any, and the re-establishment cost of the structure as a function of r ($C(r)$), increased by the depreciation allowance that could not be calculated because of the earlier damage to the original structure. In theory, the minimum cost occurs at the point where the first derivative of r is equal to zero. In order to facilitate practical applications, an approximate formula has been derived based on real data, expressing the value of optimal risk as a function of the so-called damage ratio (Mistéth 1977):

$$\frac{1}{r} = \frac{C}{50 \cdot D} \tag{4.2}$$

where C denotes the cost of establishment of the structure.

The capacity is optimal when the damage caused by the unfulfilled requirements is at the most equal to the costs of re-establishment.

Some guidance on the C/D ratio has been given by Mistéth (1981) as follows: with threat to human life, $C/D \approx 1/200$; in the case of possible injury caused to users, $C/D \approx 1/20$; when no bodily injury is probable, $C/D \approx 1/2$ to 1.

The acceptable risk for engineering structures designed for 50 years of working life is on the order of 10^{-5}–10^{-4} for the ultimate limit state (ULS) when loss of human life can be expected; for serviceability limit state (SLS), essentially impermissible deformation, the risk taken is generally around $1/r \approx 10^{-3}$.

The optimal values of these probabilities are higher for furniture, due to a shorter design working life and to the nature of damages that structural failure may cause.

In order to derive the corresponding numbers, several case studies of failures of furniture in use have been evaluated (Kovács 1989). The results justify an increase of the level of acceptable risk to $1/r < 5 \times 10^{-3}$ for ultimate limit state and to $1/r \approx 2 \times 10^{-2}$ for serviceability (deformation) limit state.

Since structural failure of fenestration products may result in loss of human life, the risk levels used for engineering structures should apply.

4.4.2 SSI (Stress–Strength Interference) and SST (Stress–Strength–Time) Models Applied to Design for Lifetime Performance of Furniture

Hereafter, we relate the question of optimal risk to load-carrying capacity. In this context, the technical demand is defined by external actions (loads) that cause internal forces or moments and deformations as load effects. Traditionally, load-carrying capacity is considered a marginal design objective besides comfort, practicality and aesthetics in the case of furniture. Therefore, no explicit design procedures have taken over in practice; instead, rules evolved by experience prevail along with occasional technical considerations, in spite of the abundance of pertinent publications. The same holds partly true for fenestration products.

Determination of the probability of failure caused by one single effect is called the stress–strength interference (SSI) method. Both load (*S*) and load-carrying capacity (*R*), i.e. the magnitude of load resisted by the structure, generally are random variables characterized with their density functions. Figure 4.10 shows a demonstration of the SSI model, where the shaded area of interference of the two density functions just illustrates but is not proportional to the probability of failure (*F*), that is the probability of the demand being higher than the capacity. Failure probability depends on the mean values μ_{S} and μ_{R}, the standard deviations σ_S and σ_R. When the variables are not normally distributed, parameters of the distribution in question have to be used.

The probability of failure can be determined by introducing a new variable $Y = R - S$. For the simple case of normally distributed loads and strength, the mean value of *Y* will be $\mu_Y = \mu_R - \mu_S$ and the variance of this new normal variable is $\sigma_Y^2 = \sigma_R^2 + \sigma_S^2$.

A metric of the reserve of load-carrying capacity, the safety margin (SM) is the negative of the *z*-score of the random variable *Y* for $R = S$:

$$Z = \frac{R - S - (\mu_R - \mu_S)}{\sqrt{(\sigma_R^2 + \sigma_S^2)}} = \frac{-(\mu_R - \mu_S)}{\sqrt{(\sigma_R^2 + \sigma_S^2)}} = -\mathrm{SM} \tag{4.3}$$

The probability corresponding to SM as a standard normal variable is the probability of survival *R*. For the probability of failure *F*, the relationship $F = 1 - R$ holds, therefore, with Φ_{SND} denoting the standard normal distribution function, and it can be written as

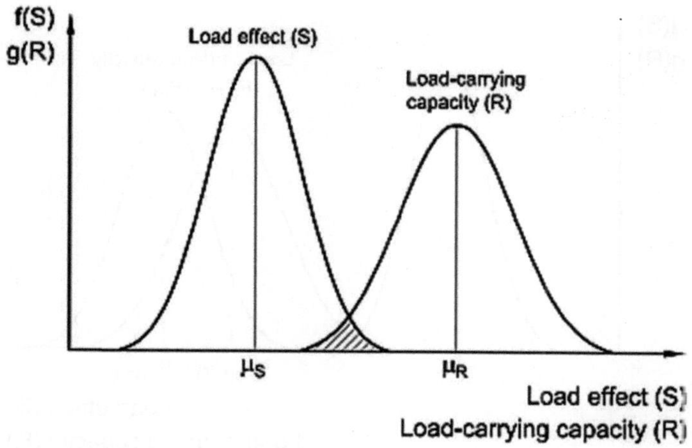

Fig. 4.10 SSI model for probabilistic strength design

$$R = \Phi_{\mathrm{SND}}(\mathrm{SM}) = \Phi_{\mathrm{SND}}(-Z) \tag{4.4}$$

and

$$F = \Phi_{\mathrm{SND}}(Z) \tag{4.5}$$

The above formula for SM is the basic relationship for dimensioning, applicable even for repeated load effects as long as the distribution of *S* and *R*, respectively, do not change with time. However, this is seldom the case. With timely variation, the stress–strength–time (SST) model gives the right solution. One example is when the load distribution is constant or has an increasing trend and the strength is decreasing. This may be typical of many wood structures, including furniture. With increasing service life, the occurrence of higher loads has an increasing probability, which explains the trend. This may not be significant during the planned service life of furniture, which is typically 5–15 years. On the other hand, a decrease in strength cannot be neglected in design, especially for structural joints.

The discrete reliability function according to the SST model for such cases is (Petrik 1985):

$$R_n(t_n) = \begin{bmatrix} 0, \text{ if } & y(t_n) \le x(t_n) \\ 1, \text{ if } & y(t_n) > x(t_n) \end{bmatrix} \tag{4.6}$$

where *y*(*t*) is the time-dependent strength and *x*(*t*) is the time-dependent load, and the relationship relates the *n*th occurrence of load. Reliability *R*(t) is defined by the strength preserved at the time of the last occurrence of load.

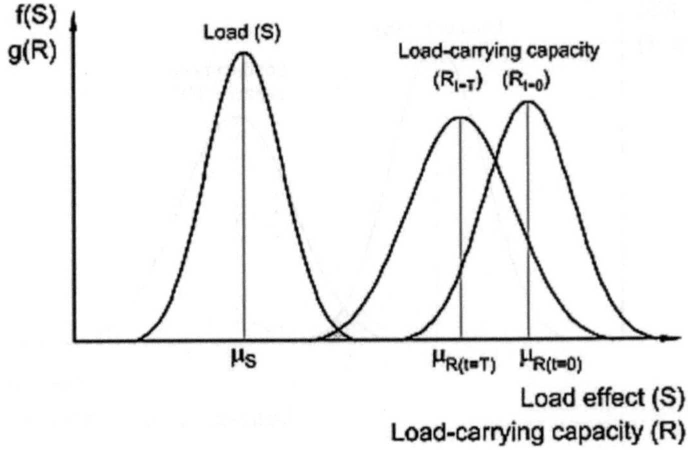

Fig. 4.11 SST model for reliability based strength design

From these considerations, the SST model is applicable for the strength design of furniture and other, non-structural wood products and can be demonstrated as shown in Fig. 4.11.

The figure shows that the reliability level is determined by the distribution of the strength at the end of the planned service life, characterized by a lower mean value and a larger standard deviation than at the beginning of use. No change of the distribution of load values is taken into consideration. Clearly, this is identical to the SSI model with a probability density function of the load-carrying capacity valid at the end of the planned service life.

4.5 Engineering Design of Furniture—Strength Design

In the course of the strength design of furniture, one encounters three questions. The first is the level of safety required and the design methodology through which it can be enforced. This question has been partly dealt with in the previous section. In this section, the adaptation of limit state design, as contrasted by permissible stress design, will be explained. Furthermore, the proposed way of determining design values for loads and for resistances of structural members and joints to comply with the chosen safety level for the whole of service life will be described.

The second question is what methods of structural analysis are suitable for furniture. This section deals with the possibilities of structural modelling of frame-type furniture, case (panel structure) furniture as well as furniture with combined frame and panel structural systems. Special consideration is devoted to the behaviour and modelling of joints.

The third question is what type of member or joint resistance values or material strength properties are suitable to compare with action effects in order to assess conformance to expectations. This will be discussed for wood and wood-based materials as well as for furniture joints.

Finally, the proposed methodology of structural design of furniture will be demonstrated on examples.

4.5.1 *Considerations on the Theory of Strength Design*

Design for permissible stresses is typically used in mechanical engineering; limit state design has taken over for building constructions. One of the differences between the two is that the concept of the former one is based on stress calculations and comparison. For some structural members and for most connections in buildings, this would be difficult or less reliable. It also holds true for furniture as a load-carrying structure. In limit state design, principally, resistance of members and joints is calculated and compared with the results of analysis of the structure under load, taking into consideration the variability of geometrical data.

A combination of several simultaneous actions may be composed of loads of different duration and statistical distribution in building structures. Therefore, the overall lifetime safety requires differentiated treatment of individual components of a load combination on the level of design values. This is not the case in machine design, where design values for all load components are derived in the same way.

One can state that the difference between the two approaches is only formal, provided that stresses in members and connections can be realistically calculated, the structure is not subjected to load combinations with components of different duration and different statistical distribution, and variability of member cross-sections is negligible.

In furniture strength design, the above conditions often hold, but not always. For a chair, different loading situations (load cases) have to be taken into consideration. Some are combined loads, like the simultaneous vertical load on the seat and horizontal load on the back. These two loads, however, are of the same duration and statistical distribution, both deriving from the weight of human body.

Load combination is different when a packed cupboard or movable bookshelf is being pushed, since the stored mass acts as a dead load and the pushing action is of short duration (or impact load), the distribution of which is that of physical force in any particular direction.

The concept of permissible stresses does not take into consideration the duration of load (except for fatigue effects) and service conditions (essentially moisture uptake) in the design values of material properties. Consequently, the concept of permissible stresses does not explicitly involve a safety level in the probabilistic sense, even if permissible stress corresponds to the 0.01 percentile of strength distribution, with design load being the nominal operative value.

It follows from the above discussion that the concept of limit state design is the right choice for strength design methodology for furniture and other non-structural wood products. For load-carrying structures, the individual parts of the Structural Eurocode published in European Standards (EN 1990- to EN 1999) stipulate limit state design rules and methods consistent with the required level of reliability in buildings and other objects of civil engineering. Among them, Eurocode 5 (EN 1995:2008) applies to timber structures.

When referring to Eurocode, one has to acknowledge the much simpler loading situations of furniture as compared to those of building structures, allowing for major simplification when adapting the basic concept to furniture design, as will be explained later. This simplification sometimes results in calculation processes formally identical with permissible stress design.

As in structural design in general, two types of limit states have to be dealt with. Ultimate limit states concern the safety of people and/or the safety of structure on account of rupture, excess deformation, and sudden loss of rigidity or stability. Serviceability limit states apply to the functioning, comfort and visual appearance under normal use, due to deformations. These deformations may be reversible or irreversible, making the identification of different load combinations necessary in terms of duration of load.

The model of strength design process has three logical steps as follows:

1. Determination of actions, their distribution, duration and representative value; identification of loading situations,
2. Setting up the structural model and solving for action effects,
3. Comparison of action effects with design criteria; assessment of the design, identification of necessary alterations if any, and go back to step 2.

The Eurocodes specify the verification of the required structural reliability by the use of the partial factor method (probabilistic-based safety factors on both the action and the resistance side) as a semi-probabilistic method. At the same time, it allows a design based on direct reliability calculations.

The comparison is made between the design value of the effect of action (E_d) and the design value of the corresponding resistance (R_d) or serviceability criterion (C_d). It shall be verified that $E_d \leq R_d$ for strength and $E_d \leq C_d$ for serviceability. The partial factor method uses the characteristic values of actions and material or product properties to derive design values of action effects and resistances. The characteristic values generally correspond to the 0.95 percentile of actions and the 0.05 percentile of material properties in their respective distribution functions. Applying the partial factors given in the relevant standard results in the target reliability, corresponding to a probability of failure $P_f = 10^{-4}$ for building constructions. These partial factors—γ_M for material and γ_F for actions—account for the variability of actions and material properties, as well as for model uncertainties and dimensional variations. A separate conversion factor on the material side takes into consideration the volume and scale effects, moisture and temperature effects, and load duration. From the probabilistic point of view, the variability parts of the partial factors are of interest and it may be useful to deal with them separately when different target reliability is to be attained.

The effect of dimensional variability can be neglected in the case of furniture. The simplified relationship to illustrate the role of these factors may be written as

$$E\{F_k \cdot \gamma_F; a_d\} \leq \eta \cdot \frac{R_k}{\gamma_M} \tag{4.7}$$

or

$$E\{F_k \cdot \gamma_f \cdot \gamma_{Sd}; a_d\} \leq \eta \cdot \frac{R_k}{\gamma_m \cdot \gamma_{Rd}} \tag{4.7a}$$

where the left-hand side of the equation is the design effect value of the action F as a function of the design geometrical data a_d. The lower case f and m subscripts denote the variability-dependent parts of the partial factors, and subscript k refers to characteristic value. The subscripts Sd and Rd denote the model uncertainty factors for action effect and resistance, respectively; η is the material property conversion factor.

For designs based on direct reliability calculation, EN 1990:2002 specifies a method of calibrating the design values for the target reliability. For reliability verification however, it is proposed to use these design values in an indirect way, through partial factors obtained by relating these design values to their respective characteristic or representative value.

For furniture strength design, one can adapt the partial factor method with the recalculation of the partial factors as shown in the next two sections. Wherever sufficient information is at hand, direct reliability calculations may be made as shown in Sects. 4.5.2 and 4.5.3.

Verification of strength is stipulated in EN 1995:2004, except for stability problems and for connectors, on the level of stresses rather than of internal forces or load-carrying capacity. Specialities inherent in furniture design include the need to deal with combined state of stress in orthotropic bodies on account of the fact that unlike in timber structures, quite often, structural members of furniture are shaped without sufficient strength considerations.

4.5.2 Design Values of Actions and Action Effects

For simplicity, we will refer to actions as loads on furniture. Pieces of furniture are loaded during their service life in different ways with varying intensity and duration of load. Therefore, a number of various load models can be associated with a piece of furniture. It is the designer's (and manufacturer's) responsibility to take into consideration what load model(s) if any should be used in the product development process. Load models for design must take into account the way the furniture is used and should represent the structurally important elements of the load spectrum during its service life. The service conditions for which a piece of furniture is intended can be classified from light to very heavy categories of use with different load levels and/or frequencies of occurrence, and a higher category sometimes calls for extra load models. Therefore, design values in a load model should reflect the level of target resistance, or target user group in other words. Moreover, design load levels depend on design working life. Manufacturers may reconsider if customers are willing to pay the cost of more durable furniture. However, it would not be honest to sell furniture that implies longer durability than it actually has.

Loads acting on furniture have been investigated for performance testing rather than for engineering design. Eckelman studied the strengths and weaknesses of performance tests for assessing the fitness of a piece of furniture for its intended use (Eckelman 1988a, b, 1999). He generally found that the ability of current test methods was not sufficient to distinguish between differences of performance or to explore the true potential strength of a piece of furniture. Likewise, there was only a meagre chance to gain engineering design information through performance tests. In order to overcome most of these shortcomings, a universal test method on the basis of the concept of multiple-level acceptance with cyclic load models was developed (FNEA 1998; Eckelman and Erdil 2001a, b). The test method has proved to be satisfactory in different areas of use of furniture (Eckelman 1995; Diler et al. 2017).

From the point of view of design loads, loadings used in performance tests as specified in the respective standards must be handled with due discretion. Load models proposed in the above-mentioned "cyclic stepped load" method are more meaningful in simulating real service loads. Whether relying on performance test stipulations or on any surveys relative to loads that furniture in use is exposed to, establishment of the best representative values needs thorough consideration, especially because of the lack of information on probabilities associated with the values considered.

The magnitude of the loads that act on furniture derive from the weight of a human body, or from the bodily force of human in different positions and directions, or from the weight of objects stored, or weight of the furniture itself, but very often from the combinations of them. Statistical distributions for a probabilistic approach are currently scares, except for human body weights which must be differentiated geographically, and according to age groups if relevant. Moreover, some loads occur repeatedly with varying frequencies and durations.

Therefore, if we want to derive design load values consistent with the target reliability of the structure, we have two possibilities. One is to take load levels required by performance test standards as deterministic, rather than stochastic values, until more information on their stochastic nature will have accumulated. Often, load values in standards are high enough so that they are not expected to be exceeded in use.

Alternatively, it may seem more appropriate to take the load given in a performance test as a nominal value. When this is the case, one may proceed with the partial factor design as exemplified in the Eurocodes, however, adhering to the target reliability for furniture.

The other possibility is to assess the mean value and standard deviation of the type of load in question and then use the first-order reliability method (FORM) stipulated as Level II procedure in Annex C "Basis for Partial Factor Design and Reliability Analysis" of EN 1990:2002.

According to this annex, the design values of action effects E_d and resistances R_d should be defined such that the probability of having a more unfavourable value is as follows:

$$P(E > E_{\mathrm{d}}) = \Phi(+\alpha_{\mathrm{E}} \cdot \beta) \tag{4.8}$$

$$P(R \leq R_{\mathrm{d}}) = \Phi(-\alpha_{\mathrm{R}} \cdot \beta) \tag{4.9}$$

where Φ denotes the standard normal distribution function,
β is the target reliability index (denoted as *SM* in Sect. 4.4.2),
α_{E} and α_{E}, with $|\alpha| \leq 1$, are the values of the FORM sensitivity factors; the value of α is negative for action effects, and positive for resistances.

Both partial factor design and design by using Eqs. (4.8) and (4.9) have to use target reliability based on load effects calculated from design loads, and the uncertainty of the load and load effect model have to be covered by the design value of load effects. These uncertainties are included in the values of the partial factor γ_{F} given in the Eurocode standards; nevertheless, when applying Level II procedure (direct reliability calculation), either loads or load effects have to be increased by a factor γ_{Sd} of 1.05–1.15 to account for the same. For furniture design, $\gamma_{Sd} = 1.1$ can be assumed. The use of this factor is also necessary when the partial factors are recalculated for achieving different target reliability.

As set forth in Sect. 4.4.1, the acceptable risk for ultimate limit states of furniture is 5×10^{-3} with the corresponding target reliability index of $\beta = 2.576$. (In Eurocode for building structures $\beta = 3.8$.)

When the statistical distribution of a load (e.g. a load stipulated in a performance test standard) is not known and is considered as a nominal value, it should be, according to EN 1990, taken as characteristic value. Assuming the same to be equivalent to the 0.95-fractile, we may take the value of the partial factor $\gamma_{\mathrm{f}} = 1$, hence $\gamma_{\mathrm{F}} = 1 \cdot \gamma_{\mathrm{Sd}}$ for defining the design value of load and load effect.

When the mean and standard deviation of a type of load can be estimated, the expressions below, based on Table C3 in EN 1990, can be used to calculate design load value. For normally distributed loads

$$F_{\mathrm{d}} = \gamma_{\mathrm{Sd}}\left(\bar{F} - \alpha_{\mathrm{E}} \cdot \beta \cdot \sigma\right) \tag{4.10}$$

or, when extremal distribution can be assumed

$$F_{\mathrm{d}} = \gamma_{\mathrm{SSd}}\left(u - \frac{1}{a}\ln\{-\ln\Phi(-\alpha_{\mathrm{E}} \cdot \beta)\}\right) \tag{4.11}$$

In the above expressions, F-bar and σ denote the mean value and the standard deviation of the given variable; these should be based on the same reference period,

Table 4.6 Probability level and formulae for calculating ULS design values of load according to the level of information on load

Load information	Formula for calculating design load value for ultimate limit states	Probability level p
Deterministic value	$\gamma_{Sd} \cdot F_{det}$	–
Nominal value; statistical distribution not known	$\gamma_{Sd} \cdot F_{nom}$	0.95
Known mean, SD (and type of distribution)	Equation (4.10) for normal distribution Equation (4.11) for extremal (Gumbel) distribution	0.98

Table 4.7 Probability level and formulae for calculating SLS design values of load according to the level of information on load

Load information	Formula for calculating design load value for serviceability limit states	Probability level p
Deterministic value	$\gamma_{Sd} \cdot F_{det}$	–
Nominal value, statistical distribution not known	$\gamma_{Sd} \cdot F_{nom}$	0.95
Known mean, SD (and type of distribution)	Equation (4.10) for normal distribution Equation (4.11) for extremal (Gumbel) distribution	0.95

as β. Φ is the cumulative distribution function of the standard normal distribution, furthermore

$$u = \bar{F} - \frac{0.577}{a} \text{ and } a = \frac{\pi}{\sigma\sqrt{6}}$$

By taking the form sensitivity factor $\alpha_E = -0.7$, the values in the parenthesis of the above equations approximatively correspond to the 0.98-fractile.

Determination of design load values for *ultimate limit states* (ULS) for these three situations of load information is summarized in Table 4.6.

For *serviceability limit states* (SLS), the acceptable probability of an occurrence is $P_f \approx 2 \times 10^{-2}$ with the corresponding target reliability index $\beta = 2.05$.

The partial factor for actions, according to EN 1990, should be taken as 1, unless differently specified in the respective parts of Eurocode. This means that the characteristic load values are design values, which, for permanent loads (self-weight), are equal to the mean value. For variable loads (loads imposed on furniture), the characteristic value corresponds to the 0.95-fractile.

With the three levels of information on loads, design values for serviceability limit state can be summarized as shown in Table 4.7. It should be kept in mind that

$\beta = 2.05$ has to be used in Eqs. (4.8)–(4.11), which, with $\alpha_E = -0.8$ gives values that approximatively correspond to the 0.95-fractile.

Some of the loads relevant for either ULS or SLS design may be combinations of loads of different natures (such as permanent, variable or accidental) occurring simultaneously. In such cases, the combination of actions has to be defined in accordance with Sects. 4.7.4 and 4.7.5 of EN 1990:2002 for ultimate and serviceability limit states, respectively. When more than one variable load is present, design combination values of accompanying variable actions are used, the latter multiplied by a $\Psi \leq 1$ combination factor.

A few examples of load cases specified in performance requirement standards or otherwise thought to be useful for design purposes are given in Table 4.8 for illustration only, with reference to Figs. 4.12, 4.13 and 4.14. Load duration class as defined in EN 1995:2004, Table 2.1, as well as load combination where relevant is also indicated. The "+" sign in these cases implies "to be combined with". Duration class or number of cycles of durability tests will be taken into consideration in the design values of material and product properties and resistances through the k_{mod} factor. When a load combination consists of actions of different load duration classes, a value of k_{mod} corresponding to the load with the shortest duration shall be used. For impact loadings, such as a person falling heavily onto a bed or chair, a dynamic factor of $\varphi = 2.0$ shall be applied to the load. The same is suitable for reaction forces caused by collision or dropping. For shocks caused by thrust or pull, a dynamic factor of $\varphi = 1.3$ is reasonable.

4.5.3 Design Values of Material Properties and Member Resistance

Reasonable design stresses for woods used in furniture had been extensively studied (Eckelman 1974). The author concluded that one-third of the average bending strength to rupture at 12% moisture content in the *Wood Handbook* (*W.H.*) can be recommended for hardwoods. Eckelman based his recommendation on test results of more than 900 full-size furniture parts that he converted to 12% moisture content, so that comparisons could be made with *W.H.* values. There were 24 hardwood and a single softwood species involved. Less than 5% of the samples failed at a bending stress lower than this threshold with just two exceptions. In analysing the results, Eckelman calculated the 0.05-fractile at $\alpha = 0.95$ confidence level, which showed a generally good agreement with the "one-third of the average" convention. He evaluated this finding as a statistical justification of the convention of one-third of average being reasonable.

Eckelman's findings for probability-based limit state design are important as explained below. First, he tested real size pieces made of furniture-grade wood. The samples had an allowable degree of defects. Secondly, samples came from a wide range of wood sources. Therefore, coefficients of variation that ranged from

Table 4.8 Examples of loads applied in performance tests of furniture and their use in design

Seatings Load case	Duration	Load(s)	Level 1	Level 2	Probability	Dyn. factor	Load combination
Seat and back static load test, Fig. 4.12	Instantaneous	Seat load Back load	1600 N 560 N	2000 N 700 N	Deterministic	–	–
Seat and back static load test, Fig. 4.12	Cyclic	Seat load Back load Cycles	10,000 300 N 100,000	1000 N 300 N 200,000	Deterministic	–	–
Drop from the height of a table, Fig. 4.13	Impact	Drop height	600 mm	600 mm	Deterministic	2.0	–
Storage furniture Test for structure and underframes, Fig. 4.14	Impact Long term	Side force $Q1$ Stored mass $Q2$	350 N Full load	– –	Charact. 0.95	1.3	USL: $G_d + Q_{d,1} + \Psi_0 \cdot Q_{d,2}$ SLS: $G_k + Q_{k,1} + Q_{k,2}$

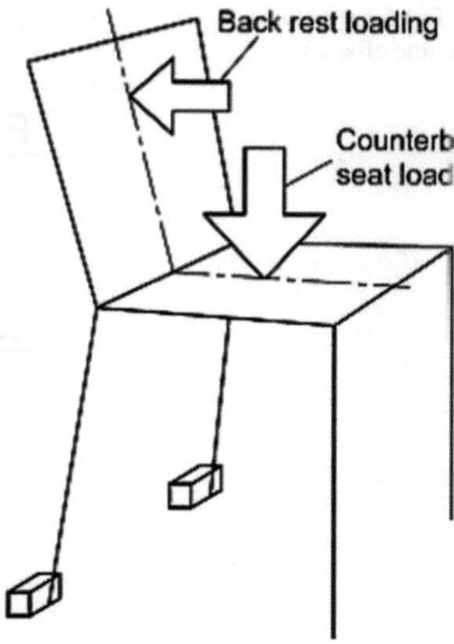

Fig. 4.12 Seat and back static loading and durability test

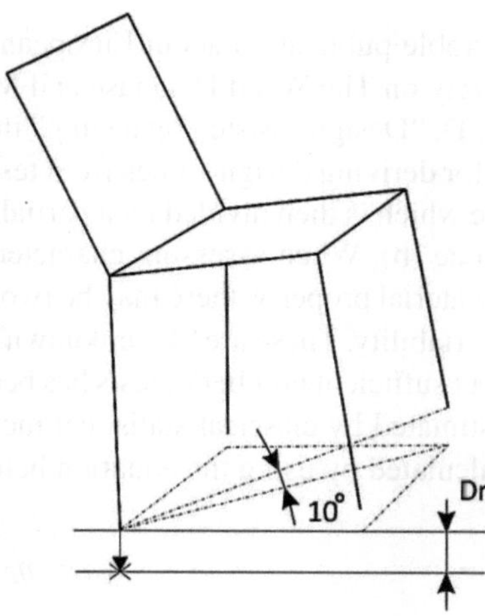

Fig. 4.13 Drop test of chair from the height of a table

17.1 to 38.0% for the individual hardwood species include the variability of a species bending strength over various geographical ranges, as well as the scattering effect of features not accepted in small clear specimens. The coefficient of variation for samples of a single softwood species (Southern Yellow Pine) coming from one source reached 40.6% This can be expected when the stock is not stress graded. The results allow the derivation of characteristic values and design values of a required probability level, valid for the size and grade typical to furniture parts. These values can be related to the mean values published in the Wood Handbook, determined on small clear specimens at 12% moisture content. Unfortunately, the Wood Handbook lists a limited number of wood species used for furniture in Europe, and currently, there is

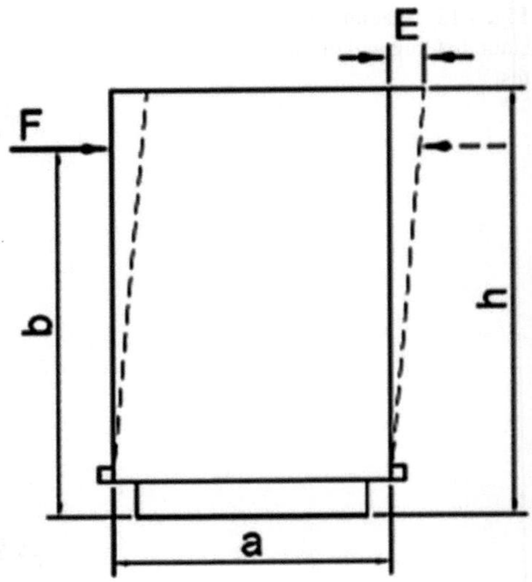

Fig. 4.14 Test for structure and underframe of storage furniture

no comparable publication about European wood species. In some cases, however, one may rely on The Wood Database and MatWeb available on the net.

Annex D, "Design assisted by testing" in EN 1990:2002 describes the applicable methods for deriving design values from tests in two ways: by assessing a characteristic value which is then divided by a partial factor (a), or by direct assessment of the design value (b). When assessing characteristic or design value of structural resistance or material property, there may be two cases depending on the prior knowledge on their variability. These are "V_x unknown" and "V_x known". When "V_x unknown" is valid, but sufficient number of tests has been performed, the coefficient of variation can be estimated by classical statistical methods. The design value of a property X can be calculated by using the equation below:

$$X_d = \eta_d \frac{X_{k(n)}}{\gamma_m} \tag{4.12}$$

The characteristic value $X_{k(n)}$ is given by the expression:

$$X_{k(n)} = \bar{X}(1 - k_n \cdot V_x) \tag{4.13}$$

In Eq. (4.13), k_n is the characteristic (5%) fractile factor, for calculating a lower confidence limit at 75% confidence level. Its value is tabulated for both cases of "V_x known" and "V_x unknown" as a function of the sample size n, with the assumption that there is no prior knowledge about the value of the mean. The Greek letters γ_m and η_d denote the partial material factor and design value of the possible conversion factor, respectively, when this latter is not included in the partial factor for resistance. The same Annex states it preferable to use the case "V_x known" together with a conservative upper estimate of V_x.

For direct assessment of the design value, Annex D in EN 1990:2002 gives the equation below:

$$X_d = \eta_d \cdot \bar{X} \cdot \left(1 - k_{d,n} \cdot V_x\right) \tag{4.14}$$

where $k_{d,n}$ is the design fractile factor, and η_d should cover all uncertainties not covered by the test.

The value for $k_{d,n}$ is again given in a tabular form depending on the sample size n. Values in the table give a probability of observing a lower than design value of about 0.1%, which corresponds to the probability belonging to a resistance sensitivity factor of $\alpha_R = 0.8$ and target reliability index of $\beta = 3.8$; see Eq. (4.9). For furniture, this probability will be different.

The European standard EN 14 358:2016 specifies a method based on non-central t-distribution that gives more conservative estimates of lower confidence limits for percentile values. Applying this method to Eckelman's test results, characteristic values at $\alpha = 0.75$ confidence level have been calculated based on mean and standard deviations assessed from samples. The 5% lower tolerance limits with 95% confidence were re-examined and found to correspond approximatively to the 2.8th percentile value in the distribution function of the bending strength.

It might be interesting to see how the characteristic values with $\alpha = 0.75$ confidence level and the 2.8th percentile values can be derived from the means determined on small clear specimens for the individual species of wood. That will be examined next.

Working on design stresses for limit state design of wood structures, based on sufficient experimental data, Mistéth established relationships between bending strength distribution parameters of selected defect-free stock and wood with allowable growth features in different quality grades as below (Mistéth 1986):

$$\bar{f}_b^i = \bar{f}_b^0 \cdot 0.95^i \tag{4.15}$$

$$\sigma_{fb}^i = \sigma_{fb}^0 \cdot 1.25^i \tag{4.16}$$

where $\bar{f}_b$ is the mean bending strength and σ_{fb} is the standard deviation, and the subscript $i = 1, 2, 3$ denotes the quality class.

Assuming the wood of furniture parts to be in class I, i is assigned the value 1. Furthermore, a coefficient of size effect can be assessed on the basis of EN 1995:2004 to vary from 1.08 to 1.38 for depths of member cross-section between 30 and 100 mm, with a value of 1.15 for depth of 50 mm; results of small clear specimens have to be divided by this factor of size effect.

In this manner, the 5th percentile value of bending strength $\bar{f}_{b05}$ for a species of wood can be derived from its mean value $\bar{f}_{b,sc}$, given in an authentic publication for 12% moisture content, by using the expression below, with the subscript "sc" referring to small clear specimen:

$$f_{b05} = \bar{f}_{b,sc} \cdot 0.95 \cdot (1/1.15) \cdot (1 - 1.645 \cdot V_{SC} \cdot 1.25) \tag{4.17}$$

When using this expression, we assume both the mean $\bar{f}_{b,sc}$ and the coefficient of variation V_{sc} to be known. If we want to apply this expression to the wood species Eckelman used in his study, we have to take notice of the fact that the coefficients of variation were determined from the samples, and their values were obtained for the quality class, which the tested furniture parts correspond to. Therefore, the multiplying factor of 1.25 has to be omitted; at the same time, the ratio of the upper 75% confidence limit to the observed standard deviations has to be applied as a factor to increase the observed value. The weighted average of these latter factors for the species of wood represented by more than 10 samples turned out to be 1.06. As a result, the above expression applies to the species of wood in the study as follows:

$$f_{b05} = \bar{f}_{b,WH} \cdot 1.95 \cdot (1/1.15) \cdot \left(1 - 1.645 \cdot V_{exp} \cdot 1.06\right) \tag{4.18}$$

where $\bar{f}_{b,WH}$ denotes the mean value of the bending strength of small clear specimens at 12% moisture content, for the given species in the Wood Handbook, and V_{exp} is the coefficient of variation obtained in testing furniture parts.

Assessing the 5th percentile value of bending strength for true-size pieces of wood of the grade used in furniture by using the method specified in EN 1990:2002, Annex D and also by the method described above allowed us to see how the results of the two approaches agree. In fact, the characteristic values obtained with the two methods are quite close for the 12 species of wood with sample sizes above 10. The weighted ratio of EN-based to WH-based characteristic strength values happened to be 0.975 for the 11 hardwoods and 1.005 overall, with the corresponding coefficients of variation of 0.134 and 0.123, respectively. The same ratio for the 2.8th percentile values turned out to be 0.975, likewise close to 1. This comparison indicates that Eq. (4.17) gives a good estimate of a given percentile value of the strength of solid wood furniture parts. In other words, the approach proposed by Mistéth is useable for deriving characteristic or design strength values for solid wood, when mean values defined on small clear specimens are known and a conservative upper estimate of the coefficient of variation is available.

Equation (4.9) (design based on direct reliability methods, Annex C of EN 1990:2002) applied to material properties of normal distribution can be re-written as below:

$$X_d = \frac{1}{\gamma_{Rd}}\left(\bar{X} - \alpha_R \cdot \beta \cdot \sigma_x\right) \tag{4.19}$$

In Annex C, the material property conversion factor η is implicitly taken into account in the value of $\overline{X}$. In the expressions where this is not the case, one has to use a conversion factor, called k_{mod} for timber structures; its values are given in EN1995:2004 for the different service classes (ranges of equilibrium moisture content) and load duration classes. The factor γ_{Rd} in Eq. (4.19) is a partial factor

Table 4.9 Proposed values of material strength property modification factor k_{mod} for furniture strength design

Material	Service class	Long-term action	Medium-term action	Short-term action	Instantaneous action
Solid wood	Indoor	0.75	0.85	0.95	1.15
	Outdoor	0.67	0.77	0.85	1.05
Particleboard	Indoor	0.50	0.70	0.90	1.10
	Outdoor	0.35	0.50	0.65	0.85
MDF	Indoor	0.45	0.65	0.85	1.10
	Outdoor	–	–	0.50	0.85
Hardboard	Indoor	0.50	0.70	0.90	1.10
	Outdoor	0.35	0.50	0.65	0.85

associated with the uncertainties in the resistance model, which for standard tests may be assigned a value of 1.05. For furniture joint resistance, $\gamma_{Rd} = 1.15$ can be used.

As compared to building structures, the ranges of equilibrium moisture content of wood furniture and buildings are different. Therefore, service classes and the values of the modification factor k_{mod} have to be taken different, on the basis of the change in strength properties as a function of moisture content, as shown in Table 4.9.

Structural resistance cannot be directly derived from material properties for furniture joints; see Sect. 4.5.6. Therefore, service classes should expediently be applied at the structural resistance level. Joints, especially glued ones, are more sensitive to repetition of loads than to their duration itself; hence, joints typically subject to cyclic loading during their service life have to be treated with due attention to the effect of cumulative damage (Table 4.10).

Table 4.10 Proposed values of structural resistance modification factor k_{mod} for glued furniture joints

Loading conditions	Service class	Long-term action	Medium-term action	Short-term action	Instantaneous action
Glued joints, no cyclic user loads	Indoor	0.50	0.65	0.90	1.05
	Outdoor	0.40	0.55	0.80	0.90
Glued joints, typically cyclic user loads	Indoor	0.33	0.50	0.90	1.05
	Outdoor	0.25	0.40	0.80	0.90

With reference to Eq. (4.14), direct assessment of the design value of a strength property is possible by using Eq. (4.19), re-written in the form of Eq. (4.20), in conformity with D.7 in EN 1990:2002 (Eq. 4.13).

$$X_d = \frac{k_{\text{mod}}}{\gamma_{\text{Rd}}} \cdot \bar{X}_{\text{SC}} \cdot 0.95 \cdot (1/1.15) \cdot \left(1 - u_{(1-p)} \cdot V_{\text{SC}} \cdot 1.25\right) \tag{4.20}$$

In this equation, the product $k_{\text{mod}} \cdot \overline{X}_{\text{sc}} \cdot 0.95 \cdot (1/1.15)$ corresponds to $\overline{X}$ in Eq. (4.19), and $V_{\text{sc}} \cdot 1.25$ to $\sigma_x/\overline{X}$, while $u_{(1-p)}$ to $\propto_{\text{R}} \cdot\beta$, where $u_{(1-p)}$ is the $(1 - p)$ percentile of the standard normal distribution function.

When neither V_{sc} nor σ_x is known and has to be assessed via testing, the $u_{(1-p)} \cdot V_{\text{sc}}$ term will be replaced by $k_{d,n} \cdot V_x$ from Eq. (4.14). The design fractile factor $k_{d,n}$ is the α-percentile in a non-central t-distribution with $n - 1$ degrees of freedom and $\lambda = u_{(1-p)} \cdot \sqrt{n}$ non-centrality parameter, divided by the square root of n, where $\alpha = (1 - p)$.

$$X_d = \frac{k_{\text{mod}}}{\gamma_{\text{Rd}}} \cdot \bar{X}_{\text{SC}} \cdot 0.95 \cdot (1/1.15) \cdot \left(1 - k_{d,n} \cdot V_x \cdot 1.25\right) \tag{4.20a}$$

Keeping in mind that the selected value of reliability index for furniture in the ultimate limit states is $\beta = 2.576$, the probability level pertinent to the design strength values can be calculated by using the formula given for the safety margin SM, Eq. (4.3). For this calculation, when design values are derived from known mean values of small clear specimens, the coefficient of variation for hardwood material strength properties can be assumed as $V = 0.20 \cdot 1.25 = 0.25$ for full-size, furniture-grade pieces.

In wood composites, mean values of the individual material strength properties for given product types are seldom known. Instead, quality control criteria of strength and stiffness are given in product specifications, which can be regarded as normative, from the probabilistic point of view 5th percentile characteristic values. When strength design of furniture is based on such information, Eqs. (4.20 and 4.20a) still apply, but instead of $\overline{X}_{\text{sc}}$, mean value $\overline{X}$ derived from the product specification value shall be used. With a conservative estimate of coefficient of variation of $V = 0.1$, this mean value of strength is calculated as:

$$\bar{\sigma} = \frac{\sigma_{\text{norm}}}{(1 - 1.645 \cdot 0.1)} = 1.2 \cdot \sigma_{\text{norm}} \tag{4.21}$$

No information on size effect is currently available for the calculation of design value. Though thickness of test pieces and structural members is identical, in order to remain conservative, a factor of 1.1 for both the mean and the standard deviation may be appropriate. At the same time, no such term as quality grade effect is applicable.

When calculating joint resistance values, $\overline{F}$ or $\overline{M}$ takes the place of $\overline{X}_{\text{sc}}$; neither quality grade factor nor size effect factor is applicable, and a coefficient of varia-

Table 4.11 Probability level and formulae for calculating ULS design values of material strength according to the level of information on load

Load information	Formula for calculating design value of material strength properties in ultimate limit states	Probability level, p
Deterministic value	For normal distribution: Equation (4.20) with $u_{(1-0.005)} =$ 2.576, or Equation (4.20a) with $k_{d,n,005}$ taken from Table 4.12	0.005
Nominal value; statistical distribution not known	For normal distribution: Equation (4.20) with $u_{(1-0.02)} =$ 2.054, or Equation (4.20a) with $k_{d,n,02}$ taken from Table 4.12	0.02
Known mean, SD and type of distribution	For normal distribution: Equation (4.20) with $u_{(1-0.02)} =$ 2.054, or Equation (4.20a) with $k_{d,n,02}$ taken from Table 4.12	0.02

Table 4.12 Values of the design fractile factor, $k_{d,n}$ for different sample sizes

n	3	4	5	6	8	10	15	20	30	50	100
$k_{d,n,005}$	4.86	3.77	3.23	3.06	2.98	2.89	2.81	2.73	4.86	3.77	3.23
$k_{d,n,02}$	3.90	3.31	3.04	2.89	2.70	2.60	2.46	2.39	2.32	2.25	2.18

tion of 0.22 can be used in Eq. (4.20), accounting for the additional variability of workmanship precision.

In the three cases of load information discussed above, the design value of a material strength property shall correspond to the percentile value in the standard normal distribution function as shown in Table 4.11.

According to EN 1995:2008, the design values of the stiffness properties (E_{d}, G_{d}) used in carrying out the analysis have to be fitted to the time-dependent behaviour of the structure. When all members and connections have the same time-dependent properties, and therefore internal forces are not effected by stiffness distribution, mean values shall be used. Otherwise, final mean values, adjusted to the load component causing the largest stress, are used. In furniture, rotational stiffness of semi-rigid joints may change more rapidly under sustained or repeated loading, than member stiffness values. This may justify the use of final means of stiffness properties, including joint rotational stiffness K (Nm/rad), calculated as

$$E_{\mathrm{mean,fin}} = \frac{E_{\mathrm{mean}}}{(1 + \psi_2 \cdot k_{\mathrm{def}})}, \text{ and } G_{\mathrm{mean,fin}} = \frac{G_{\mathrm{mean}}}{(1 + \psi_2 \cdot k_{\mathrm{def}})}$$

$$\text{and } K_{\mathrm{mean,fin}} = \frac{K_{\mathrm{mean}}}{(1 + \psi_2 \cdot f_{\mathrm{def}})} \tag{4.22}$$

The values of k_{def} given in EN 1995:2004 as time-dependent increment of deformation refer to a working life of 50 years, which now have to be adjusted to 10–15 years. No k_{def} values for furniture joints are stipulated; relevant research reports may be consulted. Recommended values of the quasi-permanent load-combination factor ψ_2 for buildings are listed in EN 1990:2002; appropriate value may be selected on the basis of the similarity of their action. If the load causing the largest stress is permanent, $\psi_2 = 1$ shall be used.

In special cases like buckling, the 5th percentile value of stiffness properties shall be used for furniture, as it is used for building structures.

For serviceability limit states, typical loading situations of different pieces of furniture can be categorized as either characteristic or quasi-permanent load combinations as defined in EN 1995:2004. For a characteristic combination, instantaneous deformation u_{inst} is calculated by using the mean values of stiffness properties. The final deformation u_{fin} shall be calculated for the quasi-permanent combination of loads. In this case, when internal forces are not effected by stiffness distribution, final deformation can be calculated in a simplified way as

$$u_{fin} = u_{inst} \cdot (1 + k_{def}) \tag{4.23}$$

for all the loads of the quasi-permanent combination and added together.

When internal forces are affected by stiffness distribution (due to different creep properties of the components), final stiffness properties are used as follows:

$$E_{mean,fin} = \frac{E_{mean}}{(1 + k_{def})}, \text{ and } G_{mean,fin} = \frac{G_{mean}}{(1 + k_{def})} \text{ and } K_{mean,fin} = \frac{K_{mean}}{(1 + k_{def})} \tag{4.24}$$

4.5.4 Setting Up the Mathematical Model—Modelling Tools

This section demonstrates some important modelling aspects specific to furniture to calculate its responses to external loads. Special situations occur for several reasons. First of all, the structure of a piece of furniture is seldom governed by rules of statics and this holds true for its individual structural members. It often follows that additional parts are connected to the primary load-carrying part in a statically undefined way, resulting in somewhat arbitrary load sharing. Secondly, joints between members are semi-rigid with generally nonlinear behaviour that cannot always be approached satisfactorily with a linear model in a range of use conditions. Thirdly, support conditions in the different loading situations are often difficult to realistically model.

In the literature, one can find valuable examples of the use of computational models for the internal forces and deformations of frame-type and case furniture. They will be referred to at the right places, and additional procedures will be proposed.

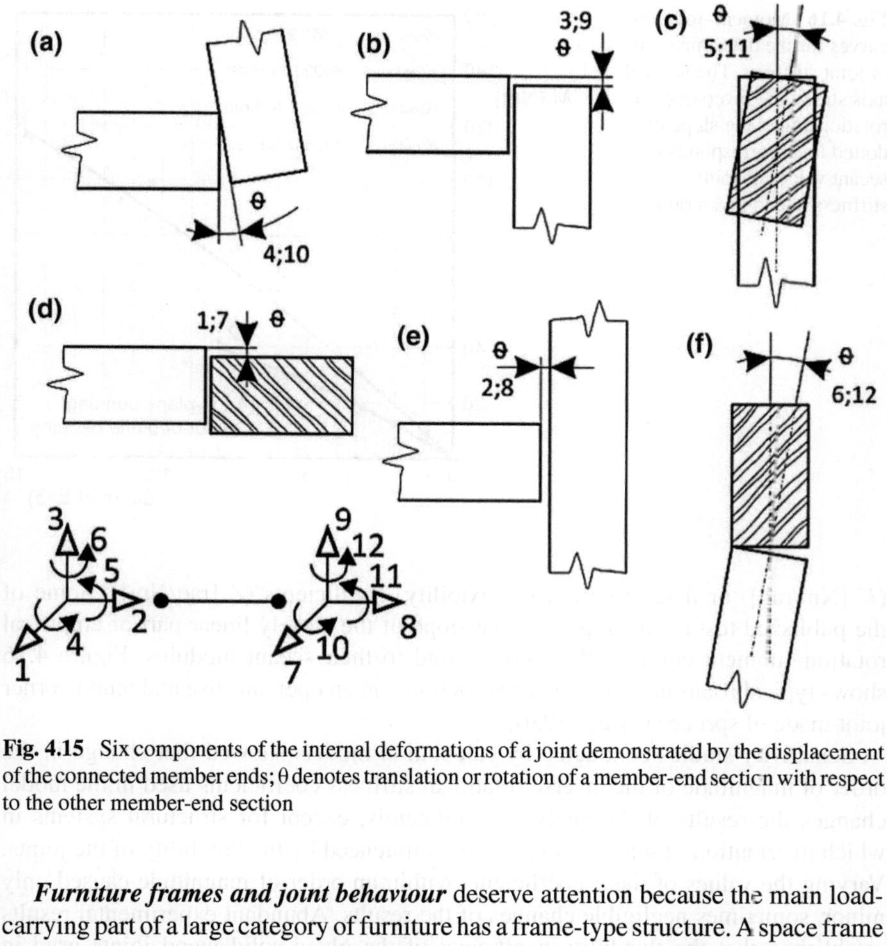

Fig. 4.15 Six components of the internal deformations of a joint demonstrated by the displacement of the connected member ends; θ denotes translation or rotation of a member-end section with respect to the other member-end section

Furniture frames and joint behaviour deserve attention because the main load-carrying part of a large category of furniture has a frame-type structure. A space frame is most often used, but in some cases, planar subunits can be defined and analysed. Joints between members intended to be fix are semi-rigid due to internal deformations within the joint domain. The overall deformation of a joint can be decomposed to components corresponding to the six independent external forces in a Cartesian coordinate system as illustrated in Fig. 4.15 with exaggerated displacements after Eckelman (1970b).

The semi-rigid behaviour of a joint can be characterized by force–translation as well as in-plane, out-of-plane and torsional moment–rotation curves. The displacements of the joints in the axial and lateral directions (*X*, *Y* and *Z* in the figure) and their effect on the distribution of internal forces in the structure are generally negligible.

Finite rigidity of furniture joints and joints of solid wood frames have been studied by a number of authors (Eckelman 1971, 2005; Erdil et al. 2005; Hill and Eckelman 1973; Hajdarević and Martinović 2014; Hüsing 1986; İmirzi et al. 2015; Kovács 1986; Márkus 1990; Palotai 1990; Vassiliou et al. 2016; Wambier and Wilczyński 2000; Wilczyński and Wambier 2003). The authors defined stiffness coefficients

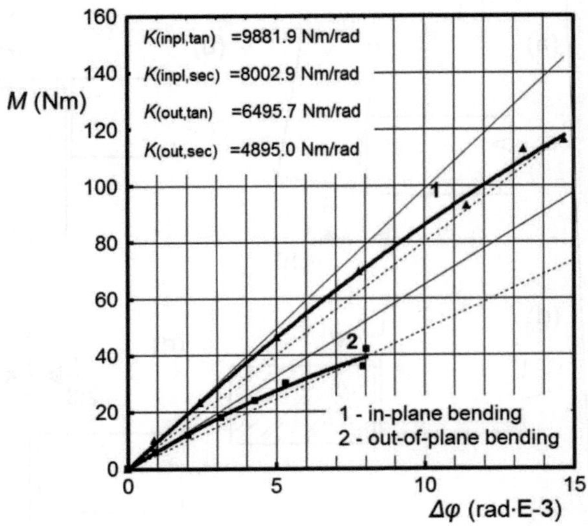

Fig. 4.16 Moment–rotation curves for the determination of joint stiffness. The vertical axis shows the inverse of the rotation angle; the slope of dotted lines corresponds to secant values of joint stiffness over a given range

(*K* [Nm/rad]) or inverse stiffness (flexibility) coefficients (*Z* [rad/Nm]). Some of the published test results represent the slope of the closely linear part of empirical rotation–moment curves; others correspond to their secant modulus. Figure 4.16 shows typical rotation–moment curves defined on an open mortise and tenon corner joint made of spruce (Kovács 1986).

Sensitivity studies conducted by calculation models showed that change in the order of magnitude of the inverse rotational stiffness coefficients used in the model changes the results of the analysis significantly, except for structural systems in which distribution of internal forces is not influenced by the flexibility of the joints. Varying the values of these coefficients within an order of magnitude caused only minor, sometimes negligible changes of the results. Abundant experimental results established that the flexibility coefficient of the glued solid wood joints used in furniture and joinery is of the order of magnitude of 10^{-5}–10^{-4} rad/Nm for in-plane bending and 10^{-4}–10^{-3} rad/Nm for out-of-plane bending.

Approximate values for flexibility of the two types of joints of different sizes have been tabulated as a guidance for structural analysis of furniture frames (Kovács 1989) as shown in Table 4.13. When applying these flexibility coefficients in the analysis of frames, it is important to understand that the experimentally established values mean the rotation of one of the member axes with respect to the other one at their intersection point under a unit moment. In the structural model however, *Z*-values are assigned to ends of the member rather than to joints; consequently, joint flexibility has to be treated as the displacement of both adjoining member ends with respect to the node, as implied by Fig. 4.17. The proportion of sharing the total deformation should depend on the bending stiffness of the respective members (Kovács 1984).

For frame members meeting at a right angle, out-of-plane bending of one member results in torque acting on the other member and vice versa. The corresponding

Table 4.13 Typical values of rotational flexibility applicable to furniture joints

Two-pin dowel joint			
Distance of axes (mm)	Diameter (mm)	In-plane Z_1 (rad/Nm)	Out-of-plane Z_2 (rad/Nm)
18	8	(1.4–1.5) e−4	5.0 e−4 to 1.0 e−3
21	8	(1.0–1.3) e−4	5.0 e−4 to 1.0 e−3
30	8	3.0 e−5	5.0 e−4 to 1.0 e−3
45	8	(1.5–3.0) e−5	5.0 e−4 to 1.0 e−3
Tenon and mortise joint			
Tenon width (mm)	Tenon length (mm)	In-plane Z_1 (rad/Nm)	Out-of-plane Z_2 (rad/Nm)
50	12	7.3 e−5	5.0 e−4
50	20	6.3 e−5	5.0 e−4
50	25	5.7 e−5	5.0 e−4
50	38	5.5 e−5	5.0 e−4
50	50	5.4 e−5	5.0 e−4
13	25	2.5 e−4	1.0 e−3
25	25	1.5 e−4	1.0 e−3
38	25	7.0 e−5	5.0 e−4
50	25	4.0 e−5	5.0 e−4
64	25	3.0 e−5	5.0 e−4
76	25	2.0 e−5	5.0 e−4
40	28	1.2 e−4	1.0 e−3
40	30	5.5 e−5	5.0 e−4
44	30	5.0 e−5	5.0 e−4
Open tenon and mortise joint			
Type	Wood species	In-plane Z_1 (rad/Nm)	Out-of-plane Z_2 (rad/Nm)
Simple	Softwood	(5.0–7.0) e−5	5.0 e−4
Simple	Hardwood	(3.0–5.0) e−5	5.0 e−4
Double	Softwood	(4.0–6.0) e−5	5.0 e−4
Double	Hardwood	(2.0–5.0) e−5	5.0 e−4

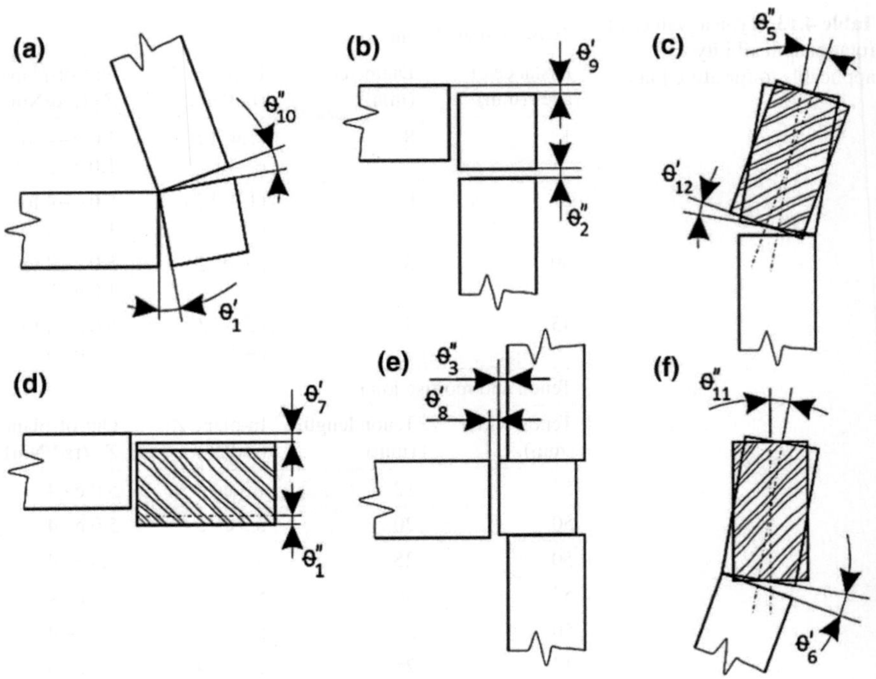

Fig. 4.17 Components of a joint's internal deformation treated as member and displacements with respect to the undeformed joint

internal distortion of the joint includes both out-of-plane bending and torsional deformation. Therefore, the total rotation one can measure is theoretically the sum of two components; the flexibility coefficient has to be shared and part of it assigned to the respective member end as torsional flexibility of the connection.

Unlike timber structures in buildings, furniture frames are often built up of members of irregular cross-sections. This in fact results in out-of-plane effects even in the case of plane frames with in-plane loads; when their magnitude cannot be neglected, a three-dimensional computational model is needed. Typically, chairs can be treated as frame-type structures. Classical methods of structural analysis, such as stiffness (displacement) method, are useful for calculating internal forces and deformations in furniture frames. The flexibility (action) method as an alternative to the stiffness method is less effective with multi-member frames because of the high degree of internal indeterminacy of the structure. For simpler frames, the flexibility method was used by several authors (Dziegelewski and Smardzewski 1990; Dziuba 1990). An early development of a computer-based stiffness method in matrix form for furniture frame analysis is CODOFF (Eckelman 1970a, b), treating space frames with members of rectangular cross-section oriented arbitrarily around their axis, connecting into nodes with final rigidity. Gustaffson used the stiffness matrix method for the analysis of planar side frames with experimental verification of the model (Gustaffson

1995, 1996). Recently, Hajdarević and Busuladžić used the stiffness method with a rigid and a semi-rigid joint model for a planar side frame of a chair in order to relate displacement results to those obtained by using FEM with isotropic and orthotropic properties. The results of the stiffness method with semi-rigid joints were identical with those from using FEM with orthotropic material model (Hajdarević and Busuladžić 2014).

Structural members in the *computational model for frames* are represented by their geometrical axis and cross-sectional data. The basis of the structural model is a two- or three-dimensional network of member axes, in which each node (axis line intersection) corresponds to a joint. The topology of axis lines and the definition of the orientation of member cross-sections with respect to a reference coordinate system complete the geometrical model. For stiffness properties of the members, modulus of elasticity (E) in bending along the grain direction and the corresponding modulus of rigidity (G) as well as Poisson's ratio (μ) should be used. Linearized rotational stiffness coefficients of the joints can be added to the model so that the computing algorithm modifies the stiffness matrix of the members connecting to a joint. Markedly nonlinear behaviour of the joints in the range of design load values may make analysis results invalid. In such cases, the true results may be approached by adding the total external load in increments with a corresponding update of joint stiffness coefficients. Loads can be external forces and moments acting either on nodes, or along member lengths; the latter will be converted to nodal actions. Boundary conditions can be defined for nodes. The analysis produces nodal displacements and rotations in the global coordinate system and member-end forces (axial force, in-plane and out-of-plane shear force) and moments (in-plane and out-of-plane bending moment and torsional moment) transferred into the members' coordinate systems.

It is important to note that the methods directly based on the theory of structures give an exact solution for the model—not necessarily for the object modelled though. Member cross-sections often change along their length. In addition, member axes may be curvilinear, sometimes 3D curves with changing cross-section. Setting up stiffness or flexibility matrices for such members is feasible in the planar case (Vanderbild 1974), but it often may remain only approximate. A simpler way of approximating the true structure is to divide a member into shorter ones having a straight axis and constant cross-section by inserting additional rigid nodes. Gradually changing cross-section and/or slope of axis can be assigned to the individual segments obtained in this way. The number of segments within a member should depend on the magnitude of error of the results that can be accepted for such a member. It has to be noted that errors of approximation in a frame member affect overall nodal displacements and member-end forces of other members to a lesser extent. This approach gives a reasonable approximation (Kovács 1989).

To keep an allowable ratio of the larger to the smaller moment of inertia belonging to the prevailing bending moment of a segment (I_{i+1}/I_i), the lengths of consecutive sections in a tapered member should be calculated starting from the narrow end by the relationship

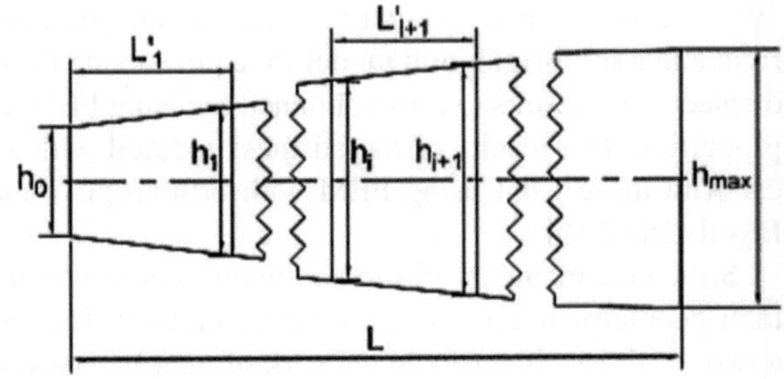

Fig. 4.18 Notation used in Eq. (4.25) for tapered frame members

$$L_i' = r^i L \cdot h_0 \cdot \left(1 - r^{-1}\right) / \left(h_{\max} - h_0\right) \tag{4.25}$$

and the segment end cross-sections grow according to the relationship

$$h_i = h_0 \cdot r^i$$

In the above equations, L is the total length of the tapered member, L_i' is the length of the ith segment, and h_0 and h_{max} denote the height of cross-section of the narrower and the wider member end, respectively; see Fig. 4.18. The ratio r of the larger to smaller cross-sectional height of a section is the cube root of the ratios of the respective moments of inertia, which can be freely chosen in order to arrive at a convenient number of segments within the tapered member. As an example, a tapered member 400 mm long, with $h_0 = 20$ mm and $h_{max} = 40$ mm, choosing the quotient of $I_{i+1}/I_i = 1.35$, the ratio of increase of section heights will be $r = 1.35^{1/3} = 1.1052$, resulting in seven segments along the length of the member. The smaller moment of inertia of a section is 1.17 times less, and the larger one 1.16 times more than that belonging to the mid-length cross-section. Choosing $I_{i+1}/I_i = 1.5$, division of the total length to five segments, these latter quotients will increase to 1.22 and 1.23, respectively.

To keep the angle between the axes of two adjacent segments under 10°, the straight segment axes cannot be longer than 0.175 times the radius of curvature R of the original curvilinear axis line; see Fig. 4.19.

Side frames of chairs sometimes are slanted, and/or the two sides are not parallel. When the angle of inclination is small, however, the geometrical model can

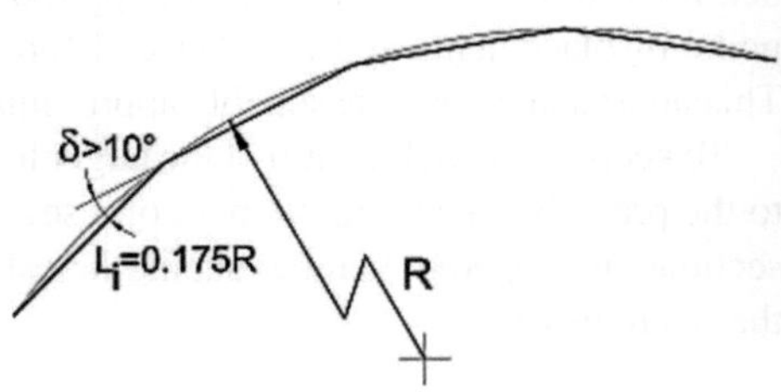

Fig. 4.19 Definition of allowable length of straight segments replacing an arc

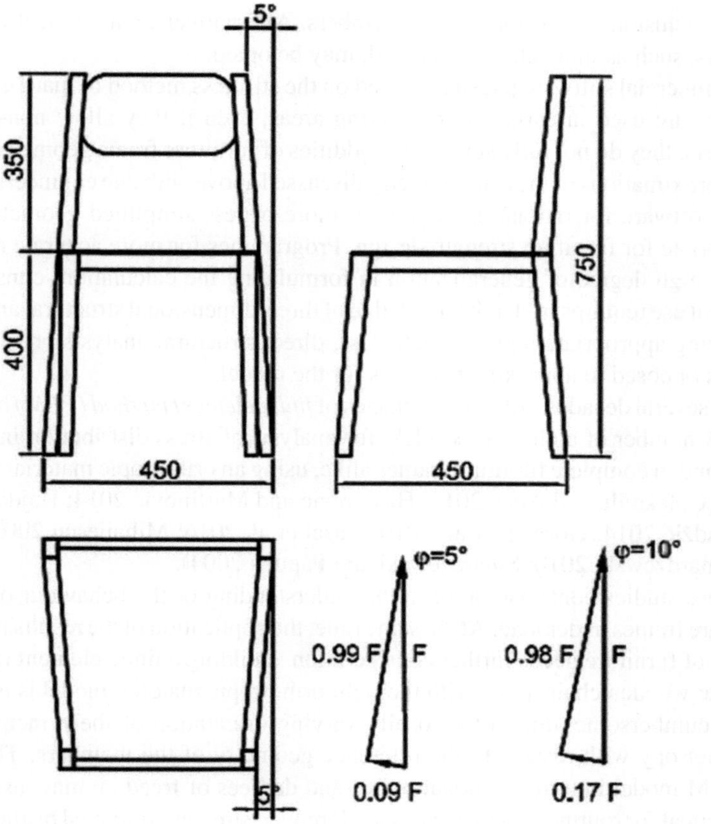

Fig. 4.20 Effect of tilt angle of chair frame parts on the decomposition of external loads acting parallel with the plane of symmetry

be simplified a lot when loads are symmetrical. As demonstrated in Fig. 4.20, such simplifications up to an angle of inclination of 5° neglect sideway load components that are not larger than 7% of the total load, and overestimate in-plane loads by 1%. For small angles, the magnitude of these errors is closely proportional to the angle; see the figure. Similar consideration may be useful concerning the angle of the principal axis of a symmetrical cross-section formed with the global coordinate planes, or with the face of the member for cross-sections without axis of symmetry.

While non-structural parts of frame-type furniture (i.e. seating and backrest of a chair) are attached to the load-carrying frame and may influence the way the frame parts function together, the frame model is only capable of simulating their effect through prescribing the modality of external load application on the structure. By neglecting their load-sharing effect, the analysis remains on the safe side. When the contribution of these parts cannot be overlooked, their effect can be approximated by the inclusion of fictitious frame members. Alternatively, a numerical method of analysis, such as finite element method, may be opted.

Commercial software packages based on the stiffness method of matrix structural analysis are used in classical engineering areas; even if they allow non-prismatic members, they do not fully support the oddities of furniture frame geometries. Using the approximations of the true geometry discussed above with due engineering judgement, software for rational structures of more or less simplified geometry can be appropriate for furniture strength design. Programmes for more accurate modelling need a high degree of generalization in formulating the calculation; consequently, practical use requires in-depth knowledge of three-dimensional structural analysis for producing appropriate input. Nonetheless, direct structural analysis provides accurate, as opposed to approximate results for the model.

For several decades, software packages of *finite element method (FEM)* have taken over. A number of authors used FEM for analysis of stress distribution in furniture joints and in complete furniture frames alike, using an orthotropic material model for wood (Çolakoğlu and Apay 2012; Hajdarević and Martinović 2014; Hajdarević and Busuladžić 2014; Horman et al. 2010; Kasal et al. 2016; Mihailescu 2003; Prekrat and Smardzewski 2010; Smardzewski and Papuga 2004).

These studies contribute a lot to the understanding of the behaviour of wooden furniture frames under load. At the same time, the application of the results in strength design of furniture needs further consideration. Building a finite element model of a realistic wooden chair frame with the right orthotropic material model is most often rather cumbersome, due to the locally varying orientation of the principal planes of orthotropy with respect to the reference geometry of the members. The size of the FEM model in terms of nodal points and degrees of freedom may also become impractical for routine design purposes. Moreover, stresses computed by the software are not readily comparable with design strength values of the species of wood for combined states of stress typical to joints.

As is verified in a study (Hajdarević and Busuladžić 2014), a reasonable approach of structural modelling for design may be either the use of frame models and direct stiffness method of matrix structural analysis with semi-rigid joint properties, or *the use of beam elements in the FEM model*. In this latter case, semi-rigid behaviour of the joints can be modelled in an indirect way. Joints in a beam element model have a non-real dimension. They may be extended along the axes of the joining members by beam sections of appropriate stiffness and length, so that the required internal joint deformation results. The concept is demonstrated in an example (see Fig. 4.21), in which two frame members (member 1–4 and member 1–5, 30 mm by 50 mm cross-section, made of spruce, MOE = 10,000 N/mm^2) join at one end with Z_1 = 1.0 E−4 in-plane rotational flexibility shared evenly between the two members with bending stiffness of 10,000 × 50^3 × 30/12 = 3.13E+9 Nmm^2. In the frame model, the flexible joint of finite dimensions is represented by splitting these members using intermediate members 1–2 of length L_1 and 1–3 of length L_2 joining rigidly at Node 1. Treating both of them as cantilever beams with fix support at Node 1, their bending stiffness has to be calculated so that a unit (1 Nm) in-plane moment applied at the free ends (node 2 and node 3), the resulting cross-sectional rotation is equal to the

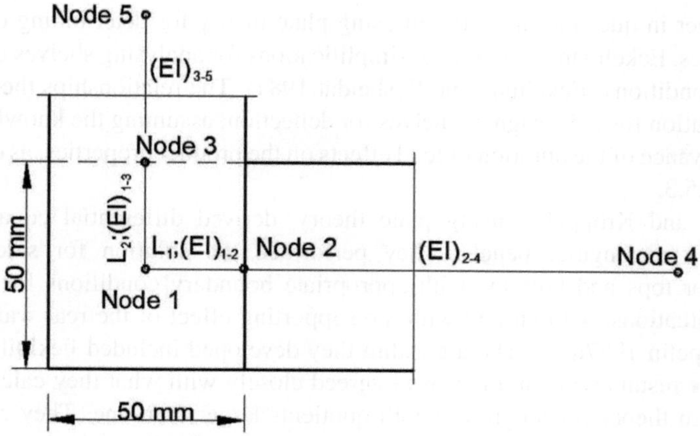

Fig. 4.21 Representation of a flexible joint in frame models

sum of the rotation of the original member at length L_1 and L_2, respectively, and an additional rotation of $1 \times 0.5 \times Z_1 = 5.0\text{ e}{-5}$ rad. For the intermediate member 1–2, this would result in a bending stiffness of

$$(\mathrm{EI})_{1-2} = \frac{L_1 \cdot (\mathrm{EI})_{2-4}}{L_1 + 0.5 \cdot Z_1(\mathrm{EI})_{2-4}} \tag{4.26}$$

which, with $L_1 = 25$ mm, results in $(\mathrm{EI})_{1-2} = 7.58\mathrm{E}{+}8\ \mathrm{Nmm}^2$ that corresponds to a height of cross-section of 31 mm instead of 50 mm.

Adopting this approach of structural modelling, the computed internal forces and moments at member ends are directly compared to the design resistance values of joints. If the state of stress within a joint domain is needed, a finite element model of the joint itself with known forces imposed on the sections joining the frame members can be built and analysed.

Case furniture is commonly used nowadays for storage. This type of furniture is built up of panels manufactured from wood-based sheet materials like particleboard, medium density fibreboard (MDF), plywood and edge-glued solid wood. The individual panels are connected one to the other along an edge of one member or along one of the edges on both members generally in a more or less flexible way, forming the overall load-carrying structure. The box-form structure may or may not have a rear wall. Simply supported horizontal members (shelves) if any contribute to the distribution of internal forces in certain loading situations. Several authors studied the stiffness of joints with different connectors in case furniture (Füchter 1984; Hüsing 1986).

The following problems are relevant in calculating resistance to external loads: damage to connectors, deflection of horizontal elements, buckling of sidewalls, shear buckling of the rear wall and stability against overturning, which will not be dealt with here. Evaluation of *load-carrying capacity of shelves is* possible by modelling the member in question in itself and using plate theory for determining deflection and stresses. Eckelman used rational simplifications for analysing shelves of various support conditions (Eckelman and Resheidat 1984). The relationships they derived give a solution for the design of shelves for deflection, assuming the knowledge and due observance of the duration of load effects on the product properties, as discussed in Sect. 4.5.3.

Kühne and Kröppelin, using plate theory, derived differential equations for orthotropic rectangular panels. They performed the solution for sidewalls as well as for tops and bottoms with appropriate boundary conditions for different loading situations, without and with the supporting effect of the rear wall (Kühne and Kröppelin 1987a, b). The algorithm they developed included flexibility of the joints. The results without a rear wall agreed closely with what they calculated by using beam theory for length-to-width quotients larger than one. They concluded that the proposed computation method could be useful for dimensioning with exact knowledge of the orthotropic product properties and their time dependency.

Satisfactorily accurate assessment of stability of sidewalls and rear walls is also possible by modelling such members in themselves, with approximate boundary conditions and loads determined from the whole structure (Curtu and Ghelmeziu 1984; Smardzewski and Dzięgelevski 1993).

Structural models for determining forces acting on connectors have been developed by a few authors (Ganowicz and Kiatkowski 1978; Ganowicz et al. 1978; Eckelman and Rabiej 1985; Dzięgelevski and Smardzewski 1995; Smardzewski 2015). These models build on the torsional rigidity of members and subunits of the body of case furniture and on the compatibility of the displacements due to torsion (twist) for the whole body, deeming edges of the individual panels to be connected by rotation-free (pinned) joints in their corner. Eckelman suggests corrections for semi-rigid edge joints (Lin and Eckelman 1987; Eckelman and Munz 1987; Eckelman and Resheidat 1985) when calculating torsional deformation of the case. Smardziewski calculated components of internal forces acting on connectors on the basis of the *torsional deformation model of case furniture*, using reasonable simplifying assumptions. He demonstrated his calculation method in full details in textbooks referred to above that suggests a complete solution for strength dimensioning of joints in case furniture. This solution extends to the combined load case of side thrust force and full loads on storing surfaces and rails, superimposing the load effects of the horizontal and vertical loads.

An alternative modelling tool for case furniture is *finite element analysis (FEA)*. Examples for finite element analysis of panel corner joints can be found in the literature. No precedents exist for a whole of a piece of furniture. We performed tentative FEM modelling of a bookcase with a rear wall for demonstration purposes, using software package Sofistik (Kovács 1995). Figure 4.22a demonstrates the deflection of horizontal members under the same load. The upper shelf is simply supported, the middle one is connected to the walls by means of dowels modelled as semi-rigid joints, and the lowest one is rigidly connected along the shorter edges. Figure 4.22b shows the same bookcase under a horizontal load acting at the front upper corner on

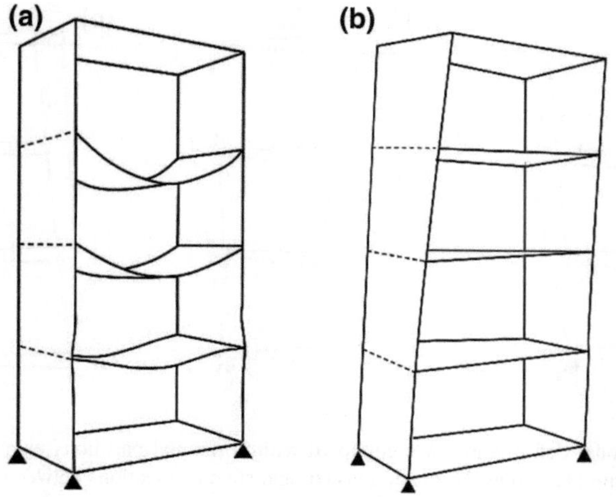

Fig. 4.22 Finite element simulation of the deformation of a piece of case furniture containing simply supported (upper), semi-rigidly (middle) and rigidly connected (lower) horizontal members, with vertical loads on shelves (**a**), and sideway thrust on the top corner (**b**)

the left side, causing torsional deformation of the case. The enlarged deformations in these figures illustrate the effect of the rigidity of joints.

While Smardzewski conducted an in-depth theoretical and experimental study on the strength of dowel joints and on the stiffness and strength of the commonly used demountable connectors, it seems that in evaluating their performance in case furniture, he did not involve bending moments acting on semi-rigid corner joints. Their magnitude depends on the finite rotational stiffness of the connectors, which, when neglected, results in overestimating case distortion. This is shown in examples in Sect. 4.5.7. Bending moments are of important magnitude in the connectors of the vertically loaded fixed horizontal members.

On the basis of the sample analyses using FEM, we will conclude a few additions to the suggested calculation method in Smardzewski (2015) based on superposition.

The results of finite element analysis illustrate how the distortion of a case is influenced by the stiffness of panel joints; see Fig. 4.23. For semi-rigid connections, typical stiffness of dowel joints along the edges is used in the calculation model. Figure 4.23 shows the deformed shape of a case of 868 mm by 1450 mm by 355 mm under the effect of a side thrust force and shelf load. The case is made of 15-mm-thick particleboard elements with a 5-mm-thick hardboard rear wall. Deformations for the three configurations of pinned, flexible and rigid joints are shown applying the same scale.

As can be expected, the pure twist model is not applicable for cases, where the five peripheral elements (top, bottom, the two sides and the rear wall) do not form a continuous surface. This may happen when extension of the sidewalls forms the foot part, and the bottom board is in an elevated position, irrespective of whether the back wall continues all along the sides or just down to the bottom board. In

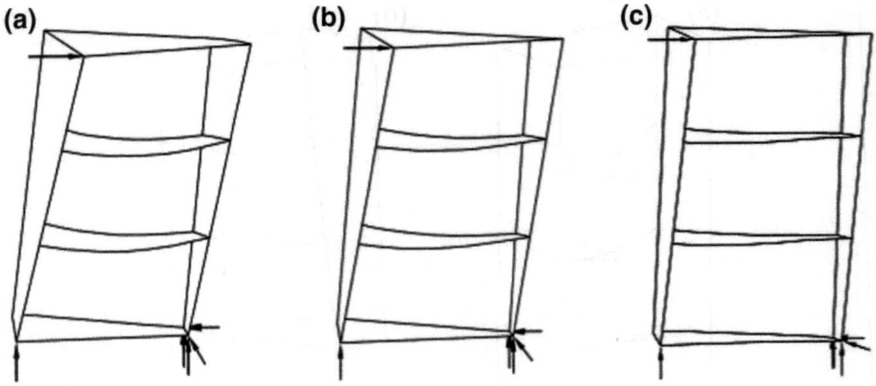

Fig. 4.23 Torsional deformation of a cupboard with horizontal partitions using pinned connections (**a**), semi-rigid (e.g. dowel) connections (**b**) and rigid connections between the sides and the horizontal elements (**c**); deformation scale: 5

Fig. 4.24 Deformation of a cupboard with pinned connections with continuous arrangement of the peripheral parts (**a**); with elevated bottom (**b**)

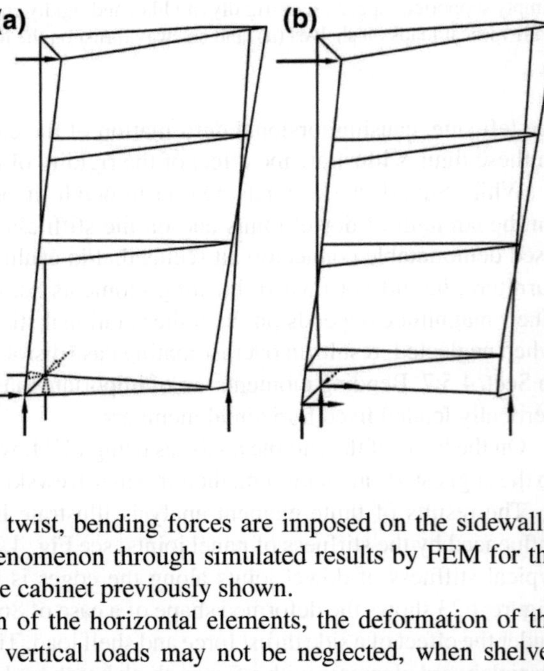

such situations, in addition to twist, bending forces are imposed on the sidewalls. Figure 4.24 illustrates this phenomenon through simulated results by FEM for the pin-jointed configuration of the cabinet previously shown.

When measuring deflection of the horizontal elements, the deformation of the sidewalls under the effect of vertical loads may not be neglected, when shelves are not simply supported. The more rigid the connection is, the more the sidewalls influence the deflection, as illustrated by Fig. 4.25.

Some types of furniture have a plane or space frame as supporting part and a body of surface or shell structure. This is quite common in tables and chairs and sometimes occurs with beds and cabinet furniture as well. From the structural analysis point of view, we may call them ***mixed structures***.

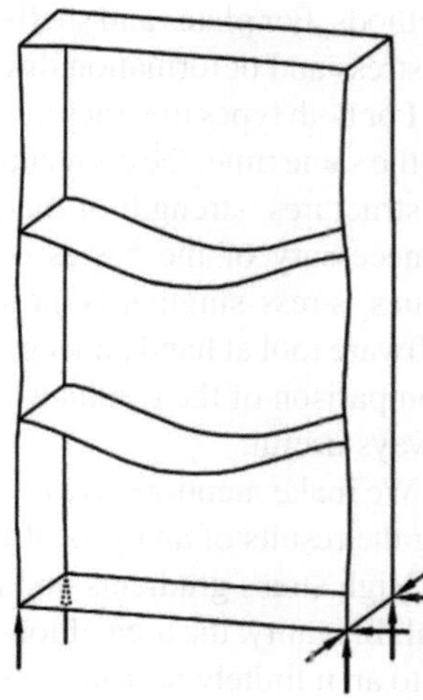

Fig. 4.25 Effect of the rigidity of the connection between sides and horizontal elements on the deformation of a cupboard with load on its shelves

When the deformations of the surface or shell part do not have an important influence on the load effects in the frame part or vice versa, the two parts can be modelled independently, or just the frame part needs to be analysed. Good engineering judgement is needed to decide when this is the case. A realistic assessment of the boundary conditions that the two parts provide to each other is also a question of engineering judgement.

Otherwise, it is necessary to account for the way the two parts of different structural features work together. The approximation of the structural influence of the non-frame part by a set of frame members is one possibility that may work in a few cases but not generally. Frame structure for the replacement of plate or shell structure can only be made equivalent either for torsional or bending effects but not for both at the same time. FEM can offer a generally valid modelling opportunity, using uniformly solid meshing for the whole of the structure or mixed meshing, including beam and solid, beam and shell, or shell and solid meshing. The choice may depend on the types of connections which are possible between the two differently meshed parts in the FEM software at hand.

The concept will be demonstrated in an example in Sect. 4.5.7.

4.5.5 Interpretation of the Results of Solving the Mathematical Model

Analytical solutions of the mathematical model of frame structures yield internal forces in the form of resultants of stresses for member cross-sections. Calculation of stresses can be performed using section properties. For the domain of a joint however, derivation of stresses is not as straightforward and may require the use of numerical methods. For plates and shells, analytical solutions generally aim at defining the state of stress and deformation directly that can be compared with failure criteria.

For both types of structures, the use of numerical tools such as FEM is widespread. At the same time, the correct interpretation of the results of analysis needs knowledge of structures, strength of materials and theory of elasticity. In addition, knowledge is necessary of the basics of approximation by numerical methods, convergence issues, stress singularity problems, modelling capabilities and deficiencies of the software tool at hand and a good understanding of the effect of model modifications. Comparison of the results with known or attainable solutions of similar problems is always useful.

We make mention here of some of the common issues to consider when interpreting the results of analysis of the strength of furniture using FEM.

High stress gradients and peaks at points of load application or restraint are not real. In reality, the area of load application is not confined to an infinitely small point or to an infinitely narrow edge, and supports of a piece of furniture emerge on flat or curved surfaces or on rounded edges, contacting a flexible rather than a rigid floor.

Using standard connectors such as pins and bolts offered by the software for defining the conditions of mutual translation or rotation of assembly parts instead of modelling the real connecting piece (e.g. dowels) causes the reported stress in the one-diameter vicinity of the connector to be higher than the actual stress.

Re-entrant corners such as edges of drilled holes and other cut-out features may create space for stress singularity; i.e., the theoretical solution for stress is infinite value. Edges in reality always have a finite radius of rounding or are purposefully provided with a fillet/chamfer, which is often neglected in the FEM model.

Depending on the geometry of the parts, stresses calculated in the global coordinate system cannot be evaluated. The use of a local coordinate system of members or parts thereof may become necessary, and occasionally polar coordinates may be more useful than the Cartesian system.

4.5.6 Comparison of Model Computational Results and the Limiting Values

The question whether a solution of a model in terms of stresses or internal forces is needed for comparison depends on our knowledge of the limiting values of the state of stress or the forces/moments of a part or connection.

The mechanical behaviour of pieces of wood and wood-based materials follows the orthotropic material model. In a uniaxial state of stress, design values of strength in the respective direction can be taken for comparison; the modification of design values in the case of deviation from the longitudinal (grain) direction is given in EN 1995.

Combined states of stress, when the magnitude of stresses in at least two of the three mutually perpendicular directions is not negligible, call for a proper failure criterion in order to derive values for comparison. The next section addresses this issue.

When different possible failure modes exist for a member or joint, and the analysis yields internal forces rather than a distribution of stresses, it is necessary to check member or joint strength by failure modes. Sometimes confirming adequacy for all failure modes independently does not guarantee satisfactory strength. Section 4.5.7 discusses these situations.

Failure criteria for wood-based materials present a delicate question. A few authors studied failure criteria applicable for wood. Among them, Szalai applied Ashkenazi's orthotropic strength criterion to wood and wood-based materials and formulated the same using tensor theory (Szalai 2001):

$$\frac{\sqrt{\frac{1}{2}\left[\left(\sigma^{ij}\delta_{ij}\right)^2+\sigma^{ij}\sigma_{ij}\right]}}{t_{ijkl}\sigma^{ij}\sigma^{kl}} \geq n \quad i,j,k,l = L, R, T, (1,2,3) \tag{4.27}$$

where t_{ijkl}—elements of the four-dimensional strength tensor of the orthotropic material, computed from the technical strength values determined experimentally,
σ_{ij}, σ^{ij}, σ^{kl}—components of stress state at the given point,
δ^{ij}—Kronecker delta; and n = factor of safety.

The technical strength values of solid wood needed for computing the elements of the strength tensor are as follows, adhering to the notation used by Szalai:

$$\mathrm{f}_{\mathrm{L}}^{+}, \mathrm{f}_{\mathrm{R}}^{+}, \mathrm{f}_{\mathrm{T}}^{+}, \mathrm{f}_{\mathrm{LR}}^{\mathrm{T(45)+}}, \mathrm{f}_{\mathrm{LT}}^{\mathrm{R(45)+}}, \mathrm{f}_{\mathrm{RT}}^{\mathrm{L(45)+}}, \mathrm{f}_{\mathrm{L}}^{-}, \mathrm{f}_{\mathrm{R}}^{-}, \mathrm{f}_{\mathrm{T}}^{-}, \mathrm{f}_{\mathrm{LR}}^{\mathrm{T(45)-}}, \mathrm{f}_{\mathrm{LT}}^{\mathrm{R(45)-}},$$
$$\mathrm{f}_{\mathrm{RT}}^{\mathrm{L(45)-}}, \mathrm{t}_{\mathrm{LR}}, \mathrm{t}_{\mathrm{LT}}, \mathrm{t}_{\mathrm{RT}}, \mathrm{t}_{\mathrm{LR}}^{\mathrm{T(45)+}}, \mathrm{t}_{\mathrm{LT}}^{\mathrm{R(45)+}}, \mathrm{t}_{\mathrm{RT}}^{\mathrm{L(45)+}}, \mathrm{t}_{\mathrm{LR}}^{\mathrm{T(45)-}}, \mathrm{t}_{\mathrm{LT}}^{\mathrm{R(45)-}}, \mathrm{t}_{\mathrm{RT}}^{\mathrm{L(45)-}}.$$

The letter f denotes normal strength in the anatomical direction shown in the subscript; t denotes shear strength with the direction of the normal of the plane of shear and the direction of stress in that plane. The upper case letter with the value in brackets in the superscript marks the direction of the force at an angle of 45° to one of the anatomical directions as follows. In normal stress, the force acts in a plane indicated by the letters in the subscript, in the direction of the bisector. With shear stress, the same is true for the normal of the plane of shear, while the force acts in a plane containing the directions in the subscript. The positive and negative signs in all cases denote tension and compression, respectively.

Technical strength values were determined by tests conducted on large samples (several hundreds of test pieces per type of test) for some commonly used species of wood and wood-based materials and were published along with similar data available in the literature (Szalai 2001; Szalai 2004).

In order to test the applicability of the strength criterion for wood products, we analysed test results obtained in a recent project (Elek 2018; Horvath 2018; Kungl 2017). Defect-free samples of European ash (Fraxinus excelsior) were used to make test pieces of corner joints. MOE and bending strength (MOR) of the material condi-

Table 4.14 Orthotropic elastic properties used in the material model for finite element analysis of test pieces

Property	N/mm² (MPa)
Modulus of elasticity E_L	15630.0
Modulus of elasticity E_T	1250.4
Modulus of elasticity E_R	1953.8
Poisson's ratio μ_{LT}	0.440
Poisson's ratio μ_{TR}	0.360
Poisson's ratio in μ_{LR}	0.371
Shear modulus in G_{LT}	1703.7
Shear modulus in G_{TR}	171.9
Shear modulus in G_{LR}	1203.5

tioned at 12% moisture content were determined according to the relevant European standard. The mean values (standard deviations in parentheses) were as below:

$$\mathrm{MOE} = 15{,}630\,\mathrm{N/mm^2}\left(1087\,\mathrm{N/mm^2}\right)$$
$$\mathrm{MOR} = 107.5\,\mathrm{N/mm^2}\left(18.6\,\mathrm{N/mm^2}\right)$$

The values reported as an overall average for the same species of wood in The Wood Database are

$$\mathrm{MOE} = 12{,}310\,\mathrm{N/mm^2}$$
$$\mathrm{MOR} = 103.6\,\mathrm{N/mm^2}$$

Compression strength in the longitudinal direction: 51.0 N/mm².

Using the tables published in the Wood Handbook Chap. 4, we made an assessment of the orthotropic material properties needed for finite element analysis as shown in Table 4.14.

The dimensions of pieces of corner joints for numerical analysis were 20 mm by 31.3 mm by 250 mm, with tenon and open mortise, respectively, at one end; see Fig. 4.26a. The assembled joints were tested by compressive force until failure, as illustrated in Fig. 4.26b.

The average breaking force in the group of test pieces with a nominal 0.05-mm-thick glue line was 679.6 N with an average vertical displacement of 4.1 mm of the upper corner.

In the finite element model, the glue line was assigned an MOE value of 465 N/mm² and a Poisson's ratio of 0.29 after (Smardzewski 2015). The model with a vertical load of 679.6 N had a displacement of 4.55 mm of the upper corner, which is just 10% higher than the experimental value.

Since the FEM software reported unrealistic stress peaks in the close vicinity of re-entering corners even when a filet with radius of 1 mm was applied, we tested convergence of the stresses around such features in the geometry by gradually refining

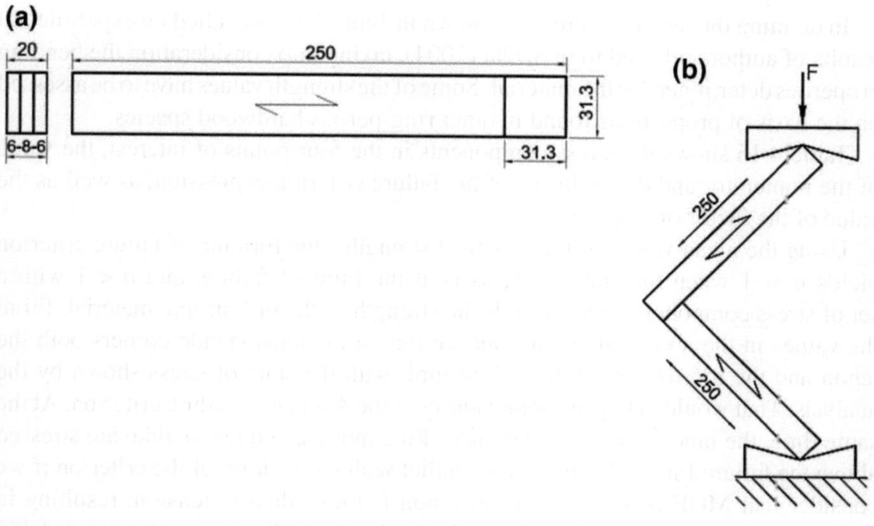

Fig. 4.26 Dimensions of parts of corner joints (a); test arrangement (b)

the mesh. By reducing the element size, a tendency of divergence of these peaks was observed, while at locations 1–2 mm apart stress values converged. Moreover, within 1 mm distance from the peak locations gradual decrease of stresses was observed with mesh refinement, indicating an increase of the stress gradient towards infinity. Therefore, we opted to read stress values at a distance of 1.0 mm from the edges. Two critical locations were identified on both the tenon and the mortise parts, one on the tension side and another on the compression side, as indicated in Fig. 4.27.

Fig. 4.27 Distribution of the longitudinal normal stresses SX in the tenon and the locations of stress inquiry

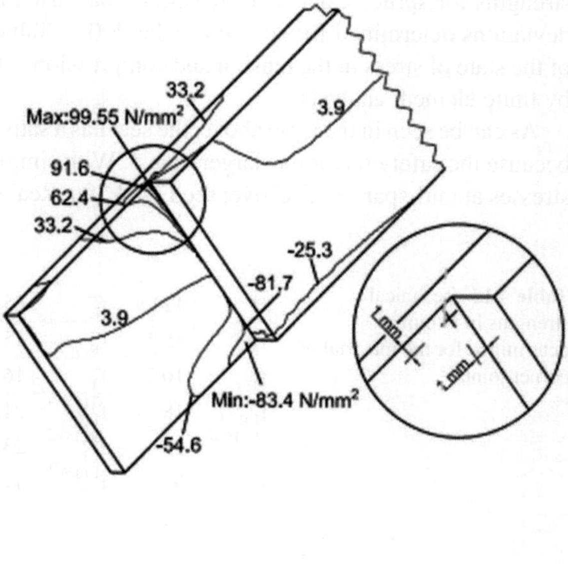

In defining the technical strengths shown in Table 4.15, we relied on experimental results of authors referred to in Szalai (2004), taking into consideration the bending properties determined for the material. Some of the strength values have to be assessed on the basis of proportions found in other ring-porous hardwood species.

Table 4.16 shows the stress components in the four points of interest, the value of the numerator and denominator of the failure criterion expression, as well as the value of the factor of safety, n.

Using the mean values of the technical strengths, the formula of failure criterion yields $n = 1$ when the state of stress is at the limit of failure, and $n < 1$ with a set of stress components that exceeds the strength of the orthotropic material. From the values in the above table, one can see that in the tension side corners both the tenon and the mortise are at the failure limit with the state of stress shown by the analysis. That would verify the applicability of the Ashkenazi failure criterion. At the same time, the inner, compressed corner of the mortise and tenon alike are stressed above the failure limit. This does not conflict with the validity of the criterion if we consider that MOE of wood in compression is lower than in tension, resulting in lower compressive stresses than those obtained with the linear orthotropic model.

It is possible to use the above failure criterion in strength design. In limit state design, the design values of technical strengths have to be applied in the failure criterion formula. Design values may have a varying proportion of the mean values depending on the standard deviation with which these means could be determined.

As an example, let us consider the solid wood seat of a chair, shown in Fig. 4.28. The seat is made of spruce and is supported by rails at both sides. The load case under consideration is the user's falling heavily onto the seat. Applying a dynamic factor of 1.2 results in an equivalent static load of 1300 N acting vertically. Assuming simple support along the sides the location of critical state of stress exhibits itself along the plane of symmetry close to the front edge of the seat where maximum bending stresses along the grain direction occur. Table 4.17 shows design values of technical strengths for spruce calculated by (Eq. 4.20a) using the mean values and standard deviations determined for spruce (Szalai 2004). Table 4.18 shows the components of the state of stress in the tension and compression side, respectively, as determined by finite element analysis.

As can be seen in the table above, the seat has a satisfactory load-carrying capacity because the safety factor n is larger than 1. With simple support conditions, bending stresses at mid-span are decisive; see Fig. 4.29a. Real support conditions can vary in

Table 4.15 Technical strengths in N/mm^2 determined for the material of corner joints

f_L^+	124.8	f_L^-	65.4	t_{LR}	13.2	$t_{LR}^{T(45)-}$	9.1
f_R^+	15.1	f_R^-	15	t_{LT}	15.5	$t_{LT}^{R(45)-}$	7.5
f_T^+	10.2	f_T^-	16.2	t_{RT}	4.9	$t_{RT}^{L(45)-}$	3.8
$f_{LR}^{T(45)+}$	18	$f_{LR}^{T(45)-}$	21.3	$t_{LR}^{T(45)+}$	8.9		
$f_{LT}^{R(45)+}$	13	$f_{LT}^{R(45)-}$	23.3	$t_{LT}^{R(45)+}$	7.1		
$f_{RT}^{L(45)+}$	9.1	$f_{RT}^{L(45)-}$	11	$t_{RT}^{L(45)+}$	3.7		

Table 4.16 Stress components and safety factors determined in the critical locations of the test piece by FEA; stresses are given in N/mm^2

Location	σ_L	σ_R	σ_T	τ_{RT}	τ_{LT}	τ_{LR}	Stress state	Stress limit	n
Tenon tension side	99.55	1.15	−1.88	3.54	−4.71	−3.78	108.60	99.44	0.92
Tenon compression side	−83.4	−17.84	−5.05	1.78	0.22	1.09	188.26	96.45	0.51
Mortise tension side	78.76	0.56	−2.78	2.4	4.05	0.11	71.65	77.83	1.09
Mortise compression side	−82.34	−1.83	−5.12	1.22	−1.34	−1.91	123.98	86.01	0.69

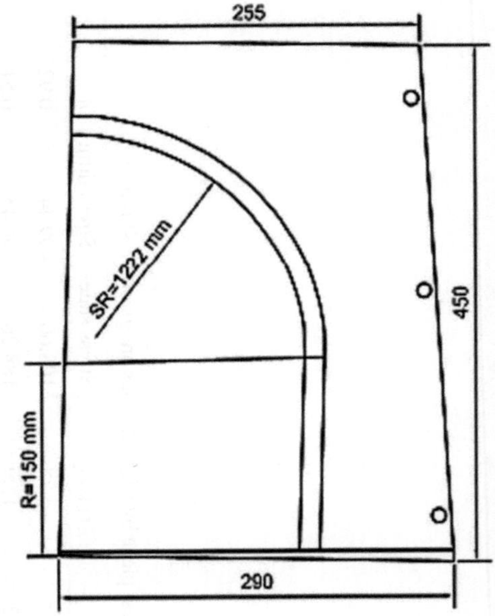

Fig. 4.28 Symmetric half of the seat analysed in the example

Table 4.17 Design values of technical strengths in N/mm^2, derived from the average values for Norway spruce (*Picea abies*)

f_L^+	26.56	f_L^-	25.88	t_{LR}	1.90	$t_{LR}^{T(45)-}$	1.34
f_R^+	1.97	f_R^-	1.54	t_{LT}	3.27	$t_{LT}^{R(45)-}$	2.39
f_T^+	1.03	f_T^-	3.37	t_{RT}	0.65	$t_{RT}^{L(45)-}$	0.47
$f_{LR}^{T(45)+}$	2.97	$f_{LR}^{T(45)-}$	3.47	$t_{LR}^{T(45)+}$	1.34		
$f_{LT}^{R(45)+}$	2.62	$f_{LT}^{R(45)-}$	7.05	$t_{LT}^{R(45)+}$	2.39		
$f_{RT}^{L(45)+}$	1.90	$f_{RT}^{L(45)-}$	1.45	$t_{RT}^{L(45)+}$	0.47		

Table 4.18 Stress components and safety factors determined in the critical locations of the seat by FEA; stresses are given in N/mm^2

Location	σ_L	σ_R	σ_T	τ_{RT}	τ_{LT}	τ_{LR}	Stress state	Stress limit	n
Seat tension side	22.08	0.03	0.00	0.00	0.00	0.00	18.56	22.10	1.19
Seat tension side	−25.36	0.00	−0.02	0.00	0.04	−0.01	24.81	25.36	1.02

practice and are seldom close to an ideally simple support. Often the seat bottom is glued and/or fastened by screws to the rail top. These solutions result in a reduction of the maximum bending stresses at mid-span and a corresponding increase of the safety factors at these locations (1.46–2.11 in our example); see Fig. 4.29b. At the

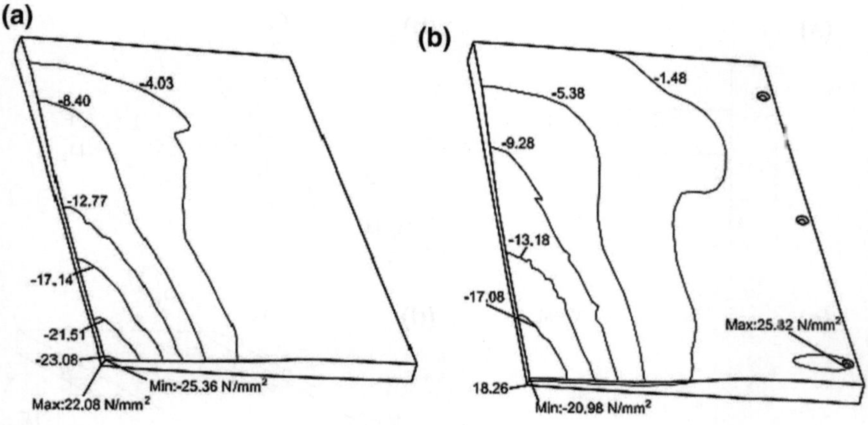

Fig. 4.29 Distribution of normal stresses acting in the direction of the front edge; simply supported seat (**a**); seat screwed on the side rail (**b**). Magnitude and location of maximum values change

same time, important stresses perpendicular to the bottom plane of seat (in tangential direction) develop in the interface of rail and seat bottom, which exceed the strength of wood and that of the glue line in this direction. When glued, tensile stresses across the glue line are critical. Fastened by screws only, the seat undergoes serious compression along the contact line with the rail. It indicates that gluing cannot be relied on; when applied with screwing, screws will hold alone, with the risk of local compression failure along the inside edge of the rail that in reality will not lead to explicit failure of the parts, on account of plastic deformation and stress relaxation.

Assessment of load-carrying capacity of joints it is of utmost importance. They are the most vulnerable structural parts in a piece of furniture; therefore, we have to be able to predict their load-bearing capacity with sufficient reliability. For the practice, the question is what type and of what geometry the joints in a piece of furniture should be in order to resist forces imposed on it in the course of its service life. The type of material (species of wood) to be used may also be a question. Since the geometry (detail dimensions) of a joint is generally related to the overall joint dimensions, i.e. thickness and depth of the joining members, the question may also be what size of a member should be used with a specific species of wood.

Loads acting on pieces of furniture cause internal forces in joints. In a general case, six (in-plane structures with in-plane loads just three) components of internal forces can be distinguished as illustrated in Fig. 4.30a–d.

In use conditions, several of these six components have nonzero value. However, assessment of load-carrying capacity of joints is usually made with respect to one of those components at a time.

One can make use of one of the four options for assessing or checking these resistance values for strength design:

1. study of values obtained by experiments (similarity search),
2. use of simplified theoretical model for stress calculation,

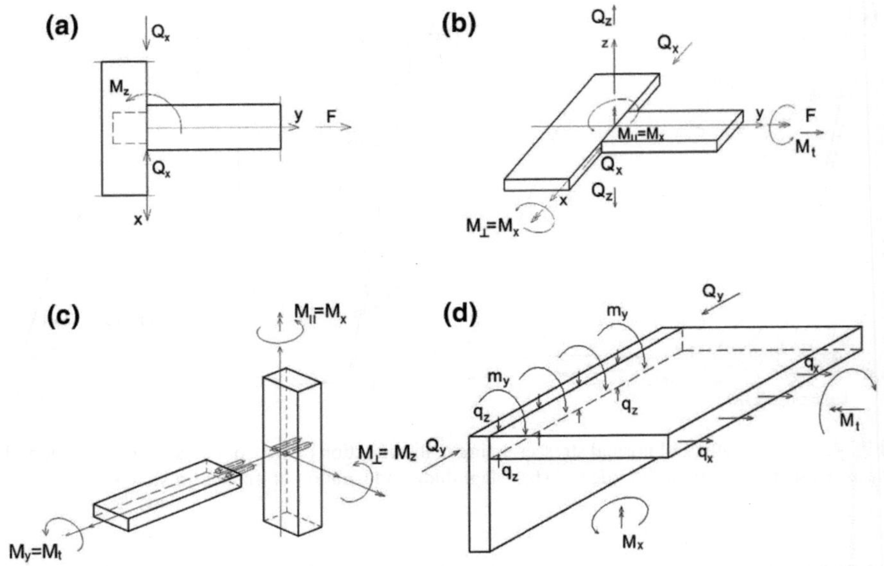

Fig. 4.30 Internal force components of furniture joints in coordinate systems connected to the joining members; plane frame with in-plane loads (**a**); plane frame with spatial loads (**b**); column and rail joint arranged to produce out-of-plane (flatwise) bending of the rail (**c**), panel joint (**d**)

3. use of predicting equations based on measured values by regression,
4. computation of combined state of stress by numerical methods (e.g. FEM).

The first approach has produced sufficient data so far, and it continues to be used. At the same time, these experiments are restricted to simple loading cases aiming at one single type of internal forces. Often the test arrangement leads to the appearance of an "accompanying" internal force component, the magnitude of which is somewhat arbitrary because the lack of consistency in testing conditions of joints of dissimilar sizes.

Examples of computing stresses based on simplified models can be found in the literature in relation to relatively simple, in-plane loading conditions (Dziengielewski and Smardzewski 1995). A thorough analytical process based on theory of elasticity is impossible to carry out due to the complexity of the geometry. Taking into account the effect of time-dependent behaviour is a further difficulty.

Among others, Eckelman and Kamenicky have derived regression-based predicting equations (Eckelman 1971, 1978, 2003; Kamenicky 1975; Kamenicky and Paulenkova 1984). This approach may be an opportunity for an extension of the first approach. For a proper use of such predictions in strength design, it is expedient to identify the influencing factors and to classify them into two groups shown below.

1. Predictors (design variables):
   - Type of joint
   - Dimensions
   - Material properties (for wood, defined by species of wood, panel type)
     - shear strength,
     - bending strength,
     - normal strength perpendicular to the grain direction.
   - Characteristics of manufacturing
     - tightness of fit,
     - surface geometry (grooved etc.),
     - glue type.
2. Workmanship quality (noise factors)
   - precision of part manufacture,
   - surface quality (roughness, wettability),
   - quantity of glue,
   - wood moisture content,
   - precision of assembly.

Clearly, the first three options support the assessment of resistance values one by one for the individual internal force components of interest, and design value of resistance (limit value) can be defined separately for each. The question remains of how to treat the simultaneous presence of the different components in order to compare the design resistance and design load action. One reasonable way is to add the squared ratios of actual to design values together, whereby the square root of the sum should not exceed 1.00, according to the expression

$$\sqrt{\left(\frac{A_d}{A_L}\right)^2+\left(\frac{Q_{i,d}}{Q_{i,L}}\right)^2+\left(\frac{Q_{o,d}}{Q_{o,L}}\right)^2+\left(\frac{M_{i,d}}{M_{i,L}}\right)^2+\left(\frac{M_{o,d}}{M_{o,L}}\right)^2+\left(\frac{T_d}{T_L}\right)^2}\leq 1 \quad (4.28)$$

where A—axial force (N),
Q—shear force (N),
M—bending moment (Nm),
T—torsional moment (Nm).

and the subscript d refers to design value, L to limit value, i to in-plane, o to out-of-plane force or moment.

In using the above relationship, it is not always necessary to use all six terms together depending on the load case and possible failure modes. Those producing their stress maximum apart from the critical locations are irrelevant. However, when some or all the terms depend on the same component of the state of stress in the same location, ratios rather than squared ratios are added together as shown in the example of Eq. (4.28a).

$$\sqrt{\left(\frac{A_d}{A_L}+\frac{M_{i,d}}{M_{i,L}}+\frac{M_{o,d}}{M_{o,L}}\right)^2+\left(\frac{Q_{i,d}}{Q_{i,L}}\right)^2+\left(\frac{Q_{o,d}}{Q_{o,L}}\right)^2+\left(\frac{T_d}{T_L}\right)^2}\leq 1 \qquad (4.28a)$$

The simultaneous presence of several internal force components does not raise these questions when numerical methods of stress calculation are used, providing that sufficient information is at hand about the orthotropic strength of the material. Comparing the design value of effects to the design (limit) value of resistance is done on the level of stresses.

Consideration of the influencing factors of the resistance of joints calls for a further approach to assessment of their load-carrying capacity. Applying the method of design of experiments (DOE), a mathematical model describing the results of experiments in terms of the setting levels of all the influencing factors is obtained. This model, called response surface, is capable of predicting joint resistance as a function of the design parameters within a defined confidence interval that can be computed from the experimental results. Noise factors contribute to the size of the confidence interval. Therefore, it is important to make them act at their characteristic level in the course of experimentation. Numerical simulation of the experiments with validated simulation models may reduce the need for physical experiments.

4.5.7 Examples of Furniture Strength Design

In this section, examples are shown for model building, solving and interpreting the results of analysis for strength design. The examples include two varieties of a cupboard, as well as a chair comprising a frame and load-sharing surface elements.

As a type of ***case furniture, a cupboard with its bottom at the extremity of the sidewalls*** is analysed. This piece of furniture along with the coordinate directions for forces is illustrated in Fig. 4.31. The sides and horizontal elements are made of 16-mm-thick particleboard with a modulus of elasticity MOE = 3500 N/mm^2, a modulus of rigidity $G = 800$ N/mm^2, and Poisson's ratio $\mu = 0.39$. The rear wall is made of hardboard; its MOE = 4000 N/mm^2, modulus of rigidity $G = 1500$ N/mm^2, and Poisson's ratio $\mu = 0.39$. One dowel per corner connects the horizontal elements to the sidewalls, and the rear wall is connected to the rest of the case by screws spaced at 288 mm intervals.

The cupboard standing on the floor or on four base supports under its corners is loaded on its shelves. The intensity of the shelf load is 1500 N/m^2 as stipulated in the European performance test standards; it causes a uniformly distributed load of 438.5 N on both shelves. The weight of the cupboard is added as a gravitation force of 281 N. The cupboard is also loaded by a side thrust force of $F_s = 349$ N at the upper left corner in the front plane. This horizontal force produces a rotating moment equal to that produced by the loads on the shelves and the weight of the cupboard itself around the axis through the right bottom corners, but rotating in the opposite direction.

In the static model for determining the global stiffness of the furniture body, three of the bottom corners are restrained as is shown in the figure. The right-side bottom corner in the front may rise up unhindered. Assuming pinned joints in all corners instead of the real connectors, we can calculate the displacement of the ccrner and the forces in the dowels according to Smardzewski (2015, pp. 481–485, 529–531).

The torsional rigidity of the case calculated according to Smardzewski is $k = 9.31$ N/mm. The magnitude of the internal force acting on the connecting pins perpendicular to the plane of the horizontal elements is calculated below:

$$F_y = \frac{G_p \cdot d_p^3}{3 \cdot a_x \cdot b_y} \cdot \left(\frac{a_x \cdot b_y}{b \cdot c}\right) \cdot \frac{F_s}{k} = \frac{800 \cdot 16^3}{3 \cdot 836 \cdot 350} \cdot \left(\frac{800 \cdot 16^3}{3 \cdot 836 \cdot 350}\right) \cdot \frac{349}{9.31}$$
$$= 79.6\,\mathrm{N} \tag{4.29}$$

where a, b and c denote the width, depth and height of the cupboard in mm; the subscripts x, y and z refer to the dimensions of the individual parts in the indicated coordinate directions, see Fig. 4.31; d is the panel thickness; the subscript p refers to particleboard.

This shear force shows upward in the left front and right back corner. There is also a force of the same magnitude, acting perpendicular to the plane of the joining sidewall. In those corners however, where there is an external load on the front upper and lower corner on the left sidewall, the actual horizontal load on the connector is the difference of the internal and the external force, which in our case is $F_b = 355 - 80.9 = 269.4$ N.

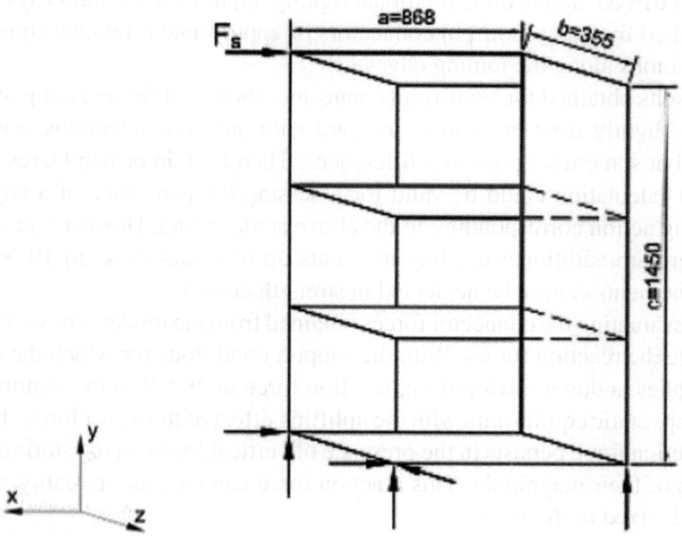

Fig. 4.31 Construction, overall dimensions, supports and load of the cupboard analysed in an example

The horizontal force acting on the connectors along the joining edges of the peripheral panels causes flatwise shear. Its value is

$$F_z = \frac{b}{c} \cdot F_s = \frac{355}{1450} \cdot 349 = 85.4\,\mathrm{N} \tag{4.30}$$

We analysed the cupboard by using finite element method, first defining pinned connections in the corners of each joining member. Connector forces with side thrust loading calculated by the FEM model agree satisfactorily with the results of the simple calculation model. The less than 4.0% differences can be attributed to minor deviations of the FEM model from the ideal model because of the available modelling techniques. Point-like pinned and semi-rigid joints could not be used in the model; furthermore, support and joint positions do not fall in the middle plane of the individual panels. The value of F_x in the corners with exterior force agrees well with that obtained by the simplified calculation, differing only 3.0%. The FEM model produces higher values for both F_x and F_y.

An exception is the edgewise horizontal force F_z. The connectors at the two ends of the horizontal parts share the force F_z unevenly, unlike in the theoretical twist model, where half of it acts on each corner. The higher absolute value on the left side appears in the front corners, while on the right-side rear corners are most loaded. The highest ratio of larger to smaller absolute value varies from 2.7 to 14, the two forces along an edge having different signs. Their sum over an edge is essentially constant, close to the value of $F_z = 85.4$ N calculated above.

The analysis was repeated by modelling rotational rigidity of $C = 465$ Nm/rad per dowel corresponding to a flexibility coefficient of $Z = 21.5\mathrm{e}{-7}$ rad/Nmm (Hüsing 1986). In the FEM model, finite rotational rigidity imparted to the joints by the dowels was modelled by using short pin connectors of appropriate rotational rigidity at the dowel locations along the joining edges.

The results obtained for semi-rigid connectors show that forces acting on connectors differ slightly from those in a cupboard with pinned connections, with mainly negative (but sometimes positive) differences. Therefore, in principle, results of the simplified calculation could be valid for assessing the resistance of a cupboard to the twisting action corresponding to the above static model. However, in semi-rigid joints, there are additional bending moments up to values close to 10 Nm. These bending moments cannot be neglected in strength design.

When evaluating the connector forces obtained from the models above, it is important to note the reaction forces. With the support conditions for which the case twist model applies, a downwards pulling reaction force of 593 N in the bottom left rear corner keeps static equilibrium with the uplifting effect of the thrust force. The downwards reaction force persists in the presence of vertical loads on the storing surfaces, regardless of their magnitude. This reaction force cannot exist in reality, unless the cupboard is fixed to the floor.

The textbook (Smardzewski 2015) calculates connector forces of loaded horizontal parts using superposition of the internal forces due to vertical loads under different support conditions (all four corners restrained in the vertical direction).

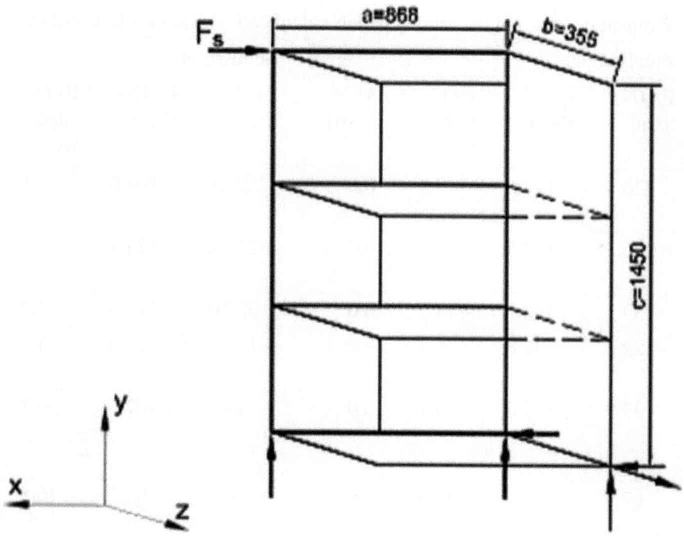

Fig. 4.32 Changed support conditions of the sample cupboard

Using unchanged support conditions for the combined load case of horizontal and vertical forces produces connector forces that are different from those calculated according to the above sourcebook.

Therefore, a static model with such support conditions can only be useful for comparing overall body stiffness of the individual pieces of case furniture. The model can be made more realistic with a different set of restraints applied on the bottom corners, when sideway translation (Ux) of the two corners on the opposite side of thrust load application is hindered and the vertical translation (uplifting) of the left rear corner is free, as can be seen in Fig. 4.32. With these support conditions, the vertical reaction force disappears since the magnitude of the side thrust force was calculated accordingly. Table 4.19 lists the internal forces acting on the connectors for both the rotation-free and semi-rigid configuration.

Note that the results of the analysis depend much on the support conditions used in the computation model. Displacement of the left top corner in the direction of the applied thrust load with the support conditions used in the twist model (Smardzewski 2015) is 14.9 mm with vertical loads (25.2 mm without vertical loads) for pinned connectors and 12.5 mm with vertical loads (20.7 mm without vertical loads) for semi-rigid connectors. The corresponding values with the changed support conditions are 69.0 mm with vertical loads (92.1 mm without vertical loads) for pinned connectors and 57.0 mm with vertical loads (72.7 mm without vertical loads) for semi-rigid connectors. Clearly, horizontal displacement values of a fully loaded cupboard are smaller than what have been calculated for pure twist. This is because vertical loads in the actual support conditions cause displacements along the *X*-axis in the opposite direction, showing the furniture body stiffer for torsional forces.

Table 4.19 Connector forces of the joining panels calculated for rotation-free and semi-rigid joints

Location	Freely rotate				Semi-rigid			
	Fx (N) axial	Fy (N) shear	Fz (N) edge-wise	*M* (Nm)	Fx (N) axial	Fy (N) shear	Fz (N) edge-wise	*M* (Nm)
Bottom left	120.5	112.6	9.3	0.0	121.1	109.6	11.1	−7.2
Bottom right	−226.3	−127.5	1.6	0.0	−217.8	−124.4	5.3	−7.6
Top left	239.2	112.7	−74.2	0.0	237.0	111.6	−75.3	−8.0
Top right	−122.1	−127.8	6.4	0.0	−132.8	−126.6	5.3	−8.3
Bottom left back	37.9	−127.6	34.1	0.0	19.8	−107.1	30.1	0.4
Bottom right back	67.9	112.6	−45.8	0.0	76.8	92.1	−46.5	−0.1
Top left back	−55.9	−127.6	25.6	0.0	−45.2	−106.9	31.4	0.4
Top right back	−61.4	112.8	42.1	0.0	−59.1	92.0	38.6	−0.5
Shelf left	−2.4	3.6	−2.4	0.0	18.6	5.0	6.9	−3.6
Shelf right	2.8	−237.8	3.8	0.0	2.9	−238.3	6.5	−14.1
Shelf back left	0.7	−237.8	2.2	0.0	−5.9	−218.3	−15.5	5.8
Shelf back right	−1.0	3.7	−3.7	0.0	−15.5	−16.4	2.1	−5.5
Reaction left back	0.0	−9.9	0.0		0.0	−9.9	0.0	
Reaction left front		597.0				597.0		
Reaction right back	349.0	559.0	0.0		349.0	559.0	0.0	

In Table 4.19, forces are given in the global coordinate system with the horizontal X-axis showing left, vertical Y-axis showing upwards and the Z-axis towards the back. F_x axial force acts along the axis of the dowels and its negative value on the left side and positive value on the right side shows a pulling out force. It can be seen that a slight negative downwards reaction force still occurs as a rounding error, but this influences the results by less than 1%.

Connector forces with a full shelf load were determined for the boundary conditions used in the sourcebook but for brevity are not shown here. Comparing the results obtained for the two different support conditions, absolute values of the vertical shear forces acting on the peripheral connectors are about 3–4 times as high as in the boundary conditions shown in Fig. 4.31. This increase of internal forces acting on the corners at a right angle to the plane of the panel is related to the reduction of the global stiffness of the furniture body with the changed boundary conditions. It follows that maximum axial force in the peripheral connectors significantly decreases, in the actual case by a factor of 1.3–1.45. Maximum vertical forces on the connectors of the shelves increase about 1.5 times. With semi-rigid connectors, internal forces are not significantly different from those obtained for pinned joints; additional bending moments in the connectors of the shelves amount to 14.1 Nm in those corners where the moments due to distortion and to bending combine.

The results of the above analyses show that for the purposes of strength design of case furniture, the simultaneous action of vertical loads (weight of stored goods and weight of furniture) and a side thrust force should be taken into account with assumed support conditions that can prevail in reality. These conditions should allow static equilibrium of the furniture as a rigid body without a downwards reaction force. The global stiffness of the furniture body will be different from what can be determined with the support conditions given in Smardzewski (2015).

We can now summarize our results by comparing them with those obtained by the simplified calculation method (Smardzewski 2015) in Table 4.20.

These numbers say that the calculation method of the internal forces of case furniture based on superposition of the results of the two loading cases, with the exception of withdrawal forces, may underestimate the load effects. Depending on the geometry of the piece of furniture, this underestimation may be quite important. In the sample piece of furniture with the load case described, representative values of internal forces for strength design according to (Smardzewski 2015) are calculated as follows and are compared with the results of the FEM models in Table 4.21. (See also calculated values in the forepart of this section.)

$F_1 = (79.6^2 + (85.4/2)^2)^{0.5} = 90.2$ N shear force to be resisted by the cross-section of the connector(s) at a corner.

$F_2 = 79.6$ N vertical shear force to be resisted by the sheared area of $h \cdot l_1$ within the sidewall, with the thickness of bottom board of h and embedment length of l_1 of the dowel in the sidewall. This force would amount to $F_2 + F_{\text{shelf}}/4 = 189.2$ N when the bottom panel is loaded.

Table 4.20 Comparison of the internal forces of connectors calculated according to the simplified superposition method and obtained by FEM using realistic support conditions

	Unit	Superposition	FEM model	Difference %
Vertical shear force in peripheral connectors Fy	N	79.6	126.6	59.1
Axial force in peripheral connectors Fx	N	269.4	237.0	−12.0
Vertical shear force in shelf connectors Fy	N	189.2	238.3	26.0
Edgewise horizontal shear in peripheral connectors Fz	N	42.7	75.3	76.3
Bending moment in peripheral connectors Mz	Nm	0.0	8.3	N.A.
Bending moment in shelf connectors Mz	Nm	0.0	14.1	N.A.
Corner displacement ux	mm	37.5	57.0	51.9
Shelf deflection Δy (mm)	mm	8.0	6.3	−21.3

Table 4.21 Comparison of the internal forces of connectors calculated for strength design according to the simplified superposition method and obtained by FEM using realistic support conditions

	Unit	Superposition	FEM model pinned joints	FEM model semi-rigid joints
Shear force acting across the dowel, F_1	N	90.2	134.9	134.6
Interlaminar shear force F_2	N	79.6	127.8	126.6
Tensile force across the panel F_3	N	189.2	237.4	238.3
Withdrawal force F_4	N	269.4	239.2	237.0

$F_3 = 189.2$ N vertical force to be resisted by an area of $l_2^2 \cdot 2 \cdot \pi$ loaded for tension perpendicular to the plane of the shelf material, l_2 being the embedment length of the dowel on the horizontal element.
$F_4 = 349.0 - 79.6 = 269.4$ N pull-out force to be resisted by one dowel.

Another way of analysing the state of stress and deformation of the cupboard with the given load conditions is to include the dowels as further parts in the FEM model. The displacement of the top left corner in the direction of the applied thrust force is then 39.93 mm, which when compared to the corresponding value of 57.0 mm obtained with semi-rigid joints indicates that the actual rigidity of the dowel joints in the members of given mechanical properties is somewhat higher than what was assumed in the semi-rigid model. Figure 4.33 illustrates the deformed shape of the cupboard modelled with dowels.

Cupboard with extended sidewalls forming feet is analysed next. The piece of furniture is shown in Fig. 4.34.

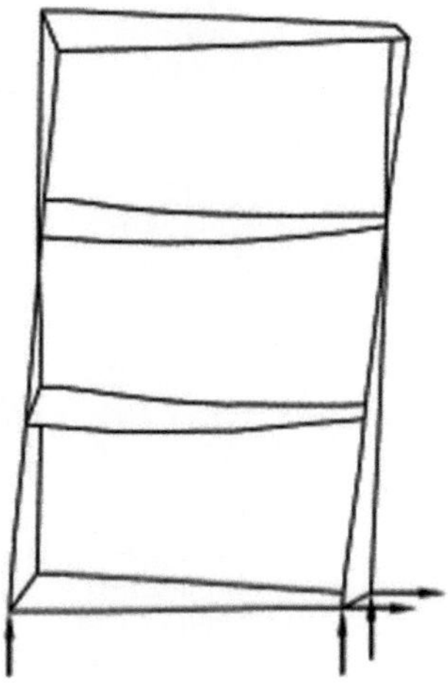

Fig. 4.33 Deformation of the cupboard modelled with one dowel at each corner; deformation scale: 5

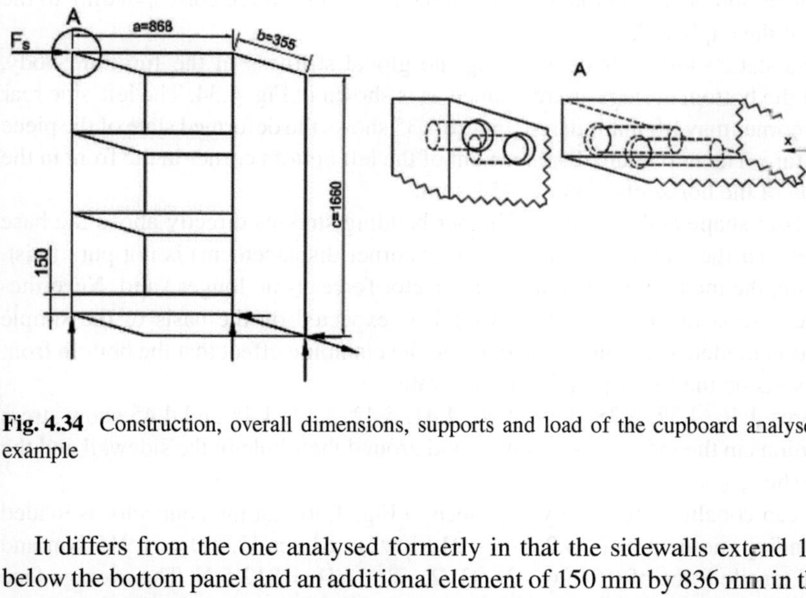

Fig. 4.34 Construction, overall dimensions, supports and load of the cupboard analysed in the example

It differs from the one analysed formerly in that the sidewalls extend 150 mm below the bottom panel and an additional element of 150 mm by 836 mm in the front plane is added; the rear wall does not extend downwards below the bottom board. As demonstrated in Sect. 4.5.4, the case twist model does not apply because there is no bottom peripheral element. One dowel made of beech wood per corner connects horizontal elements to the sidewalls. The narrow front element is connected in the same way to the sides and to the bottom; the rear wall is connected to the rest of the case by screws spaced at 288 mm.

The side thrust force is 436 N, having a rotating moment equal to that produced by the vertical loads in the opposite direction around the axis defined by the support conditions. The vertical loads include 1500 N/m^2 uniformly distributed load on the

Fig. 4.35 Deformed shape, scale of deformation: 6

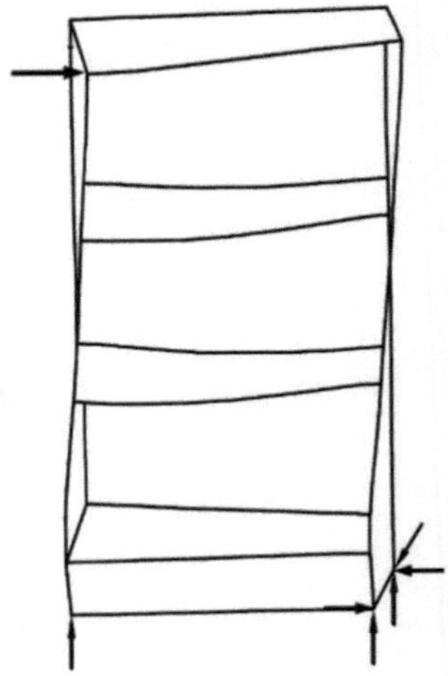

two shelves and on the bottom panel, and the gravitation force corresponding to the weight of the cupboard.

In the static model for determining the global stiffness of the furniture body, three of the bottom corners are restrained as is shown in Fig. 4.34. The left-side rear bottom corner may lift unhindered. Figure 4.35 shows the deformed state of the piece of furniture. The maximum displacement of the left bottom corner in the front in the direction of the horizontal force is 45.2 mm.

The bent shape and presence of higher bending stresses directly above the base indicates that the wall deformation, (upper corner displacement) is not pure twist. Therefore, the method of calculating connector forces is no longer valid. Nevertheless, the case is much stiffer, than would be expected on the basis of the simple calculation model. This can be attributed to the clamping effect that the bottom front panel exerts on the lower part on the sidewalls.

Figures 4.36, 4.37, 4.38, 4.39, 4.40, 4.41, 4.42, 4.43, 4.44 and 4.45 show stress distributions in the most loaded dowels and around their hole in the sidewall and the bottom shelf.

One can conclude from the values seen in Fig. 4.36 that the connector is loaded by a bending moment of $((54.26 - (-21.93))/2) \cdot 8^3 \cdot \pi/32{,}000 = 1.915$ Nm and an axial tensile force of $((54.26—21.93)/2) \cdot 8^2 \cdot \pi/4 = 812.5$ N. The average face withdrawal strength of one dowel of the plane surface embedded $L = 12$ mm in particleboard (internal bond strength IB $= 0.76$ N/mm^2) was 1913 N (Eckelman 1985). The functional relationship proposed in the same study to predict the face withdrawal force in N is

$$F = 303 \cdot (\mathrm{IB})^{0.85} \cdot L^{0.85} \tag{4.31}$$

where IB is the internal bond (tensile strength perpendicular to the plane) of particleboard in N/mm^2,

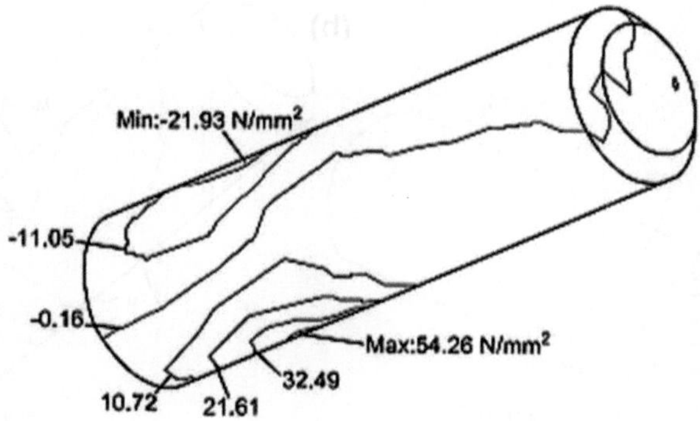

Fig. 4.36 Axial normal stress distribution in the dowel at the top left corner

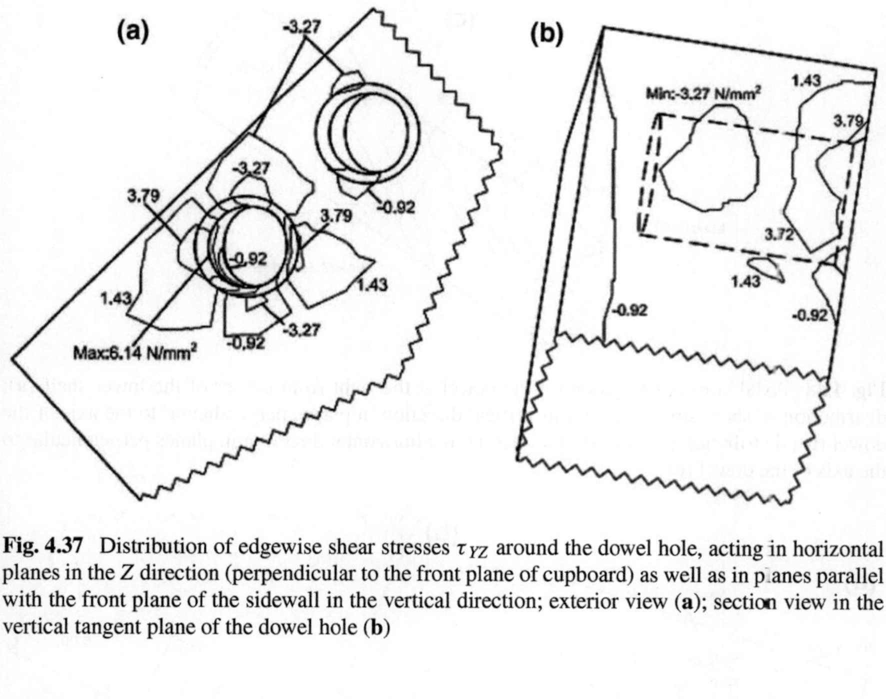

Fig. 4.37 Distribution of edgewise shear stresses τ_{YZ} around the dowel hole, acting in horizontal planes in the Z direction (perpendicular to the front plane of cupboard) as well as in planes parallel with the front plane of the sidewall in the vertical direction; exterior view (**a**); section view in the vertical tangent plane of the dowel hole (**b**)

L is the embedment length of the dowel in mm.

The design value corresponding to the $p = 0.02$ probability level can be derived on the basis of either this predicting formula or directly from test results by using Eqs. (4.20 and 4.20a). In that example, we take $k_{\text{mod}} = 1.00$ for all materials and components, recognizing that the side load applied is in fact instantaneous but in test conditions is repeated 10 times; in Eq. (4.20), $V_X = 0.22$ is used, while in Eq. (4.20a), $V = 0.1$ obtained from the test along with $k_{d,n} = 2.59$ for sample size between 10 and 15 is applied. The predicting formula yields $F = 2078.8$ N, which, when substituted into Eq. (4.20) results $F_d = 1139.4$ N. Using Eq. (4.20a) with the

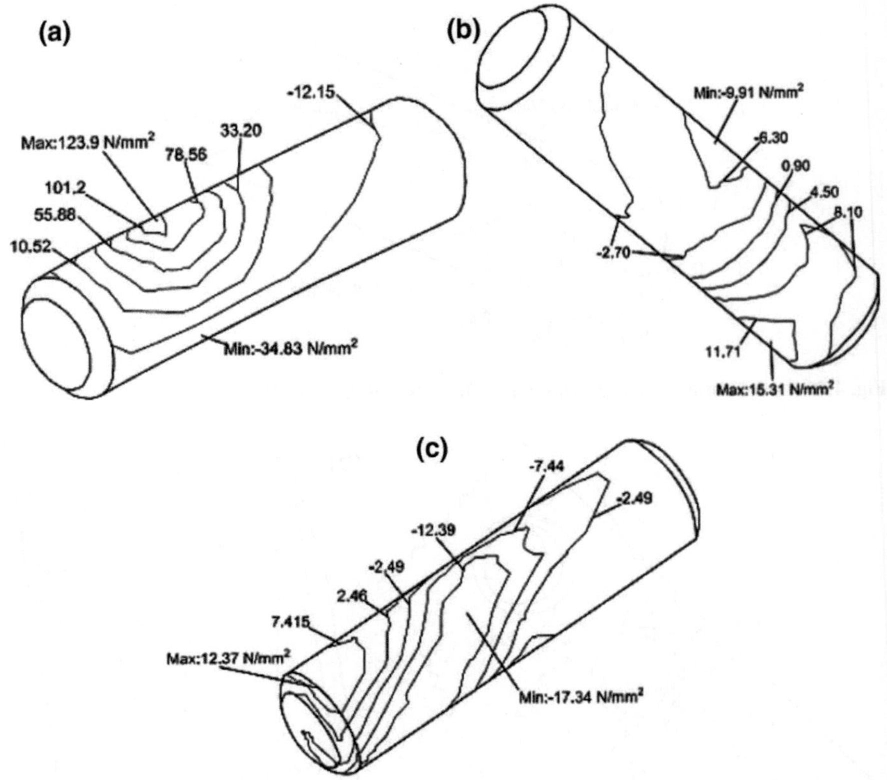

Fig. 4.38 Axial stress distribution on the dowel at the right front corner of the lower shelf (**a**); distribution of shear stresses acting in vertical direction in planes perpendicular to the axis of the dowel (**b**); distribution of shear stresses acting in a horizontal direction in planes perpendicular to the axis of the dowel (**c**)

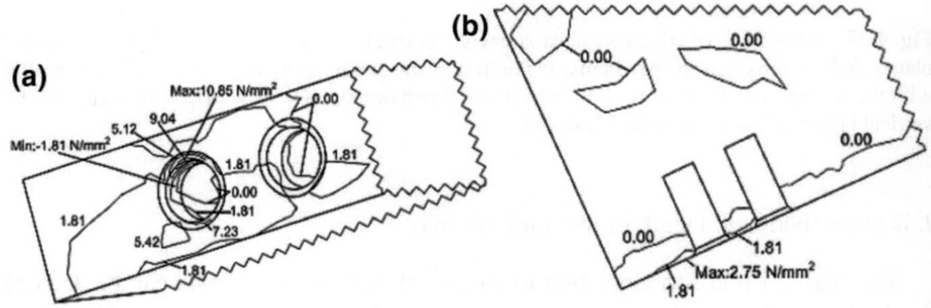

Fig. 4.39 Distribution of σ_Y normal stresses perpendicular to the plane of shelf; exterior view (**a**); mid-plane section view (**b**)

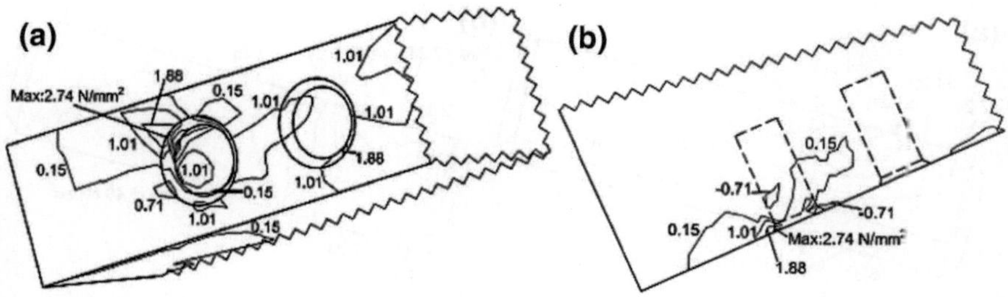

Fig. 4.40 Distribution of τ_{YZ} interlaminar shear stresses; exterior view (**a**); section view, horizontal tangent plane of the dowels (**b**)

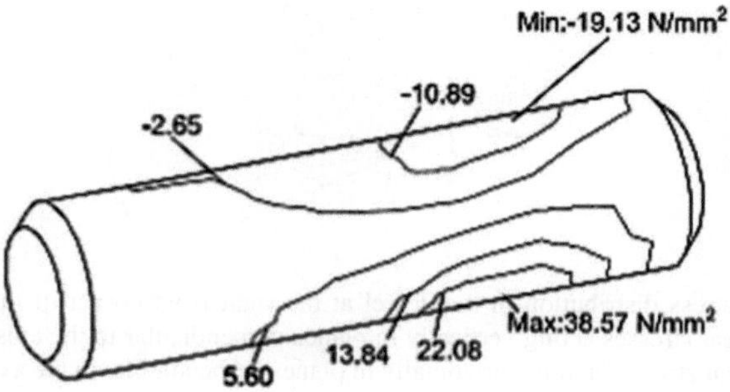

Fig. 4.41 Axial normal stress distribution in the dowel at the top left corner

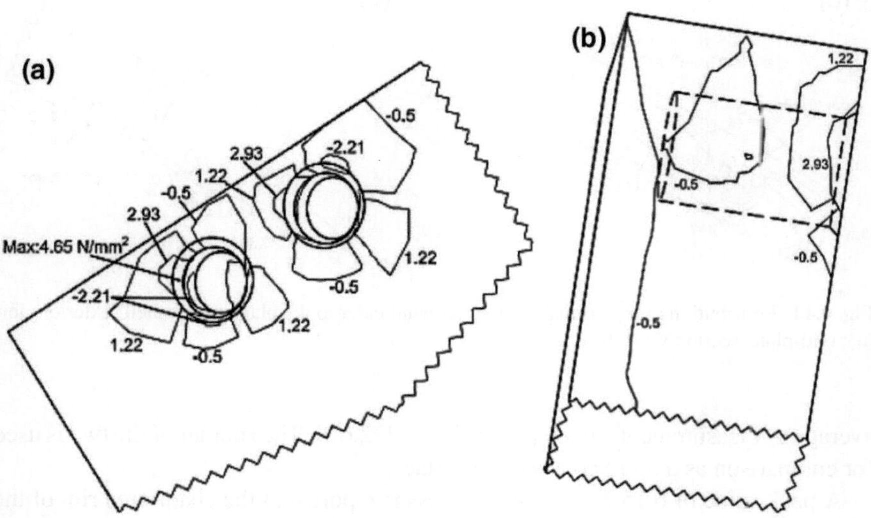

Fig. 4.42 Distribution of edgewise shear stresses τ_{YZ} around the dowel hole, acting in horizontal planes in the Z direction (perpendicular to the front plane of cupboard) as well as vertically in planes parallel with the front plane of the sidewall; exterior view (**a**); section view, the vertical tangent plane of the dowel hole (**b**)

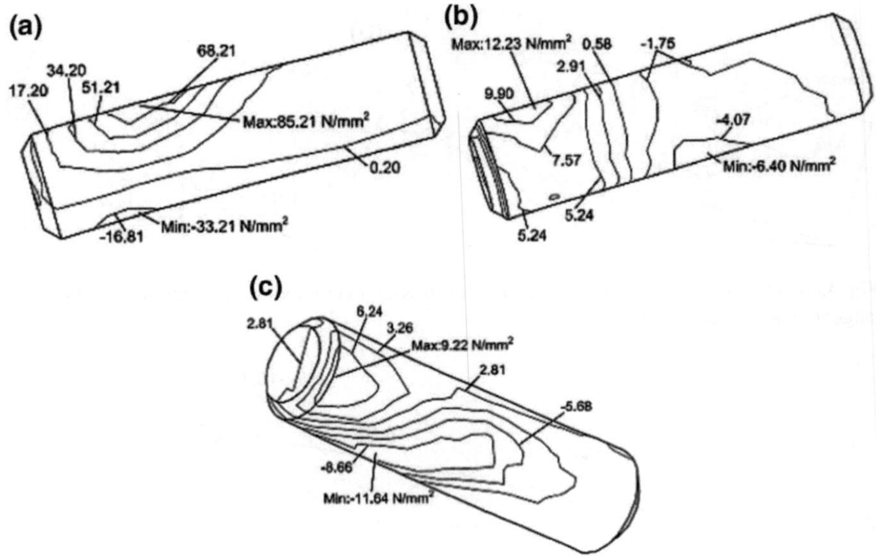

Fig. 4.43 Axial stress distribution in the dowel at the right front corner of the lower shelf (**a**); distribution of shear stresses acting vertically in planes perpendicular to the axis of the dowel (**b**); distribution of shear stresses acting horizontally in planes perpendicular to the axis of the dowel (**c**)

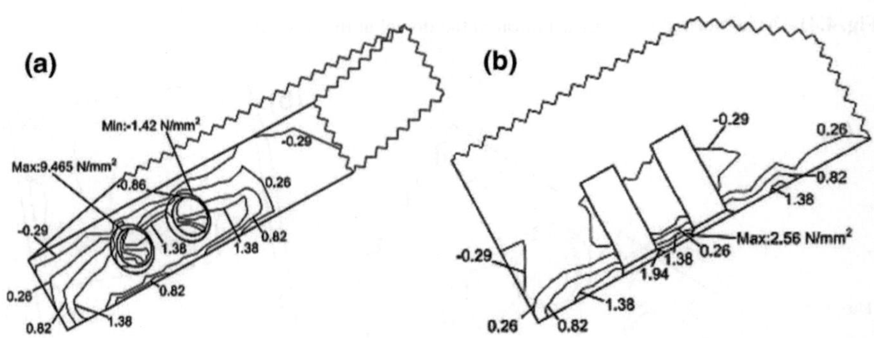

Fig. 4.44 Distribution of σ_Y normal stresses perpendicular to the plane of the shelf; exterior view (**a**); mid-plane section view (**b**)

average of measurement values yields $F_d = 1232.6$ N. The smaller of the two is used for comparison as a more conservative value.

A peak value of 6.15 N/mm^2 shear stress is reported at the chamfered rim of the drilled hole in the side board. Edgewise shear strength values of particleboard can be found in a Forest Product Laboratory report (FPL 1973). Based on a mean value of 7.8 N/mm^2 and $V_x = 0.077$ from the test results of 12 specimens, the limit strength value calculated using Eq. (4.20a) again is 5.55 N/mm^2, which is exceeded by the actual peak value.

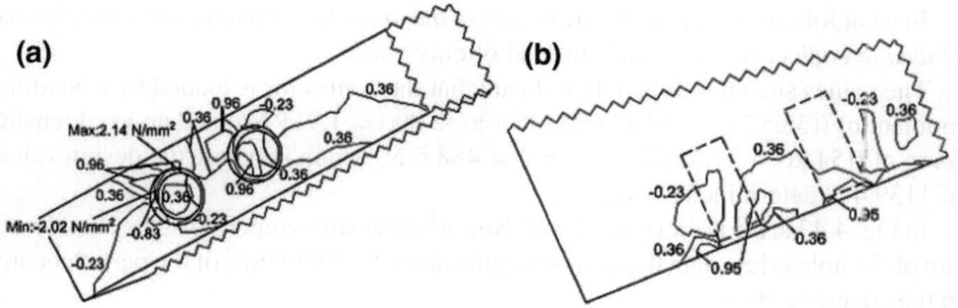

Fig. 4.45 Distribution of τ_{YZ} interlaminar shear stresses; exterior view (**a**); section view, horizontal tangent plane of the dowels (**b**)

Figure 4.38 demonstrates stress distributions in the front-side dowel of the lower shelf.

From the values shown in the above figure, a bending moment of ((123.9 − (−34.83))/2) · 8^3 · π/32,000 = 3.572 Nm and an axial tensile force of ((76.74 + 38.98)/2) · 8^2 · π/4 = 2238.6 N, nearly double of the design value of 1139.4 N can be calculated. Moreover, tensile stresses in the exterior fibre of the dowel are high above the limit value of $\sigma_{t,d} = 50.67$ N/mm^2 (quality grade and size effect do not apply to dowels).

The resultant of the shear stresses τ_{XY} and τ_{XZ} (τ_{LR} and τ_{LT}) amounts to a maximum of 17.39 N/mm^2 which can be related to the design value of the shear strength of beech wood along the grain. Taking the mean value of $\tau_{LT,sc} = 16.85$ N/mm^2 (Szalai 2004) the design value of $\tau_{LT,d} = 6.45$ N/mm^2 is found from Eq. (4.20) which is several times exceeded by the actual value.

Figure 4.39 shows stress distribution in the drilled region of the above dowel in the shelf.

Stress perpendicular to the plane of the board is several times higher in many places than the internal bond strength of about 0.4–0.94 N/mm^2 of the composite material (FPL 1973; Eckelman 1985; Semple and Smith 2006). Taking IB = 0.76 N/mm^2 as the mean value of the internal bond strength as above, the design value of $\sigma_{d,\mathrm{perp}} =$ 0.46 N/mm^2 can be derived with a coefficient of variation of 0.15 for this property of particleboards.

Figure 4.40 shows the distribution of the interlaminar shear stresses at the same location of this shelf.

Extreme values of the interlaminar shear stress do not exceed the strength of particleboard, typically around 3.0 N/mm^2 (FPL 1973), but are above the design value of 1.99 N/mm^2, calculated from Eq. (4.20).

The results of the analysis demonstrated above show that the high withdrawal force acting on dowels and possibly high tensile stresses perpendicular to the plane of the particleboard, along with interlaminar shear stresses exceeding the limit values of the board material necessitate an increase in the number of dowels along the edges of the horizontal elements.

In what follows we illustrate the *results of the analysis conducted with two dowels placed at each corner* of panels instead of only one.

The values shown in Fig. 4.41 indicate that the connector is loaded by a bending moment of ((38.57 − (−19.13))/2) · 8^3 · π/32 000 = 1.915 Nm and an axial tensile force of ((54.26 − 21.93)/2) · 8^2 · π/4 = 488.6 N, which is below the design value of 1139.4 N determined above.

In Fig. 4.42 a, the peak value of 4.65 N/mm^2 shear stress reported at the chamfered rim of the hole is less than the limit strength value of 5.55 N/mm^2 of the particleboard in the edgewise shear.

Figure 4.43 shows stress distribution in the same dowel at the front side of the lower shelf as with one single dowel per corner.

The values shown in Fig. 4.43 suggest a bending moment of ((85.21 − (−33.81))/2) · 8^3 · π/32,000 = 2.990 Nm and an axial tensile force of ((76.74 + 38.98)/2) · 8^2 · π/4 = 1291.8 N to act on the dowel which is still 13.4% above the limit value.

The resultant of the shear stresses τ_{XY} and τ_{XZ} (τ_{LR} and τ_{LT}) peaks at 12.85 N/mm^2, double the limit value of 6.45 N/mm^2.

Figure 4.44 shows stress distribution in the region of the hole of the above dowel in the shelf.

Stress perpendicular to the plane of the board in restricted regions is again several times the internal bond strength of the material. Around the mid-plane of the board between the drill holes the highest value is 2.56 N/mm^2.

Extreme values of the interlaminar shear stress (2.143 N/mm^2 and −2.016 N/mm^2) shown in Fig. 4.45 are below the mean strength of particleboard; apart from these local peaks they do not exceed the design value of 1.99 N/mm^2.

Comparing the results of analysis of the cupboard with single and double dowels at the corners of horizontal members, it can be stated that doubling the number of connectors significantly reduces the magnitude of stresses in the connectors and around their embedment but by far not proportionally to the increase of the number of connectors.

Finally, the load case under study corresponds to a standard test procedure. It simulates the situation that can happen in real use of the cupboard, when shelf loads act as medium- or long-term loads in a load combination with an instantaneous side load. To calculate design load effects, see Table 4.6, Sect. 4.5.2.

A chair frame with seat and backrest will be analysed as a whole by using FEM; additionally, internal forces of the supporting frame will be determined by the classical method of structural analysis for comparison.

The chair shown in Fig. 4.46 (Multiplex © János Vásárhelyi) has a wooden and a metal frame supporting each other and the seat plate. The frame constructed of stainless steel pipe (Φ22 mm, wall 2 mm thick) comprises the back legs and armrests; its two sides are connected by a frame member under the wooden rails to which it is fastened on their inside. The chair can be checked for load-carrying capacity in different ways. When loads are parallel with the plane of symmetry, such as a seat and back load, it can be assumed that the contribution of these flat parts to the overall strength and rigidity is negligible; they only transfer the loads to the frame

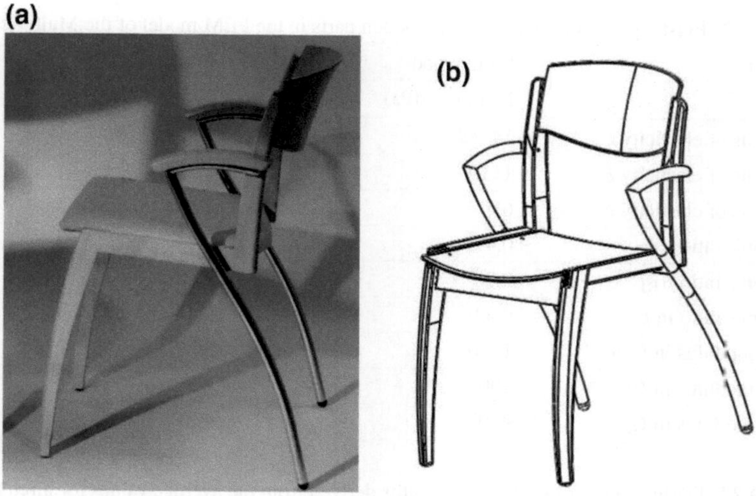

Fig. 4.46 Chair analysed for strength; real furniture (**a**); CAD model for strength analysis (**b**)

construction. Therefore a frame model solved by matrix structural analysis can be appropriate and the flat parts can be checked independently if necessary. Another possibility is to prepare a finite element model for the whole of the structure, treating the frame parts as solid bodies. Elaboration of the details of the joints in the model (tenons and mortises) presupposes the use of an orthotropic material model for the wooden parts and generally leads to FEM models which are too large to be practical. A further possibility is to use the results of matrix structural analysis to finite element models of joints only in order to define stress fields within the joints.

In his thesis G. Nagy, tentatively analysed the chair by FEM (Nagy 2017). We conducted a more in-depth study using the CAD model that he prepared (see Fig. 4.46b). We prepared an FEM model of the complete chair with an orthotropic material model of the wooden parts. While the front legs, rails and backrest of the original chair were made from European ash wood (*Fraxinus excelsior*) we opted to predict strength when using European birch (*Betula pendula*) for the solid wood parts; birch plywood was assumed for the plates. Table 4.22 shows the orthotropic material constants taken from pertinent sources (Smardzewski 2015; Szalai 2004). The chair was loaded according to EN 1728:2012 6.4 Seat and back static load test, Load level L2, that stipulates 2000 N vertical seat load and 700 N force perpendicular $(90 \pm 10)°$ to the backrest plane.

Results of the analysis were evaluated by calculating the Ashkenazi–Szalai failure criterion for the most dangerously loaded locations within the joint between the side rail and the back frame member. Table 4.23 shows the design values of the technical strengths of birch calculated from the mean values and standard deviations found in a publication (Ashkenazi 1978). In Table 4.24 the stress components at the critical location as well as the numerator and denominator of the failure criterion are given.

Table 4.22 Elastic properties used for the wooden parts in the FEM model of the Multiplex chair

Property	Solid wood	Plywood
	N/mm^2 (MPa)	N/mm^2 (MPa)
Modulus of elasticity E_L	14,000	5000
Modulus of elasticity E_R	1110	6500
Modulus of elasticity E_T	610	4000
Poisson's ratio μ_{LR}	0.49	0.44
Poisson's ratio μ_{RT}	0.78	0.22
Poisson's ratio in μ_{LT}	0.43	0.44
Shear modulus in G_{LR}	1180	800
Shear modulus in G_{RT}	190	400
Shear modulus in G_{LT}	910	400

Table 4.23 Design values of technical strengths derived from the average values for birch (*Betula pendula*)

f_L^+	77.78	f_L^-	52.84	t_{LR}	6.32	$t_{LR}^{T(45)-}$	4.32
f_R^+	4.32	f_R^-	8.92	t_{LT}	6.32	$t_{LT}^{R(45)-}$	3.45
f_T^+	5.20	f_T^-	3.72	t_{RT}	2.18	$t_{RT}^{L(45)-}$	1.80
$f_{LR}^{T(45)+}$	10.05	$f_{LR}^{T(45)-}$	5.53	$t_{LR}^{T(45)+}$	4.08		
$f_{LT}^{R(45)+}$	8.5	$f_{LT}^{R(45)-}$	8.06	$t_{LT}^{R(45)+}$	3.25		
$f_{RT}^{L(45)+}$	3.47	$f_{RT}^{L(45)-}$	4.89	$t_{RT}^{L(45)+}$	1.75		

Table 4.24 Stress components and safety factors determined by FEA in the critical locations of the side rail—back joint; stresses are given in N/mm^2

Location	σ_L	σ_R	σ_T	τ_{RT}	τ_{LT}	τ_{LR}	Stress state	Stress limit	n
Rail, point 1	48.16	4.76	4.23	2.07	−0.77	1.33	67.13	53.10	0.79
Rail, point 2	48.43	4.17	3.14	−0.29	−3.51	4.87	60.79	52.69	0.87
Rail, point 3	24.53	3.80	2.68	1.58	4.02	2.63	26.62	28.60	1.07
Rail, point 4	21.42	3.77	1.76	0.22	1.08	5.33	19.93	25.11	1.26
Back, point 1	41.84	4.23	4.93	1.00	−1.77	4.65	66.02	47.74	0.72
Back, point 2	43.95	5.23	4.65	2.11	2.84	1.31	66.36	49.53	0.75
Back, point 3	22.21	5.23	2.37	2.50	4.02	1.59	28.02	27.06	0.97
Back, point 4	13.84	4.90	3.87	0.30	1.85	4.20	24.83	19.80	0.80

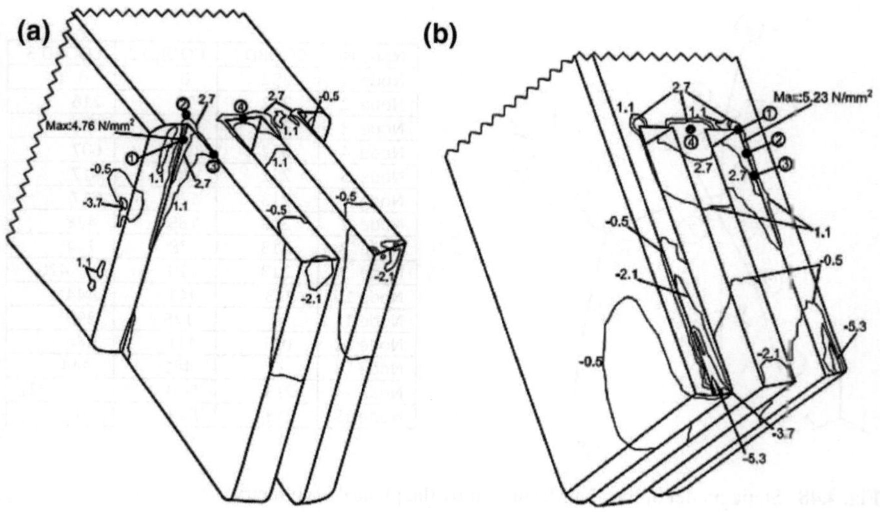

Fig. 4.47 Locations of critical state of stress on the tenoned and mortised end of chair frame members and isolines of the distribution of normal stresses σ_R acting parallel to the tenon faces

Unlike testing the strength of joints by in-plane bending forces as described in Sect. 4.5.6, when longitudinal tensile or compressive stresses exhibited unrealistic values along re-entrant corners, joints in chairs are subject to a general state of stress in which, due to out-of-plane effects, in-plane stress components alone are seldom critical.

The stress values in the table are generally nodal values. At locations where an unjustifiable rate of stress concentration occured, either the element values or the mean values of all immediate neighbouring nodes were used whichever the most unfavourable.

Figure 4.47 illustrates the locations on the tenon end and mortise end members, where stress readings were taken.

The n values in Table 4.22 indicate that the tension side of the joint in the back piece is loaded above the required safety level at the most critical location by $100 \cdot ((1/0.72) - 1) = 38.9\%$.

The analysis of the chair was redone excluding the seat and backrest plate from the analysis. The seat load was shared on the surfaces of contact with the rest of the structure and the back load on a length corresponding to the joint of the back plate in such a way that the vertical reaction at the end of the front legs remained the same as in the analysis of the whole structure. Moreover, since the seat under load contacts the metal stretcher underneath, to which it directly transfers some of its load, part of the seat load in the model was placed on this member, with maximum intensity in the midpoint, so that its bending stresses attained similar magnitude as in the analysis with the seat plate on. Under these conditions, the maximum displacement of the back frame member was close to the value obtained in the analysis of the complete chair.

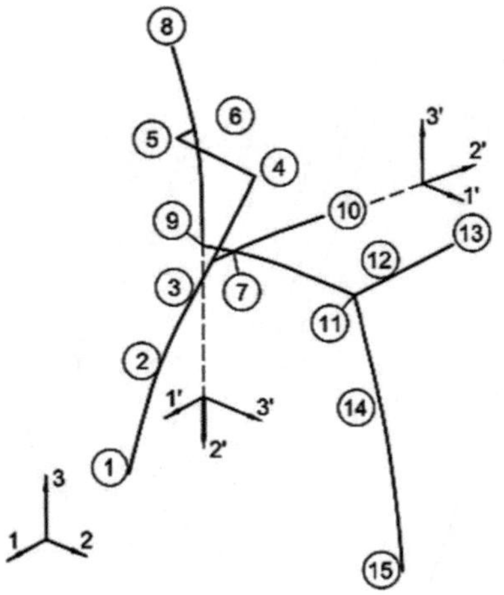

Node No	COORD 1	COORD 2	COORD 3
Node 1	252	0	0
Node 2	252	73	216
Node 3	252	185	428
Node 4	252	281	607
Node 5	252	120	617
Node 6	213	120	617
Node 7	213	185	428
Node 8	213	78	744
Node 9	213	129	420
Node 10	213	441	444
Node 11	0	175	399
Node 12	106	441	444
Node 13	0	441	444
Node 14	213	500	222
Node 15	213	524	0

Fig. 4.48 Static model of the chair frame cut by the plane of symmetry

Similarly, the magnitudes of the individual stress components at the critical points of the joint, and the extreme values of the members were fairly close to those defined in the previous analysis. All these justify our presumption that the contribution of plate members to the overall rigidity of the chair is negligible and the load bearing of the structure can be analysed on a frame model. However, it is important to note that the results of the analysis depend much on the distribution of loads in the model, which may not be self-evident and needs trials and good engineering judgement. An unproved load application on a simplified model may easily lead to an erroneous assessment of adequacy of the chair for its intended use.

An analysis of the chair based on matrix structural analysis of the frame was also performed with the aim of applying the results to the individual wood joints with an orthotropic material model in finite element analysis. The static model consisting of 15 members and 15 nodes is shown in Fig. 4.48, making use of the symmetry of structure and loading. The Modulus of elasticity of the members is given the value of $E = 14{,}000\ \text{N/mm}^2$, and the modulus of rigidity is $G = 1050\ \text{N/mm}^2$.

Local coordinate directions $1'$, $2'$ and $3'$ are illustrated for members 6–9 and 7–10, respectively. The model pictures curved parts but the sections between nodes are considered as straight lines. Coordinates of nodes are shown in tabular form. Table 4.25 contains section properties of the individual frame members along with section orientation and connection flexibility.

Table 4.26 shows the member-end forces and moments computed in the respective coordinate system for a few beams including member 6–9 and 7–9. These two join at node No. 9. The node is not loaded by outside forces; thereby forces and moments acting in the joining cross-sections of both beams keep the node in static equilibrium.

Stresses within the joint domain can be computed using FEA by making the member-end forces of both beams at end 9 act with negative sign on the respective cross-sections of the truncated joint (see Fig. 4.50). When doing that, it has to be

Table 4.25 Section properties and connectivity of the frame members in the static model

BEAM	$I1$ mm^4	$I3$ mm^4	J mm^4	A mm^2	Beta rad
BEAM 1–2	9.500E+04	9.500E+04	1.000E+06	1.260E+02	0.000E+00
BEAM 2–3	9.500E+04	9.500E+04	1.000E+06	1.260E+02	0.000E+00
BEAM 3–4	9.500E+04	9.500E+04	1.000E+06	1.260E+02	0.000E+00
BEAM 4–5	9.500E+04	9.500E+04	1.000E+06	1.260E+02	0.000E+00
BEAM 5–6	9.500E+04	9.500E+04	1.000E+06	1.260E+02	0.000E+00
BEAM 6–8	7.400E+04	5.200E+04	1.100E+05	8.640E+02	1.570E+00
BEAM 6–9	2.210E+05	9.520E+04	1.950E+05	1.330E+03	1.570E+00
BEAM 7–9	1.290E+05	8.230E+04	1.720E+05	1.113E+03	1.570E+00
BEAM 7–10	1.290E+05	8.230E+04	1.720E+05	1.113E+03	1.570E+00
BEAM 10–14	1.730E+05	7.680E+04	1.910E+05	1.176E+03	1.570E+00
BEAM 14–15	3.520E+04	3.000E+04	5.820E+04	6.240E+02	1.570E+00
BEAM 10–12	4.800E+05	5.500E+04	1.690E+05	1.400E+03	1.258E+00
BEAM 12–13	1.090E+05	3.460E+04	9.450E+04	8.580E+02	1.258E+00
BEAM 3–7	5.280E+05	5.280E+05	5.560E+06	3.520E+02	0.000E+00
BEAM 7–11	9.510E+04	9.510E+04	1.000E+06	1.260E+02	0.000E+00
	BEAM END		$Z4$ rad/Nm	$Z5$ rad/Nm	$Z6$ rad/Nm
BEAM 6–9	Node 9		1.0E−05	1.0E−05	1.0E−05
BEAM 7–9	Node 9		1.0E−05	1.0E−05	1.0E−05
BEAM 7–10	Node 10		1.0E−05	1.0E−05	1.0E−05
BEAM 10–14	Node 10		1.0E−05	1.0E−05	1.0E−05
BEAM 10–12	Node 10		1.0E−05	1.0E−05	1.0E−05

taken into consideration that nodes formed by the intersection of beam axes in the theoretical (node and axis) model are of infinitely small dimension, while real joints have finite dimensions. Therefore the length of a member in the theoretical model comprises roughly half of the width of the joining member. Consequently, when member forces vary along the axis of the member, the theoretical values of member-end forces have to be adjusted to the distance from the node, corresponding to the location of the cross-sections in the FEA model of the joint. Since section forces and moments as exterior loads applied on a member cross-section give rise to local effects in their close vicinity, end sections in the model have to be positioned at some distance from the boundaries of the joint to let these disturbances to die off. Figure 4.49 lists the adjusted values of member forces applied on the model.

Results of the analysis were evaluated by calculating the Ashkenazi–Szalai failure criterion for the most dangerously loaded location within the joint. Table 4.27 gives the stress components at the critical location as well as the numerator and denominator of the failure criterion. The critical locations for the mortise and the tenon end, respectively, may be seen in Fig. 4.50.

Table 4.26 Member-end forces and moments in member coordinate directions

Member-end forces	*F*1 kN	*F*2 kN	*F*3 kN	*M*1 Nm	*M*2 Nm	*M*3 Nm
BEAM 6–9 End 6	0.0021	0.2134	−0.1547	8.8084	−1.8112	3.9848
BEAM 6–9 End 9	−0.0021	−0.2134	0.1547	−39.3133	1.8112	−4.3966
BEAM 7–9 End 7	0.0021	−0.1662	0.2947	49.4956	4.4416	−1.7858
BEAM 7–9 End 9	−0.0021	0.1601	−0.2093	−39.3133	−4.4416	1.6978
BEAM 7–10 End 7	−0.0464	0.0477	0.6138	92.7017	−0.3184	5.2073
BEAM 7–10 End 10	0.0464	−0.0237	−0.3053	10.2735	0.3184	7.3528
BEAM 10–14 End 10	−0.0000	0.1196	−0.0318	−10.2735	−0.0000	0.0000
BEAM 10–14 End 14	0.0000	−0.1196	0.0318	2.9706	0.0000	−0.0000
BEAM 10–12 End 10	0.1324	0.0464	0.4091	−2.0237	−0.0000	−7.0760
BEAM 10–12 End 12	−0.0659	−0.0464	−0.2036	36.7705	0.0000	−3.5292
BEAM 12–13 End 12	0.0659	0.0464	0.2036	−36.7705	−0.0000	3.5292
BEAM 12–13 End 13	−0.0000	−0.0464	0.0000	47.5614	0.0000	−7.0203
BEAM 3–11 End 3	0.1759	−0.0015	0.0003	−2.6398	−0.5226	6.1090
BEAM 3–11 End 3	−0.1759	0.0015	−0.0003	2.7131	0.5226	−43.8279

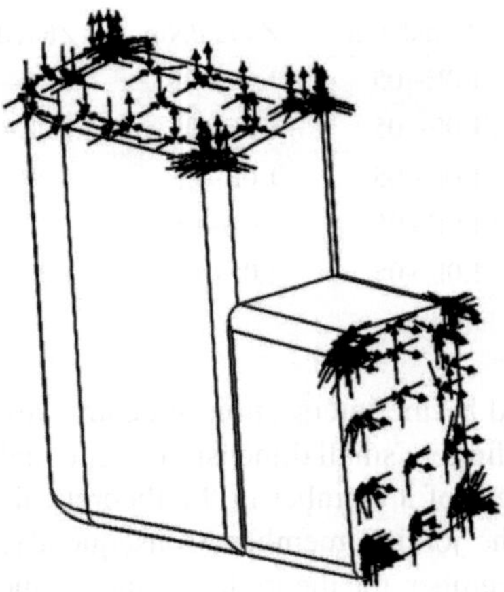

	Back	Side rail
F1 (kN)	0.0021	0.0021
F2 (kN)	0.2143	-0.1061
F3 (kN)	- 0.1547	0.2093
M1 (kNm)	34.3100	49.4900
M2 (kNm)	-1.8112	4.4416
M3 (kNm)	4.3966	-1.6978

Fig. 4.49 Truncated model of the joint of a backrest and side rail at the right side of the chair frame with internal forces applied as exterior load on the end sections

In conclusion, the joint between the backrest and side rail in the Multiplex chair, when the wooden parts are made from common birch, is not strong enough when the chair is loaded according to the load case considered. The state of stress at several critical points is beyond the failure conditions.

An alternative way of checking the joint strength could be *direct comparison of the calculated member-end forces with design values of joint resistance* predicted by any of the options 1, 2 and 3 listed in Sect. 4.5.6.

Table 4.27 Stress components and values of safety factor n determined in the critical locations of the joint modelled by FEA; stresses are given in N/mm^2

Location	σ_L	σ_R	σ_T	τ_{RT}	τ_{LT}	τ_{LR}	Stress state	Stress limit	n
Back, point 1	41.70	3.03	3.75	0.65	1.07	5.23	51.17	45.66	0.89
Back, point 2 (vis-a-vis rear)	−15.63	−0.91	−0.49	−0.33	−3.21	−3.79	15.24	17.10	1.12
Back, point 3	25.33	4.28	4.52	−0.35	−1.80	0.44	34.04	30.43	0.89
Back, point 4 (rear side)	20.95	4.85	1.99	−2.19	−0.14	2.05	21.33	25.07	1.18
Back, point 5	2.49	3.40	2.57	1.58	1.77	−3.31	10.40	8.03	0.77
Back, point 6 (rear edge)	−2.36	−5.23	−1.95	0.25	0.23	0.03	9.59	8.00	0.83
Back, point 7	24.00	3.03	3.58	−0.33	−1.67	5.54	29.14	28.30	0.97
Rail, point 1	−18.25	−0.74	−0.31	−0.20	2.82	−0.98	13.96	19.03	1.36
Rail, point 2	42.08	3.98	3.31	0.00	−4.20	−4.61	52.44	46.43	0.89
Rail, point 3	7.53	2.78	3.31	1.72	−4.26	−3.98	16.42	12.94	0.79
Rail, point 4	37.53	3.13	3.37	0.03	−2.61	−4.61	42.91	41.38	0.96
Rail, point 5	19.52	2.35	2.64	−2.07	4.22	−0.40	19.33	22.79	1.18
Rail, point 6	41.38	3.97	1.57	−2.20	0.06	1.13	36.52	44.42	1.22

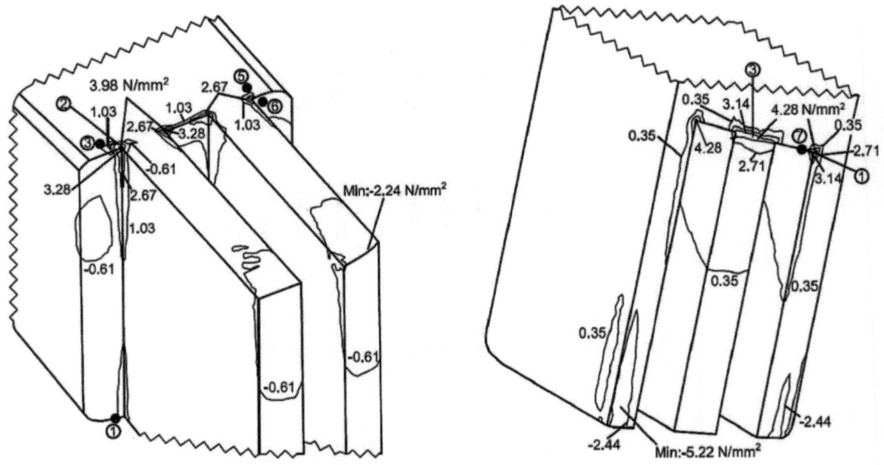

Fig. 4.50 Locations of critical state of stress on the tenoned and mortised part of joining members in the truncated model and isolines of the distribution of normal stresses σ_R acting parallel to the tenon faces

The chair under study is unique enough so that similarity search is not promising. Strength predicting formulas established by regression analysis of experimental results cannot be found in the literature for open tenon (bridle) joints. Therefore, the only available option is to use simplified theoretical models. Member-end forces from Table 4.26, on the point of the side rail with the back are listed below:

In-plane shear force (F2 for beam 7–9)	$F_{\text{sh1}} = 160.1$ N
In-plane shear force (F3 for beam 7–9)	$F_{\text{sh2}} = 209.3$ N
Out-of-plane shear force (F1 for both beam 6–9 and 7–9)	$F_{\text{sh,out}} = 2.1$ N
In-plane bending moment (M1 for both beam 6–9 and 7–9)	$M_{\text{in}} = 39.313$ Nm
Out-of-plane bending moment (M3 for beam 7–9)	$M_{\text{out}} = 1.6978$ Nm
Torsional moment (M2 for beam 7–9)	$M_t = 4.442$ Nm.

We assessed the load-carrying capacity of the joint for the individual load components, applying the concept given in Smardzewski's textbook (Smardzewski 2015) for in-plane bending of open tenon joints. The effect of the rest of the load components either on the tenoned or on the mortised end was considered in a conservative way. Out-of-plane shear force $F_{\text{sh,out}} = 2.1$ N is negligible.

We assumed that glue bonds fail at shear stresses that reach the shear strength of the wood in the respective direction. Wood may fail from shear, torsion or bending. Design values of strength for shear and torsion were taken from Table 4.23: $\tau_{\text{RL},d} = \tau_{\text{TL},d} = 6.32$ N/mm^2, $\tau_{\text{RT},d} = 2.18$ N/mm^2. The design value of bending strength was derived from the mean value of 114.3 N/mm^2 at 12% M.C., given in The Wood Database. By using Eq. (4.20) and assuming COV = 0.28, we obtained $\sigma_{b,d} =$ 38.25 N/mm^2.

In-plane bending capacity was calculated with the following considerations: The joint comprises four parallel rectangular glue lines, each with a surface area of 39 mm by 46 mm. In-plane bending moment loads the glue lines (and glued surfaces) in torsion. The maximum value of the developed shear stresses, $\tau_{max,t}$ occurs in the corners (see Fig. 4.51) acting perpendicularly to the diagonal of the rectangle and is given in N/mm^2 by the expression:

$$\tau_{max,t} = \frac{M_{in}}{J} \frac{\sqrt{\left(\frac{a}{2}\right)^2 + \left(\frac{b}{2}\right)^2}}{n} \tag{4.32}$$

where a is the shorter dimension of the rectangular glue-line surface in mm,
b is the longer dimension of the rectangular glue line in mm,
M is the bending moment in Nmm,
n is number of glue lines,
J is the torsional moment of inertia of the glue-line surface area in mm^4.

Replacing $\tau_{max,t}$ by the design strength value of wood, we can rearrange the expression to get the design value of the in-plane bending capacity, $M_{in,d}$ in Nmm.

$$M_{in,d} = \tau_{t,d} \frac{J \cdot n}{\sqrt{\left(\frac{a}{2}\right)^2 + \left(\frac{b}{2}\right)^2}} \tag{4.33}$$

The orientation of the maximum torsional shear stress with respect to the anatomical directions of wood is between that of τ_{LR} and τ_{RT}, or τ_{LT} and τ_{TR}; its dual pair acts perpendicularly to the tenon face, hence across the grain. In order to remain on the safe side, in this expression we use the shear strength $\tau_{RT,d} = \tau_{t,d} = 2.18$ N/mm^2 as design strength value.

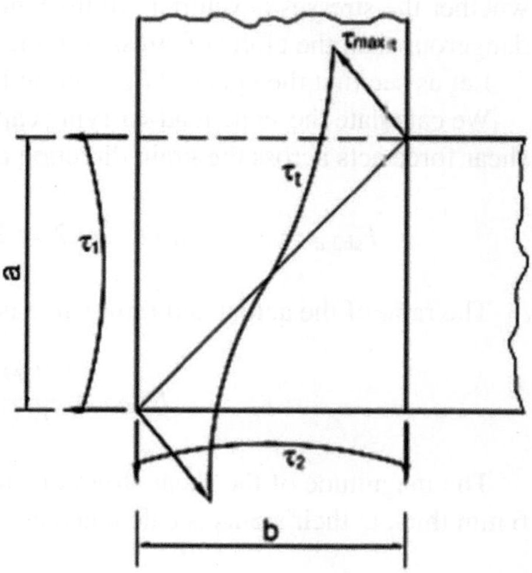

Fig. 4.51 Shear stress components acting in the plane of glued surfaces; τ_1 and τ_2 denote shear stresses due to the shear forces F_{sh1} and F_{sh2}, τ_t is shear stress due to torsion

With these values Eq. (4.33) yields $M_{\mathrm{in},d} = 126\,718$ Nmm. The ratio of the acting bending moment to the limit capacity is

$$R_{M,\mathrm{in}} = \frac{39{,}313}{126{,}718} = 0.310$$

The effect of the *in-plane shear force* F_{sh1} is counteracted by the glued faces of the tenons and mortises. The corresponding shear stresses act as τ_{RT} or τ_{TR} in the mortise faces with the average value of

$$\tau_1 = \frac{F_{\mathrm{sh1}}}{n \cdot a \cdot b} \cdot \frac{160.1}{4 \cdot 39 \cdot 46} = 0.022\,\mathrm{N/mm^2}$$

The glue line acts as a lap joint and stress concentration occurs at the ends by a factor of 2, resulting $\tau_{2\max} = 0.044$ N/mm^2. The limit (design) value of $\tau_{\mathrm{RT},d} = \tau_{\mathrm{TR},d}$ = 2.18 N/mm^2 should be used for these stresses.

The load-carrying capacity of the joint is

$$F_{\mathrm{sh1},d} = \tau_{\mathrm{RT},d} \cdot n \cdot a \cdot b/2 = 2.18 \cdot 4 \cdot 39 \cdot 46/2 = 7821.8\,\mathrm{N}$$

The ratio of the acting and limit force is

$$R_{\mathrm{Fsh1}} = \frac{160.1}{7821.8} = 0.020$$

The evaluation of the *effects of the in-plane shear force* F_{sh2} needs some further consideration. While the shear force F_{sh1} pulls the two members apart, the force F_{sh2} presses them against each other on the 6 mm wide edges of the tenons, resulting in shear stresses at the stems of the tenons with their relative direction between τ_{RL} and τ_{TL}, depending on the orientation of the annual rings. Therefore we examined whether the stresses developing in the tenon cross-sections at their stems are more dangerous than the effect of stresses in the glue line.

Let us see first the *effect of* F_{sh2} acting in the planes of glued faces.

We calculate the same load-carrying capacity of the joint as for F_{sh1}, because the shear force acts across the grain direction on the glued tenon faces.

$$F_{\mathrm{sh2},d} = \tau_{\mathrm{RT},d} \cdot n \cdot a \cdot b/2 = 2.18 \cdot 4 \cdot 39 \cdot 46/2 = 7821.8$$

The ratio of the actual and limit value is

$$R_{F\mathrm{sh2}} = \frac{209.3}{7821.8} = 0.027$$

The magnitude of the shear stress due to F_{sh2} in the cross-section of tenons $d =$ 6 mm thick at their stems is calculated as

$$\tau_2 = \frac{F_{\mathrm{sh2}}}{n' \cdot d \cdot a}$$

where d is the tenon thickness in mm, and n' denotes the number of tenons. The shear stress τ_2 corresponds to τ_{RL} or τ_{RT}.

The load-carrying capacity of the joint for that mode of failure is

$$F'_{\mathrm{sh2},d} = \tau_{\mathrm{RT},d} \cdot n' \cdot d \cdot a = 6.32 \cdot 2 \cdot 6 \cdot 39 = 2957.8\,\mathrm{N}$$

The ratio of the maximum and limit value is

$$R'_{\mathrm{Fsh2}} = \frac{209.3}{2957.8} = 0.071$$

which is clearly more unfavourable than the former quotient of $R_{\mathrm{Fsh2}} = 0.027$.

Using the three quotients corresponding to the effect of bending moment, shear force F_{sh1} and F_{sh2} we get the share of joint load-carrying capacity consumed by the in-plane forces under the more unfavourable supposition, expressed as

$$R_{\mathrm{inplane}} = \sqrt{(R_{M,\mathrm{in}})^2 + (R_{F,\mathrm{sh1}})^2 + (R'_{F,\mathrm{sh2}})^2} = \sqrt{0.310^2 + 0.20^2 + (0.71)^2}$$
$$= 0.338$$

The *effect of the out-of-plane bending moment* can be considered in two ways. First, it causes normal (bending) stresses along the grain in the tenons, and the moment resisting capacity can be calculated as

$$M_{\mathrm{out},d} = K \cdot \sigma_{b,d}$$

where $\sigma_{b,d}$ denotes the design bending strength of wood and K is the modulus of resistance of a cross-section through the two tenons; see Fig. 4.52.

This latter is calculated from the moment of inertia I_3 and the distance e of the exterior fibre from the neutral axis of the cross-section as

$$K = \frac{\left((3 \cdot d)^3 - d^3\right) \cdot a}{12 \cdot e} = \frac{\left((3 \cdot 6)^3 - 6^3\right) \cdot 39}{12 \cdot 1.5 \cdot d} = 2028\,\mathrm{mm}^3$$

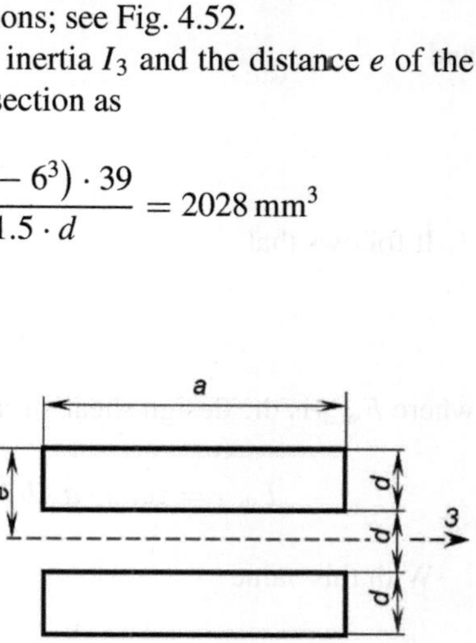

Fig. 4.52 Cross-section of the tenoned end, the out-of-plane bending moment acts around the axis 3

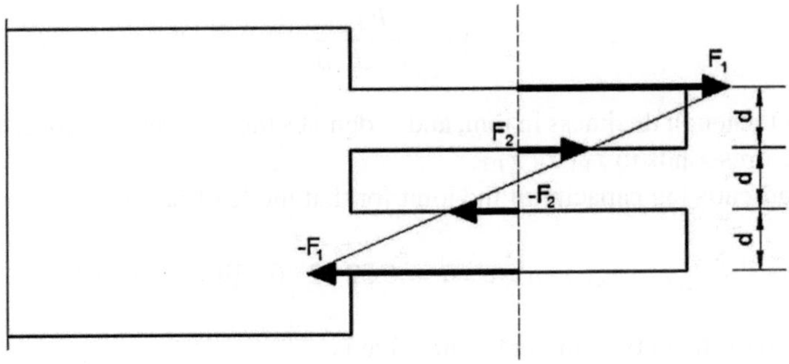

Fig. 4.53 Shear forces developing in the planes of the glued tenon surfaces due to out-of-plane bending moment

With the value of design bending strength given above, this is expressed as

$$M_{\text{out}.d} = 2028 \cdot 38.25 = 77{,}571\,\text{Nmm}$$

The corresponding ratio of utilization of the load-carrying capacity is

$$R_{\text{out}} = \frac{1698}{77{,}751} = 0.022$$

The other way of considering the effect of out-of-plane bending is to calculate the shear forces on the glued surfaces the moment of which counteracts the moment acting at the base of tenons; see Fig. 4.53.

The equations of the equality of moments and ratio of forces are

$$3 \cdot d \cdot F_1 + d \cdot F_2 = M_{\text{out}}$$

and

$$\frac{F_1}{1.5 \cdot d} = \frac{F_2}{d}$$

It follows that

$$M_{\text{out}.d} = 3.67 \cdot d \cdot F_{\text{sh},d}$$

where $F_{\text{sh},d}$ is the design shear strength of one glued face of a tenon.

$$F_{\text{sh},d} = \tau_{\text{RT},d} \cdot a \cdot b/2 = 2.18 \cdot 39 \cdot 46/2 = 1955.5\,\text{N}$$

With this value

$$M_{\text{out},d} = 3.67 \cdot d \cdot F_{\text{sh},d} = 43{,}059\,\text{Nmm}$$

The ratio of the out-of-plane bending moment to its limit value is

$$R'_{\text{out}} = \frac{1698}{43{,}059} = 0.039$$

This value is lower than that calculated for bending stresses, therefore this latter will be used.

Finally, *the torsional moment* of $M_t = 4.4412$ Nm has to be accountec for. The torsional section modulus of the cross-section of the two tenons can be calculated as (Muttnyánszki 1980):

$$K_t = 0.33 \cdot d^2 \cdot 2 \cdot (b + d)$$

This yields $K = 1080$ mm^3. The corresponding shear stresses are $\tau_{\text{R}\_}$ and τ_{TL} with the same design strength value of 6.32 N/mm^2.

The design value of torsional resistance of the two tenons is

$$M_{t,d} = K_t \cdot \tau_{\text{RL},d} = 1080 \cdot 6.32 = 6826\,\text{Nmm}$$

The ratio of the acting torque to the limiting value is

$$R_t = \frac{4442}{6826} = 0.651$$

Using the calculated partial quotients of load-carrying capacity utilization it has to be taken into consideration that the shear force $F_{\text{Sh},1}$ and the out-of-plane bending moment produce the same type of shear stresses in the glue line and add up directly in the failure mode. Likewise, the shear force $F_{\text{Sh},2}$ and the torsional moment acting on the tenoned end causes an identical type of shear stresses in the stem of tenons. Applying Eq. (4.28) accordingly we get

$$R_{\text{joint}} = \sqrt{\left(R_{M,\text{in}}\right)^2 + \left(R_{F,\text{sh1}} + R_{M,\text{out}}\right)^2 + \left(R'_{F,\text{sh2}} + R_{M,t}\right)^2}$$

Finally, it is proper to account for an increased uncertainty associated with the simplified theoretical models applied in these calculations. In Sect. 4.5.2 we proposed the model uncertainty factor of $\gamma_{\text{Sd}} = 1.1$ for furniture design which is supposed to be included in the design load values. Applying a higher value $\gamma_{\text{Sd}} = 1.25$ for the unusual degree of uncertainties needs multiplying the partial quotients by a factor of $1.25/1.1 = 1.14$. Using these increased ratios yields

$$R_{joint} = \sqrt{0.310^2 + (0.020 + 0.039)^2 + (0.071 + 0.651)^2} = 0.898$$

This indicates that the joint between the backrest and side rail is loaded below but close to its limiting load-carrying capacity. We arrived at more unfavourable conclusion from the results of FEM analysis of both the complete structure and the isolated joint using the Ashkenazi–Szalai failure criterion. It has to be noted that both direct comparison by load components and FEM analysis suggested a major contribution of shear stress components to the safety factor n at the critical points in the joint. Occasionally, across-the-grain tensile stresses also proved to be a concern; we could not derive a simplified theoretical model to calculate these stresses for the partial quotient approach.

4.6 Description of Wood Joint Strength and Stiffness by Using Similarity Relationships

As stated in Sect. 4.5.6, assessment of the load-carrying capacity of furniture joints is not straightforward enough. All woodworking joints, furniture joints included, obey the laws of physics when subjected to mechanical loads. The physical phenomena taking place are often complicated, being difficult, if not impossible to describe analytically. In such cases, the method of dimensional analysis (Langhaar 1951) is helpful. Having identified the key influencing factors (parameters) relevant to the kind of joint and external action at hand, it is possible to establish relationships that govern the behaviour of joints in terms of generally valid dimensionless quantities.

Determination of the numerical constants in these relationships needs experimental results. A great number of historical and current research results on furniture joint strength are now available in the literature. They may provide sufficient experimental data to process, provided that they encompass the ranges of variation of both the key and the accidental influencing factors.

In the ideal case, the processed experimental results, when plotted on a log–log paper, are expected to align closely along a straight line. A scatter experienced is the consequence of the inevitable experimental errors, aggravated by our ignorance of some of the influencing factors that contribute to the variation of the measured strength of test pieces.

The similarity relationships that may be determined by the use of test results allow the derivation of calculation formulas for the draft estimation of the resistance of the joint to the load in question. The numerical constants in these formulas are dimensionless, thus independent of the unit system used. However, one has to be aware that the resistance values predicted by these formulas can be surpassed or underperformed, depending on the accidental influencing factors, such as variability of material properties, correctness of fit, precision of machining, proper type and application of glue that we omit from the analysis. Deviations from the good practice regarding the geometrical details of the joint (ratio of dowel diameter or tenon thickness to rail thickness, distance of the dowel axis from the rail edge, the size of the tenon shoulder, length of embedment in the rail), unless treated as parameter, may also play a part.

Test data for dimensional analysis of joint load-carrying capacity have been collected from a number of published and unpublished sources. These are referred to in the sections dealing with the individual loading cases of the different types of joints.

Typically, average results of series of 5–15 test pieces were available; therefore, it was thought that the use of unweighted average results does not introduce appreciable error into the analysis. Weighting of test results was applied in the case of very dissimilar (generally large) sample sizes.

Our study extended to glued T-shape and L-shape (corner) joints of wood frame members with dowel, mortise and tenon and bridle (open mortise) joints, widely used in practice. We considered the withdrawal force, in-plane and out-of-plane bending strength.

In almost all experimental studies found in the literature, the applied bending loads were exercised by the application of force rather than by pure bending moment, resulting in a combined loading of bending moment and shear force. Real loading situations in a furniture frame are likewise combinations of bending and shear—and sometimes of axial load too; though arm length of the bending force varies in a range close to what is applied in laboratory tests. At the same time, in both experimental and real service conditions, bending moment is the dominant action; moreover, an important bending force is very seldom accompanied by a serious axial load. Consequently, these test results may give a good indication of the resistance of a joint to external bending loads for strength design; they yield conservative estimation rather than overestimate load-bearing capacity.

4.6.1 Choosing Parameters of the Similarity Relationships—General Remarks

When we want a generally valid relationship of resistance of joints and the governing parameters, one of the problems is to choose the proper parameters.

The key influencing factors can be grouped as

- size and geometry of the joint,
- strength properties of the wood,
- quality of the glue bond.

Details of joint geometry include thickness, width, length and shape (rectangular or rounded) of the tenon as well as its position with respect to the cross-section of the rail in mortise and tenon joints. For dowel joints, most generally made up of two dowels, the diameter and spacing of dowels and their embedded lengths both in rail and post are the factors of interest. They are chosen by the designer along with the dimensions of the cross-section of the members as design parameters.

Involvement of too many parameters in the analysis results in relationships that are inconvenient to use. Since detail dimensions are related to dimensions of member cross-sections, due to some rules of thumb of good practice, not all of these dimensions are absolutely necessary to evaluate the load-carrying capacity of a joint. Cross-sectional dimensions (width and height) generally have a decisive role. In practice however, some important deviations from the "good practice" can be expected for a number of different reasons. Therefore, omitting some of the parameters from the analysis either to make it simpler or because of lack of information, the processed data will scatter more; at the same time, the general validity of the relationship established for the derived dimensionless quantities will be kept. To sum up, cross-sectional dimensions along with a few pertinent detail dimensions should be a good choice when deriving similarity relationships of furniture joint strength.

The quality of the glue bond is influenced by a number of factors, including the type of glue, its mode and amount of application, the state of surfaces to be bonded, and tightness of fit. These influencing factors cannot be involved in the analysis because of scarcity of information. As matters of design consideration, fit tightness, glue type and dowel surface formation were rarely treated as parameters in the available studies. Moreover, fit tightness, though regarded as design parameter, may be influenced by the precision of machining to such an extent that makes it uncertain as parameter.

Glue bonds in furniture joints fail in shear; hence, their strength cannot exceed the shear strength of wood. Therefore, except for pure glue-line failure, occasionally reported in the studies, measures of glue bond quality can be reasonably replaced by the shear strength of wood along their respective anatomical planes.

This partly answers the question of which material properties should be used as parameters. As further parameters, the bending strength of wood and in some cases tensile strength perpendicular to the grain orientation may be relevant.

Species of wood used in a joint define its material properties, but one must not forget about their inherent variability. Along with test result data in the literature, some physical properties of the materials used for test specimens are sometimes, but not always reported. Whenever they were available, we used them in processing the results. In other cases, we extracted values for the necessary material properties from three sources: Wood Handbook—Wood as an Engineering Material, USDF FPL 2010, The Wood Database (www.wood-database.com) as well as S. Molnár et al.: The World's Industrial Timbers (Földünk Ipari Fái, in Hungarian) ERFARET, Sopron, 2016. The strength values that apply to wood with 12% moisture content were adjusted to the actual moisture content of test pieces when reported.

4.6.2 Dowel Joints: Withdrawal Strength, In-Plane Bending Resistance, Out-of-Plane Bending Resistance

Figure 4.54 illustrates a dowel joint with the characteristic dimensions: h—rail width, v—rail thickness, d—dowel diameter, s—dowel spacing and L—dowel embedment length in the post-member.

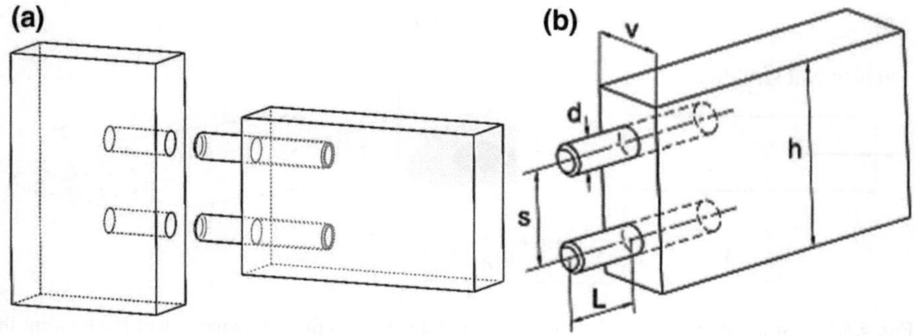

Fig. 4.54 T-shape dowel joint (**a**); characteristic dimensions (**b**)

The load-carrying capacity of T-shape dowel joints depends on a number of influencing factors. They can be purposefully grouped in two categories as follows:

design parameters

- rail member thickness,
- rail member width,
- post-member thickness,
- post-member width,
- wood species of rail and post,
- wood species of dowels,
- dowel diameter,
- dowel spacing,
- dowel surface (axially grooved, spirally grooved or smooth),
- length of embedment of the dowel in the rail member,
- length of embedment of the dowel in the post-member,
- clearance/interference of dowel and hole diameter,
- type of glue.

manufacturing parameters

- precision of hole diameter,
- precision of dowel diameter,
- precision of spacing,
- precision of assembly (size of gap between rail and post, if any),
- quantity of glue.

Dimensions of the cross-section of a post-member can be ignored in practice since a post's thickness is at least equal to that of the rail. Its width may limit the embedment length of the dowel; however, when this is not the case, a larger dimension does not improve the withdrawal strength.

A testing set-up for withdrawal resistance is demonstrated in Fig. 4.55 for flat frame members (a) as well as for frame members standing on the plane of frame (b).

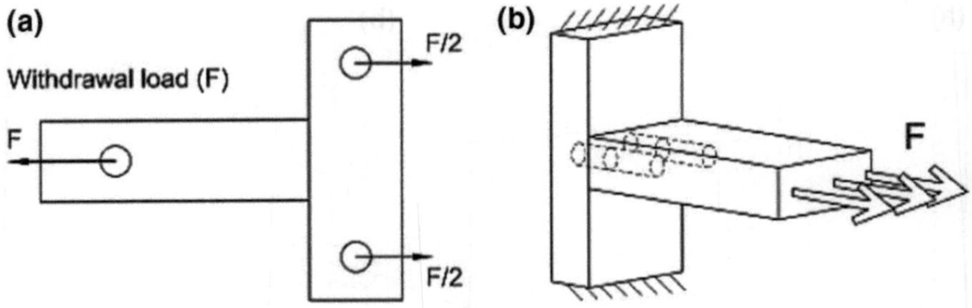

Fig. 4.55 Test arrangement for determining withdrawal strength of T-shape dowel joints; joint in flat frame (**a**), dowel joint of members with wider faces oriented perpendicular to the plane of frame (**b**)

Test data found in the literature do not always record the spacing of dowels. Likewise, the surface form of the dowels (smooth, axially or spirally grooved) is seldom given. Other data that are often missing are clearance/interference of the dowel and hole diameter and sometimes the type of glue as well. Therefore, in order to arrive at a more general treatment of the withdrawal capacity, we regard these mentioned parameters as noise factors contributing to the variability of test results, along with the manufacturing parameters listed, which are also often unknown. Embedment length in the rail member matters only when it is too short. Good manufacturing practice would not allow this situation to occur.

Withdrawal force of T-shape dowel joints is analysed next.

Seven variables were considered in processing the data on dowel joint withdrawal force when applying the method of dimensional analysis. These are: withdrawal force F [N], rail width h [m], rail thickness v [m], dowel diameter d [m], dowel embedment length L [m] in the post-member, shear strength τ_1 [N/m^2] of the wood species used for the post-member and shear strength of the dowel τ_2 [N/m^2].

By using the standard method of dimensional analysis, and after purposeful rearrangement of the original relationship obtained, the dimensionless quantity for withdrawal strength has the following form:

$$\frac{F}{L \cdot d \cdot \tau_1}$$

This ratio contains a meaningful expression in the denominator; $2 \cdot \pi \cdot L \cdot d$ is the surface area of the glued dowels which, multiplied by the shear strength results in force. The calculated values of the similarity number for withdrawal resistance range from 1.62 to 6.97, with an average value of 3.83. However, if the denominator was the correct predictor for withdrawal resistance, the similarity number would average at $2 \cdot \pi = 6.28$ indicating the need for other parameters.

Dimensional analysis supplies the following similarity equation:

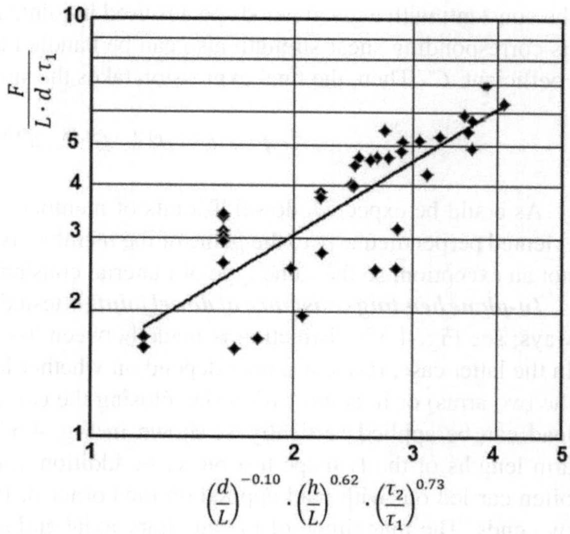

Fig. 4.56 Similarity relationship of dowel joint withdrawal force

$$\frac{F}{L \cdot d \cdot \tau_1} = C \cdot \left(\frac{d}{L}\right)^k \cdot \left(\frac{h}{L}\right)^l \cdot \left(\frac{v}{L}\right)^m \cdot \left(\frac{\tau_2}{\tau_1}\right)^n \tag{4.34}$$

The test data of 39 samples found in the literature were used for determining the value of the coefficient C, as well as the exponents k–n, (Eckelman 1979; Kamperidou and Vassiliou 2011; Najafi 2013; Sagmeister 1982; Wilczyński 1999). The values of variables evaluated in the analysis covered the ranges as below.

Rail width $h = 0.052$–0.080 m; rail thickness $v = 0.020$–0.025 m; dowel diameter $d = 0.008$–0.012 m; dowel embedment length $L = 0.012$–0.040 m; member shear strength $\tau_I = 5.60\text{E}+6$ to $16.01\text{E}+6$ N/m^2; dowel shear strength $\tau_1 = 11.00\text{E}+6$ to $16.01\text{E}+6$ N/m^2.

The analysis yielded the numerical values of the parameters in the similarity relationship as shown below:

$$\frac{F}{L \cdot d \cdot \tau_1} = 1.5243 \cdot \left(\frac{d}{L}\right)^{-0.10} \cdot \left(\frac{h}{L}\right)^{0.62} \cdot \left(\frac{\tau_2}{\tau_1}\right)^{0.73} \tag{4.35}$$

Figure 4.56 shows the representation of Eq. (4.35) with experimentally obtained results; the coefficient of determination R^2 turned out to be 0.7014.

From the above similarity equation, one may derive the formula predicting the withdrawal force of T-shape dowel joints:

$$F = 1.5243 \cdot d^{0.9} \cdot L^{0.48} \cdot h^{0.62} \cdot \tau_1^{0.27} \cdot \tau_2^{0.73} \tag{4.36}$$

In a great many cases, dowels are made of the same species of wood, namely beech; thus, the last power term of the equation can be numerically incorporated into the constant; with a given wood species used in jointed members, the power term of its corresponding shear strength also can be handled this way, resulting in the new coefficient, C'. Then, the final expression takes the simpler form

$$F = C' \cdot d^{0.9} \cdot L^{0.48} \cdot h^{0.62} \tag{4.36a}$$

As could be expected, dowel T-joints of members, the wider faces of which are oriented perpendicularly to the plane of the member axes as shown in Fig. 4.55b, are not an exception, so the same type of material constants is valid.

In-plane bending resistance of dowel joints is tested in practice in several different ways; see Fig. 4.57; distinction is made between T-shape and L-shape test pieces. In the latter case, resistance may depend on whether loaded in tension (i.e. opening the two arms) or in compression (i.e. closing the two arms). An opening or closing load can be applied vertically as shown in Fig. 4.57b), or diagonally with equal arm lengths of the L-shape test piece. In addition, a tension-type (opening) test is often carried out with load applied on the corner of the test piece supported at the two ends. The magnitude of the auxiliary axial and shear forces acting beside the bending moment depends on the type of loading and on the moment arm k, none of which are standardized.

For the objective of our study, it was expedient to treat these influencing effects as factors contributing to the scatter of the experimentally obtained results.

The following variables were involved in the dimensional analysis for establishing a useable relationship for in-plane bending resistance of dowel joints: ultimate bending moment M [Nm], rail width h [m], rail thickness v [m], dowel embedment length L [m] in the post-member, spacing of dowels s [m], shear strength of the wood species used for the post-member τ_1 [N/m^2], shear strength of the dowel τ_2 [N/m^2] and bending strength of the dowel σ [N/m^2].

The dimensionless quantity obtained for in-plane bending resistance by using dimensional analysis has the form

$$\frac{M}{d^2 \cdot h \cdot \sigma}$$

The rationale behind this formula is that the denominator's physical meaning is normal force time moment arm, ignoring again the multiplier $\pi/4$. The calculated values of the similarity number for in-plane bending resistance of the dowel joints ranged from 0.083 to 0.667, with an average value of 0.375. Dimensional analysis led to the following similarity equation:

$$\frac{M}{d^2 \cdot h \cdot \sigma} = C\left(\frac{v \cdot s \cdot L}{d^3}\right)^k \cdot \left(\frac{\tau_1}{\sigma}\right)^l \cdot \left(\frac{\tau_2}{\sigma}\right)^m \tag{4.37}$$

In order to determine the value of the coefficient C and the exponents k, l and m, data of 33 samples of different test piece configurations found in the literature were used (Eckelman 1971; Imizri et al. 2015; Kamperidou and Vassiliou 2012; Kaygin

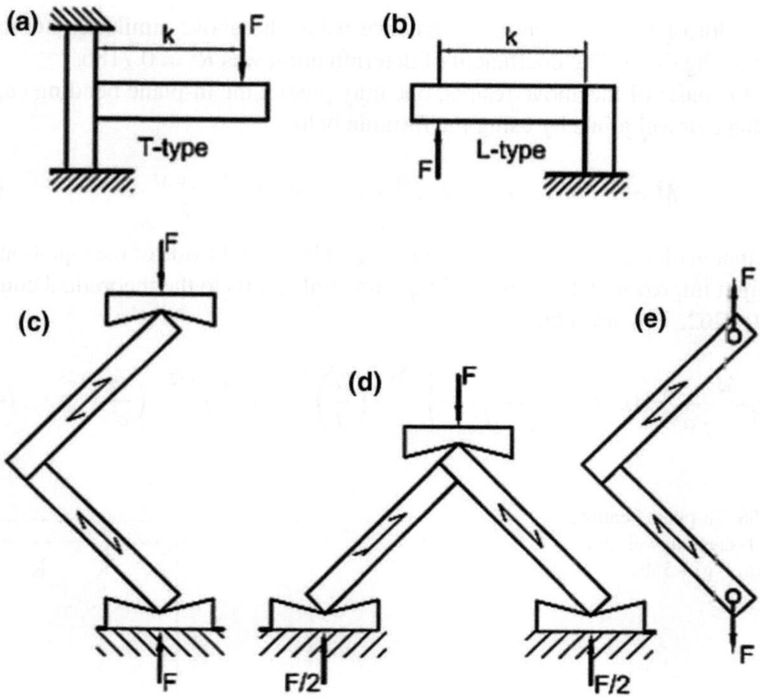

Fig. 4.57 In-plane bending test arrangements for T-shape and L-shape (corner) joints; bending load on T-shape joint (**a**), bending load on corner joint (**b**), bending of corner joint in compression (**c**), bending of corner joint in tension (**d**), (**e**)

et al. 2016; Palotai 1990; Sagmeister 1982; Vassiliou et al. 2016). The variables involved in the analysis covered the ranges as below:

Rail width $h = 0.043$–0.089 m; rail thickness $v = 0.019$–0.042 m; dowel diameter $d = 0.008$–0.011 m; dowel spacing $s = 0.0127$–0.0637; dowel embedment length $L = 0.0165$–0.0254 m; member shear strength $\tau_1 = 2.00\text{E}+6$ to $16.01\text{E}+6$ N/m^2; dowel shear strength τ_1: 11.00E+6 to 16.01E+6 N/m^2; dowel bending strength = 108.3E+6 to 125E+6 [N/m^2].

The kind of dowel T-joints shown in Fig. 4.55b when loaded in bending is expected to perform differently than dowel joints of flat members, because the plane of dowel axes is oriented across rather than along the grain orientation in the adjoining member. Indeed, using the shear strength in the plane across the grain direction rather than in the planes parallel with the grain for the parameter τ_1 aligns the data points of the test pieces with those of the bulk of the results (Fig. 4.58).

The relationship with the numerical values that we established for the dimensionless quantities is as follows:

$$\frac{M}{d^2 \cdot h \cdot \sigma} = 0.3635 \cdot \left(\frac{v \cdot s \cdot L}{d^3}\right)^{0.58} \cdot \left(\frac{\tau_1}{\sigma}\right)^{0.41} \cdot \left(\frac{\tau_2}{\sigma}\right)^{0.37} \tag{4.38}$$

The plot of in-plane bending resistance using the above similarity numbers is shown in Fig. 4.59. The coefficient of determination was $R^2 = 0.7186$.

On the basis of the above results, one may predict the in-plane bending capacity of T-shape dowel joints by using the formula below:

$$M = 0.3635 \cdot (v \cdot s \cdot L)^{0.58} \cdot d^{0.27} \cdot h \cdot \tau_1^{0.41} \cdot \tau_2^{0.37} \cdot \sigma^{0.22} \tag{4.39}$$

Inclusion of the dimensionless number (*h*/*d*) in the right side of the equation leads to a slight improvement of the fit of experimental results to the theoretical equation; $R^2 = 0.7262$. The new equation reads

$$\frac{M}{d^2 \cdot h \cdot \sigma} = 0.2458 \cdot \left(\frac{v \cdot s \cdot L}{d^3}\right)^{0.52} \cdot \left(\frac{h}{d}\right)^{0.22} \cdot \left(\frac{\tau_1}{\sigma}\right)^{0.32} \cdot \left(\frac{\tau_2}{\sigma}\right)^{0.33} \tag{4.38a}$$

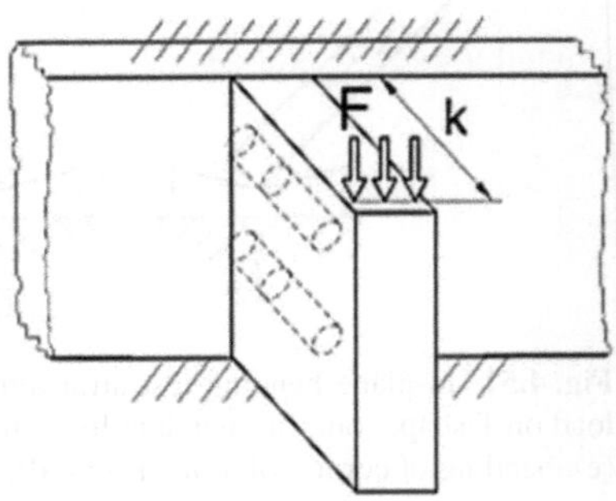

Fig. 4.58 In-plane bending test of T-shape dowel joint shown in Fig. 4.55b

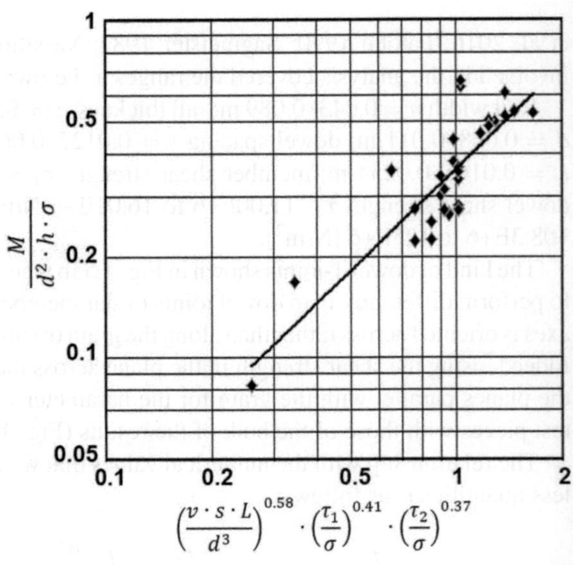

Fig. 4.59 Similarity relationship of dowel joint in-plane bending resistance

This relationship is illustrated in Fig. 4.60.

The new prediction formula is

$$M = 0.2458 \cdot (v \cdot s \cdot L)^{0.52} \cdot d^{0.44} \cdot h \cdot \tau_1^{0.37} \cdot \tau_2^{0.33} \cdot \sigma^{0.30} \quad (4.39a)$$

The shear strength and bending strength terms become constant for each wood species and can be handled by the resulting new coefficient, C'. The final expression then takes the form:

$$M = C' \cdot (v \cdot s \cdot L)^{0.58} \cdot d^{0.27} \cdot h \quad (4.39b)$$

Often a dowel joint undergoing in-plane bending fails as a result of withdrawal of the dowel loaded in tension, as a result of failure in shear. With this fact in view, a different dimensionless number for bending moment capacity was derived, leading to the similarity relationship with the dataset, as seen below, with $R^2 = 0.7797$.

The calculated values of the similarity number $M/(d \cdot L \cdot h \cdot \tau_1)$ cover the range of 0.87–5.81, averaging at 1.69.

$$\frac{M}{d \cdot L \cdot h \cdot \tau_1} = 2.1747 \cdot \left(\frac{v}{h}\right)^{0.46} \cdot \left(\frac{s}{h}\right)^{1.07} \cdot \left(\frac{L}{h}\right)^{-0.61} \cdot \left(\frac{\tau_2}{\tau_1}\right)^{0.61} \quad (4.40)$$

The plot of processed data from which this equation follows is shown in Fig. 4.61. The prediction formula that follows from Eq. (4.40) takes the form

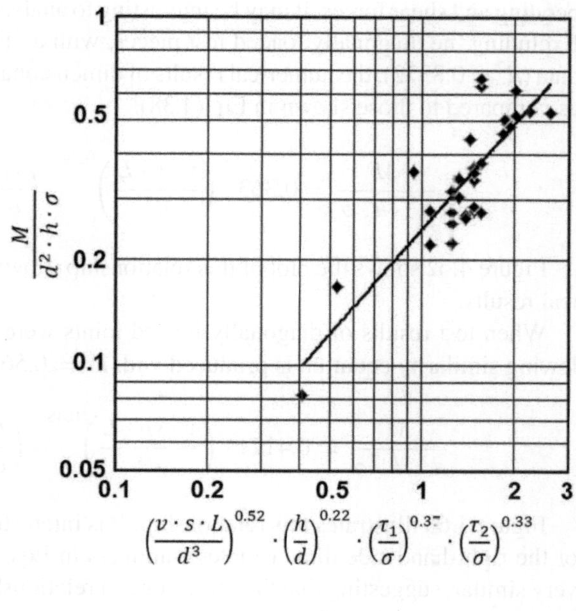

Fig. 4.60 Plot of the relationship of dimensionless quantities derived for in-plane bending resistance of dowel joints

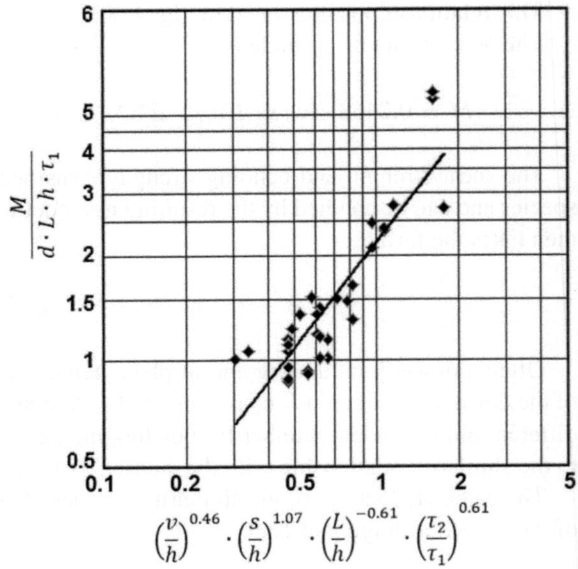

Fig. 4.61 Similarity relationship of dowel joint in-plane bending resistance when the expected failure mode is withdrawal of the dowel in tension

$$M = 2.1747 \cdot (v)^{0.46} \cdot d \cdot L^{0.39} \cdot h^{0.08} \cdot \tau_1^{0.39} \cdot \tau_2^{0.61} \quad (4.41)$$

Furthermore, it has to be noted that part of the dataset relates to tests in which the L-shaped test pieces were loaded diagonally, by a force tending to either open or close the corner joint. Since this type of loading results in a different combination of bending and shear forces, it may be interesting to analyse these test results separately. Excluding the diagonally loaded test pieces, with a fairly close fit to the processed data ($R^2 = 0.8372$), the numerical results of dimensional analysis change just slightly as compared to those shown in Eq. (4.38):

$$\frac{M}{d^2 \cdot h \cdot \sigma} = 0.363 \cdot \left(\frac{v \cdot s \cdot L}{d^3}\right)^{0.54} \cdot \left(\frac{\tau_1}{\sigma}\right)^{0.39} \cdot \left(\frac{\tau_2}{\sigma}\right)^{0.35} \quad (4.42)$$

Figure 4.62 shows the plot of this relationship showing experimental and theoretical results.

When test results of diagonally loaded joints were analysed separately, the following similarity equation is produced with $R^2 = 0.5636$:

$$\frac{M}{d^2 \cdot h \cdot \sigma} = 0.4171 \cdot \left(\frac{v \cdot sp \cdot L}{d^3}\right)^{0.48} \cdot \left(\frac{\tau_1}{\sigma}\right)^{0.37} \cdot \left(\frac{\tau_2}{\sigma}\right)^{0.32} \quad (4.43)$$

Figure 4.63 illustrates this relationship. It is interesting to note that the exponents of the right-hand side dimensionless numbers in Eqs. (4.38), (4.42) and (4.43) are very similar, suggesting that the same general relationship is valid irrespective of the loading configuration (Fig. 4.63).

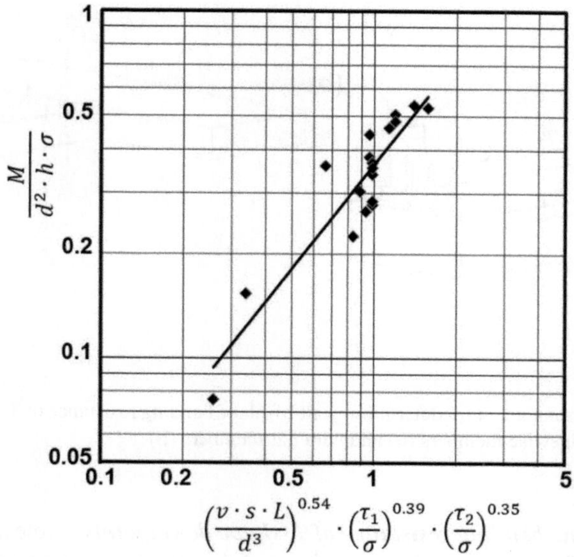

Fig. 4.62 Similarity relationship of dowel joint in-plane bending resistance, not including the test results of diagonally loaded joints

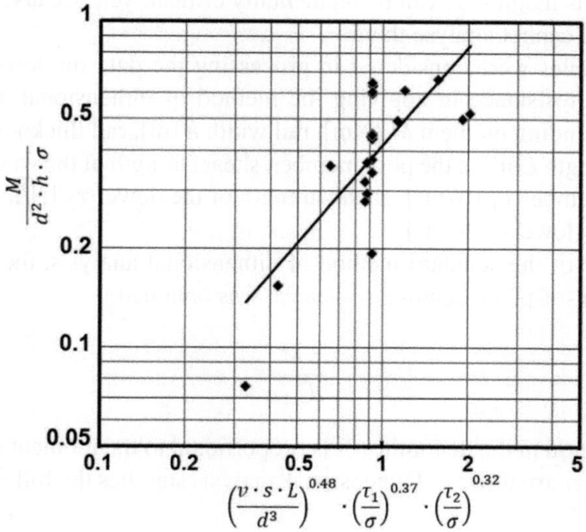

Fig. 4.63 Similarity relationship of dowel joint in-plane bending resistance in the case of diagonal loading

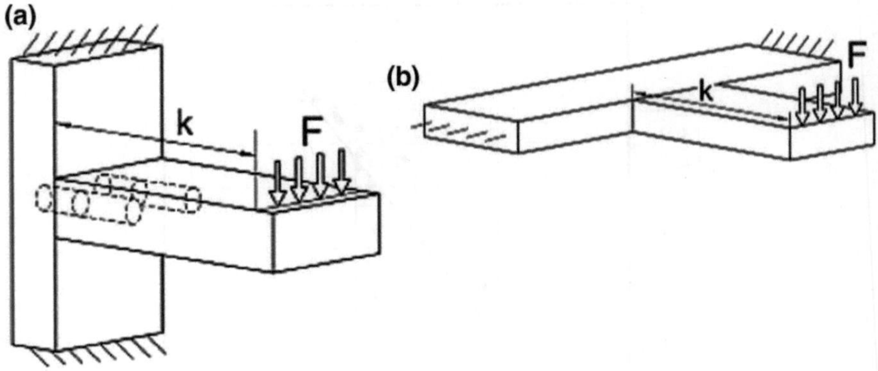

Fig. 4.64 Test arrangement for determining out-of-plane bending resistance of T-shape dowel joint in frames with edgewise members (**a**) and with flat members (**b**)

Out-of-plane bending resistance of T-shape dowel joints is relevant to sofa side frames or bed frames where the wider faces of the members are perpendicular to the plane of frame and bending forces act on the wider faces, as shown in Fig. 4.64.

Out-of-plane bending of joints of flat furniture frames occurs in a real loading situation, and its magnitude can be incidentally critical; yet, because of the scarcity of test data, we cannot analyse them.

Seven variables were considered in processing the data on dowel joint out-of-plane bending resistance in applying the method of dimensional analysis. These are: ultimate bending moment M [Nm], rail width h [m], rail thickness v [m], dowel embedment length L[m] in the post-member, shear strength of the wood species used for the post-member τ_1 [N/m^2], shear strength of the dowel τ_2 [N/m^2] and bending strength of the dowel σ [N/m^2].

Making use of the standard method of dimensional analysis, the dimensionless quantity for out-of-plane bending resistance was obtained:

$$\frac{M}{d^3 \cdot \sigma}$$

The expression in the denominator is proportional to the moment of resistance of the dowel time normal stress. Dimensional analysis supplies the following similarity equation:

$$\frac{M}{d^3 \cdot \sigma} = C \cdot \left(\frac{v}{d}\right)^k \cdot \left(\frac{L}{d}\right)^l \cdot \left(\frac{h}{d}\right)^m \cdot \left(\frac{\tau_1}{\sigma}\right)^n \cdot \left(\frac{\tau_2}{\sigma}\right)^o \qquad (4.44)$$

The numerical values for the constant C and the exponents k through o were determined using 6 pertinent test data found in the literature (Eckelman 1979b; Kamperidou et al. 2011). The variables among the test pieces covered these ranges:

Rail width $h = 0.050–0.076$ m; rail thickness $v = 0.025–0.038$ m; dowel diameter $d = 0.0095–0.0120$ m; dowel embedment length $L = 0.019–0.021$ m; member shear strength $\tau_1 = 6.83\text{E}+6$ to $13.86\text{E}+6$ N/m^2; dowel shear strength $\tau_1 = 13.86\text{E}+6$ to $16.01\text{E}+6$ N/m^2; dowel bending strength $\sigma = 108.3\text{E}+6$ to $134\text{E}+6$ [N/m^2].

The relationship with the numerical values of the exponents of dimensionless quantities and the coefficient is:

$$\frac{M}{d^3 \cdot \sigma} = 0.7026 \cdot \left(\frac{v}{d}\right)^{1.48} \cdot \left(\frac{L}{d}\right)^{0.64} \cdot \left(\frac{h}{d}\right)^{0} \cdot \left(\frac{\tau_1}{\sigma}\right)^{0.38} \cdot \left(\frac{\tau_2}{\sigma}\right)^{0.30} \tag{4.45}$$

The calculated values of the similarity number for out-of-plane bending resistance of the dowel joints analysed ranged from 0.406 to 1.65.

The plot of out-of-plane bending capacity using the above similarity numbers is shown in Fig. 4.65.

The coefficient of determination is $R^2 = 0.7332$. The prediction equation takes the form

$$M = 0.7026 \cdot d^{0.88} \cdot v^{1.48} \cdot L^{0.64} \cdot \tau_1^{0.38} \cdot \tau_2^{0.30} \cdot \sigma^{0.32} \tag{4.46}$$

We demonstrate the results of another analysis with a different similarity number of

$$\frac{M}{v^2 \cdot d \cdot \sigma}$$

The similarity equation that was derived using the test result is shown below (Eq. 4.47).

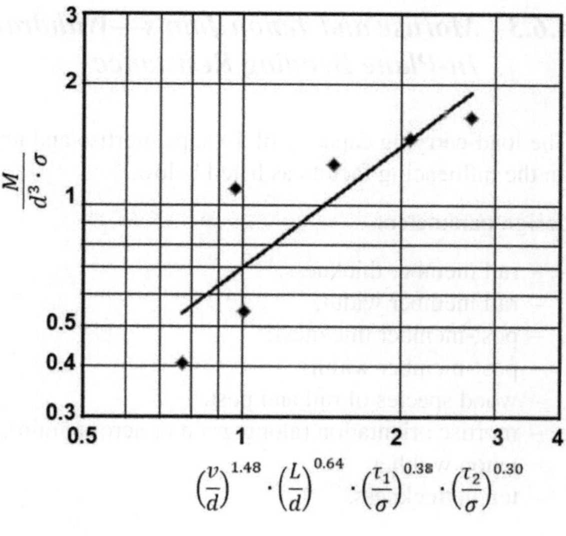

Fig. 4.65 Similarity relationship of dowel joint out-of-plane bending resistance

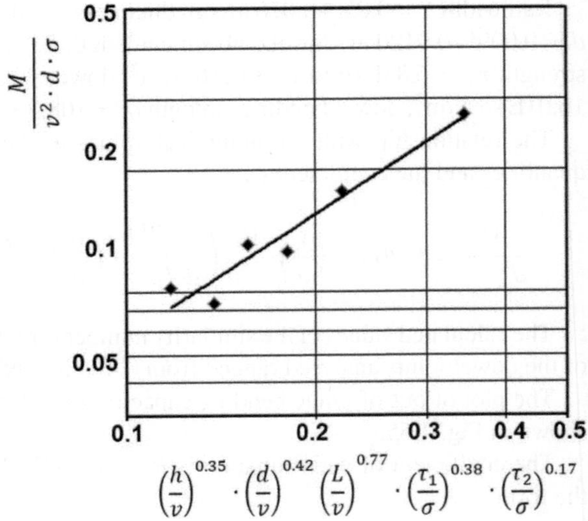

Fig. 4.66 Similarity relationship of dowel joint out-of-plane bending resistance

$$\frac{M}{v^2 \cdot d \cdot \sigma} = 0.7805 \cdot \left(\frac{h}{v}\right)^{0.35} \cdot \left(\frac{d}{v}\right)^{0.42} \cdot \left(\frac{L}{v}\right)^{0.77} \cdot \left(\frac{\tau_1}{\sigma}\right)^{0.38} \cdot \left(\frac{\tau_2}{\sigma}\right)^{0.17} \quad (4.47)$$

The relationship is illustrated in Fig. 4.66 as a plot of the experimental results and the line corresponding to the theoretical equation; $R^2 = 0.9353$. Equation (4.48) was derived for the prediction of the out-of-plane bending resistance of dowel joints.

$$M = 0.7805 \cdot h^{0.35} \cdot d^{1.42} \cdot v^{1.23} \cdot L^{0.77} \cdot \tau_1^{0.38} \cdot \tau_2^{0.17} \cdot \sigma^{0.45} \quad (4.48)$$

4.6.3 Mortise and Tenon Joints—Withdrawal Strength, In-Plane Bending Resistance

The load-carrying capacity of T-shape mortise and tenon joints (Fig. 4.67) depends on the influencing factors as listed below.

design parameters

- rail member thickness,
- rail member width,
- post-member thickness,
- post-member width,
- wood species of rail and post,
- mortise orientation (along grain or across grain),
- tenon width,
- tenon thickness,

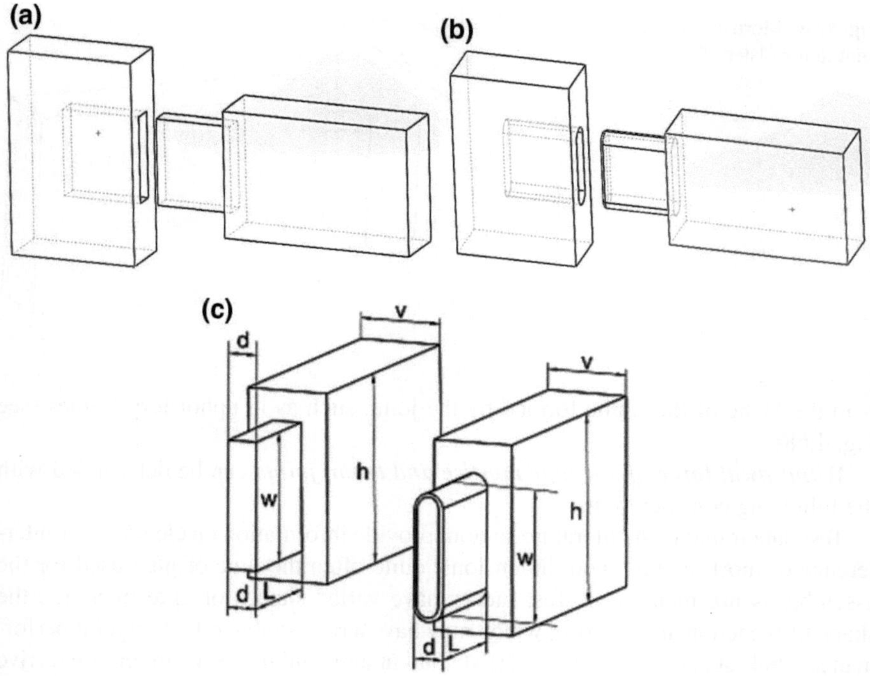

Fig. 4.67 Mortise and tenon joints; square edge mortise and tenon (**a**); rounded edge mortise and tenon (**b**); characteristic dimensions (**c**)

- length of the tenon,
- through mortise or blind mortise,
- shape of tenon (rectangular or rounded),
- shape of mortise (rectangular or rounded),
- clearance/interference of mortise and tenon thickness,
- type of glue (Fig. 4.67).

manufacturing parameters

- precision of tenon thickness and width,
- precision of mortise thickness and width,
- precision of assembly (size of gap between rail and post if any),
- quantity of glue.

Dimensions of the post-member cross-section do not matter in the analysis since the thickness of a post is at least equal to that of the rail. Its width may limit the length of the tenon; on the other hand, extra larger dimension does not improve its strength. The orientation of the mortise must be dealt with when the mortise width is measured in the direction perpendicular to the principal plane of the joint. This situation may occur when the smaller cross-sectional dimension of the joint member

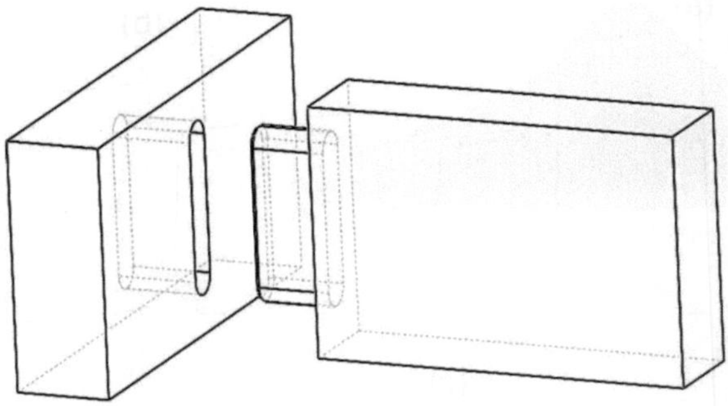

Fig. 4.68 Mortise and tenon joint in upholstery frames

is in the plane of the frame formed by the joint, such as in upholstery frames (see Fig. 4.68).

Withdrawal force of T-shape mortise and tenon joints can be determined with the following considerations.

Test data found in the literature seldom provide information on clearance or interference of mortise and tenon dimensions; quite often the type of glue used for the assembly is not mentioned. Test pieces have varied shapes of glue lines, i.e. the shape of the tenon and mortise, which may have a remarkable effect on joint performance (Prekrat and Smardzewski 2010; Tankut and Tankut 2004). In order to arrive at a more general treatment of the withdrawal capacity, we regard these mentioned parameters as noise factors contributing to the variability of test results, along with the manufacturing parameters listed, which are also generally unknown.

On the basis of these considerations, we used eight variables in applying the method of dimensional analysis with the collected data on mortise and tenon joint withdrawal force. They were: withdrawal force F [N], rail width h [m], rail thickness v [m], tenon thickness d [m], tenon width w [m], tenon length L [m], shear strength τ [N/m^2] and tensile strength across the grain direction $\sigma_\perp$ [N/m^2] of the wood. For the type of T-shape joints shown in Fig. 4.68 shear strength in the plane across the grain direction is used because of the position of the mortise in the post-member with respect to the grain direction.

The use of dimensional analysis led to the similarity number of tenon withdrawal strength as follows:

$$\frac{F}{h \cdot v \cdot \sigma_\perp}$$

The denominator of this formula is the area times the normal strength resulting normal force. With this dimensionless number of withdrawal force, the similarity equation takes the form

$$\frac{F}{h \cdot v \cdot \sigma_\perp} = C \cdot \left(\frac{d}{v}\right)^k \cdot \left(\frac{w}{v}\right)^l \cdot \left(\frac{L}{v}\right)^m \cdot \left(\frac{\tau}{\sigma_\perp}\right)^n \tag{4.49}$$

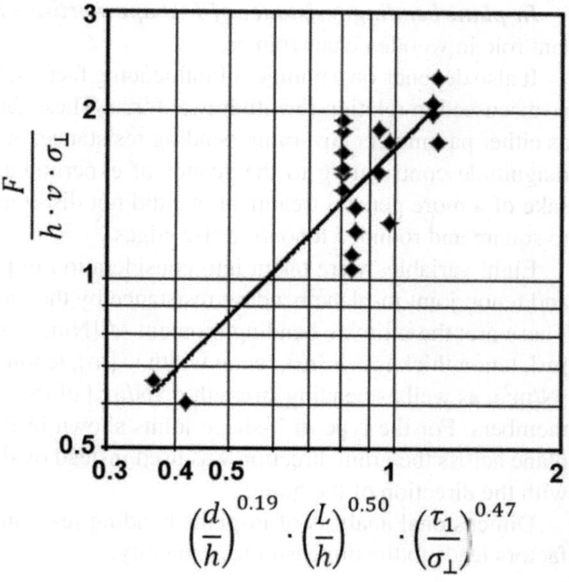

Fig. 4.69 Similarity relationship of mortise and tenon joint withdrawal force

For determining the numerical constants in Eq. (4.49), we used data from 20 relevant test series found in the literature (Acar et al. 2016; Kamperidou and Vassiliou 2010; Sagmeister 1982). The parameters of the test pieces involved in the analysis covered the ranges as below:

Rail width $h = 0.043–0.050$ m; rail thickness $v = 0.022$ m; tenon thickness $d = 0.009–0.013$ m; tenon width $w = 0.023–0.050$ m; tenon length $L = 0.020–0.035$ m; shear strength $\tau = 2.0\text{E}+6$ to $13.9\text{E}+6$ N/m^2; tensile strength across the grain $\sigma_\perp = 2.1\text{E}+6$ to $7.5\text{E}+6$ N/m^2.

The calculated values of the similarity number for withdrawal resistance of the mortise and tenon joints analysed ranged from 0.33 to 2.29, with an average value of 1.43. Evaluation of experimental results yielded the numerical values of the parameters in the similarity relation with $R^2 = 0.7313$ as follows.

$$\frac{F}{h \cdot v \cdot \sigma_\perp} = 1.7212 \cdot \left(\frac{d}{h}\right)^{0.19} \cdot \left(\frac{L}{h}\right)^{0.50} \cdot \left(\frac{\tau}{\sigma_\perp}\right)^{0.47} \qquad (4.50)$$

The plot of withdrawal strength by using the above similarity numbers is shown in Fig. 4.69.

Based on the above results, one may predict the withdrawal force of T-shape mortise and tenon joints by using Eq. (4.51), derived from Eq. (4.50).

$$F = 1.7212 \cdot h^{0.31} \cdot v \cdot L^{0.50} \cdot d^{0.19} \cdot \tau^{0.47} \cdot \sigma_\perp^{0.53} \qquad (4.51)$$

In-plane bending resistance of T-shape mortise and tenon joints plays an important role in wooden chair frames.

It also depends on a number of influencing factors, which are essentially the same as discussed in relation to withdrawal force. These factors were expediently treated as either parameters governing bending resistance, or factors of known or unknown magnitude contributing to the scatter of experimentally obtained results. For the sake of a more general treatment, we did not distinguish between test data relating to square and rounded tenon/mortise edges.

Eight variables were taken into consideration in processing the data on mortise and tenon joint in-plane bending resistance by the method of dimensional analysis. These are: the ultimate bending moment M [Nm], rail width h [m], rail thickness v [m], tenon thickness d [m], tenon width w [m], tenon length L [m], shear strength τ [N/m^2], as well as bending strength σ [N/m^2] of the wood species used for the joint members. For the type of T-shape joints shown in Fig. 4.68, shear strength in the plane across the grain direction was used instead of shear strength in planes parallel with the direction of the grain.

Dimensional analysis of in-plane bending resistance and the related influencing factors leads to the dimensionless quantity:

$$\frac{M}{h^2 \cdot d \cdot \sigma}$$

In this formula, the physical meaning of the denominator is the section moment of resistance time normal stress, resulting in bending moment. Dimensional analysis supplies the similarity equation as follows:

$$\frac{M}{h^2 \cdot d \cdot \sigma} = C \cdot \left(\frac{v}{h}\right)^k \cdot \left(\frac{w}{h}\right)^l \cdot \left(\frac{L}{h}\right)^m \cdot \left(\frac{\sigma}{\tau}\right)^n \quad (4.52)$$

Test data of 100 different joint configurations were taken from the literature (Atar et al. 2017; Bardak et al. 2017; Boadu and Antwi-Boasiako 2017; Eckelman 1971; Erdil et al. 2005; Imirzi et al. 2015; Kamperidou and Vassiliou 2012; Kasal et al. 2015; Márkus 1990; Palotai 1990; Ratnasingam et al. 2013; Ratnasingam and Ioras 2013; Sagmeister 1982; Sparkes 1968; Tankut and Tankut 2005; Vassiliou et al. 2016) in order to determine the numerical values of the coefficient C, as well as the exponents k through n. The variables among the test pieces involved in the analysis covered the ranges below:

Rail width $h = 0.030$–0.076 m; rail thickness $v = 0.019$–0.030 m; tenon thickness $d = 0.007$–0.012 m; tenon width $w = 0.013$–0.076 m; tenon length $L = 0.013$–0.050 m; member shear strength τ = 9.00E+6 to 16.01E+6 N/m^2; dowel shear strength τ_1 = 8.20E+6 to 16.07E+6 N/m^2; bending strength σ = 59E+6 to 123E+6 N/m^2.

Using the numerical values determined from the test results, the similarity equation is

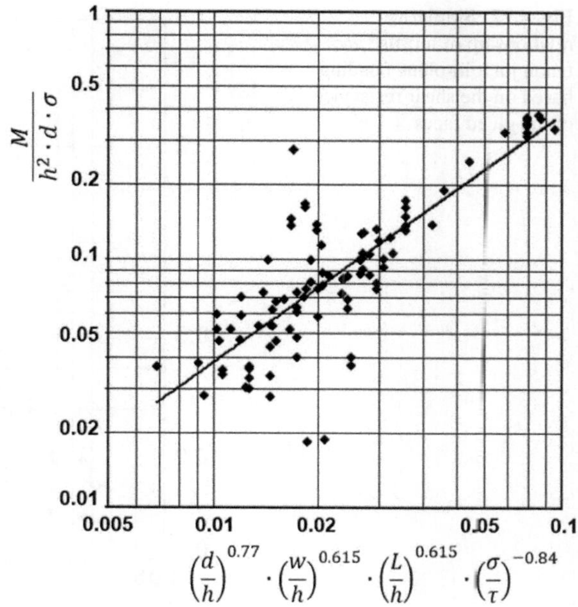

Fig. 4.70 Similarity relationship of mortise and tenon joint in-plane bending resistance demonstrated by the log–log plot of experimental and theoretical values

$$\frac{M}{h^2 \cdot d \cdot \sigma} = 3.8576\left(\frac{d}{h}\right)^{0.77} \cdot \left(\frac{w}{h}\right)^{0.615} \cdot \left(\frac{L}{h}\right)^{0.615} \cdot \left(\frac{\sigma}{\tau}\right)^{-0.84} \tag{4.53}$$

The values of the similarity number calculated from test results for in-plane bending resistance of mortise and tenon joints range from 0.016 to 0.383, with an average value of 0.116.

The plot of similarity numbers determined from the analysis of test data is shown in Fig. 4.70; the coefficient of determination R^2 turned out to be 0.6587.

On the basis of the above results, one may predict the in-plane bending capacity of T-shape dowel joints by using the formula below:

$$M = 3.8576 \cdot d^{0.77} \cdot w^{0.615} \cdot L^{0.615} \cdot \sigma^{0.16} \cdot \tau^{0.84} \tag{4.54}$$

Because the dimensionless number of joint resistance used in Eq. (4.52) had been derived on the basis of the bending strength of wood, an attempt was made to use a different similarity number, based on the shear strength of glue line. The new similarity equation determined from the test results by dimensional analysis reads

$$\frac{M}{L^2 \cdot w \cdot \tau} = 2.1534 \cdot \left(\frac{v}{L}\right)^{1.4} \cdot \left(\frac{w}{L}\right)^{-0.61} \cdot \left(\frac{d}{L}\right)^{0.7} \tag{4.55}$$

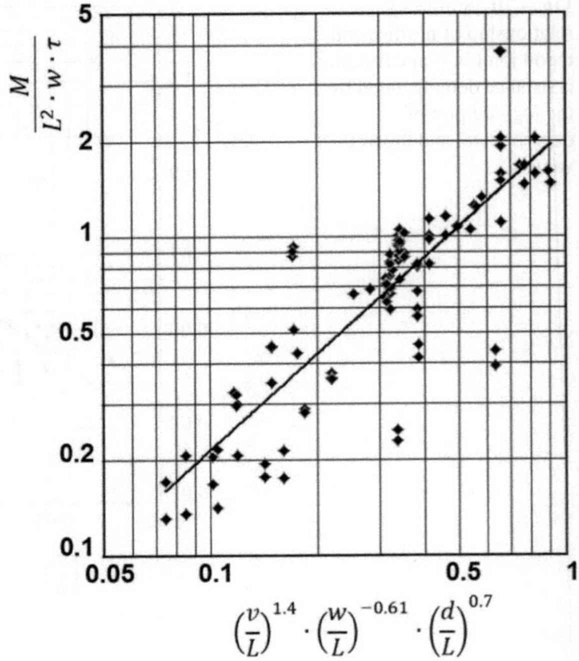

Fig. 4.71 Similarity relationship of mortise and tenon joint in-plane bending based on the shear resistance of the glued faces

Figure 4.71 shows the fit of test results to the theoretical relationship; in this case, the coefficient of determination increases, $R^2 = 0.7153$. The new relationship to assess in-plane resistance of a mortise and tenon joint takes the form

$$M = 2.1534 \cdot v^{1.4} \cdot w^{0.39} \cdot L^{0.51} \cdot d^{0.7} \cdot \tau \tag{4.56}$$

4.6.4 *Open Mortise and Tenon Corner Joint In-Plane Bending Resistance*

Load-carrying capacity of open mortise corner joints depends on the influencing factors as listed below.

design parameters

- rail member thickness,
- rail member width,
- post-member thickness,
- post-member width,
- wood species of rail and post,
- number of tenons,

– tenon thickness(es),
– clearance/interference of mortise and tenon thickness,
– type of glue.

manufacturing parameters

– precision of tenon thickness and width,
– precision of mortise thickness and width,
– precision of assembly (size of gap if any),
– quantity of glue.

In the above list of parameters, post refers to the member ending an open mortise, while the rail is the one with a tenon. The length of the tenon corresponds to the width of the post, while mortise depth is equal to rail width; see Fig. 4.72.

In-plane bending resistance of open mortise and tenon corner joints (bridle joints) is typically important in-plane frames such as window sashes, cabinet door frames or chair side frames. Test setups used in practice are the same as for L-shape or corner joints, demonstrated in Fig. 4.57c–e.

For the sake of a more general treatment of the corner joint performance, the following factors were treated as noise contributing to the variability of test results: clearance or interference data of mortise and tenon fit, the type of glue used for the assembly, as well as the generally unknown precision of manufacturing.

Eight variables were used to process the data on open mortise corner joints when applying the method of dimensional analysis. These are: ultimate bending moment M [N/m], rail width h [m], rail thickness v [m], tenon thickness d [m], tenon length L [m], number of glue lines n [–], shear strength τ [N/m^2], tensile strength across the grain direction σ_{perp} [N/m^2] as well bending strength σ [N/m^2] of the wood species.

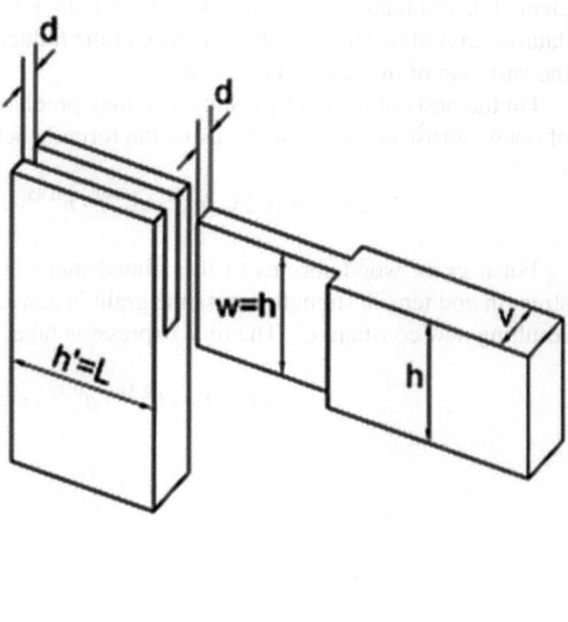

Fig. 4.72 Open mortise and tenon corner joint and its characteristic dimensions

By using the standard method of dimensional analysis, the dimensionless quantity for bending capacity was obtained as

$$\frac{M}{h^3 \cdot \tau}$$

The denominator in this fraction is a quantity proportional to the torsional moment of resistance of the glued surface of the tenon times shear strength. That results torque in planes parallel to the glued surfaces. The form of the similarity equation derived is the following:

$$\frac{M}{h^3 \cdot \tau} = C \cdot \left(\frac{v}{h}\right)^k \cdot \left(\frac{d}{h}\right)^l \cdot (n)^m \cdot \left(\frac{\sigma}{\tau}\right)^n \tag{4.57}$$

In the test pieces, the parameters of the equation had values within the following range: rail width = tenon width $h = 0.042$–0.050 m; rail thickness $v = 0.03$–0.034 m; tenon thickness $d = 0.007$–0.011 m; post width = tenon length $L = 0.042$–0.050 m; number of glue lines $n = 2, 3, 4, 6$; shear strength $\tau = 9.0\text{E+6}$ to 14.0E+6 N/m^2; bending strength $\sigma = 59.8\text{E+6}$ to 110.0E+6 N/m^2; tensile strength across the grain $\sigma_\perp = 4.2\text{E} + 6$ to $8.0\text{E} + 6\,\text{N/m}^2$.

Evaluation of the available test results yielded the following relationship:

$$\frac{M}{h^3 \cdot \tau} = 0.1023 \cdot \left(\frac{v}{h}\right)^0 \cdot \left(\frac{d_1}{h}\right)^{0.49} \cdot (n)^{0.85} \cdot \left(\frac{\sigma}{\tau}\right)^{0.57} \cdot \left(\frac{\sigma_\perp}{\tau}\right)^{0.32} \tag{4.58}$$

The calculated values of the similarity number for in-plane bending resistance of the open corner joints ranged from 0.108 to 0.556, with an average value of 0.369.

The similarity relationship is illustrated by the plot of experimental results with theoretical function appearing as a straight line on log scale, in Fig. 4.73. The coefficient of determination, R^2 turned out to be 0.4923, reflecting the scatter in the limited dataset, caused by the ignorance of some of the influencing factors that contribute to the variation of the strength of test pieces.

On the basis of the above results, one may predict the in-plane bending capacity of open mortise corner joints by using the formula below:

$$M = 0.1023 \cdot h^{2.51} \cdot d_1^{0.49} \cdot n^{0.85} \cdot \sigma_\perp^{0.32} \cdot \tau^{0.11} \tag{4.59}$$

For a given wood species of the jointed members the shear strength, bending strength and tensile strength across the grain is constant and can be handled by the resulting new constant, C'. The final expression takes the form:

$$M = C' \cdot h^{2.51} \cdot d^{0.49} \cdot n^{0.85} \tag{4.59a}$$

Fig. 4.73 Similarity relationship of open mortise and tenon joint in-plane bending resistance

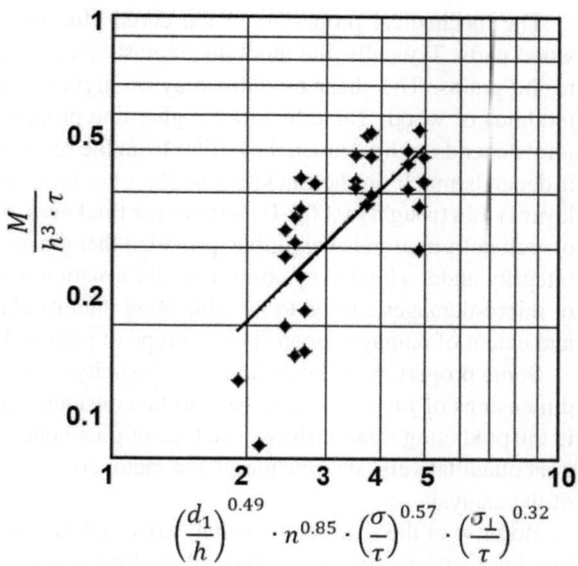

4.6.5 *Influencing Factors of Stiffness of Furniture Joints—General Remarks*

As stated in Sect. 4.5.4, furniture joints, and wood joints in general, cannot be assumed to be rigid even when that is intended. Rather, they exhibit finite stiffness.

The factors contributing to the deformability of glued wood joints can be grouped as

- properties of wood,
- orientation of grain direction in the joint,
- glue-line properties.

The orthotropic nature of wood is the main reason of the finite stiffness of glued furniture joints. Material contained in the joint domain is much easier to deform in the across the grain directions than along the fibres, and because the grain direction of one of the joining members is always across (typically at a right angle or close to a right angle) to the member-end forces, the internal deformation of the joint area is larger than with isotropic materials. Another reason is that the moduli of rigidity ($G_{i,j}$ shear moduli, $i, j = 1, 2, 3$) of wood are much lower compared to the moduli of elasticity (E_i, $i = 1, 2, 3$) than in most structural materials. For example, the E–G ratio of steel is 2.65–1; for hardwoods, $E_L/G_{LT} \approx 18$, $E_L/G_{TR} \approx 50$ and $E_L/E_T \approx 20$ (Bodig 1983). It also follows that torsional deformability of wood around the grain direction is especially low.

The orientation of the grain direction in the different parts of a joint is defined by the type and the dimensional details of the joint.

The mechanical properties of the cured glue line are different of those of the wood parts. Typically, the modulus of elasticity is inferior to that of wood, parallel to the grains. The shear modulus may be higher or lower than the relevant shear modulus of wood. The effect of the glue-line properties on the deformation of the joint depends on how much they differ from the respective wood properties. Besides, it depends much on the thickness of the glue line. In general, because of the glue layer is thin (roughly 0.002–0.04 times the thickness of wood), its contribution to the overall deformation is negligible, provided that the glue bond perfectly preserves its integrity under a load. Any observable deformation of the glue line is more the result of micro-damages caused to the glue bond than its elastic deformation. The chance and extent of damage depends on the type of glue and conditions of gluing.

Some properties of the orthotropic elasticity of wood along with the appropriate dimensions of joint geometry have to be considered as relevant influencing factors in the predicting relationships. The type of glue bond is difficult, if not impossible to treat quantitatively. We consider it as a factor contributing to the generality of results of the analysis.

Because of the scarcity of available research results other than for in-plane bending, joint stiffness under this type of loading was analysed for dowel joints as well as mortise and tenon joints.

4.6.6 In-Plane Bending Stiffness of Dowel Joints

In-plane bending stiffness of dowel joints has been investigated by a number of researchers. We analysed the test data published by several of them (Eckelman 1971; Imirzi et al. 2015; Vassiliou et al. 2016; Wambier and Wilczyński 2000). We found that when applying the method of dimensional analysis, the modulus of elasticity in tension across the grain direction E_{perp} is a useful parameter of the material. The modulus of elasticity E in bending of the dowel could probably be of interest with dowels made of different species; in the available datasets, there were only beech and sugar maple, with similar E-values so we had to omit this parameter from the analysis. The necessary joint dimensions were: dowel diameter d, rail width h, dowel spacing s and embedment length L of the dowel into the post.

The ranges these variables covered in the 74 test piece variants were: modulus of elasticity in tension perpendicular to the grain direction (average of E_R and E_T) E_{perp} = 5.5 E+8 to 1.6 E+9 N/m^2; dowel diameter d = 0.006–0.012 m; dowel spacing s = 0.00127–0.0635 m; rail width h = 0.0381–0.0889 m; dowel embedment length L = 0.016–0.032 m.

The dimensionless expression established for the dowel joint rotational rigidity is

$$\frac{K}{d^2 \cdot h \cdot E_\perp}$$

where K (Nm/rad) is the rotational stiffness of the joint.

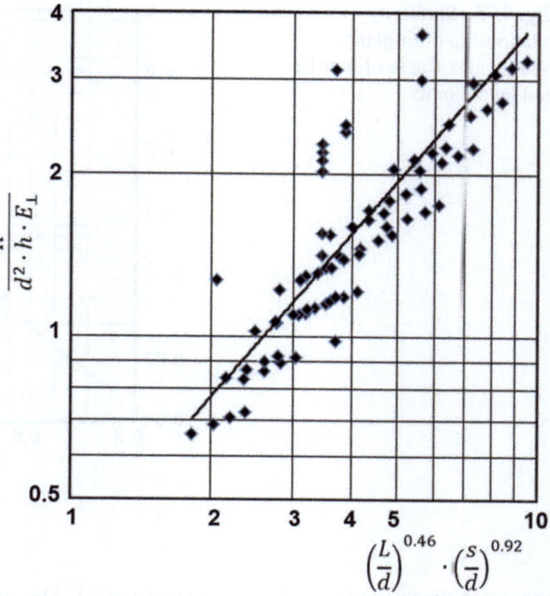

Fig. 4.74 Similarity relationship of in-plane rotational stiffness of dowel joints

The following similarity equation was derived from the test results:

$$\frac{K}{d^2 \cdot h \cdot E_\perp} = 0.3876 \cdot \left(\frac{L}{d}\right)^{0.46} \cdot \left(\frac{s}{d}\right)^{0.92} \quad (4.60)$$

The constants in Eq. (4.60) were determined with $R^2 = 0.7289$. From this similarity equation, the formula for the assessment of the rotational stiffness of a dowel joint is

$$K = 0.3876 \cdot E_\perp \cdot h \cdot d^{0.62} \cdot L^{0.46} \cdot s^{0.92}. \quad (4.61)$$

Figure 4.74 shows the plot of the test results with the fitted straight line, using a log–log scale.

4.6.7 In-Plane Bending Stiffness of Mortise and Tenon Joints

We collected available test results on the in-plane rotational stiffness of mortise and tenon joints for analysis, not including open mortise corner joints (bridle joints). The data of 126 samples of varying parameters were published as referred here (Hajdarević and Martinović 2014; Erdil et al. 2005; Imirzi et al. 2015; Vassiliou et al. 2016; Wilczyński and Wambier 2003). In applying the dimension analysis, the relevant parameters were: moduli of elasticity both in bending and in tension across

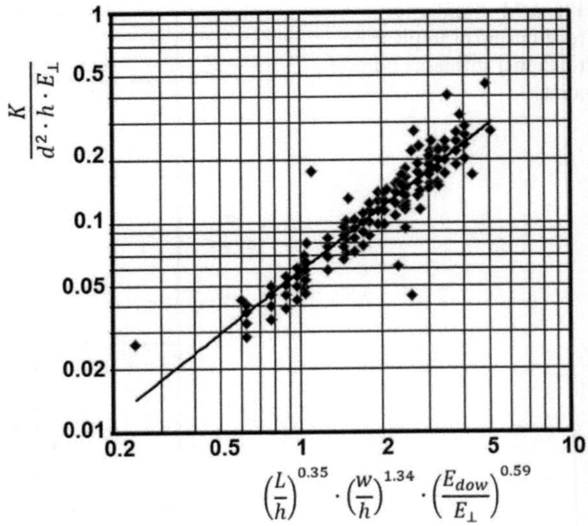

Fig. 4.75 Similarity relationship of in-plane rotational stiffness of mortise and tenon joints

the grain direction (*E* and $E_{perp,}$ respectively). The important joint dimensions were: rail width *h*, rail thickness *v*, tenon width *w* and tenon length *L*.

These variables occurred in the following ranges: modulus of elasticity in bending $E = 7.27$ E+9 to 1.425 E+10 N/m^2; modulus of elasticity in tension perpendicular to the grain direction (average of E_R and E_T) $E_{perp} = 5.5$ E+8 to 1.38 E+9 N/m^2; rail width $h = 0.05$–0.079 m; rail thickness $v = 0.02$–0.03; tenon length $L = 0.01$–0.051 m; tenon width $w = 0.18$–0.0635 m.

For the in-plane rotational stiffness of the mortise and tenon joints, the similarity number has the following form:

$$\frac{K}{v \cdot h^2 \cdot E_\perp}$$

where *K* (Nm/rad) is the rotational stiffness of the joint.

The similarity equation that resulted from the analysis of the test data takes the form and numerical values of the coefficient and exponents below:

$$\frac{K}{v \cdot h^2 \cdot E_\perp} = 0.0592 \cdot \left(\frac{L}{h}\right)^{0.35} \cdot \left(\frac{w}{h}\right)^{1.34} \cdot \left(\frac{E_{\text{dow}}}{E_\perp}\right)^{0.59} \tag{4.62}$$

Figure 4.75 demonstrates the relationship in log–log scale showing experimental and theoretical results. The latter represented by the straight line was obtained with $R^2 = 0.8284$. The expression for the prediction of in-plane rotational stiffness of mortise joints reads

$$K = 0.0592 \cdot E^{0.59} \cdot E_\perp^{0.41} \cdot h^{0.31} \cdot L^{0.35} \cdot w^{1.34} \tag{4.63}$$

4.7 Engineering Design of Wood-Based Products—Designing Capable and Reliable Products

To improve customer satisfaction and business competitiveness, companies need to reduce the levels of non-conformance and the failure costs associated with poor product design and development. Failure costs that occur during production, or when the product is in service with the customer, are the largest cost in a manufacturing business and include those attributable to rework, scrap, warranty claims, product recall and product liability claims (Russel and Taylor 1995). Attention needs to be focused on the quality and reliability of the design as early as possible in the product development process. This can be achieved by understanding the potential for variability in design parameters and the likely failure consequences when the product is used. These potentials are present in manufacture, assembly and service conditions.

The inherent variability of processes and products needs thinking in terms of tolerances. There are two kinds of tolerances in product design:

- product tolerances,
- manufacturing tolerances.

Product tolerances relate properties of products that are linked to functional and aesthetic requirements. Manufacturing tolerances have to be specified in relation to part or assembly properties as results of production processes to keep product properties within acceptable limits with a reasonable probability; in other words, to produce *capable products*.

The functional requirements of the design become detailed into dimensional tolerances or into attributes of the components or assembly. The ability of the manufacturing process to consistently provide dimensions within tolerance should be an integral part of the design. The two basic process capability indices C_p and C_{pk} have been most commonly used to measure this ability (ISO 3534-2 1993).

4.7.1 Tolerance and Machining Accuracy

For the sake of a more general treatment of the problem, instead of the conventional indices of process/machine capability C_p and C_{pk} we used a unique metric C that expresses utilization of the capability of processes or machines.

Tolerance, machining accuracy and numbers of unacceptable parts due to machining errors are closely correlated.

Tolerance width, ΔT, (see Fig. 4.76) may be:

- symmetric ($\pm\Delta T/2$),
- asymmetric ($-T_1$ and $+T_2$),
- one-sided ($-T_1$ and 0, or 0 and $+T_2$).

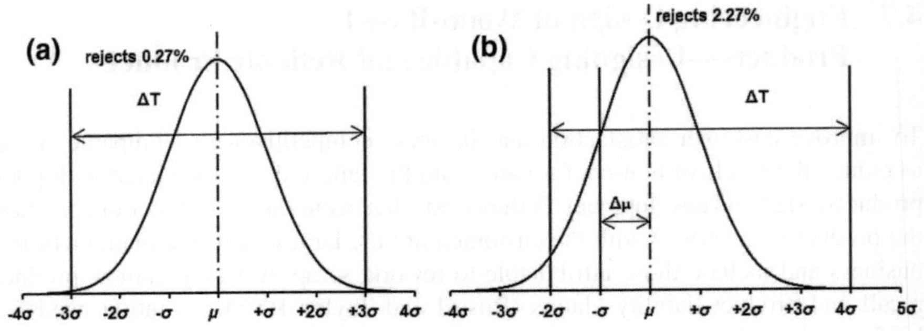

Fig. 4.76 Symmetric tolerance (**a**) and asymmetric tolerance with shifting of the design value (**b**) and their interaction with machining accuracy

Machining accuracy (machine capability) is characterized by

- stochastic error (standard deviation, σ),
- systematic error (typically shifting of the design value, $\pm\Delta\mu$).

Reasons for systematic errors typically include setting error and tool wear, causing a constant or continuously growing shift in the process.

In order to be able to utilize machine capability, the machining process has to be made stable first by identifying and eliminating assignable causes of variation; then, stability has to be maintained by keeping the process under control. The tools of statistical process control (SPC) serve these objectives.

The accuracy of machining depends on many influencing factors:

- machine rigidity and damping,
- accuracy of spindle running, bearings,
- clamping of workpiece,
- sharpness of tool,
- running circle accuracy,
- tooth bite,
- feed speed,
- relative workpiece mass (g/cm) in through-feed machines,
- mechanical properties of the processed material,
- spring constant of press rolls,
- resolution of setting mechanism.

These factors contribute to the stochastic error, except for the last one which manifests itself in systematic error.

Variation of the machined dimensions with respect to the mean follows normal distribution when a number of causes are acting in a random way, each with small effect as related to the total variation. This is the case of the stochastic error, also called common-cause variation in the literature of quality engineering. Because of the great number and small individual effect of these causes, there is no point of trying to mitigate or eliminate some of them. They belong to the machining process and define the smallest possible variation which is regarded as quasi-stationary with unchanged machining parameters in the context of quality control. The standard deviation σ as a measure of this variation may be different depending on the processed material and dimensions of the workpieces.

In the ideal case, only common-cause variation exists when the machining process is stable, meaning that neither the process mean nor the standard deviation changes over time. That is, machining accuracy is predictable. However, when causes of variation other than those inseparable from the process are present, the process will not be constant; the so-called assignable cause variation, manifesting itself as systematic error and/or increasing influence of some factors mentioned above essentially changes the statistical population. Assignable cause variation means that it is possible to find the cause because of its relatively important effect on variation; moreover, it can be eliminated by a reasonable effort. Assignable cause variation may become evident through a varying shift of the process mean, or change—generally an increase—of the standard deviation, or both. The process is no longer stable; the lack of stability may also appear through a non-normal distribution of the machined dimensions.

Some of the assignable causes of variation resulting in impaired accuracy of machining of wood:

- change or alteration of the quality of raw material (batches of different material properties or quality grade),
- improperly performed or missed prior treatment, operation, such as seasoning, jointing,
- omitting of prior checking where necessary,
- mixing of workpieces coming from different sources,
- incorrect positioning of the workpiece (e.g. bark side up vs. down),
- chips or sawdust left on the workpiece or machine table,
- imperfect workpiece clamping or pressing.

To utilize machine capability, the machining process has to be made stable first by identifying and eliminating assignable causes of variation; then, stability has to be maintained by continuously keeping the process under control. The tools of statistical process control (SPC) serve these objectives.

Interaction of tolerance width* ΔT *with machining accuracy is treated here assuming machining processes are in control, thus characterized by the least feasible standard deviation of machined dimensions of the processed material.

Simple case: symmetric (bilateral) tolerance ($\Delta\mu = 0$); see Fig. 4.76a. Utilization of machine capability is given by

$$C = \frac{\Delta T}{3 \cdot \sigma} \tag{4.64}$$

General case: with asymmetric tolerance and shifting the design value, the above equation reads:

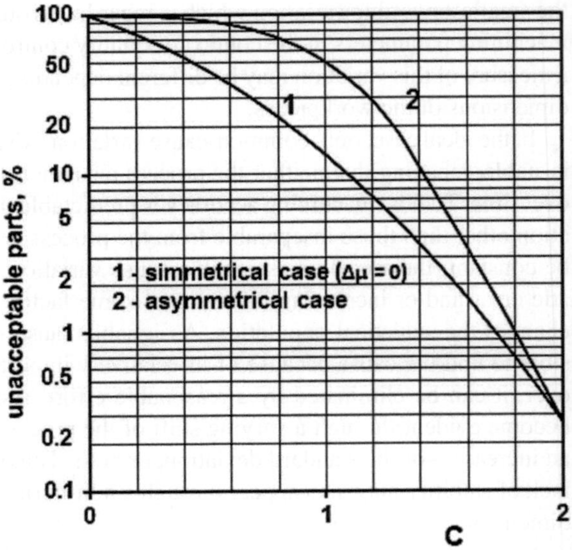

Fig. 4.77 Expected number of rejected parts as a function of relative machine capability coefficient C

$$C = \frac{\Delta T \mp \Delta T_{\mathrm{as}} \pm \Delta \mu}{3 \cdot \sigma} \tag{4.65}$$

where ΔT_{as} is the shifting of the tolerance limit in relation to the symmetric case ($\Delta T_{\mathrm{as}} \approx -\Delta T/3$ in Fig. 4.76b).

If the tolerance width ΔT equals the $\pm 3 \cdot \sigma$ range, then $C = 2$ and the ratio of unacceptable parts (rejects) on the two sides is 0.27% (see Fig. 4.77). If the value of C decreases, the number of rejects increases. Asymmetric tolerance and shifting of the design value mean a worse case compared to the symmetrical case with no shifting ($\Delta\mu = 0$).

The asymmetric tolerance can be compensated easily however, setting an accurate shifting $\Delta\mu$. If $\Delta T_{\mathrm{as}} = \Delta\mu$, then we return to the more favourable symmetric case of Eq. (4.64).

4.7.2 Experimental Study of Machining Accuracy

Machining accuracy and feasible tolerances of planing/shaping final cross-section was studied in a university–industry joint project, aiming at enhancing the efficiency and conformance of component manufacture in furniture plants (Kovacs et al. 2012). Precision of cross-sectional machining on an up-to-date four-sided moulder was studied. 500-mm-long workpieces of three hardwood species (beech, oak and black locust), 60 of each were machined to width and thickness, shown in Table 4.28. Thickness and width of the machined workpieces were measured to two decimal places at three points along their length, by using a digital slide gauge. Primary

Table 4.28 Nominal cross-sectional dimensions of workpieces used in process capability study

Species	European beech		English oak		Black locust	
Dimension	Width (mm)	Thickness (mm)	Width (mm)	Thickness (mm)	Width (mm)	Thickness (mm)
Raw	57.0	48.0	68.0	38.0	61.0	27.0
Final	51.5	44.0	62.5	34.5	55.5	21.5
SD mm	0.084	0.078	0.091	0.0517	0.088	0.037
Density g/cm^3	0.7	0.7	0.71	0.71	0.8	0.8

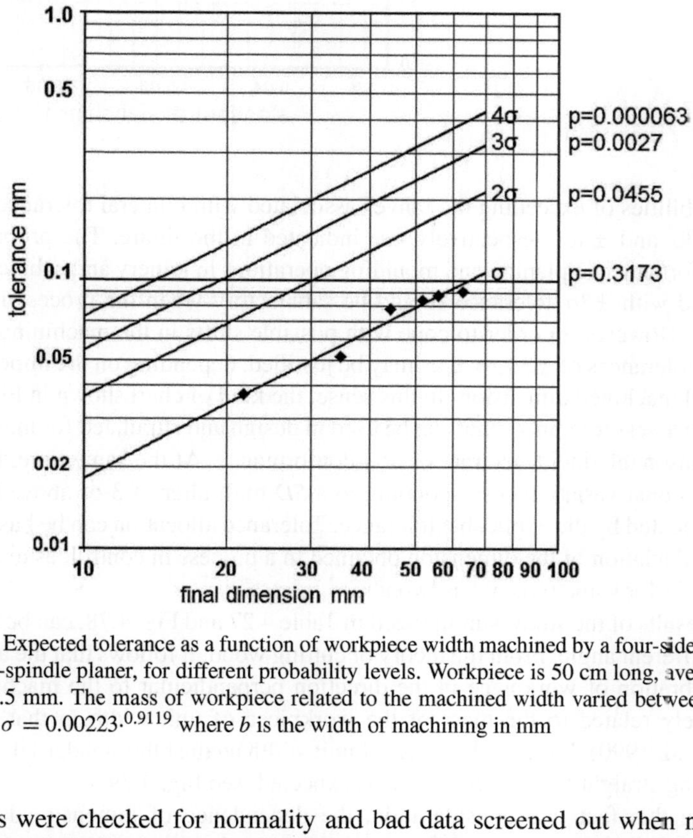

Fig. 4.78 Expected tolerance as a function of workpiece width machined by a four-sided through-feed multi-spindle planer, for different probability levels. Workpiece is 50 cm long, average depth of cut is 2.5 mm. The mass of workpiece related to the machined width varied between 104 and 243 g/cm. $\sigma = 0.00223^{\cdot 0.9119}$ where b is the width of machining in mm

databases were checked for normality and bad data screened out when needed in order to obtain standard deviations (SD) characterizing processes when in statistical control. These values are also given in the table.

Standard deviations versus machined dimensions were plotted on a log–log chart in which the data points aligned along a straight line, suggesting a power function relationship. The exponent turned out to be close to 1.0. In order to evaluate process capability for the individual dimension groups, straight lines corresponding to 2 sigma, 3 sigma and 4 sigma were drawn in the chart; see Fig. 4.78.

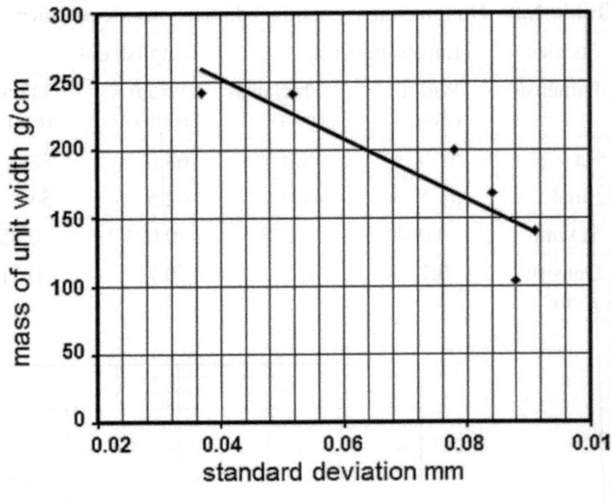

Fig. 4.79 Dependence of variation of machined dimension on the relative mass of the workpiece

Probabilities of exceeding the ranges associated with bilateral tolerances of $\pm\sigma$, $\pm 2\sigma$, $\pm 3\sigma$ and $\pm 4\sigma$, respectively, are indicated in the figure. The probability of non-conformance in planing and moulding operations in joinery and cabinet making associated with $\pm 3\sigma$ tolerance should be satisfactory when the process is not out of centre. However, in order to cope with possible shifts in the machining process, bilateral tolerances of $\pm 3\sigma$ to $\pm 4\sigma$ may be justified, depending on the importance of the actual machined dimension. In this sense, the kind of chart shown in Fig. 4.78 is helpful to assess tolerances that can be used in design and stipulated for manufacture with a reasonable incidence rate of non-conformance. At the same time, the range of dimensional variation corresponding to a *SD* multiplier of 3 or above has to be accommodated by the applicable tolerance. Tolerance allocation can be based on the standard deviation of the dimension obtained in a process in control, as exemplified in Fig. 4.78 for some typical hardwoods.

The results of the study, summarized in Table 4.27 and Fig. 4.78, can be looked at from a different angle. From the theory of cutting wood, it follows that the amplitude of the vibration of workpiece in the direction perpendicular to the machined face is inversely related to the mass of the workpiece of unit width in that direction (Sitkei et al. 1990). Plotting this mass of unit width against the standard deviation, a descending straight line can be drawn as expected; see Fig. 4.79.

Taking this fact into consideration, the dependence of dimensional variation explored in cross-sectional machining relates to random sizes of the workpieces in the direction across the machined one. Therefore, it would be interesting to study the standard deviation of the same machined dimension with changing cross-sectional size in the other direction for the assessment of feasible tolerances.

Machining accuracy and feasible tolerances of tenoning was the topic of a capability study conducted on the single end tenoning unit of a wooden window machining centre (Kovacs et al. 2011). Window frame parts of pine wood (*Pinus sylvestris*) with tenons 8, 12 and 30 mm thick at the end were prepared, 30 pieces of

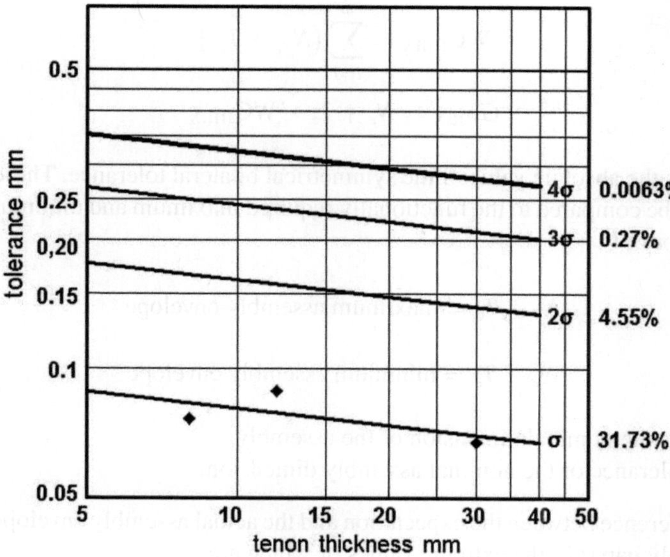

Fig. 4.80 Expected tolerance as a function of tenon thickness machined by the single end tenoning unit of window machining centre, for different probability levels

each. Tenon thickness was measured at mid-length to two decimal places using digital slide gauge, and the standard deviation was calculated for the individual tenon thickness groups. It was found that the standard deviation decreases with increased tenon thickness. The process capability map as shown below was constructed (Fig. 4.80).

These preliminary results, showing a reverse tendency may be explained by the less efficient support conditions of the machined end of workpieces on a tenon making machine.

4.7.3 Tolerance Design

Three methods of tolerance design and analysis will be shortly reviewed:

- worst-case method,
- statistical tolerance design,
- cost of quality-based tolerance design.

Worst-case method is the simplest most conservative method of tolerance design. Assembly dimensions and tolerances are determined from extreme part dimensions given by nominal dimensions and (N_{pi}) and part tolerances (T_{pi}). The maximum and minimum worst-case conditions can be expressed as

$$\text{WC}_{\text{max}} = \sum_{i=1}^{m} \left(N_{p_i} + T_{p_i}\right) \tag{4.66}$$

$$G_{\text{max}} = N_e + T_e - \text{WC}_{\text{max}} \tag{4.67}$$

T_{pi} denotes the absolute value of the symmetrical bilateral tolerance. These extreme values can be compared to the functionally required maximum and minimum assembly envelopes of

$$N_e + T_e = \text{maximum assembly envelope}$$

$$N_e - T_e = \text{minimum assembly envelope}$$

where N_e is the nominal dimension of the assembly,
T_e is the tolerance of the nominal assembly dimension.

The difference between the expectation and the actual assembly envelope is called the assembly gap (G), the extreme values of which are

$$G_{\text{max}} = N_e + T_e - \text{WC}_{\text{max}} \tag{4.68}$$

$$G_{\text{min}} = N_e - T_e - \text{WC}_{\text{min}} \tag{4.69}$$

and it has the nominal value of

$$G_{\text{nom}} = \sum_{i=1}^{m} \left(N_{p_i}\right) \tag{4.70}$$

The maximum and minimum worst-case assembly gap can also be written in the following way:

$$G_{\text{max}} = G_{\text{nom}} + T_e - \Sigma T_p, \tag{4.71}$$

and

$$G_{\text{min}} = G_{\text{nom}} - T_e + \Sigma T_p \tag{4.72}$$

These formulae suggest that the planned clearance or interference of a fit can increase or diminish with the cumulative part tolerances and assembly inaccuracies. On the other side, regarding minimum and maximum assembly envelopes as specification limits for assembly stack, part tolerances and/or dimensions can be assigned in order to avoid incompatibilities in the assembly or unwanted sizes of clearances/interferences.

Statistical tolerance design takes into consideration the distribution of the variation of part dimensions postulating normal distribution for the same. Dimension variability is described by means of the adjusted standard deviation derived from the process capability index C_p (ISO 3534-2 1993) or C as shown below:

$$\sigma_{\text{adj}} = \frac{T}{3 \cdot C_p} \tag{4.73}$$

or

$$\sigma_{\text{adj}} = \frac{\Delta T}{3 \cdot C} \tag{4.73a}$$

where T is the bilateral tolerance ($\pm T$) and ΔT is the tolerance width.

With a process capability $C_p = 1.0$ or $C = 2.0$, the probability of exceeding either of the lower and upper tolerance limits is 0.00135 or 0.135%, as was pointed out in Sect. 4.7.1.

The assembly envelope standard deviation (resultant standard deviation) can be computed as the square root of the sum of squares of the component variances:

$$\sigma_{\text{res}} = \sqrt{\sigma_{p1}^2 + \sigma_{p2} + \ldots . \sigma_{pn}^2} \tag{4.74}$$

The same relationship applies to the resultant assembly tolerance (T_{res}) and the component tolerances when all component dimensions are produced by the same process capability C_p:

$$T_{res} = \sqrt{T_{p1}^2 + T_{p2}^2 + \ldots . T_{pn}^2} \tag{4.75}$$

We may also account for the different process capabilities associated with the individual component dimensions. Then, for the assembly tolerance T_a that we are to specify, it must hold that

$$z_a \cdot \left[\sum_{i=1}^{n} \left(\frac{T_{pi}}{z_{pi}} \right)^2 \right]^{0.5} \leq T_a \tag{4.76}$$

Using the notion of assembly gap, its standard deviation can be computed as

$$\sigma_{\text{gap}} = \sqrt{\left(\frac{T_e}{3C_p} \right)^2 + \sum_{i=1}^{m} \left(\frac{T_{pi}}{3C_{p_i}} \right)^2} \tag{4.77}$$

From the formula above, the standard deviation of the assembly gap is expressed as the square root of the pooled variance made up of the envelope variance the assembly is to fit within and the sum of the component variances.

The upper and lower 0.135 percentile values ($\pm 3 \cdot \sigma$ limits) of the assembly envelope will be

$$\mathrm{UL}_{0.135} = \sum_{i=1}^{n} N_{pi} + 3 \cdot \left(\sum_{i=1}^{n} \left(\frac{T_{pi}}{3 \cdot C_{pi}} \right)^2 \right)^{0.5} = \sum_{i=1}^{n} N_{pi} + \left(\sum_{i=1}^{n} \left(\frac{T_{pi}}{C_{pi}} \right)^2 \right)^{0.5} \tag{4.78}$$

and

$$\mathrm{LL}_{0.135} = \sum_{i=1}^{n} N_{pi} - \left(\sum_{i=1}^{n} \left(\frac{T_{pi}}{C_{pi}} \right)^2 \right)^{0.5} \tag{4.79}$$

where UL denotes upper limit and LL denotes lower limit.

With similar reasoning, the upper and lower limits of the assembly gap belonging to the probabilities of $p = (1 - 0.00135)$ and $p = 0.00135$, respectively, can be calculated as follows:

$$G_{\max} = N_e + T_e - \left(\sum_{i=1}^{n} (N_{pi}) + \sum_{i=1}^{n} \left(\left(\frac{T_{pi}}{C_{pi}} \right)^2 \right) \right)^{0.5} \tag{4.80}$$

and

$$G_{\min} = N_e - T_e - \left(\sum_{i=1}^{n} (N_{pi}) - \sum_{i=1}^{n} \left(\left(\frac{T_{pi}}{C_{pi}} \right)^2 \right) \right)^{0.5} \tag{4.81}$$

and the nominal value is the same as in the case of the worst-case method:

$$G_{\mathrm{nom}} = N_e - \sum_{i=1}^{m} (N_{P_i}) \tag{4.82}$$

Since standard deviation and tolerance are related by the process capability, the probability of assembly gaps up to any size occurring can be expressed by the probabilities of the standard normal distribution. In order to determine the probability associated with an assembly gap of Q, one calculates the standard deviation multiplier (z-score) as below.

$$z_Q = \frac{Q - G_{\mathrm{nom}}}{\sqrt{\left(\frac{T_e}{3C_p} \right)^2 + \sum_{i=1}^{m} \left(\frac{T_{P_i}}{3C_{P_i}} \right)^2}} \tag{4.83}$$

Using the above expression, one can calculate the Z-values at the gap limits by replacing Q with G_{min} and G_{max} for the conditions in which we exceed the assembly gap. If we take $Q = 0$, the resulting z is the number of standard deviations away from the nominal gap that relates to the condition of line-to-line contact as the starting point of interference fit. The corresponding probability can be looked up in a table of z-distribution.

The cost of quality approach determines the design of economic tolerances. The concept of economic tolerances is based on the recognition that quality costs are optimal (i.e. minimum) when failure costs are equal to the cost of achievement of quality. Ideally, besides the response of customers and further results of the QFD analysis, the output of parameter design of the Taguchi's approach to quality engineering serves as input of this method. These include the optimum setting of design variables, initial tolerance estimates and material grades.

Quality loss is the kernel of the Taguchi method. Deviation of any perceived property y of a product from its target value m causes loss $L(y)$ to the customer and to the business because of the impaired product performance. The relationship between the deviation $y - m$ and the ensuing cost can be described by a Tailor series that most often will be reduced to a quadratic function of the form

$$L(y) = k(y - m)^2 = \frac{A_0}{(\Delta_0)^2}(y - m)^2 \quad (4.84)$$

In the above equation, $\pm\Delta_0$ denotes the deviation from the target beyond which the product performance is intolerable to the customer; therefore, it is called customer tolerance. A_0 is the loss occurring at the tolerance limit. It is essentially the cost that a malfunction or failure causes to the customer.

The objective is to relate these customer tolerance limits to the technological characteristics of the product. In order to achieve this, firstly, one needs to determine the cost to the business to adjust the off-target performance values back onto target during the manufacturing process. Secondly, the sensitivity between the customer tolerance and manufacturing tolerances has to be clarified. Once the function and its limits are established, the engineering team needs to determine the safety factor, ϕ, to prevent off-target performance values. The company also needs to quantify how much they are willing to spend to remedy the off-target performance. Therefore, the safety factor can be described as the square root of the average loss to customer when a product characteristic exceeds customer tolerance limits (A_0), over the average loss to business (A) when engineering parameters determining the characteristics perceivable to a customer exceed the manufacturing and/or design tolerance limits:

$$\varphi = \sqrt{\frac{A_0}{A}} \quad (4.85)$$

Sensitivity (β) is the change in the high-level perception of a characteristic which can be observed by a customer (or a product-level engineering characteristic), y, when a unit change occurs from the target set point of the manufacturing characteristic x. It can be written that

$$A = \frac{A_0}{\Delta}[\beta(x - m)]^2 \tag{4.86}$$

Then the manufacturing tolerance can be determined by the safety factor and sensitivity:

$$\Delta = \sqrt{\frac{A_0}{A}}\left(\frac{\Delta_0}{\beta}\right) \tag{4.87}$$

4.7.4 Worked Examples of Tolerance Design

Worst-case design is illustrated through the problem of tolerance allocation of tenon and mortise. Let us see as an example a tenon and mortise joint with a nominal thickness of 10.0 mm. The joint is to be made with an interference fit; therefore, the nominal values of the two parts (N_d for the tenon and N_D for the mortise) are different. The difference of part sizes is equal to the size of interference of 0.1 mm, that is

$$N_d = 10.0\,\text{mm}$$
$$N_D = 9.9\,\text{mm, and}$$
$$N_e = -0.1\,\text{mm.}$$

Specifying symmetrical bilateral part tolerances of ± 0.2 mm, the permissible sizes are:

for the tenon $9.8\,\text{mm} < d \leq 10.2\,\text{mm}$, and
for the mortise $9.7\,\text{mm} < D \leq 10.1\,\text{mm}$.

The right-side face of the tenon is selected as the starting point to build up the assembly envelope; see Fig. 4.81. The assembly stack-up is $\Sigma N_p = -N_d + N_D = -0.1$ mm and is shown along with the tolerance stack-up in tabular form (Table 4.29).

Table 4.29 Mating assembly nominal dimension and tolerance stack-up

	Assembly stack-up	Tolerance stack-up (mm)
N_d	−10.0	±0.2
N_D	9.9	±0.2
Σ	−0.1	±0.4

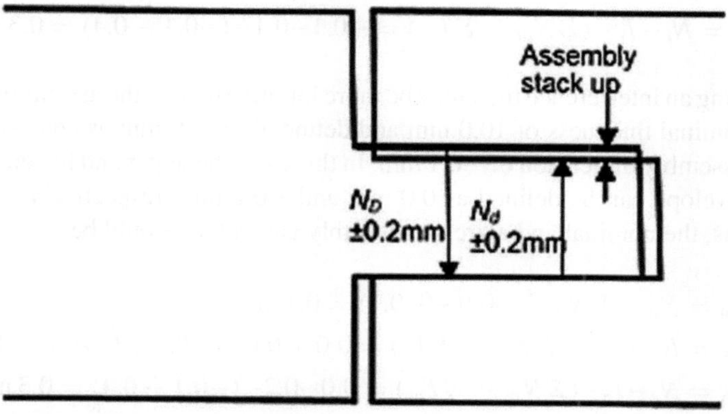

Fig. 4.81 Mortise and tenon assembly with interference fit

The value of −0.1 is the nominal distance (with nominal part dimensions) between the right faces of the mortise and the tenon, when tenon and mortise faces on the left-hand side are in contact. A positive distance indicates clearance, a negative one interference, the amount of which is shared between the two sides in reality.

The maximum and minimum worst-case situation will be

$$\begin{aligned} \mathrm{WC_{max}} &= -N_d + N_D + \left(T_{p1} + T_{p2}\right) = -10.0 + 9.9 + (0.2 + 0.2) \\ &= 0.3\,\mathrm{mm}\ \text{(clearance of 0.3 mm)} \\ \mathrm{WC_{min}} &= -N_d + N_D - \left(T_{p1} + T_{p2}\right) = -10.0 + 9.9 - (0.2 + 0.2) \\ &= -0.5\,\mathrm{mm}\ \text{(interference of 0.5 mm).} \end{aligned}$$

Though the interference of 0.5 mm is halved on both sides, occurrence of size differences of this magnitude cannot be allowed from the functional point of view. However, the worst-case model does not take probabilities into account, whereas the chance of both extreme values occurring simultaneously is very low, as we will see later (in the case of process capability $C_{pk} \geq 1.0$, this probability is $p \leq (0.00135)^2 = 1.82 \cdot 10^{-6}$).

Let us write up nominal and extreme values of the assembly gap. For this, we need to define the maximum and minimum assembly envelopes $N_e + T_e$ and $N_e - T_e$, respectively. Let us specify

$$T_e = \pm 0.1\,\mathrm{mm}$$

Then, we can write

$$\begin{aligned} G_{\mathrm{nom}} &= N_e - \Sigma N_p = -0.1 - (-0.1) = 0.0\,\mathrm{mm} \\ G_{\max} &= N_e + T_e - (\Sigma N_{pi} + \Sigma T_p) = -0.1 + 0.1 - (-0.1 + 0.4) = -0.3\,\mathrm{mm} \end{aligned}$$

$$G_{\min} = N_e - T_e - (\Sigma N_{pi} - \Sigma T_{pi}) = -0.1 - 0.1 - (-0.1 - 0.4) = 0.3\,\text{mm}$$

Planning an interference fit, it may be more intuitive to base the assembly envelope on the nominal thickness of 10.0 mm and define $N_e = 0.0$ mm as contrasted to the mating assembly dimension of −0.1 mm. In this case, the upper and lower tolerances of the envelope can be defined as 0.0 mm and −0.2 mm, respectively. With these conditions, the nominal and extreme assembly gap values would be

$$G_{\text{nom}} = N_e - \Sigma N_p = -0.0 - (-0.1) = 0.1\,\text{mm}$$
$$G_{\max} = N_e + T_e - (\Sigma N_{pi} + \Sigma T_p) = 0.0 + 0.0 - (-0.1 + 0.4) = -0.3\,\text{mm}$$
$$G_{\min} = N_e - T_e - (\Sigma N_{pi} - \Sigma T_{pi}) = 0.0 - 0.2 - (-0.1 - 0.4) = 0.3\,\text{mm}$$

We can see that the maximum and minimum assembly gap values have not changed. Either way of writing up maximum and minimum assembly gaps tells us that the upper envelope of the actual mating assembly is 0.3 mm higher, while the lower envelope is 0.3 mm lower than the corresponding tolerance limits. Indeed, these are the same results that the worst-case values show: the maximum clearance is 0.3 mm as opposed to the allowable value of 0.0 mm and the allowable interference of 0.2 mm increases to 0.5 mm.

This same problem could be regarded in a simpler manner. The actual assembly envelope being −0.1 mm with a tolerance stack of ±0.4 mm, nominal, maximum and minimum assembly gap, respectively, can be written up as

$$G_{\text{nom}} = -\Sigma N_{pi} = -0.1\,\text{mm}$$
$$G_{\max} = -\left(\Sigma N_{pi} + \Sigma T_{pi}\right) = -(-0.1 + 0.4) = -0.3\,\text{mm}$$
$$G_{\min} = -\left(\Sigma N_{pi} - \Sigma T_{pi}\right) = -(-0.1 - 0.4) = 0.5\,\text{mm}$$

In this case, assembly gap values are the negatives of the worst-case values, directly showing net clearance/interference values. One can directly evaluate them to decide on the necessity of changing part dimensions and/or tolerances in order to remedy the unsatisfactory situation.

We examine the same problem applying the method of ***statistical tolerance design***.

Assuming $C_{p1} = C_{p2} = 1.0$ or $C_1 = C_2 = 2.0$ process capabilities for both part dimensions, the upper and lower 0.135 percentile values ($\pm 3 \cdot \sigma$ limits) of the assembly stack will be

$$\text{UL}_{0.135} = -N_d + N_D + \left((T_{p1})^2 + (T_{p2})^2\right)^{1/2} = -10.0 + 9.9 + \left(0.2^2 + 0.2^2\right)^{1/2}$$
$$= 0.18\,\text{mm (clearance)}$$

and

$$\mathrm{LL}_{0.135} = -N_d + N_D - \left((T_{p1})^2 + (T_{p2})^2\right)^{1/2} = -10.0 + 9.9 - (0.2^2 + 0.2^2)^{1/2}$$
$$= -0.38\text{ mm (interference)}$$

These values give us a more favourable picture of the manufacturing process than the worst-case analysis did, but they still indicate higher values of both interference and clearance than what is admissible.

Namely, the probability of exceeding the limits in either direction of the admissible interval of -0.1 ± 0.1 mm by at least 0.18 mm is $p = 0.00135$.

The probability of occurrence of any clearance, or of interference tighter than -0.2 mm, is calculated as below. For larger than zero gaps

$$z_{0.0} = \frac{0 - (-0.1)}{\sqrt{\sum_{i=1}^{2} \left(\frac{0.2}{3\cdot 1}\right)^2}} = \frac{0.1}{0.09428} = 1.061$$

Likewise, for fits tighter than -0.2 mm, $z_{-0.2} = -1.061$. The corresponding probabilities are $p_{0.0} = 1 - 0.8556 = 0.1444$ and $p_{-0.02} = 0.1444$, respectively.

One may calculate the probabilities corresponding to the worst-case extreme values, 0.3 mm and -0.5 mm:

$$z_{0.3} = \frac{0.3 - (-0.1)}{\sqrt{\sum_{i=1}^{2} \left(\frac{0.2}{3\cdot 1}\right)^2}} = \frac{0.4}{0.09428} = 4.244$$

and $p = 1.98\text{E}{-5}$.

It is easy to conclude that the same probability applies to the interference of -0.5 mm.

Taguchi's design of economic tolerances is useful to apply in relation to the cost implications of tolerance allocation. In the above example of mortise and tenon joints, let the load-carrying capacity of the joint in a chair be the customer perceived characteristic. The target value is to resist the weight of 105 kg of a person using the chair with dynamic effect added. As customer tolerance, the joint should resist dynamic use of the chair by a person weighing at least 95, i.e. $\Delta_0 = 10$ kg. By analysing the load-carrying behaviour of the joint, it has been established that the correct fit of the tenon in the mortise is the most important manufacturing characteristic to keep the load-carrying capacity on target. The optimum fit is characterized by an interference of 0.1 mm. If the tolerances to be set for this engineering characteristic are exceeded, the additional cost of corrective measures would be A = US\$20 per joint. At the same time, if the joint fails in the chair used by the customer, the cost of the damage caused is A_0 = US\$50.

A shift of 0.02 mm in the fit interference decreases the allowable body weight of user; the sensitivity factor is $\beta = 2$ kg/0.02 mm = 100 kg/mm.

The safety factor is $\phi = (A_0/A)^{0.5} = (50/20)^{0.5} = 1.58$.

With these data, the manufacturing tolerance is

$$\Delta = \sqrt{\frac{50}{20} \cdot \left(\frac{10}{100}\right)} = 0.158\,\text{mm}$$

This means that the precision of manufacturing the tenon and the mortise should be specified in such a way that the resulting interference fit comply with the specification of 0.1 ± 0.158 mm. In that case the cost of deviation from the target value incurred to the business times the safety factor will be equal to the cost to the customer in case of product failure.

A different, simpler application of Taguchi's quality loss approach to tolerance design (Kemény and Deák 2000; Kovács 2002) is demonstrated next.

Target value of fit interference	$m = 0.1$ mm
Bilateral tolerance t	$t = \pm\, 0.1$ mm
Lower specification limit	LSL = 0.0 mm
Upper specification limit	USL = 0.2 mm.

Failure of the product in use by the customer incurs a cost of US$50; this much of damage occurs with high probability when double value of the manufacturing tolerance is exceeded, that is $x \geq m + 2 \cdot t = 0.1 + 2 \cdot 0.1 = 0.3$. The loss function can be written as

$$L_{(x)} = k \cdot (x{-}m)^2 = k \cdot (0.3 - 0.1)^2 = \text{US\$50}.$$

From this, the value of the loss factor k is

$$k = \text{US\$1250/mm}^2$$

The surplus cost to business caused by rejecting a unit with out-of-tolerance part is US$20 (disassembly, part exchange, remedy of damages due to disassembly). From this loss, the deviation from the target that the manufacturer may allow is determined using the loss function:

$$20 = 1250 \cdot \left(x' - m\right)^2 = 1250 \cdot (\Delta)^2$$
$$\Delta = 0.13\,\text{mm}$$

At this amount of deviation from target, the cost in production is equal to the loss that would occur if the product was delivered to the customer without the expenditure at the factory. At higher deviations, the cost of selling the product is higher than the cost of repair in work.

4.7.5 Optimization of Fit and Tolerances—Example

In relation to the example above, extensive experimentation showed that the strength of a joint in in-plane bending, tightness of fitting the tenon into the mortise is acceptable with a clearance of 0.1 mm to an interference fit of 0.35 mm overlapping, the optimum being at 0.1 mm interference with the type of glue used (Kungl et al. 2018). What should be the nominal settings for the thickness of tenon and mortise, and how strict should tolerances be prescribed for those dimensions, in order that the probability of exceeding the fit limits be minimum? This is an optimization problem with two variables, $x_1 = d - D$ and $x_2 = T$, where d is the tenon thickness, D is the mortise thickness, and T is the bilateral tolerance, all expressed in mm. For simplicity, we consider the same tolerance for both parts. On the basis of the above-mentioned study, a shift of 0.02 mm of setting is reasonable to account for. When writing up the objective and constraints functions, one should have in mind that interference fit is preferable. We want to assure identical probabilities (or frequencies) of exceeding both the clearance limit and the interference limit specified above. The shift of process setting is to be taken into consideration with the most unfavourable sign which is positive when the limit of interference (0.35 mm) is approached and is negative with respect to the clearance. The objective and constraint equations would read

$$F(x_1, x_2) = \frac{0.35 - x_1 - 0.02}{x_2} \tag{4.88}$$

$$g_1(x_1, x_2) = \left| \frac{-0.1 - x_1 + 0.02}{x_2} \right| \geq \frac{0.35 - x_1 - 0.02}{x_2} \tag{4.89}$$

$$g_2(x_1, x_2) = 0.2 \tag{4.90}$$

The first constraint can be brought to a simpler form:

$$g_1(x_1, x_2) = x_1 - 0.125 \geq 0 \tag{4.91}$$

which could have been written straight from the condition that equal probabilities belong to magnitudes of the random variable at equidistance from the mean in any symmetric distribution; therefore, the mean satisfying the first constraint has to be in the middle of the acceptance interval.

Without the second constraint, $T \to 0$ would yield optimum, but variation of part dimensions cannot be totally eliminated; thus, T cannot be equal to zero. Therefore, the smallest attainable tolerance will be the optimum, which in our example is ± 0.2 mm.

Through the formal application of the Davidon–Fletcher–Powell algorithm with the due penalty terms as described in Sect. 4.5, one can arrive at the same solution that the optimum can be found at the boundary of the feasible range of design variables.

Optimization for minimum of non-conformance of tenon and mortise joints simply means to set the nominal dimensions of the tenon and mortise with a desired difference between the two parts, and strive for the minimum variation of tenon and mortise dimensions in the manufacturing process. However, the consequences of an unwanted deviation in fit tightness may be different depending on whether it is positive or negative (i.e. makes the fit tighter or looser). In other words, we want a trade-off between the probabilities of exceeding the clearance and interference limits, respectively. In fact, we would prefer to exceed the interference value of 0.35 mm with less probability than what is associated with clearances larger than 0.1 mm. For such problems, Carrington's desirability functions are useful (Deák 2000). The general form of the desirability function for a property of one-sided limit is

$$d_i = \exp\left[-\exp(-(-y_i'))\right] \tag{4.92}$$

where y'_i is a linear (or nonlinear) function of the property y_i.

In the linear case

$$y' = b_0 + b_1 \cdot y \tag{4.93}$$

The two parameters b_0 and b_1 can be determined by means of two pairs of corresponding y- and d-values. In our case, we have two quantities: y_1 is the z-score of the limit interference value (right side of the first constraint inequality), and y_2 is the absolute value of that of limit clearance size (left side of the same inequality). In the expressions of y_1 and y_2, we take $x_2 = (0.2^2 + 0.2^2)^{0.5} = 0.28$ mm as a feasible assembly tolerance based on the study mentioned above (Kungl et al. 2018). For the more critical property, one defines a d versus y' curve of a higher slope through a proper choice of corresponding y- and d-values. The desirability function has a value of $d = 1/e \approx 0.37$ at the threshold of acceptance which belongs to the $y' = 0$ point. Another point may correspond to $d = 0.8$ at $y' = 1.5$, where the property attains its excellent level; see Fig. 4.82. For exceeding clearance limit, we take $|Z_{\mathrm{thr}}| = 1.5$ as acceptance threshold and $|Z_{\mathrm{exc}}| = 2.1$ as excellent level. The values are set higher for the interference limit; $Z_{\mathrm{thr}} = 2.0$ and $Z_{\mathrm{exc}} = 2.4$. The resulting desirability functions read

$$d_1 = \exp\left[-\exp(-(-3.75 + 2.5y_1))\right] \tag{4.94}$$

and

$$d_2 = \exp\left[-\exp(-(7.0 + 3.5 \cdot y_2))\right] \tag{4.95}$$

The complex objective function can be written as the geometric mean of the component d-values:

$$D = \sqrt[n]{d_1 \cdot d_2 \cdot \ldots . d_n} \quad D = \sqrt[n]{d_1 \cdot d_2 \cdot \ldots . d_n} \tag{4.96}$$

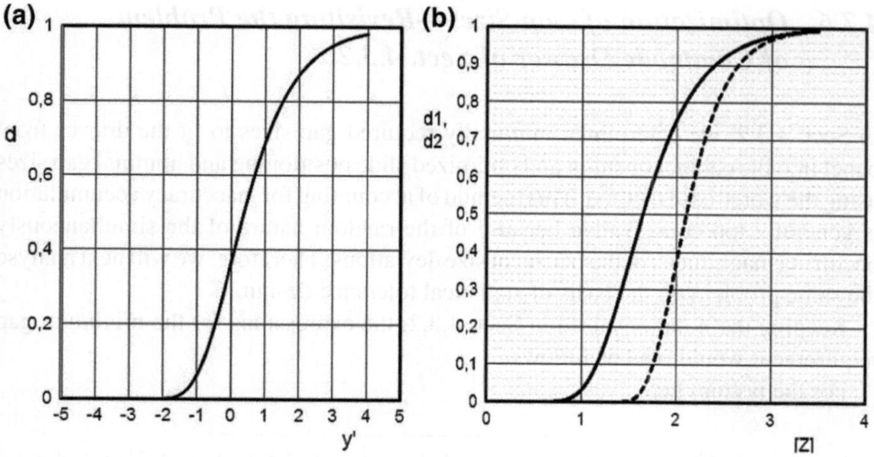

Fig. 4.82 Desirability function for one-sided acceptance limit (**a**); d-values for clearance (continuous line) and for interference (dotted line) as a function of Z-value associated with the acceptance limits (**b**)

Writing up the objective function for our d_1 and d_2 desirability functions, we can solve it numerically by calculating d_1- and d_2-values for Z_1-values changed in small increments.

The optimum value obtained for the difference of the nominal setting of tenon and mortise thickness happens to be $x_1 = d - D = 0.106$ mm, and the associated Z-values without a shift in the process are 2.61 and −2.21 for interference and clearance limit, respectively. The corresponding probabilities are 0.004 for interference and 0.014 for clearance, summing up to a total of 1.8% non-conformance. With a shift of either +0.02 mm or −0.02 mm in the process, the same trend in the relation of these probabilities still holds. With the setting of $x_1 = 0.125$ and the same tolerance of fit of 0.28 mm, the probability of non-conformance at the interference side would amount to $1 - \Phi((0.35 - 0.125)/(0.28/3)) = 1 - \Phi(2.41) = 0.008$, where Φ is the standard normal distribution function. The probability of non-conformance at the clearance side would be the same, corresponding to a total of 1.6% non-conformance. Though the rate of non-conformance is slightly less, the frequency of assemblies with a fit close to the interference limit is twice as high as when using the setting optimized by using the Carrington's compromise model. Therefore, the general quality of a production batch of joints is better.

4.7.6 *Optimization of Gap Sizes—Revisiting the Problem of Commode Drawer of Sect. 4.3.2*

In Sect. 4.3.2, we determined minimally required gap sizes to fit the drawer front panel into the cabinet opening and optimized slide positioning and nominal gap sizes using the worst-case method. This method of accounting for inaccuracy accumulation is generally too conservative because of the random nature of the simultaneously occurring magnitudes of the pertinent size deviations. Therefore, we will next analyse the same problem on the basis of statistical tolerance design.

Keeping the notations used in Sect. 4.3.2, the expressions for the minimum gap requirement would read as follows.

For the bottom gap:

$$h_{\min,B} = \sqrt{\left(\Delta x_{l,d}\right)^2 + \left(\Delta x_{l,w}\right)^2 + (\Delta x_s \cdot a)^2 + 2 \cdot \left(\Delta x_{l,h}\right)^2} + 1.3 \cdot a + \Delta x_w \tag{4.97}$$

For the top edge, in a similar way

$$h_{\min,T} = \sqrt{\left(\Delta x_{l,d}\right)^2 + \left(\Delta x_{l,w}\right)^2 + (\Delta x_s \cdot (1-a))^2 + 2 \cdot \left(\Delta x_{l,h}\right)^2} + 1.3(1-a) \tag{4.98}$$

In these expressions, the deflection under load and dimensional changes due to humidity is treated as deterministic values and their maximum value is added to the tolerance stack, because the occurrence of these maxima is highly probable. The same principle is followed in defining maximum gap sizes.

Bottom gap:

$$h_{\max,B1} = h_{\min,B} + \sqrt{\left(\Delta x_{l,d}\right)^2 + \left(\Delta x_{l,w}\right)^2 + (\Delta x_s \cdot a)^2 + 2 \cdot \left(\Delta x_{l,h}\right)^2} + 0.65 \cdot a \tag{4.99}$$

Top gap in the same circumstances (bottom-gap-based situation):

$$h_{\max,T1} = h_{\min,T} - \sqrt{\left(\Delta x_{l,d}\right)^2 + \left(\Delta x_{l,w}\right)^2} + \sqrt{(\Delta x_s \cdot (1-a))^2 + 2 \cdot \left(\Delta x_{l,h}\right)^2} + 0.65 \cdot (1-a) \tag{4.100}$$

Top gap:

$$h_{\max,T2} = h_{\min,tT} + \sqrt{\left(\Delta x_{l,d}\right)^2 + \left(\Delta x_{l,w}\right)^2 + (\Delta x_s \cdot (1-a))^2 + 2 \cdot \left(\Delta x_{l,h}\right)^2} + 0.65 \cdot (1-a) + \Delta x_w \tag{4.101}$$

Bottom gap, in the same circumstances (top-gap-based situation):

$$h_{\max,B2} = h_{\min,B} - \sqrt{(\Delta x_s \cdot a)^2 + (\Delta x_{l,h})^2} + \sqrt{(\Delta x_s \cdot a)^2 + 2 \cdot (\Delta x_{l,h})^2} + 0.65 \cdot a - \Delta x_w \quad (4.102)$$

Table 4.30 contains the minimum and maximum gap sizes calculated for the sample commode with three selected values of the ratio a of the fixing height of drawer slide, namely $a = 0$ (bottom mount or undermount), $a = 0.5$ (side mount, mid-height of drawer) and $a = 1$ (top of drawer).

The last row in the table shows mid-height positioning of the slider is the best with the ratio of bottom-to-top gap the closest to 1. It is possible to find the optimal value of the ratio a on condition of the equality of the bottom and top gap size. The use of the principle of sum of the square root in summing up tolerances leads to equations difficult to solve analytically. Numerical approximation produces the optimum mount-height ratios and gap shown in values shown in Table 4.31.

As in worst-case design, optimal values of the ratio $a = B_1–B_0$ do not vary much, being below 0.5 when criteria for the bottom gap are used and above 0.5 for top gap size. It might be the designer's choice to choose the most relevant circumstances of extreme gap evaluation.

When nominal gap sizes larger than the minimum requirements are chosen by the designer, minimum gap requirements are met, and the ratio of $a = B_1/B_0$ may be optimized in the same way as before. Numerical results of optimum values of the ratio a can be obtained and are plotted against the difference between bottom and top nominal gap size in Fig. 4.83. Square data points show a decrease of a-values with the increase of nominal bottom gap for the top-gap-based situation; triangle data points indicate the effect of increasing the nominal top gap in the case of bottom-gap-based optimization. Top-gap-based optimization leads to above the middle slide position with identical nominal sizes of the bottom and top gap; however, a nominal bottom gap 1 mm larger makes again mid-height as optimum. In contrast, bottom-gap-based optimization sets forth below the middle height slide positioning with uniform nominal gap sizes, and a top gap 1 mm larger shifts the optimum closer to mid-height. As seen in Sect. 4.3.2, this approach of optimization for similar top and bottom gap sizes allows the choice of nominal gap sizes based on the choice of the slide positioning. For example, when an undermount slide is chosen ($a = 0$), top-gap-based design (conditions include fully loaded drawer and dryer than average climate) dictates a nominal gap 3.0 mm larger at the bottom (Fig. 4.83).

Closing remarks

To date, the development of wood products does not use methods of engineering design as much as product development processes in many other branches of industry. This can be explained partly by the unique nature and variability of the raw material that have always placed knowledge accumulated by experience in the fore. Novel methods and decision-making procedures are almost excluded in the practice. Another reason is that many of the products made of wood are expected to carry aesthetics values sometimes overriding their usability.

Table 4.30 Minimum gap requirements and maximum gap sizes at the bottom and top of drawer front for three distinct positions of the slider, using the RSS method of tolerance design

$a = B_1/B_0$	0			0.5			1.0		
	h_{min} mm	$h_{max,1}$ mm	$h_{max,2}$ mm	h_{min} mm	$h_{max,1}$ mm	$h_{max,2}$ mm	h_{min} mm	$h_{max,1}$ mm	$h_{max,2}$ mm
Bottom	0.70	1.10	0.40	1.40	2.17	1.49	2.17	3.38	2.72
Top	1.87	2.72	3.38	1.10	1.49	2.17	0.40	0.40	1.10
Ratio of bottom-to-top gap		0.40	0.12		1.46	0.68		8.45	2.48

Table 4.31 Optimized slider positioning in different circumstances of gap increase

Circumstances	Bottom-gap-based, drawer full loaded	Bottom-gap-based, drawer empty	Top-gap-based, drawer full loaded	Top-gap-based, drawer empty
$a = B_1/B_0$	0.5	0.35	0.65	0.5
Gap size mm	1.84	1.83	1.84	1.84

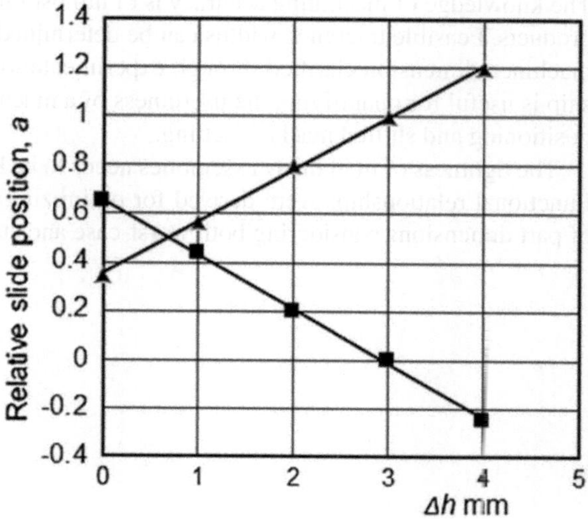

Fig. 4.83 Relationship between the difference of bottom and top nominal gap sizes *h* and the relative height of slide position *a* in the top-gap-based (square data points) and bottom-gap-based design (triangle data points), using the RSS method of design tolerance

Early detection of wrong decisions in the development process to avoid poor design necessitates the adaptation of engineering methods used successfully in other branches of industry. Such methods can be most effective when based on functional relationships that exist between the characteristics of a product and its influencing factors. Unfortunately, many of these functional relationships are still missing. However, they can be established in some simple cases through engineering thinking and can be used to make a design robust that may be insensitive to changing environmental and use conditions. Methods of quality engineering, like QFD and FMEA and DOE when used with good engineering judgement, can suite the needs of furniture design very well, even when comfort of use and aesthetics dominate the customer needs.

Design for strength and durability of pieces of furniture has been recently in the focus of a number of researchers. We proposed a concept of probability-based design of furniture using the principles of Eurocode for establishing design values for both the loads and the load-carrying capacity of furniture parts and joints. Various calculation models for load effects are available. There is increasing knowledge about the orthotropic nature of wood and behaviour of wood joint aids to minimize the uncertainties inherent in any structural and testing models.

The load-carrying capacity of furniture joints most often determines the suitability of a piece of furniture for its intended use. Generally valid dimensionless quantities and similarity relationships can be established on the basis of test results of a number of researchers for the load-carrying capacity and rigidity of dowel joints and mortise and tenon joints for different types of loading.

Quality demands that designers give tolerances to component characteristics; however, this works only when these tolerances are within process capabilities. The knowledge of machining accuracy is of utmost importance in designing quality products. Feasible tolerance widths can be determined and their dependence on the machined dimension clarified through experimentation. A generally valid relationship is useful for quantifying the usefulness of a machine as a function of tolerance positioning and shifted machine setting.

The tightness of fit in many assemblies needs to be kept within operational limits. Functional relationships were derived for optimizing design values and tolerances of part dimensions considering both worst-case and statistical tolerance stack-up.

Chapter 5
Furniture Production Processes: Theory to Practice

5.1 Introduction

The conceptual design of furniture is turned into a physical product through the various manufacturing processes. The manufacturing has individual processes (knife machining, sanding, drilling, etc.) which are limited in numbers and vary slowly in time. A purposefully selected sequence of machining processes creates a manufacturing technology which may be very different, even for the production of the same item. As a consequence, the production technologies are much more inclined to variations than the individual process.

In this chapter, therefore, a discussion of the general principles of manufacturing processes and its optimization, rather than the detailed description of the various manufacturing technologies is presented. After an overview of the global furniture industry, the general principles of optimum manufacture are discussed. Interrelations among product structure, degree of automation, inventory and batch size are also treated. The description of main woodworking operations, surface coating and finish, packaging and value-adding technologies round up this chapter.

The material discussed in this chapter is considerably supported and supplemented by the previous chapters, and cross-referenced as it is appropriate. Especially, the functional relationships concerning the basic machining processes and quality properties (surface roughness, colour and gloss) are important supports and supplements. Furthermore, Sect. 3.6 contains an array of engineering solutions to demonstrate process optimizations.

E. Csanády et al., *Optimum Design and Manufacture of Wood Products*,
https://doi.org/10.1007/978-3-030-16688-5_5

5.2 Overview of the Global Furniture Industry

5.2.1 Design Evolution

Furniture industry is an ever-evolving industry that encompasses innovation and strives continuously to evolve and meet the expectations of individuals from different tastes and lifestyle on a global platform. The evolution of design in the furniture industry can be dated back to cabinet-makers and apprentices in the 1700s to the current twenty-first century.

Before the nineteenth century, the earliest form of furniture used were stumps and rocks and then it evolved into range of styles with the increased knowledge of the craftsmen making it and the availability of new materials and techniques. A quick review of the design evolution throughout the centuries is shown in Fig. 5.1.

- Neolithic period/The Classical World—The evolution of furniture during this time (3000 B.C.–eighth century) gave the world an insight into the early human civilization where furniture was actually made out from stone and wood for the sole purpose of daily utilities such as storage and they were often adorned with gold, silver, ivory and ebony for decoration purposes (Richter 1926).
- Early Modern Europe/Medieval period—In contrast to the ancient civilizations, furniture from this period (500–1500 A.D.) mainly featured exquisite artistic designs often made with heavy oak and ornamented with carved designs. An explosion of design and renaissance of culture started a rapid development in the furniture design in this era (Cescinsky and Gribble 1922).
- Nineteenth Century—Starting in the nineteenth century (1801–1900), furniture designs started to get more detailed and artistic, defined by concurrent revival styles

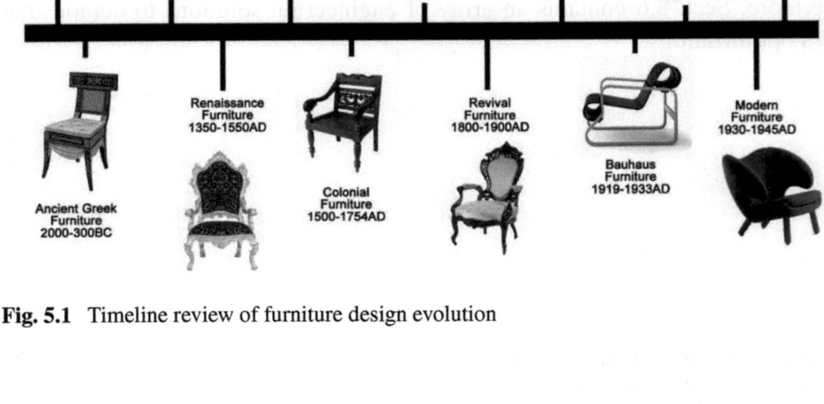

Fig. 5.1 Timeline review of furniture design evolution

including Gothic, Neoclassicism, Rococo, Arts and Crafts Movement and the Art Nouveau Movement. The designs inspired handcraft and natural designs. The furniture often had fancy cut-out designs, often used by the wealthy community.

- Early North American—In the beginning of the twentieth century, furniture from this time started to gain more importance on its functionality rather than its outer appearance filled with artistic and detailed design. A simplified design was preferred with more emphasizes on form and materials over fancy furniture.
- Modernism—In post-World War II (1945 and after), a fusion between many different styles and movements of the early twentieth century, including Bauhaus, Art Deco and futurism occurred. It emphasized more on producing furniture that embraces the quality of functional, elegant and ultra-modern (Hinchman 2009).
- Eco-design—Eco-design is a form of furniture design that is influenced by the balance between the human race and nature complimenting to their surroundings. It can be traced back to the 1920s and become increasingly popular in the current generation. The furniture produced are based on environmentally friendly design with higher sustainability without overutilizing the earth's resource.
- Contemporary—Contemporary furniture describes the outgrowth of post-modern furniture designs, from the 1970s onwards till the recent time designs from all over the world. It is characterized by the advancements, developments, styles and materials in furniture design highlighting simple, uncomplicated, minimalism designs yet remained functional with low-cost mass production (Ratnasingam et al. 1997).

5.2.2 Major Producers Around the World

The world trade of furniture has grown tremendously in the past ten years, and has been consistently contributing about 1% to the world trade of manufactures. In 2017, the global production of furniture was worth US$420 billion (CSIL 2018). This estimate is based on CSIL processing of data from official sources, both national and international, that cover the 100 most important countries.

On the global stage, there are several countries that have kept the pace with advancement and holds significant root in furniture production globally for many years. The world production of furniture can be broken down by geographical regions and is presented in Fig. 5.2.

Currently, China dominates as the largest furniture producer with 39%, followed at a distance by other major furniture manufacturing countries such as the USA, Germany, Italy, India, Poland, Japan, Vietnam, the UK and Canada.

Over the last decade, China has witnessed an unprecedented period of growth in the furniture production globally making it the world leader in furniture production. This massive growth is accompanied by the export-driven industrial production, foreign investments, abundant human capital skill and low-costs production which has enabled China to become highly competitive in the international market.

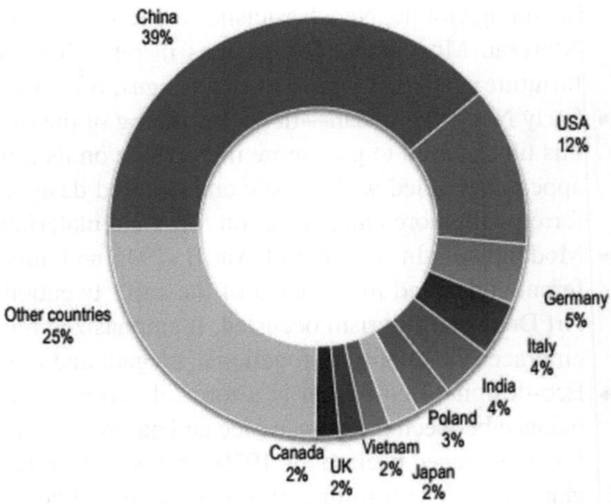

Fig. 5.2 Percentage breakdown of world furniture production, 2016. *Source* CSIL processing of data from official sources: National Statistical Offices, National Furniture manufacturers associations, Eurostat, UN

For decades, the USA has been the main engine of growth in international arena of furniture. However, with the increasing growing dependency on imports from low labour cost suppliers like Vietnam and China, has led to the closure of many furniture production plants in the country and this has caused a gradual weakening of the USA-based furniture industry.

It is interesting to note, however, that high-quality furniture made of precious timber with carvings are imported by China from the neighbouring countries such as Vietnam and Malaysia (Ratnasingam 2015).

5.2.3 Manufacturing Technology

Manufacturing technology for furniture production has faced revolutions where large machinery and computers have taken over craftsmanship and have increased the accuracy and the speed of manufacture. A good machine with defined technology can definitely increase design and production efficiencies of the furniture industry (Ratnasingam and Tanaka 2002).

Among the most common technologies applied in the furniture production field includes computer-aided design (CAD) and computer numerically controlled (CNC) manufacturing. Most manufacturers apply combination of these technologies to safeguard a good flow in their production ensuring sustainable value-added market in the furniture industry.

CAD software is a tool that helps a designer to visualize, to improve precision of various designs and enable them to produce 3D designs in the furniture industry (Ratnasingam et al. 2018). The application of CAD allows the design ideas to be viewed clearly and efficiently where the design can be rotated, colour rendered and get evaluated. Such software enhances creativity, precision and accuracy, increases the production speed and reduces the production cost.

CNC routers are machines that are guided by CAD data rather than by human beings to accommodate cutting and machining of parts that require high tolerances and finishes. CNC routers provide good solution in the modern furniture industry to conduct a wide range of shaping tasks. The CNC router turns wood stock and metal into decorative components once the desired shape is entered in the computer.

The application of robotic arms is also on the increase in the furniture industry, especially in the application of finish materials and also packaging. These operations are usually monotonous and hence, robotic arms increase production rate, while reducing the production cost (Ratnasingam 2015).

5.2.4 Manufacturing Scheme-Production Flow

There are several stages of processes involved in the production flow of furniture. The operation varies depending on the type of furniture produced. The operation stages must be followed accordingly to maximize quantity, quality and durability of the product as they hold the main key to customer satisfaction (Ratnasingam et al. 2018).

Figure 5.3 exhibits the production flow of furniture.

Fig. 5.3 Production flow of furniture factory

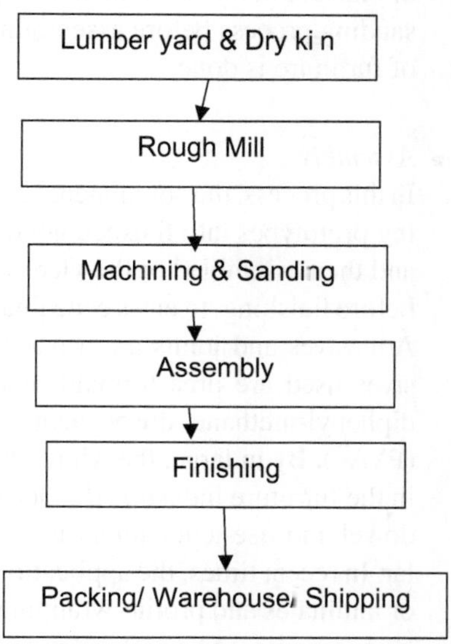

- *Lumber yard and drying kilns*
  The lumber yard functions as storage of raw material where the lumber is sorted into species, size, thickness and length; furthermore, it also acts as a medium for air drying. As the lumber has to be dried before it can be used, a kiln drying process is installed to reach an optimum moisture content level between 12 and 15% in the lumber. Kiln drying is a process of placing the lumber in a kiln where heated air is circulated with controlled relative humidity to reach the desired equilibrium moisture content (EMC) of the lumber.

- *Rough mill*
  Rough mill converts dried lumber board into components of specific dimensions where it is planed, cut, glued and processed to form panel in random width, except for the times when the components are used in the finishing room to manufacture parts such as rails or posts. Generally, the final product mostly comes in a rectangular component with a specific rough dimension of length, width and thickness. The stock removal rate at this section is usually within the range of 1.0 mm or more, as the final outcome is the production of rectangular or square blanks or white parts of the desired dimensions, with clear surfaces.

- *Machining and sanding*
  The sized lumber from the rough end are brought into the machining or machine shop to be processed into components to achieve the required or specific shapes, profiles and sizes for furniture manufacturing. As an example, Fig. 5.4 shows parts of a door. These operations include different machines as moulder, router, round end tenoner, oscillation mortiser and many more. Sanding is a process to smooth the surface of the component for finishing. All components will be sent through the sanding process before assembling, once the construction manufacturing process of furniture is done.

- *Assembly*
  In this process, the components are assembled according to the drawing and designing prototypes into finished good. Sometimes, the furniture can either be finished and then assembled or the vice versa but commonly assembling process are done before finishing, to ensure the final product has smooth surface without any defect. Adhesives and joints are usually used in the assembly process. Common adhesives used are urea-formaldehyde (UF), phenol-formaldehyde (PF), polymeric diphenyl-methane diisocyanate (PMDI), hot melt and the polyvinyl acetate (PVAc). By in large, the white glue of the PVAc is the most widely used adhesive in the furniture industry (Bandel 1995). The common furniture joints used are the dowel, mortise-tenon joints and for flat panels, finger joints are increasingly popular. In recent times, the application of edge-banding, over-laying with face veneers or laminates and profile-wrapping is also on the rise, especially on furniture made from wood-based panels (Bandel 1995).

Fig. 5.4 Parts of a door

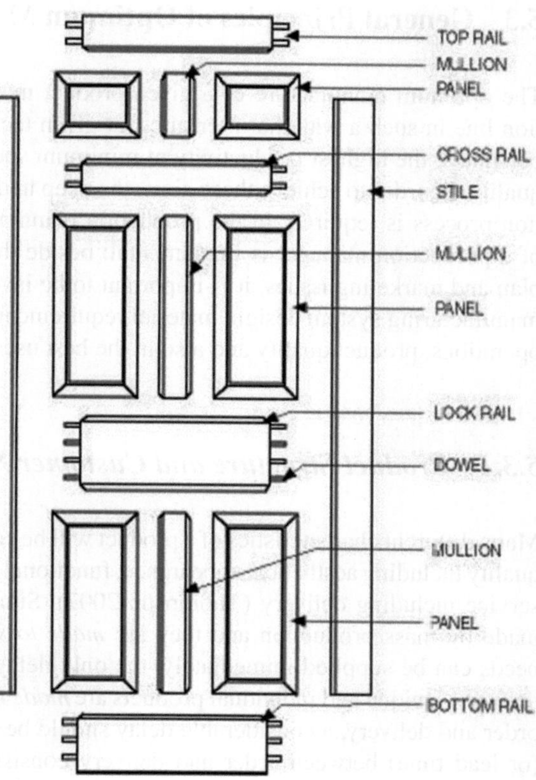

- *Finishing*
  Finishing involves a variety of coatings, such as stains, lacquers, varnishes and other required material to give the assembled good a desired appearance. The finishing material can be applied by dipping, roller-coating and spraying. Spraying technique has become the most common finishing application technique since it requires lesser time. In most furniture factories, finishing conveyors are in practice to reduce the handling time. Finishing is a crucial aspect to ensure the success of furniture sales, as the appeal of the furniture through its beautiful appearance is the trigger for sales.
- *Packing/Warehouse/Shipping*
  Packaging provides protection to the finished products from damage, i.e. impact damage and damage during transit. Corrugated carton boxes are the most common external packaging materials used, while for internal packaging plastic sheets, paper, polystyrene and polypropylene (PP) are widely practiced. The packaged finished products are then shipped to the customers on skids, pallets and cottons, or are kept in the warehouse until an order is received.

5.3 General Principles of Optimum Manufacture

The optimum manufacture of a given product means the organization of production line in such a way that it ensures, at given technical possibilities and working personals, the highest productivity at minimum specific cost and with the required quality. In order to achieve these aims, the deep understanding of the whole production process is required. In the production planning and control function, the role of a production manager is fundamental: beside the concern of financial, business plan and marketing issues, it is important to be involved in the technical aspects of manufacturing system design, material requirement planning, inventory, production operations, product quality and also in the best use of the workmen in the factory.

5.3.1 *Product Structure and Customer Needs*

Many different characteristics of a product will be concerned by the customer such as quality including aesthetical appearance, functional performance, price and customer service including delivery (Thompson 2007). Simple wooden items are generally made by mass production and they are *made to stock*. In this case, the customer needs can be supplied immediately, the only delay may be which is necessary for delivery. Unique and individual products are *made to order* and in this case, between order and delivery, a considerable delay should be reckoned with. The whole delay (or lead time) between order and delivery consists of several lead times such as order processing lead time, supplier lead time for raw material and components, manufacturing lead time and delivery lead time. The total lead time highly depends on the complexity of the ordered product and how the manufacturing company is organized. If the company keep stocks of raw materials and an array of components, the total lead time can considerably be reduced.

The product structure manufactured determines the production system and the flow of materials. In a serial production system, the raw material enters at the one end of the series of operations and at the other end, the finished product is received. This is the simplest production system. In most of the cases, the production system is more complicated, having several assembly operations.

The assembly operation takes parts and puts them together into an assembly. Not only common part but also subassemblies can be used to make assemblies. The finished product may have several assemblies which consist of subassemblies and parts, Fig. 5.5.

A bill of materials (BOM) describe how many and which parts are necessary to put together in the subassemblies, assemblies and finally the finished product. In general, assembly operations put together a large number of components into a small number of finished products. This is called a *convergent product structure*. Today, however, in order to meet customer's needs, companies tend to offer a variety of assemblies for the same basic product design. This means for the opposite to happen: a small

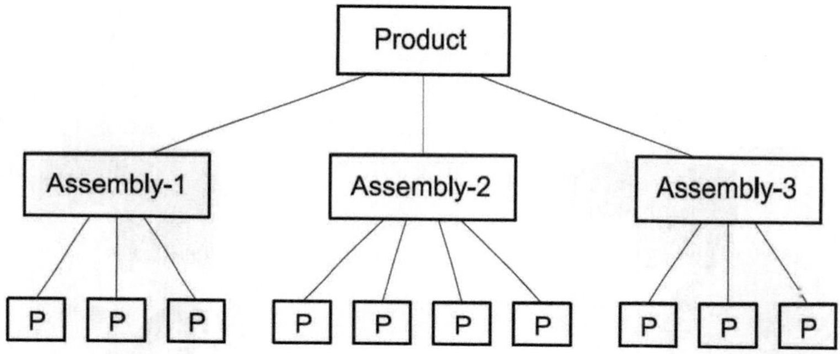

Fig. 5.5 Schematic product structure of a cabinet made of a frame structure (Assembly-1), doors (Assembly-2) and drawers (Assembly-3). All assemblies made of several parts (P)

number of basic components are combined into a large number of finished products. This is the case in the present-day furniture industry. A kitchen furniture, cabinet or sofa are offered a number of varieties depending on fabrics in different colour and quality, built-in mechanisms, number of sections in assembled cabinets and kitchen furniture. This is called the *divergent product structure* offering a variety of finishes and options. One of the most typical products is the strip flooring parquet differing only in its uncounted variety of top layer including wood species, number of rows of lamellae and finish, with the same basic design. There may also be a mix between convergent and divergent structures depending on the number of components and finished goods.

The product structure has a definite interaction with the production system. One characteristic feature is the relationship between volume and variety of products. Generally, an increasing volume of production corresponds to decreasing variety and accordingly, it relates to mass production, batch production and one-off production.

Mass production is characterized by producing a small number of different products in a large quantity. Its characteristic feature is operations are linked together in a line and product is moving directly to next operation when one operation is finished.

Batch production is characteristic for a greater variety of products in correspondingly smaller volume. The one-off production is common when individual customer requires individual and unique product. In the furniture manufacture, it is more common in cases when exceptional raw materials in smaller quantities are available which enables the manufacture of one-of-a-kind items with highest quality, aesthetical beauty and stability of their value. This production method, mostly with hand-work, is typical in Southeast Asia but it also occurs in Western Australia, thanks to the availability of precious and unique timber materials.

It should be noted that the above-mentioned three categories are often mixed in any production system. For example, in the production of strip flooring parquet, the middle lamellae (865 $\times$ 28 $\times$ 9 mm for four strips) as a common part for all end products are required in a great quantity and their manufacture may be regarded as a

Fig. 5.6 Different drilling and sawing units for milling in CNC machine

high-volume batch production. In order to combine high variety with high volume, the use of flexible manufacturing system may be a possible solution.

Generally, the physical layout of a factory follows the production process itself. Some examples to this question have already been discussed in Sects. 3.6.8 (see Table 3.5) and 3.6.9 (see Fig. 3.37). In the case of batch production, one possibility may be the use of functional layout, where similar machines are grouped together making a work-center. Mostly, CNC machines are used having the capability to change tools automatically and to use additional units for drilling, trimming and sawing at any optional angles, Fig. 5.6. The economic use of this technology highly depends on volume and variety of products, balanced workload on the machine group and the organization level of all operations.

5.3.2 Degree of Automation, CIM

In the development of manufacturing, the introduction of CNC machines and CNC machining centres has played a major role (Stonebraker and Leong 2005). They have the capability to carry out a variety of different operations required, such as milling, drilling, form cutting of boards and other sawing operations. The tools required for these operations are stored in a tool magazine and changed automatically when required. The machines are capable to work with different rotation speeds to ensure the optimum cutting speed with different tool diameters. For loading and unloading, a programmable manipulating device (robot) can be used.

The simpler automation (so-called hard automation) of machines and machine lines have been introduced several decades before. It had used a fixed equipment to automate the manufacture of a given product for longer time without any changes in design. CNC machines and machine centres, at the same time, can be used for flexible automation. These machines can be programmed to manufacture a variety of different parts through a sequence of operations. The need for flexibility is supported today by a number of reasons:

- the requirement to introduce new products in a short time. An increasing number of goods, often as fashion goods with new look, is required even in the case if there is not much real improvement in the design,
- making a variety of different products, the risk of equipment becoming redundant may be reduced due to the obsolescence of some products,
- a quick change in manufacture of product to other generally decreases the economic batch size,
- a flexible manufacture can respond more effectively to any changes in customer demand both in volume or design,
- in a flexible manufacturing system, the breakdown of a machine may more easily be replaced temporarily by another machine,
- in a flexible system, parts can be produced in order to give the possibility to assemble products in the desired quantity.

In achieving a successful flexible manufacturing system, there are also difficulties which may arise from software and control aspects. Especially, the high complexity of control tasks may cause problems. For example, tool wear and the formation of workpiece surface defects should be monitored in order to take its effect into account in the software governing the machining process or to replace the tool when necessary.

The more advanced development in the product manufacturing is the computer integrated manufacturing (CIM). In this case, the role of computer is extended to further areas such as the design, the entire production process and inventory control. The main company's functions integrated in the CIM system are shown in Fig. 5.7.

Computer-aided design (CAD) comprises the use of computer in the whole design process. The computer is used to make detailed drawings and perform strength, stability and functional calculations. In the case of furniture, the designer should

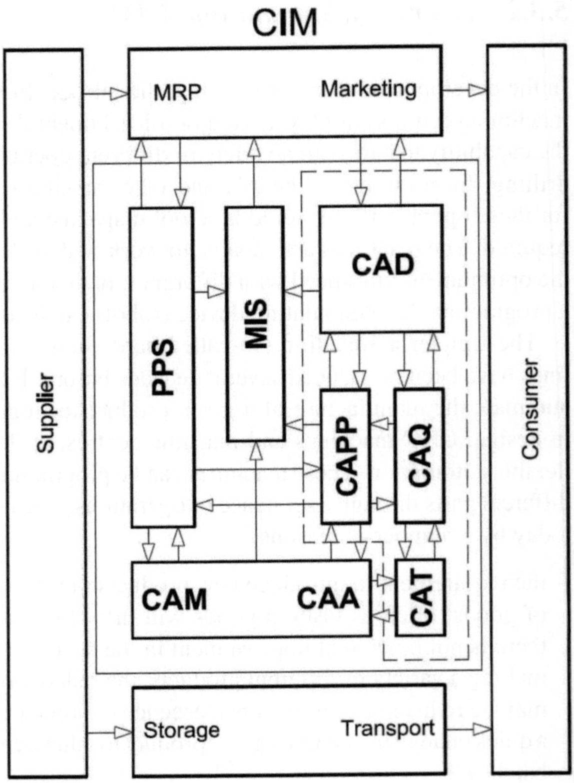

Fig. 5.7 Integrated company's functions in computer integrated manufacturing system

consider not only the mechanical aspects of the design but also its ergonomic, safety and aesthetical requirements. Furthermore, the designer may contribute to a more economic manufacture of the given product.

Computer-aided engineering (CAE) comprises, beside CAD and CAM, the process planning (CAPP), the quality assurance (CAQ) and testing (CAT).

Computer-aided manufacturing (CAM) comprises all activities associated with the use of computer control and manage the whole manufacturing process. Some of the most common activities based on the CAD design, are as the following:

- process monitoring and control for machining, transfer and assembly,
- data collection on shop floor on material movements,
- cost estimating for products and jobs in production, and
- routing for operation sequences, kitting and assembly instructions.

An important ability of the system is to command CNC machines, generated from the CAM system, in order to produce parts automatically.

As an example, Fig. 5.8 shows the drawing of a clock house made by the CAD program. The surface is quite complicated with round, ellipsoidal and trapezoid cavities. The next step is the generation of tool path using tools with given profiles and

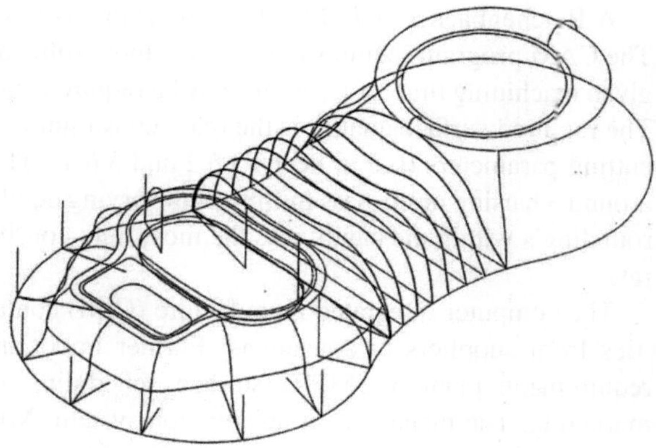

Fig. 5.8 CAD drawing of a clock house. Overall size: 650 × 300 × 100 mm

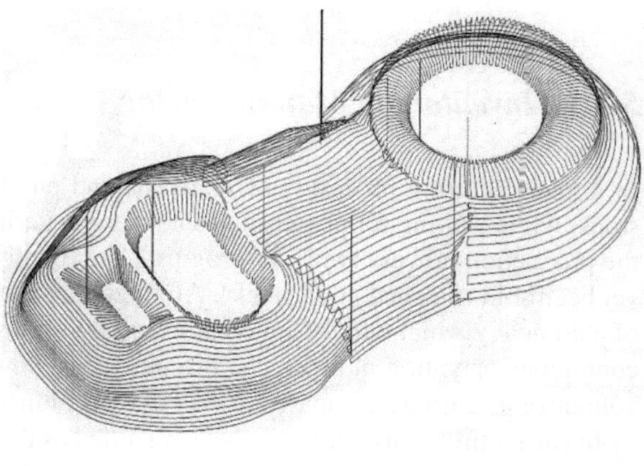

Fig. 5.9 Generated tool path of a clock house (MasterCam)

Table 5.1 Cutting parameters of milling cutters

Cutter	RPM	Feed rate (m/min)	Tooth bite (mm)	Number of edges	Diameter (mm)
Hel. finger	16,000	15	0.47	2	16
Conical	14,000	0.7	0.05	1	–
Flat	10,000	2	0.05	4	60
Ball	14,000	2	0.05	2	18
Ball	14,000	2	0.05	2	24

with constraints due to the required surface roughness (see Sect. 3.6.5 and Figs. 3.19 and 3.20). The generated tool path is shown in Fig. 5.9 using five different tools with automatic tool change. For rough milling, a helical cylindrical router of 16 mm diameter has been used.

For profile milling, a conical milling head, a flat router and two ball-milling cutters were used. The main cutting parameters used are summarized in Table 5.1.

A Reichenbacher 207AMW CNC machine with MasterCam program was used. The CAM program requires the accurate tool profile and geometric data set. After a given machining time, corrections may be required due to tool wear (see Sect. 3.6.4). The required surface quality is the main constraint for determining and limiting the cutting parameters (see in Sects. 3.6.2 and 3.6.4). The overall machining time was around 3 h using multi-pass milling with varying depth of cut. The machining of fine rounding's within the cavity was the most time-consuming operation with low feed rate.

The computer integrated manufacture (CIM) contains all informations on activities from suppliers to consumers. Further important activities are the materials requirements planning (MPR), storage, scheduling or assignment of products and marketing. The management information system (MIS) is connected to all subsystems in order to get up-to-date informations from the whole production system.

5.3.3 *Inventory in Manufacturing*

In a manufacturing process, raw materials and purchased components are held in stock, while waiting for processing. There are materials and semi-finished goods in the production process (work-in-progress) and, finally, finished goods that have not yet been sold (finished good stock). All three types of goods make up the inventory of a company which are the inherent necessity of any production system to ensure continuous operation in time. The inventory has, however, its costs. Therefore, the volume of inventories compared to the value of company's sales should be controlled.

In earlier times, inventory has been regarded as the company's asset. Present-day companies regard inventories more as an investment and the money is tied up almost without benefit. If the company hold inventories as small as possible, the money released in this way can be used for capital investments (Stonebraker and Leong 2005).

The minimum value of inventories depends on many influencing factors, with some of them being summarized below:

- the level of value adding in the production process, which is the value of purchased raw materials and components compared to the value of products sold,
- distance of supplies and their reliability, the minimum lead—time of suppliers,
- if the raw material and components are directly ordered from abroad, the transport costs for smaller volume may be relatively high and uneconomic,
- exclusive timber materials are not always available without delay and delivery time may be longer
- sensitivity of the production system to delays in delivery of a given raw material or component.

The costs of inventory have several components (Gibson et al. 1995). The major component is the cost of capital tied up. Inventories take up place often requiring a building for safe storage in order to protect from damage and deterioration due to environmental factors. There may be heating and lighting costs too. There may also be costs associated with ordering and transport which may considerably depend on the volume of ordering.

An awkward situation may appear when larger stocks of goods become meanwhile obsolete or their demand on the market are falling. The stocks are part of inventories and results in a loss for the factory. A good production control system can, however, avoid such situations.

Holding inventory brings also benefits. The main idea of inventory is that the supply of raw materials never exactly equals the demand for production. Therefore, it is necessary to hold a buffer stock for circumstances in which the demand otherwise could not be supplied. It concerns not only the raw material and component supply, but also the need for buffer stocks between work places and in assembly line.

Concerning raw materials, there is always a delivery cost associated with each delivery. Frequent deliveries mean more delivery cost and, therefore, a balance between delivery costs and the cost of holding stocks should be examined. There may often be market fluctuations which would need a production fluctuation. An operations smoothing may be more convenient and economical, in which some buffer stock of finished goods should be reckoned with.

Many companies deliver their furniture by lorry. An economic lorry load is necessary for the stock of finished goods to be delivered economically, especially in the case when the lorry is carrying of a variety of products transported to different places.

Summarizing the above discussion, it can be stated that inventory is an important and indispensable part of any production system but its volume should be held as low as possible. The successful Japanese "*Just in Time*" (JIT) technology uses reduced buffer inventories and more frequent deliveries from suppliers, to achieve lower levels of inventories in the factories (Stonebraker and Leong 2005).

5.3.4 Economic Order Quantity for Inventory

As outlined above, the inventory which consists of the raw materials and purchased components used in the production process is an inherent necessity. Its value may, however, be chosen optimally based on delivery possibilities, and all costs associated with ordering and holding on stock (Gibson et al. 1995).

In the following discussion, an example of a continuous production with constant rate, which is a common striving of furniture companies, is presented. The main cost factor is given by purchasing different timber raw materials and other materials (gluing, finishing, etc.), which is generally required in proportion to the timber materials. The following cost factors will be taken into account:

- cost of placing an order, P (\$),
- cost of holding a unit of stock, H (\$/m^3 year),
- demand in number of unit per year, A (m^3/year),

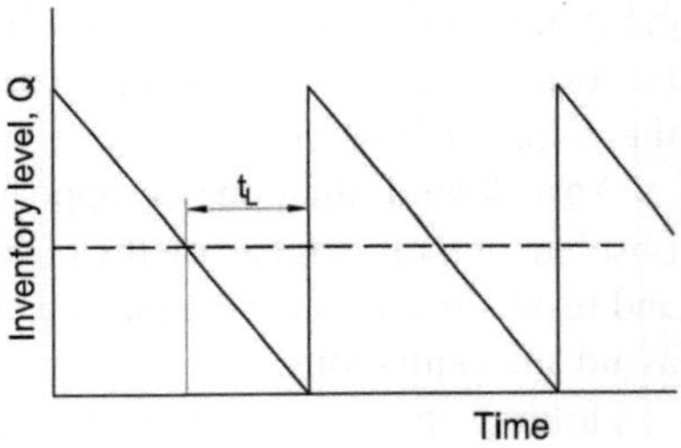

Fig. 5.10 Inventory level and lead time of supplier as a function of time

– unit price of material, M (\$/m^3).

Furthermore, if the lead time of the supplier t_L is known and the optimum order quantity Q_0 and the necessary number of ordering per year, can be established. The graph of inventory as a function of time is shown in Fig. 5.10. The total cost is summed up from the particular cost factors as follows

$$C = A \cdot M + \frac{P \cdot A}{Q} + \frac{Q \cdot H}{2} \quad \$ \tag{5.1}$$

Taking the derivative in respect to Q yields

$$\frac{dC}{dQ} = -\frac{P \cdot A}{Q^2} + \frac{H}{2} = 0$$

and the optimum order quantity is given by

$$Q_0 = \sqrt{\frac{2PA}{H}} \tag{5.2}$$

which is independent of material cost.

As an example, take a factory producing strip flooring parquet in a three-layer design and in a variety of top layers. The production capacity is around 600,000 m^2 per year and the raw material demand is composed of 4000 m^3/year for top layers, 9500 m^3/year for the middle lamellae and 1200 m^3 veneer of 2 mm thickness for the bottom layer. The total amount is 14,700 m^3 per year for an average cost of P = 200 \$, H = 4 \$/m^3/year and using Eq. (5.2), the optimum order quantity is Q = 1212.4 m^3. If this value is rounded up to 1225 m^3 then the time between two ordering is 30 days. In the case of 12 days lead time for delivery, the order should be sent at an inventory level of 490 m^3. The production system in this case has no buffer capacity for any uncertainty in the delivery.

Including a buffer into the inventory, the graph will be slightly modified (Fig. 5.11).

For calculation of optimum order quantity Eq. (5.2), also holds in this case (Anderson 1994). The holding cost increases slightly, however, due to buffer quantity. Using the above example and taking a q = 200 m^3 buffer quantity, the period of ordering

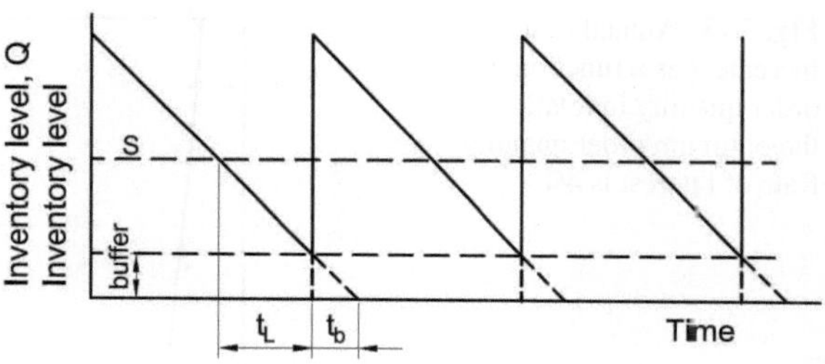

Fig. 5.11 Variation of inventory level as a function of time with buffer capacity

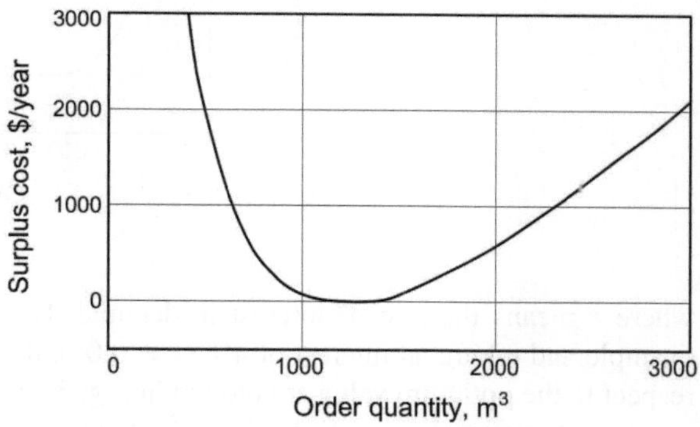

Fig. 5.12 Annual cost increment as a function of order quantity in relation to the optimum order quantity

is 30 days with $Q_0 = 1225\ m^3$. The maximum stock level is $1225 + 200 = 1425\ m^3$ and the buffer quantity covers 4.9 days of production in the case of delayed delivery.

Due to the prevailing material cost, the cost function has a flat course around its optimum value. Due to the high numerical values (over three million), it is more expressive to plot the cost differences related to the optimum as a function of order quantity, Fig. 5.12.

The order quantity, which changed between 900 and 1500 m^3 does not result in a higher increase in the annual cost. Higher deviations from the optimum case, however, leads to considerable cost increase.

Due to the high material costs, it seems to be rightful to take a given rate of interest into account for each period of ordering. In this case the material cost in Eq. (5.1) should be completed with the cost of interest (Anderson 1994). The annual cost function has now the following form:

$$C = M \cdot A + rMQ + \frac{P \cdot A}{Q} + \frac{(Q + q) \cdot H}{2} \quad \$/\text{year} \tag{5.3}$$

Taking the derivative in respect to Q, the optimum order quantity reads

$$Q_0 = \sqrt{\frac{P \cdot A}{r \cdot M + (H/2)}}\ m^3 \tag{5.4}$$

Fig. 5.13 Annual cost increment as a function of order quantity in relation to the optimum order quantity. Rate of interest is 4%

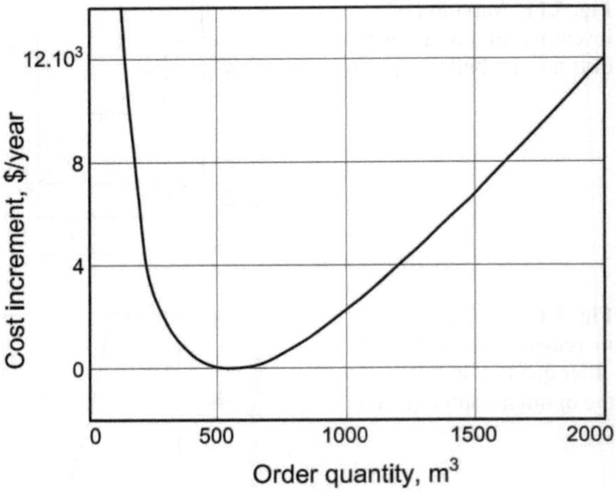

where r means the rate of interest in decimal. Using the values of the previous example and taking an interest of 4% ($r = 0.04$), the calculated cost increments in respect to the optimum value are plotted in Fig. 5.13.

The optimum order quantity is 525 m^3 which is much less than without the charge of interest. The annual cost increment within the range of the 300–900 m^3 order quantity, is tolerable, but outside this range, the cost strongly increases notably. It is important to note that, using the optimum order quantity in both cases, the annual cost is with 4475 $ higher when the interest is charged. Furthermore, if the whole annual demand would be ordered all at once, a massive surplus cost of 150,000 $ should be reckoned with mostly due to the high cost of interest of 3.3 million $ charged.

The two different approaches discussed above indicate that the financial fundamentals of the company may slightly influence the optimum order policy. The above discussion supposes a continuous operation with stabilized demands. It is not always the case. Varying demand may often occur due to several reasons. In the better case, the average demand does not vary considerably, but the demand in time randomly fluctuates around its average value. In this case, the ordering rules discussed above can be used without the risk of stock out or overstocking provided that the supplier lead time is known and the system has some buffer capacity.

In another case, a continuous rise or fall in demand can be predicted in time and the corresponding measure can be worked out to modify production rate and the optimum order quantity. A more complicated situation may arise in the demand distributed in time when a high number of varieties of products are offered, for example, top layer varieties for strip flooring parquets (20 or more). The customer choice is highly dependent on the unit price, but the choice is also a matter of taste. It is a typical batch production and the main problem is the economic batch size. The set-up for making a parquet with another top layer may be moderate when the finishing sequences are the same.

For calculating the *economic batch size*, the set-up cost and holding cost should be taken into account. The holding cost may comprise the storage and handling costs, but in the case of finished goods, interest cost may be included which will modify the economic batch size considerably (Gibson et al. 1995).

As an example, the economic batch size for the production of a less consumed strip flooring parquet is examined. If the predicted selling volume is 10,000 m^2 per year, the corresponding raw material volume is 250 m^3, the set-up cost is $P = 200$ \$ and the holding specific cost is $H = 4$ \$/m^3/year, using the simplified method with Eq. (5.2), results in the following:

$$Q_0 = \sqrt{\frac{200{,}250}{2}} = 158\ \text{m}^3$$

The average production capacity is 200 m^2/h corresponding to 5 m^3 raw material and 31.6 h production time. Taking 7.5 h net production time per shift, the required number of shifts is 4.2. The rational division is to make two batches, one with 4 shifts and the other with 3 shifts.

Holding finished goods have higher values compared to the raw materials used due to processing and labour costs, which are spent but not realized yet. This value-adding money may be charged with a given rate of interest, and in this case Eq. (5.4) will be used.

Supposing that a continuous and evenly distributed sales of finished products is realized throughout the year, the average tied-up capital is the half of the total amount.

Taking a rate of interest of 4% and a value adding of 225 \$/m^3, the optimal batch size is $Q_0 = \sqrt{\frac{200{,}225}{4.5+2}} = 87.7\ \text{m}^3$.

Which corresponds to 2.34 shifts. Again, a rational division into three batches, for example in 3 + 2 + 2 shifts, is a possible solution.

It may be concluded that any increase in holding costs decreases the economic order quantity or batch size.

In many cases, the demand varies with time and cannot be predicted with deterministic method. In such cases, the probable average demand and its standard deviation may be used. Choosing a probability level of 95%, that means that at least 95% of the customers are served in time. Therefore, as in the previous example, the predicted average demand is 10,000 m^2 strip flooring parquet per year with a standard deviation of 500 m^2. The 95% probability level corresponds to the average plus 1.64σ which amounts to 10,820 m^2.

5.3.5 *Optimized Production Technology*

The main goal of a production company is to run a system in which

– the throughput both in production and sales is maximized, to ensure the highest return on investment, in a short period,

– the inventory level is reduced to the necessary level, avoiding to have excess money tied up in inventory,
– operational expenses are reduced by using system optimization by which the money spent to convert inventory into throughput is minimized.

In order to achieve the above goals, a competitive product design, a well-planed and controlled production system, marketing and customer service are needed. A complete production system can be optimized as a whole with the advantage that the interactions among the subsystem will be considered. A substantial drawback is, however, that the procedure inherently relies on many supposed and hypothetical assumptions which may considerably lower the accuracy and reliability of the obtained results. Therefore, no one global optimization method exists which would generally be applicable to all cases.

There are several approaches and conceptions for global optimization. Zhou and Chen (2003) proposed a systematic optimized design methodology of the business process that highlights a holistic approach for the organizations to restructure their organization by focusing on the knowledge of business process reengineering (BPR), a strategic analysis and design of business and manufacturing processes on overall operation levels within an organization. The main objective of BPR is to create a revolution in the way an organization thinks and conducts its operation in order to improve its operational cost, productivity and to form a larger customer base in the industry.

Other modern optimization concept includes Lean Six Sigma System, Lean Process Principles, Just-In-Time Production, and Total Quality Management, which are focused on the responsible use of resources by eliminating waste (Sebastianelli and Tamimi 2003; Gejdoš 2015).

Lean Six Sigma is a data-driven structured approach for process improvement in an organization. The main goal of this concept is to eliminate defects, reduce production and development cost, increase profit margin and to sustain the competitive market. This is achieved by DMAIC and DMADV, the two subcategories of Six Sigma. The Six Sigma DMAIC process incorporates define, measure, analyze, improve, control stages to ensure incremental improvement on existing products which are falling below the required standard. On the other hand, the Six Sigma DMADV process incorporates the define, measure, analyse, design, verify stages as a tool to focus on improving and to develop new processes and products.

Just-in-Time concept has been in-practice since 1970s (developed in Japan), and it is a concept that was introduced to the USA by the Ford Motor Company working based on a demand-pull basis. The production of goods in this concept is to meet the number of customer demand exactly, in exact time, quantity and quality, producing what the customers wants to actual request and not to forecast. JIT generally reduces waste linked with time, labour and storage space. The benefits of JIT include reduction in the order to payment timeline, inventory cost, reduction in lead time, increase productivity and quality as well as the problems becomes easier to identify and solve it.

Total Quality Management (TQM) is also one of the optimization concepts that seek integration between all organizational functions like marketing, finance, design, engineering, and production, customer service and many more to meet market demand and organizational goals. TQM is a collective process in which improvements are made continuously in an organization by integrating the knowledge and experiences of workers, which ensures that the quality of the product is improved all the time. TQM is a powerful management tool as it not only controls but also governs the manufacturing process, providing a wholesome management tool to manage the wood products manufacturing industries. Total Quality Management can be divided into four categories such as plan, do, check and act, which are also referred to as the PDCA cycle (Feigenbaum 1991).

Lean manufacturing is a production practice that considers to apply business processes that achieve high quality, safety and worker morale, at the same time reducing cost and lead times in design, manufacturing, distribution and customer service processes. Lean manufacturing focuses on designing the processes ideally in a way that eliminates waste from all the processes, and this is accomplished by identifying all unnecessary process and designed as lean as possible targeting maximum efficiency with clearly defined responsibilities, exactly described processes and traceable ways of communication.

The above-discussed optimization technique may be regarded as global optimization methods concerning the whole production system often including sales and customer service. An array of optimization task may, however, be handled as subsystems using physically well-founded relationships and deterministic optimization approaches (see Chaps. 2 and 3). It may namely be supposed that an optimized subsystem (woodworking operation, surface finish) shall not heavily conflict with other optimized subsystems and the whole system. If necessary, the interrelations among subsystems can be examined and some corrections may be made without difficulties.

There is often a need to find a compromise between maximum production rate and minimum production cost. If the cost function around its minimum is flat (see Figs. 3.17 and 3.36), then a good compromise is always possible: against a small increase of cost, a considerable increase in production rate can be achieved. Surface quality requirements must be handled carefully: excess requirements may result in a limited production rate and higher production costs.

In many cases, the minimization of non-productive time is a major factor to arrive at optimal conditions. This requires skill and good organization of individual operations, especially in the wood industry.

In many operations, the feed speed is a fundamental factor in process productivity. If operations are linked together, a bottle-neck in the line will determine the maximum feed speed. A bottle-neck is working at its capacity and acts as a restriction on total output. Therefore, the time saved at a bottle-neck is time saved for the factory's output. It may be worth to keep a small buffer before a bottleneck machine to ensure that it does not run idle at any time and is working always at its maximum capacity.

Finally, it should be stressed that a recognition of the decisive influence of a given constraint may highly facilitate the optimization procedure and to get generalized solution in analytical form (see Sect. 3.4). Furthermore, a graphic representation of constraints often allows to select feasible operational parameters near to their optimal values (see Figs. 3.8 and 3.45) without the need to make detailed labourious calculations.

5.4 The Main Woodworking Operations

5.4.1 Introductory Remarks

Since ancient times, the sizing, shaping and surface preparation of timber products occur with mechanical woodworking operations. The common feature of mechanical woodworking is the chip formation accomplished by a cutting edge. The cutting edge itself is very similar in the different woodworking operations but the arrangement of edges on tools may be made in great diversity according to the requirements of the specific machining operation.

Due to the similarity of chip formation in the different machining operations, it is possible to find general relationships for the cutting process to facilitate the common understanding of seemingly quite different woodworking operations (Csanády and Magoss 2013). One exception is the widely used abrasive sanding process having negative rake angle and random position of its grits.

Most of the wood cutting tools utilities the so-called free-cutting mechanism without counteredge. In order to achieve a clear-cut without surface deformation, high cutting speed and sharp edge are required. Softwoods have smaller strength and they are more sensitive to lower cutting velocities with regard to the produced surface roughness (see Figs. 2.45 and 2.46). Some machines with reciprocating mechanism (frame saw, veneer knifing) can operate with limiting cutting velocity. In order to get an acceptable surface quality in veneer cutting, the use of a pressure beam is unavoidable. Moreover, the timber should be plasticized using hot water or steam.

Depending on the complexity of parts or products, different combination of machining units may be used. Simple woodworking machines perform only one operation (sawing, milling, planning, sanding, etc.). An established practice is the use of multiple blades in frame and circular saws. Multi-head planers are also in frequent use to machine all four sides of a bar or window frame in one pass, decreasing the machining time considerably.

For high capacity production, machine lines are assembled using woodworking and special machining units, for example, to manufacture edge-banded panels. Where smaller quantities but different sizes and profiles are involved, the use of CNC machines equipped with the required units (gluing unit) may be a solution. The functional relationships given in Chap. 2 may be used in all cases for system optimization either for single-purpose machines or for complex machine line.

A very important task is the proper selection of tools, their edge material and configuration, the expected tool life concerning operating parameters and the required surface quality (see Sects. 2.5 and 3.6.4). In general, it is worth to make a compromise between maximum production rate and minimum production cost (see Sect. 3.6.5). In many cases, an acceptance of slightly higher specific cost allows to increase the production rate considerably.

To ensure a good surface roughness, a smoothing pass is required using small tooth bite and sharp edges. Beside the common roughness parameter R_z, it is worth to check also the core depth R_k which is more or less, independent of wood species,

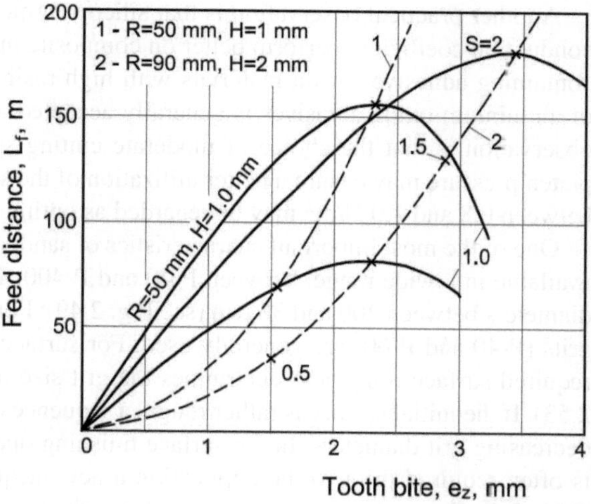

Fig. 5.14 Interrelation between optimum tooth bite and feed distance for different shardness values and the effect of tool diameter and cut of depth

i.e. of anatomical structure. At the same time, the R_k parameter is sensitive to any worsening in the surface roughness (see Figs. 2.58 and 2.59). When higher material removal is required, then multi-pass machining is the solution using a roughing pass.

Edge machining is sensitive to edge breaking which requires special care in the selection of appropriate operating parameters (see Sect. 3.6.3). It is important to remember that the edge quality is characterized by the shardness in mm^2/m.

Using Eqs. 3.32 and 3.32a, a chart can be elaborated such as shown in Fig. 5.14 which can easily be used to make some compromise in the selection of tooth bite, tool diameter and depth of cut. The feed distance curve in the vicinity of optimum tooth bite may be regarded as flat which allows some deviation from the optimum value with minimum loss in feed distance. The tool diameter and depth of cut have considerable effect on feed distance but do not influence the optimum tooth bite.

A special and important woodworking operation is the sanding. The main tasks of sanding are thickness calibration and smoothing of surfaces to a given roughness prior to coating, gluing or surface finishing. Due to the spherical profile of the sanding grits, the upper surface layer will always be crushed to a given depth depending on the grit size, and the ratio of core depth R_k to the average grit diameter (see Sect. 2.4.2 and Fig. 2.54) which is constant for all wood species. Due to the crushing effect and clogging, the measured roughness after sanding may be less than the anatomical roughness (see Fig. 2.55).

The abrasive sanding, as its name implies, is dominated by friction forces and, therefore, the energy of stock removal is high (see Sect. 2.4.2). Due to the continuous wear of the sand belt, the stock removal of the belt decreases as a function of working time. The rate of decrease is the function of operating parameters, wood species and also constructional parameters of the belts (grit material, open or closed coat). As a general rule, the open coated belts are better suited for softwoods and closed coated belts for hardwoods.

Another practical observation is that silicon carbide abrasives with its high heat conduction coefficient perform better on composite materials (particleboard, MDF) containing adhesives or on materials with high resin contents. Otherwise, the use of aluminium oxide abrasives is generally accepted and well established. Another observation is that the choice of moderate cutting speed (10–15 m/s) and higher platen pressure may ensure a better utilization of the sanding belts. Platen pressures between 0.8 and 1.0 N/m^2 may be regarded as optimum values (see Fig. 5.17).

One of the most important characteristics of sandpapers is their grit size which is available in a wide range, between P-40 and P-400. This range corresponds to grit diameters between 400 and 36 μm (see Fig. 2.49). For calibration purposes, coarser grits (P-40 and P-60) are generally used. For surface preparation (smoothing), the required surface roughness determines the grit size to be used (see Figs. 2.52 and 2.53). If the initial surface is rather rough, a sequence of several sanding is used with decreasing grit diameters. In the surface finishing operations, intermediate sanding is often required prior to the application a new lacquer layer (see Sect. 3.6.8 and Table 3.5). For such applications, quite fine grit sizes between P-300 and P-400 are commonly selected.

5.4.2 Rough Milling

Rough milling is the primary operation in the furniture manufacturing process. Once the drying process has been completed, the lumber is then sent to the rough mill for the next phase, where it will be cut, planed and possibly glued, through a group of machines. Rough milling converts the sawn lumber into blanks/white parts ready to be converted into components with correct length, width and thickness required to make the finished part, such as drawer front. It is also considered the operation that poses the strongest cost implication on the manufacturing process. Before going in further, few common terms used in this discussion should be highlighted.

- *Lumber*—Raw material procured from sawmills which are processed into beams and planks,
- *Blank*—A rough end product in a rectangular shape that goes to the machining room,
- *Rough dimension*—The dimension of the blank,
- *Finished dimension*—The final dimension of the furniture part.

A rough mill blank can be either rail stock or panel stock. Figure 5.15 shows the difference between a rail and panel stock. A rail stock is capable of producing more than one finished size rail, resulting from the rough length being a multiple of the finished lengths, whereas a panel stock is capable of producing only one part such as top of a solid case and can be used to produce a few wide rails.

As lumber enters the rough mill process in random shapes and lengths, and upon beginning the rough milling process, the company is required to decide on the quantity of blanks that is required to make the furniture part of the right size. The right size of

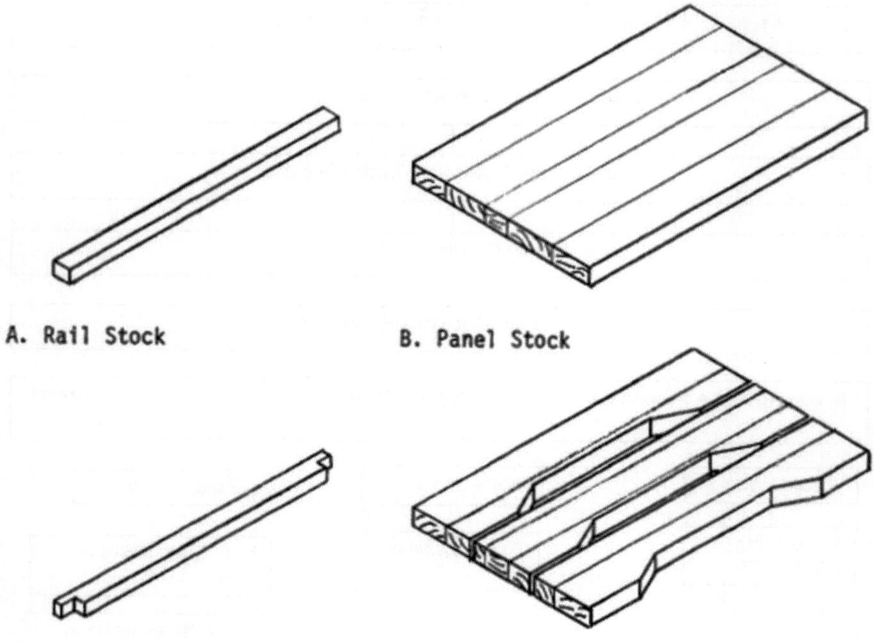

Fig. 5.15 Rail and panel stock

the furniture parts can be determined by the standard thickness in the industry that is best adapted to that particular part. Therefore, one of the main purposes of rough milling is to use lumber of ideal standard sizes and grades, and to cut the lumber into specific lengths, widths and thicknesses for each furniture part.

The second purpose of rough milling is to ensure that the furniture parts are free of natural wood defects and drying checks and/or splits. A detailed explanation of natural wood and processing defects can be found in the report by Ratnasingam et al. (1996). A rough lumber may consists of many defects such as knots, bark edges (called wane), decay and worm holes and these defects can diminish the quality and the appearance of the final product. There are some minor defects permitted in hidden part of some furniture, and there are other defects, which if allowed can ruin the final product life span and quality. In that case, the rough mill converts the lumber into desired dimension and to remove undesirable defects. Both these functions are carried out together in the furniture industry in a combined manner by maximizing the number of output of blanks, with the least amount of input material.

There are two important criteria that must be fulfilled by the blank in the rough mill department before departing to the machining department. Firstly, the blank should have the specified length, width and thickness, and secondly it should be free of defects or with allowable minimum defects in the finished parts.

There exist variety of sequences of operations, different combinations of machines and rough mill layouts commonly used to structure a successful rough mill operation.

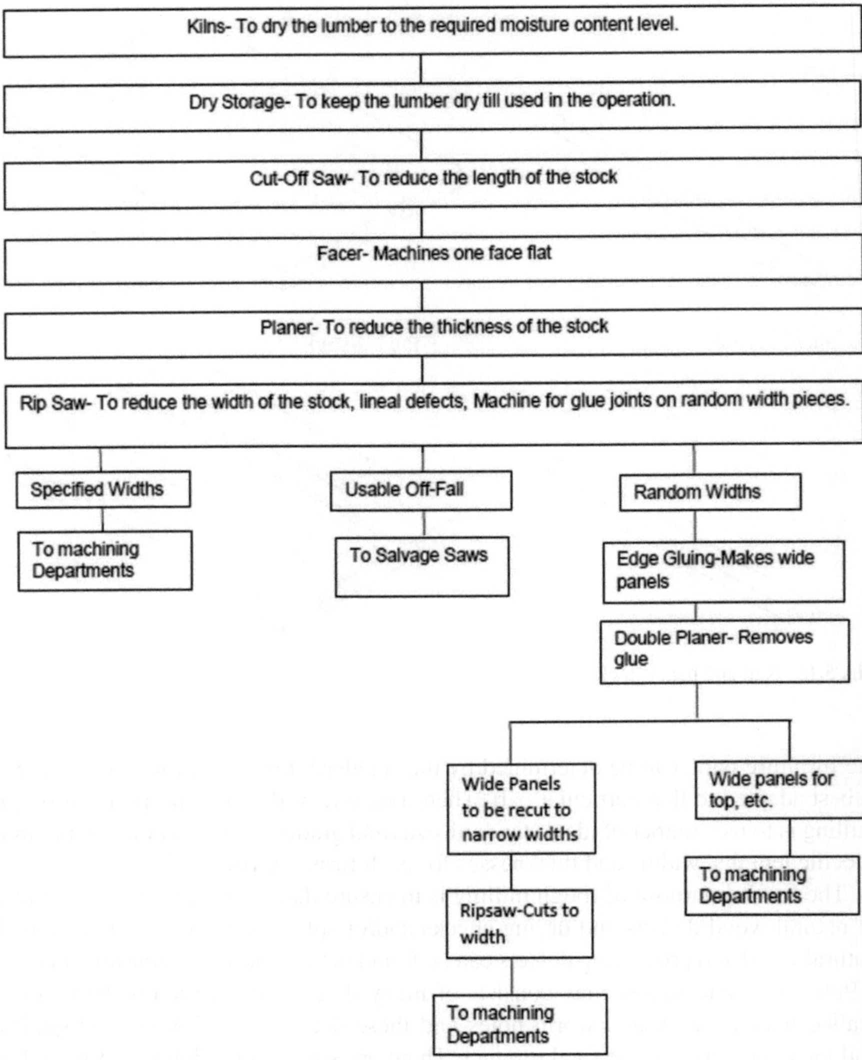

Fig. 5.16 Rough mill operation sequences

The most common sequence for rough milling process, highlighting only the major operations is outlined in Fig. 5.16.

In order to ensure an optimum production economy in time and costs, and the prescribed quality requirements, a detailed process analysis is required. The main operational characteristics of particular woodworking machines can be reliably be calculated (Csanády and Magoss 2013) and, using the quality and other limiting constraints, the optimum machining parameters can be established (see Sect. 3.6). Generally, quality requirements, production costs and production time are in conflict

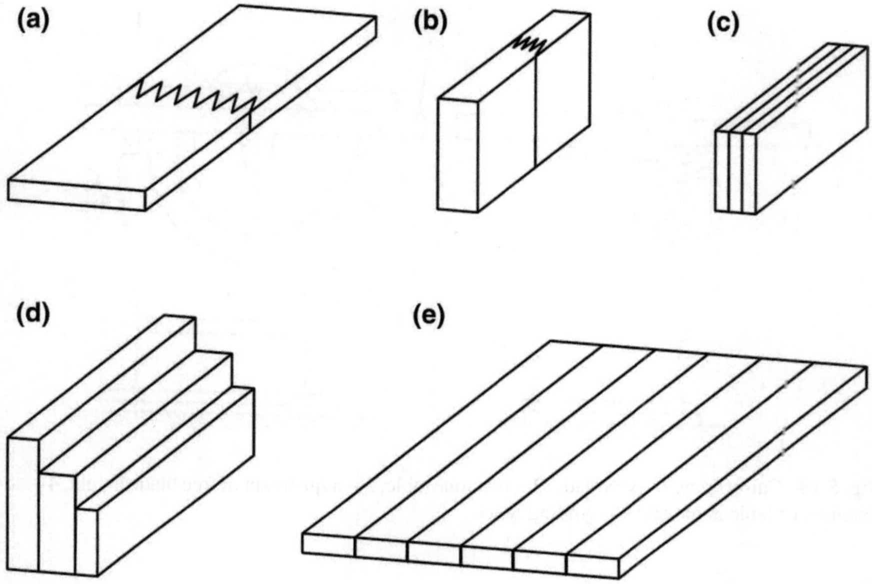

Fig. 5.17 Different methods for utilizing of small wood pieces

with each other and the optimum solution is always a rational compromise. Concerning the rough milling, the surface quality requirements are generally moderate and, therefore, they are seldom the limiting factors for optimum process design.

In order to utilize the wood raw material as much as possible, jointing of smaller pieces into boards, posts or panels are often needed. First, the short pieces are lengthened, using finger joint, into sizable boards (Fig. 5.17a, b) which may be glued together into posts of regular or irregular cross-section (i.e. for window frame) and into panels (Fig. 5.17c–e).

There are through-feed production lines with different capacity and degree of automation depending on the required capacity. Due to the special profile of window frames, the utilization of raw material is not better than 50% if it is worked out from a solid rectangular post. Making glued posts, as shown in Fig. 5.17d, can considerably increase the utilization of the wood raw material. Furthermore, the glued multi-layer post may have better form-stability compared to the one-piece when it is exposed to environmental changes.

5.4.3 Sawing Operations

The rough milling uses different woodworking machines and their combination. Most of all, the different sawing operations play an important role which will be summarized in the following discussion.

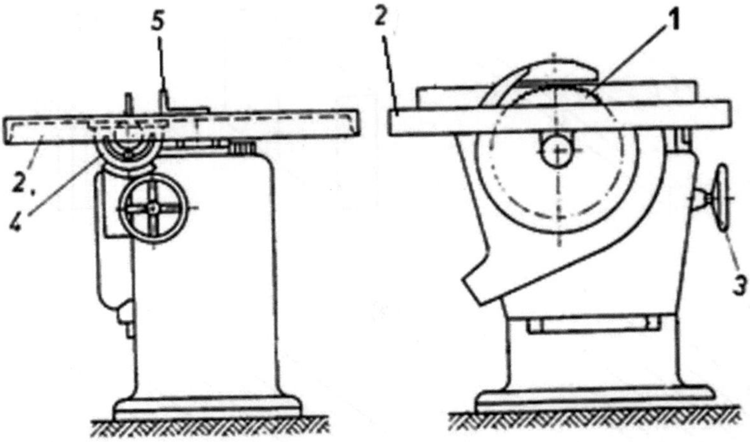

Fig. 5.18 Cut-off saw, 1—sawblade, 2—machine table, 3—adjustment of free blade height, 4—adjustment of table angle and 5—guiding fence

As outlined in the previous section, an important task is to

- cut to rough length,
- remove defects.

which are accomplished with various types of **cut-off saws**. An important economic point of view is that both operations would be conducted with minimum lumber waste and labour cost. For these purposes, lightweight cut-off saws are generally used (Fig. 5.18).

As mentioned in Sect. 5.4.2, to increase the utilization of wood raw material, may require the use the smaller pieces falling off at different cut-off operations. These smaller pieces should be cut to given width and thickness using the **salvage saws** (Fig. 5.19).

Another task may arise by using boards. The faces of the boards that are cut from a log are flat and have fairly straight edges. Once the boards start to lose their moisture during air or kiln drying, however, the board will then take various forms of crookedness due to shrinkage stresses. Therefore, a **facer** is used to produce a rectangular piece of stock, having one flat face.

This operation does not aim to cut

- one face smooth,
- the stock to uniform or specified thickness,
- the edge of the stock.

In order to cut the stock to a specified thickness and, in some cases, also to a specific width, the different **planers** are used. Planers can generally machine one surface or two surfaces, Fig. 5.20, but there are multi-head planers capable of machining also the two side edges. Planers are built in a way they can handle stock of widths up to the full size of the machine, either one piece at a time or several pieces side by side.

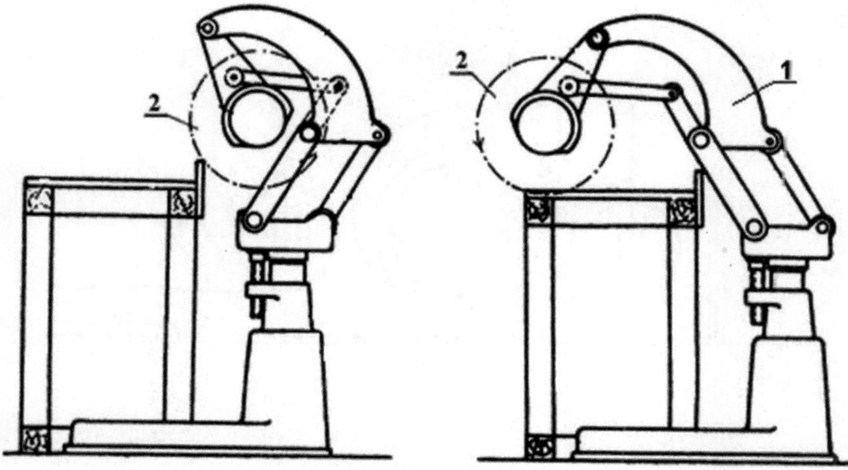

Fig. 5.19 Salvage saw, 1—mechanism for straight line motion of saw blade and 2—saw blade

A well-planed surface enables the operator to identify any small defect better than a rough surface. The planed uniform thickness makes production and quality better at the following operations such as moulding and edge gluing.

It should be remembered that the planer is a through feed machine with some consequences. The moving workpiece behaves as a vibrating system and the amplitude of vibration can substantially influence the surface roughness and size accuracy. In the shaper mode, the exciting forces (radial force component on the edge) are acting towards the rigid machine table which impedes higher vibration amplitude. In the thicker mode, however, the bottom cutterhead generates exciting forces towards the spring-loaded press-rollers and sliders which allow higher vibration amplitudes worsening the surface roughness (Csanády and Magoss 2013). Furthermore, an inherent feature of the planing operation is the wavy surface caused by the cycloid path of the tool edge. If the maximum waviness is prescribed, then it should be handled as a constraint (see Sect. 3.6.2 and Figs. 3.11 and 3.12).

The next possible operation is the edging using edging saws with feeding device. In a glue jointer edging saw, the stocks are vertically fed and the saw has straight knives to ensure the edge after cutting is at right angle (90°) to the face of the stock with a high degree of precision.

A good-quality edging operation also requires much care regarding the selection of feed per tooth and edge sharpness (see Sect. 3.6.3). Using edging saws, the overhang of the saw blade has definite influence on the edge breaking (shardness) and the edge break can substantially differ on the upper and lower edges (Csanády and Magoss 2013).

Ripsaw is the last machining operation in a typical rough mill. Ripsaw plays a key function in cutting the lumber to a given size along its length. The process involves:

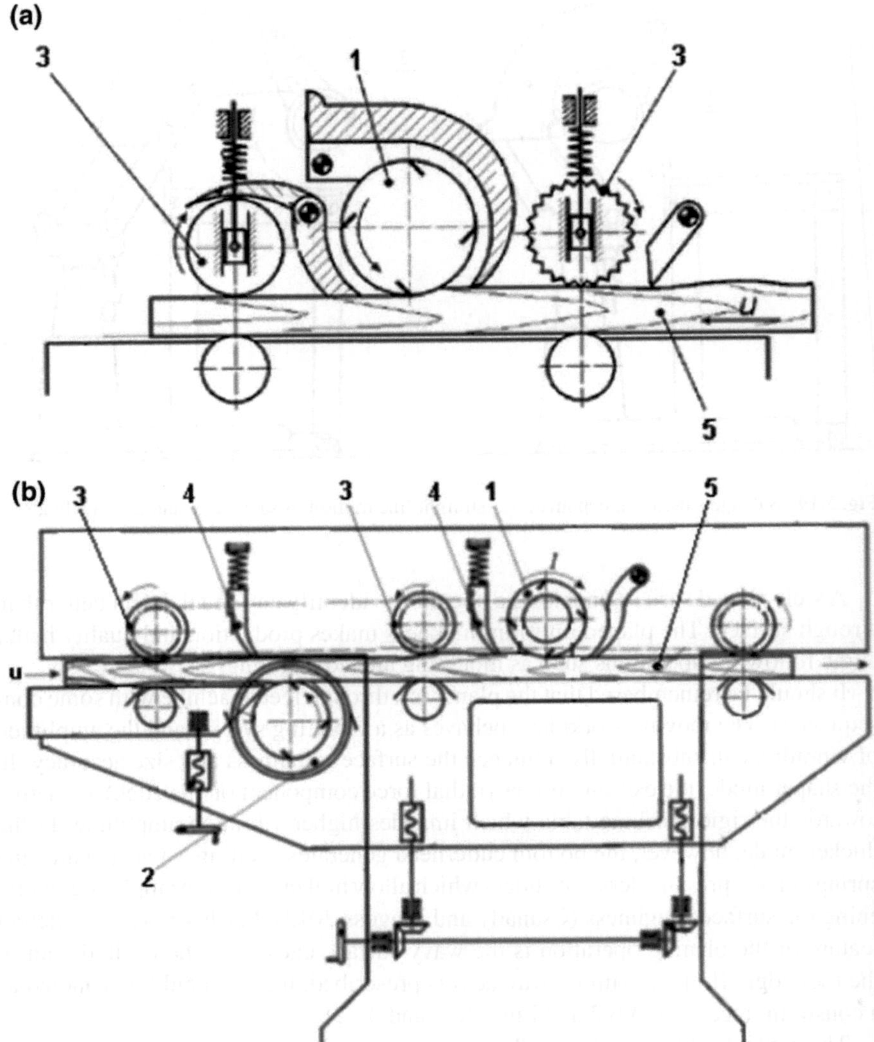

Fig. 5.20 Single planer (**a**) and double planer (**b**). 1—Upper cutterhead, 2—lower cutterhead, 3—press roll, 4—press slider and 5—workpiece

- cutting to width knows as “flat cutting”,
- cutting to thickness known as “deep cutting”,
- angle cutting in certain circumstances.

The accurate and uniform cutting width is ensured by using guiding fence.

5.4.4 Machining and Components Preparation

The blanks from the rough milling section are brought into the machining or machine room to be processed into components by some sequence of machining operations. The term “finish machining” is used to distinguish from the type of machining activities conducted in the rough milling operation.

The key operation in machining furniture part usually has one particular order in which they must go through. There exist few opportunities at times, to switch the order of the operation but, generally sticking with only one order is way better for certain problems. Thus, there are several considerations to determine the sequence to be applied in the machining operation (Ratnasingam and Tanaka 2002).

First is the reference plane. Every cut that is taken on a furniture must be taken with reference of one or two planes to serve as a guide for at subsequent processes. Next is the smoothness of the surface ensuring surface of the furniture is smooth by choosing the correct machines to avoid any unnecessary wood defects and lastly, the convenience in handling plays an important role in the sequence operations. There can be few sequences easier than other, and in order to save time we may even violate the smoothness principles but always bear in mind that the convenience methods must be able to produce the outcome which is desirable.

There are several essential processes and machines involved in machining and component preparation in furniture production. Figure 5.21 displays the key operations and machines involved.

As has been mentioned earlier, the sequences may vary according to the furniture parts and its set-up time, but mainly focusing to keep the set-up time as short as possible, factories schedule the machining operation with the correct sequence and machine. To achieve this purpose, route sheets and process charts are often used to give a quick overall look of the operation involving all the machines, while enabling the operator to make better choices in this section.

Several operations are commonly encountered in this section, which include rough cut-out, shaping, routing, joint formation like bore holes, tenon, mortise and finger and sanding.

Rough cut-out is undertaken by the narrow bandsaw. The main purpose of this process is for cutting which can be divided into two categories:

- Straight—Free hand or with the help of fence,
- Curved—Free hand or with the help of jigs/templates for repetition tasks.

Machine	Function	Illustration
Band-Saw	To perform a rough cut-out of the profile	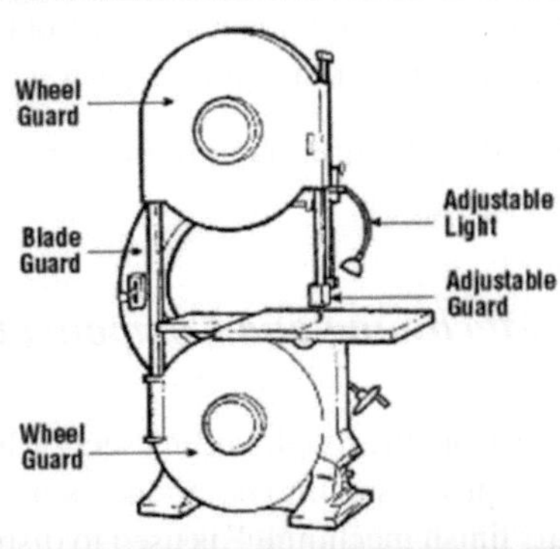
Shaper	To machine the rough cut-out by eliminating the excess material to reflect the actual profiled component	
Router	To perform the final machining of the profiled component to the exact dimensions	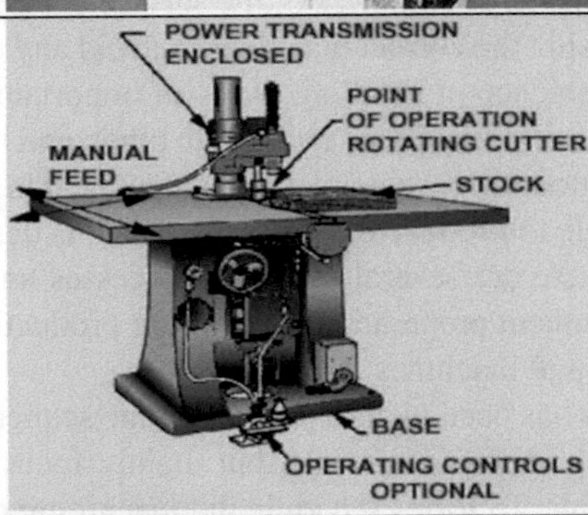
Borer	To machine borer holes for insertion of dowels according to the joint specification	

Fig. 5.21 Key operations and machines

Mortiser	To machine the mortise for the insertion of the tenon according to the joint specification	
Tenoner	To machine the tenon for insertion into the mortise according to the joint specification. In other circumstances, the tenoner can also be used to make perpendicular cuts of the stock	
Lathe	To perform turning operations for wood stock	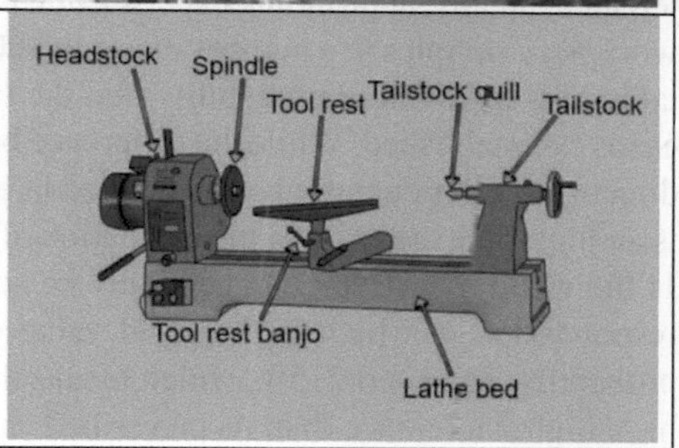

Fig. 5.21 (continued)

For details about the selection of operational parameters, energy consumption, bandsaw vibration and wash-boarding, blade tensioning, stability and cutting accuracy of bandsaw see in Sects. 2.4 and 2.5 and further in (Csanády and Magoss 2013).

Shaping is undertaken by the shaper to shape, size or mould the components of the furniture parts. A shaper is designed to make heavier cuts, making them much cleaner, handles large runs of mouldings and is able to cut larger profiles than a router.

Routing is undertaken by the router. A router is used to rout/hollow out an area in quite hard material like wood or plastic especially in making cabinetry. Cutters that are specially designed with various patterns, make cuts and edging and currently both hand- and machine-aided routers are very common in furniture production.

One famous type of router used in the industry is a CNC router. A computer numerical controlled router is programmed by the operator to control all the work sequences of the machine with the computer data provided. The CNC router is high in speed, multifunctional and works to acquire accuracy, thus permitting manufacturers to conduct a wide range of machining operations just in a single machine (see Sect. 5.4.7).

Decision on the question whether a CNC machine or a special through feed machine is a better choice is governed by the number of types and the total quantity of parts. Large number of individual parts makes the use of CNC machines or machine centres more economical, and fit well into a flexible production system.

Accuracy of machining for different woodworking operations depends on many influencing factors such as the rigidity and vibration behaviour of the machine, the excitation forces and the clamping of the workpiece. Furthermore, the cutting parameters, the accuracy of the running circle and sharpness of edges definitely influence the surface quality, size accuracy and tool life too. Blunt edges generate higher radial force component which increases the vibration amplitudes and decreases the surface quality and size accuracy (Ratnasingam et al. 1999).

The size accuracy of a machined workpiece can be characterized by the difference between the nominal dimension (design value) and the average dimension due to machining, and by the standard deviation. In order to determine these quality characteristics reliably, appropriate experimental measurements should be carried out using workpiece samples in a number of around 60–100 pieces. Processing the experimental results on a Gauss probability net, the two characteristics mentioned above can easily be established. While the difference between the design and machined average dimension can generally be maintained in a narrow range, the standard deviation σ is an important criterion in the evaluation of machining accuracy (see in Sect. 4.7.1). In the wood industry, it can generally be accepted that the allowable tolerance corresponds to 3σ which is threefold value of standard deviation. In this case, the probability level is 0.275%, which means that a quarter per cent of workpieces may have higher tolerance than its prescribed value.

Sadly, very few experimental results are known for evaluation of cutting and machining accuracy (Csanády and Magoss 2013). It may be supposed that CNC machines using pneumatic clamping have better accuracy than through feed machines with moving workpiece. A moving workpiece is more inclined to vibration, especially towards the press rolls in double planer. In this case, not only the size accuracy but also the surface roughness properties may considerably be worsening.

5.4.5 Joint Formation

Joints play a very important role in making assembled units and parts. Joint formation is undertaken by the boring machine, tenoner, mortise and many other. Joint formation is where the pieces of lumbers are joined together in order to produce more complex parts. A joint formed certainly give impacts to the furniture's strength and life. Different types of joints are used for different requirements in load-bearing capacity, repeated load and life time (Ratnasingam and Ioras 2015).

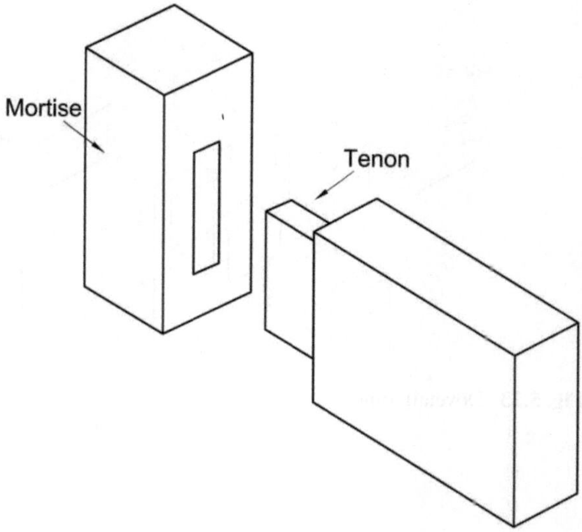

Fig. 5.22 Mortise and tenon jointer

Mortise and tenon joints, consist of the mortise which is rectangular or rounded in shape, drilled into a piece of wood to accept a tenon, which is cut to exactly fit the mortise (Fig. 5.22).

Dovetail joint, is a form of box joint, where the fingers are locked together by diagonal cuts and is extremely strong more than finger joint. This type of joints has been used previously in constructing items such as drawers and cabinets, but today they are much less seen due to their demanding and more expensive production (Fig. 5.23).

Finger joint is also a good-quality box joint in which the corner joints are with interlocking fingers and receives pressure from two directions (Fig. 5.24).

Both types of joints are made by using tapered or cylindrical finger router. The accurate dividing distance (pitch) between the fingers is assured by using a master pattern or it is programmed on a CNC machine.

Dowels are wooden pegs which are glued against another piece of wood. They are cheap, offers fairly strong joint and is quick to produce using production line machinery (Fig. 5.25). Today, CNC machine can also be equipped with drilling units (Fig. 5.30).

As outlined above, machining and component preparations in furniture production mainly have two objectives:

- Profile or geometric alteration,
- Surface modification or smoothening.

There are several operations involved in the machining department. In order to avoid duplicating set-ups and to perform in the lowest possible cost, the machining process is required to schedule its operation cautiously by involving both product and process engineering department for a better economic alternative (Ratnasingam et al. 1999).

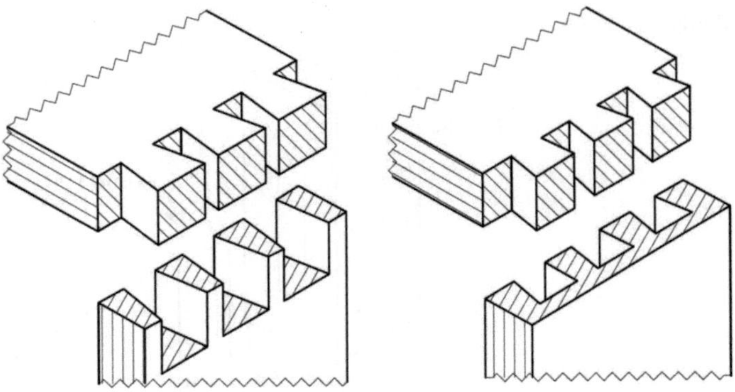

Fig. 5.23 Dovetail joint

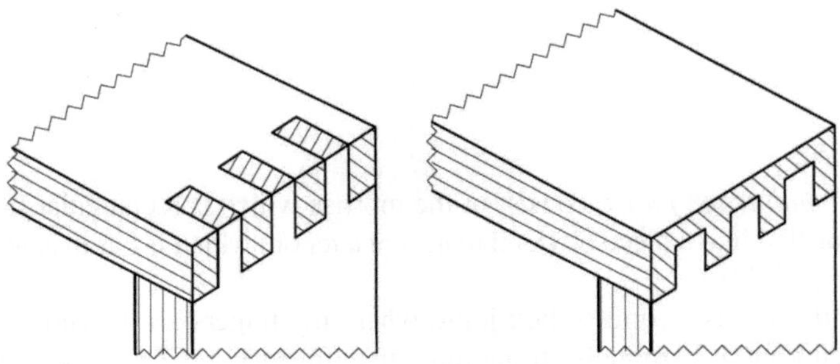

Fig. 5.24 Finger joint

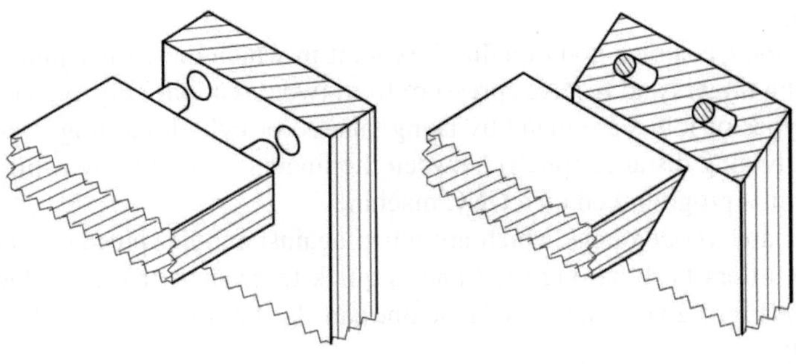

Fig. 5.25 Dowels

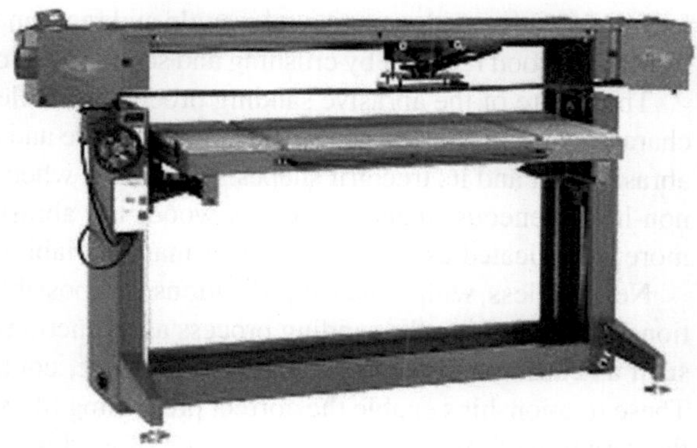

Fig. 5.26 Stroke sander

The performance (production rate) and economy of machining is increased by

- reducing the amount of stock removing,
- using higher cutting speed,
- producing smaller machining tolerances,
- reducing the amount of finer wood dust particles produced and
- enabling eco-friendly waste disposal.

In surface and components preparation, the surface quality requirements are generally higher and they may be limiting factors (constraints) in the optimum selection of operational parameters in the machining process. The required surface is determined by the subsequent surface finish whether it will be sanding, coating with veneer, laminate or any other material. The maximum production rate or minimum production cost is always a fundamental requirement (see Sect. 3.6). The necessary functional relationships for cutting forces, energy consumption, tool life and roughness parameters may be found in Chap. 2.

5.4.6 Abrasive Sanding Process

Sanding can be considered the last process in machining operation and is undertaken by various types of sanding machines. The main purpose of sanding is to remove the first wood layer, ensuring smooth and uniform surface as well as to remove blemishes from previous operations such as gluing. Sanding can also be carried out on layers of coating materials after drying, and before applying the finishing coats to smooth out any raised grains and to enhance the intercoat adhesion bond between the coating layers (see also Sect. 2.10).

Abrasive sanding process is performed as one of the most important value addition operations in the furniture manufacturing industry. The main objective of this process is to attain high-grade furniture finish and flat as well as smooth surfaces on the products. Abrasive sanding involves an orthogonal cutting process as a result of its spherical cutting edge with negative rake angle and random position of grits to each other leading to wood removal by crushing and scraping actions (Ratnasingam et al. 2004).

The nature of the abrasive sanding process is challenging as it is very difficult to characterize and analyse due to the random nature and distribution of the grits on the abrasive belt and its true grit shapes. Especially, when working with highly variable non-homogeneous material such as wood, the abrasive sanding process becomes more complicated as it has to consider many variables.

Nevertheless, with some simplifications, it is possible to elaborate the general relationships describing the sanding process as a function of main influencing variables such as cutting and feed speed, platen pressure, contact length and wood species. These relationships enable the correct processing of experimental results, especially the variation of removal rate as a function of working time (see in Sects. 2.4.2, 2.6.4 and 2.6.5).

One of the advantages of abrasive sanding is that, apart from eliminating surface wood damage associated with machine planing, there is also almost no minimum stock removal amount set to produce a smooth surface, as required in the conventional machine planing operation. However, the abrasive sanding process without a standard acceptable lifetime leads to trial-and-error process, a costly practice in the furniture industry (Ratnasingam et al. 1997, 1999).

Sanding generally involves an orthogonal cutting in which the cutting edge is perpendicular to the cutting direction of the relative motion of tool and workpiece, and the surface generated is a plane parallel to the original work surface (see Sect. 2.4.2).

Oscillation of the belt slightly modifies the orthogonal cutting principle. Moreover, an additional fine cross sanding unit (Martin 2015) and also an oblique sanding method has been developed (Kündig 2015) to increase the uniformity of gloss and to avoid visible scratches on the surface.

The sanding process can be automatic, semi-automatic and manual. The automatic sanding process involves machines that are equipped with two or three heads (usually belts) in which the first is rougher than the second and have brushing system to remove the dust particles as a health prevention step. The semi-automatic sanding machine involves machines that have dust suction system operated by one operator only, controlled manually, whereas manual sanding process is carried out with three-dimensional elements and is done by brushes, sheets or pads. It should be handled more cautiously to avoid further marks on the surface.

In the cases of abrasive machining, the process works by forcing the abrasive particles (grits), into the surface of the workpiece so that each particle removes some layer of the wood material. There are four different types of abrasive grains that are frequently used based on the machining such as aluminium oxide, garnet, silicon carbide and ceramic.

Generally, aluminium oxide grain produces long lasting and even cutting sandpaper, garnet produces a softer finish on wood although it wears out faster, and silicon carbide produces a better surface than aluminium oxide at the coarsest grit size for all the wood species. Furthermore, silicon carbide has higher heat conduction coefficient and, therefore, performs better sanding on composite surfaces with adhesive content.

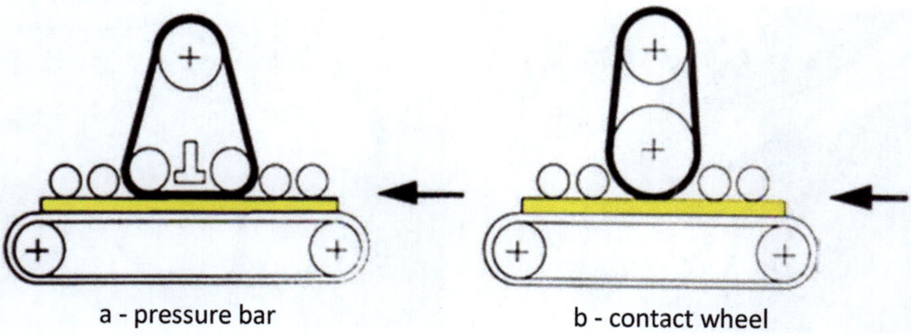

Fig. 5.27 Types of wide belt sanding machines

Sanding process is subjective according to the operator, so it varies among the professionals. However, for the best result, it is highly recommended to begin the sanding process with medium or course grade paper and gradually change the grade as the work progress. Although it is not important to use every available grits, skipping too many grits may leave scratches and it may consume more time to smooth it out. Thus, it has become important to use the correct types of abrasives, grades and grits for a productive line of tasks. A detailed description of the mechanics of the wood and panel sanding processes is given in the reports by (Ratnasingam et al. 2004, 2005).

Among the most extensively used sanding machines for plane surfaces, the stroke sanding machine is the most important one. The stroke sander is available with two variations in terms of pressure application, i.e. it uses either a pressure pad or a pressure bar (Fig. 5.26).

By and large, the wide belt sanding (WBS) machine is considered the most important sanding machine on the factory shop floor. It is regarded as the work-house, because it has a large stock removal and large production throughput rates. Figure 5.27 provides a simplified illustration of the options available in wide belt sanding machine, i.e. machines with either the contact wheel or the pressure bar.

There is much similarity between abrasive sanding of profiled geometries and abrasive sanding of plane geometries. The significant differences are however in the use of narrower abrasive sanding belts and profiled support elements. Further, profile-sanding processes require coated abrasive tools, which have multi-flex patterns, such as edge profile sander, brush sander and edge sander (Figs. 5.28 and 5.29).

In recent times, robotic sanding is also finding increasing application. With a combination of translocation and rotation axels, robots can reach every point in the workspace with precision, similar to humans. Further, robots are suitable for repetitive and monotonous tasks, and are especially useful in a surrounding that may be potentially risky to human health and safety as encountered in the sanding operation. A complete description of the sanding operations and the machines used can be found in the report by Ratnasingam et al. (2004).

A = profile sanding wheels

B = profile sanding

C = brush sander

Fig. 5.28 Edge profile sanders

Fig. 5.29 Edge sander

Investing in systematic operator training programmes is worth the while in the long-run, as a skilled operator produces twice the production of the average operator, because a skilled operator is aware of "*how good is good enough*".

Sanding has at least a magnitude higher energy consumption compared to knife machining. The abrasive belt is continuously wearing as a function of working time, producing decreasing removal rate. The economic use of belts is crucial and it depends on the cutting speed and platen pressure (see Sect. 2.4.2).

5.4.7 The Use of CNC Machines

One of the main features of the modern wood industry is the increasing use of CNC machines and machining centres (Thompson 2007). These machines have the capability to perform different operations (milling, drilling, form-cutting and sawing), change tools automatically, and can be programmed to manufacture different parts making a sequence of operations. Using manipulating devices, the loading and unloading of workpieces may also be made automatically.

The economic use of this technology depends, however, on the product structure, balanced workload and on the organization level of the factory.

The need for increased product individuality is one the main features in the globalized world, and this trend is also similar in the wood industry. This was the main reason for the rapid development in manufacturing technologies to fulfil customer desires. This situation requires maximum production flexibility in term of product variety and maximum production economy in order to keep the product costs more or less competitive with those of cheap mass suppliers.

CNC-based technologies are suitable for cases when a well-defined product or subassembly should be manufactured in a great variety (Thompson 2007). The main application fields are the following:

- panel dividing into arbitrary form and size, with subsequent trimming and drilling,
- machining of door components in great variety,
- machining of kitchen furniture components, first of all door components in the desired variety both in form and material. Using special trimming unit, the profiling of furniture fronts is possible,
- the CNC processing centres equipped with a gluing unit is capable to carry out edging of freeform components economically, first of all at low batch size. If large quantities are manufactured, however, the use of through-feed machine line is more economical,
- using 5-axis CNC machines, the setting of all axes occurs automatically for carrying out drilling, trimming and sawing at any optional angle. Special and precise fitting cavity cuts are also possible for interior fittings and
- the spindle rotation speed can be varied in a wide range allowing to use optimum cutting speed independently of tool diameter. This capability highly increases the production rate and lowers the production cost.

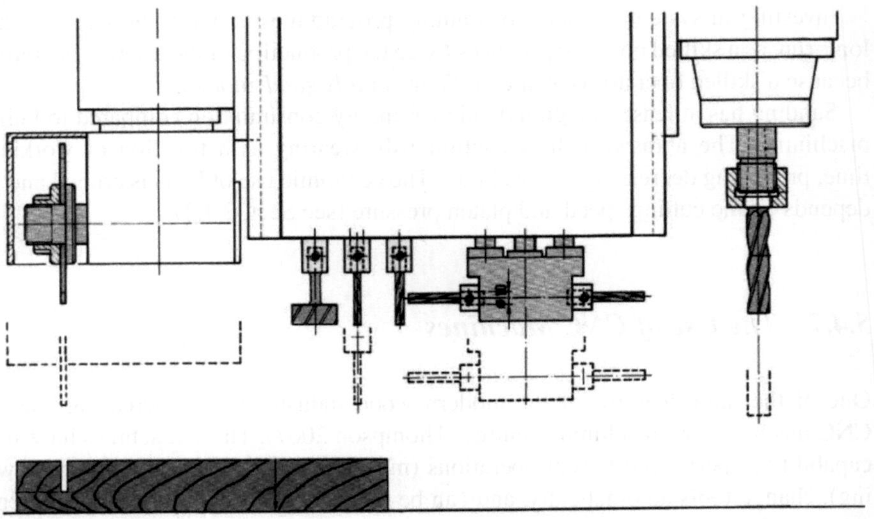

Fig. 5.30 Using different modular units in a CNC machine for milling, drilling and sawing

In order to ensure the desired economy of production system, a high level of organizational environment is unconditionally needed. All stored data must be up-to-date and even the wear of tool edges must be taken into account. In order to facilitate the economic use of these machines, the machine suppliers are offering a series of modular units to ensure functional capability, and also control units and programming system (Fig. 5.30).

In order to extend the machining capability, modular units with own variable axis are also widely used (Fig. 5.31).

It should also be mentioned that the use of CNC machines is not always practical and economical. Woodworking operations are of high diversity and, depending on the number and complexity of parts, different technologies may represent an optimum production system. Labour and production costs are quite different all over the world, and therefore different technologies may be competitive depending on local circumstances, including the availability of local raw materials. The carpenters in Southeast Asia, using *"low technology"*, are fully competitive with unbeatable master pieces. They are the main supplier of high-quality furniture for China who is, at the same time, the largest producer of mass merchandise furniture in the world (Ratnasingam 2015).

5.4.8 Through-Feed Machine Lines

In the case of mass production of similar parts or simpler end products, the high capacity through-feed machine lines are often used. As its name implies, different

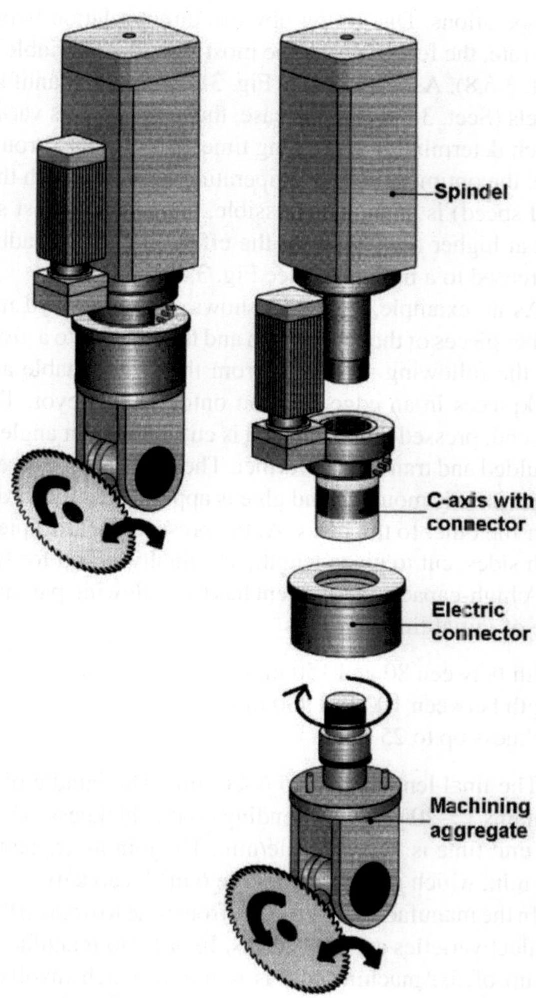

Fig. 5.31 Machining units for extending the number of axes in a CNC machine

feeding, woodworking, gluing, spraying, sanding, pressing units and conveyors are grouped together following the sequence of planned operations. The capacity of the machine line is mainly determined by the feed speed which is limited by the lowest operation speed (bottleneck). Typical applications are

– manufacture of strip flooring parquet and its finishing,
– manufacture of kitchen furniture parts with edge-banding and finishing,
– finishing of open structure parts and products (door, window),
– utilization of waste raw materials by joining into boards, posts and panels.

First, the sequence of operations is drawn up in an operation process chart (see Table 3.5 in Sect. 3.6.8). The layout of the machine line follows the sequence of operations with minimum material transportation and synchronized feed speed for all operations. Due to the obvious direct relation between feed speed and production rate, the feed speed is the most important variable concerning optimization (see Sect. 3.6.8). As an example, Fig. 3.37 shows a manufacturing line making veneered panels (Sect. 3.6.9). In this case, the main process variable is the platen temperature which determines the pressing time, and also the throughput in the unit time. Therefore, the optimum platen temperature is chosen such that the throughput (or average feed speed) is as high as possible, giving the lowest specific cost. It is remarkable that at higher temperatures, the effect of different adhesives on the specific cost is decreased to a minimum (see Fig. 3.40).

As an example, Fig. 5.32 shows a through-feed machine line for joining short timber pieces of the same width and thickness into a given length. The main operation has the following sequence: from the storage table an automatic feeder sends the workpieces in an edge position onto the conveyor. The workpieces are aligned to one end, pressed down, the end is cut-off at right angle, wedge-shaped finger joint is moulded and transported further. These workpieces are then aligned to the other end, finger joint is moulded and glue is applied, and then transported by the conveyor one after the other to the press. At the press, these workpieces are pressed, machined on both sides, cut to given length, and finally stored for further use.

A high-capacity equipment has the following parameters concerning: the possible size of initial timber pieces

width between 80 and 150 mm,
length between 100 and 900 mm,
thickness up to 25 mm.

The final length is up to 6000 mm. The bundle of workpieces handled together contains 15–20 pieces depending on the thickness. The number of pressing cycles in the unit time is 8 to 12 cycle/min. The joining capacity amounts to 150–200 pieces per min, which means an average 6 m^3/h capacity.

In the manufacturing process, from time to time on the same machine line, different product varieties are in progress. In order to manufacture a new product variety, the set-up of the machine line is required which involves a set-up time and a set-up cost. The costs due to a set-up may include the time required to change tools, set the new operational parameters such as feed speed, depth of cut, etc. Some costs are associated also with the start of a new run (scrap or rework). The set-up time and cost are independent of the size of the batch being processed. Therefore, a correct balance between the cost of holding inventory and the cost of setting up the machine line should be established.

In the present day, in order to reduce the set-up time, there are possibilities to make some part of the set-up operations, while the machine is operating on the previous batch (external set-up). The reduction of set-up time is extremely important in such cases, when the machine line is working at full capacity.

The economic order quantity and economic batch size have been discussed in detail in Sect. 5.3.4. The total cost of a batch is the sum of the batch cost itself and set-up cost. The batch cost depends, however, on the batch size which should be optimized.

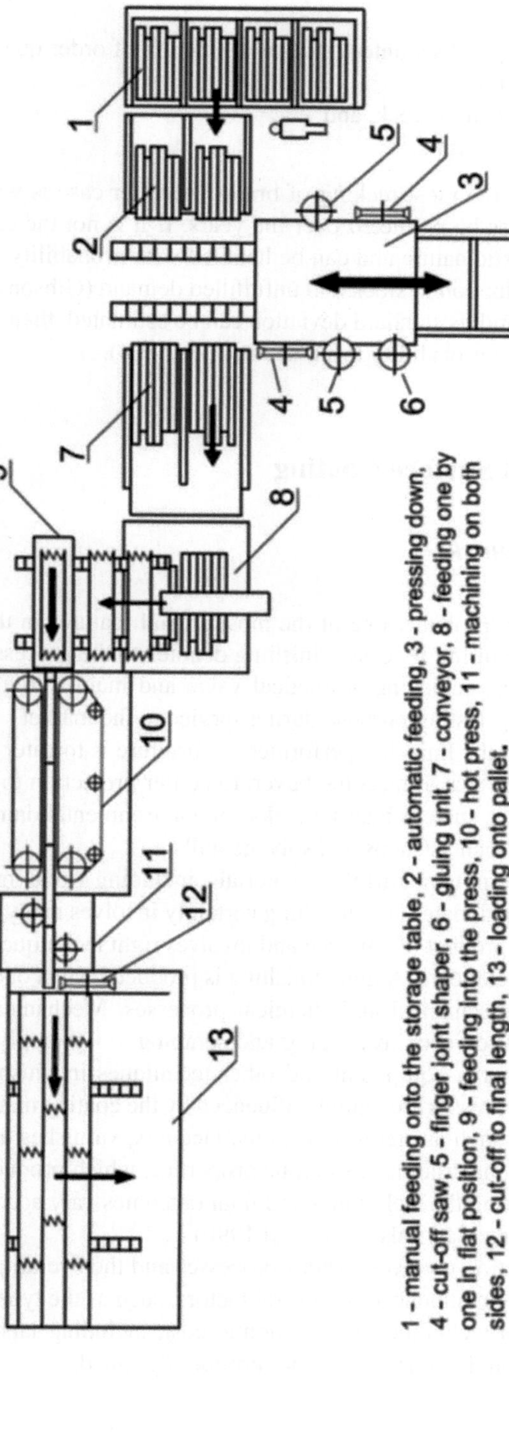

Fig. 5.32 Layout plan of a joining machine line

In practical cases, the situation may be complicated by the following circumstances:

- the supplier is offering a discounted price as a function of order quantity (incremental quantity discount),
- non-instantaneous arrival of stock, and
- the demand is varying in time.

The last case causes the most racking of brains. A better case is when varying demand can more or less be predicted over the years. If it is not the case then the problem is of probabilistic nature and can be handled with probability calculations balancing between production to stock and unfulfilled demand (Gibson et al. 1995). If the average demand and its standard deviation can be estimated, then the problem can be solved for a given probability level (see in Sect. 5.3.4).

5.5 Finishing and Surface Coating

5.5.1 General Remarks

Finishing operations are probably one of the most crucial features in the final step in the furniture manufacturing process. Finishing denotes to the process of protecting the furniture surface, achieving aesthetical value and maintaining the desired characteristics for a long-lasting purpose during service in the market.

The main purpose of the finishing performed in furniture is to cater the finished good, a good appearance that attracts the buyer, to confer protection to the product from biological damages, such as insect attacks and environmental damages due to weather and accidents, such as stains and solvent spills.

Furniture is made from many different materials, including wood, metal, plastic and many other combined materials. Finishing certainly involves many processes, a great outcome can be a product of patience and involves right techniques to perform each process with ample of care. A great finishing is produced when one takes good consideration of both mechanical and chemical processes. Mechanical processes involve surface preparation such as sanding and scraping wood, applying finishes with rags, brushes or spray equipment and other techniques in finishing. On the other hand, chemical processes are mainly influenced by the coating material chosen to use in the operation such as stains, dyes, oils, lacquers, varnishes and solvents. Every finish or coating material has its unique properties, which produces different effects and results, making the techniques and final outcomes vary according to the different finish or coating materials used (Paul 1996).

Therefore, the operation involves various processes and the overall performance of the finishing process is influenced by several factors, such as the types of finishes or coatings, application techniques and economic cost, including labour cost and production cost based on the objective of the furniture designed.

5.5.2 *Types of Coatings*

The process of furniture finishing begins with preparing the surface for the final finish which is a very important factor. A good finish will diminish if the product is not prepared for the finishing process. Hence, by avoiding pitfalls before starting the finishing process, it is possible to avoid the after works needed, to correct the finish that does not come out as planned.

The preparing of the surface usually involves some sanding, which aims to remove defects from the surface that will impact the outer appearance and the performance of the finishes that are applied subsequently. After sanding and surface preparation, all unintentional imperfections on the surface such as holes from knots or nails must be removed. This can be done by filling these gaps and holes with wood putty, wood fillers and, in cases of larger holes, a combination.

Next, in the cases of changing the colour on the product's surface, a few colouring techniques are available, ranging from wood staining to bleaching and, if the desire is to keep the surface in its natural colour, especially for wood, it would be advisable to skip this step and aim for a coloured finish, using coloured wax. Once the surface is prepared and stained, depending on the type of effect that is aimed for, the finish is applied followed by sanding to smoothen the surface again, followed by drying and curing, and finally polishing to produce a shiny finish.

The gloss of the finished surface is also an important aesthetical property which may be achieved in different ways. The main influencing factors are the machining, the wood density and the finish itself. Knife machining gives higher natural gloss but the variation of gloss locally may be higher. The highest natural gloss is obtained using the Japanese planer (see Figs. 2.135 and 2.136). Sanding, especially perpendicular to the grains, results in a more uniform gloss distribution with low gloss value. However, using lacquer with the desired gloss, a uniform glossiness may be achieved (see Sect. 2.12). Further, the natural gloss of wood surfaces varies in time. Shellac and lacquers preserve the gloss for a much long time. Generally, the dry-film thickness of the finish or coating material should be in the range of 120 µm, to give the best possible optical properties to the finish film (Ratnasingam and Scholz 2006).

The whole finishing process planned will be assorted in a *process chart* containing all specific operations to be carried out in a sequence.

Different finish yields different measures of durability, protection, ease of application and performance, thus it is essential to pick the most suitable finish that will impart the long-lasting look.

5.5.3 *Properties of Coating Materials*

There are several types of finish with different characteristics and aesthetical values, used in the furniture industry (Paul 1996).

Varnish is a film that is applied on finished good surface to protect the surface from heat, water and solvents. It is naturally transparent allowing to be used on stained surface or coloured, hardwearing and function as decorative and preservative, providing a tough, durable finish which is resistant to most household chemicals. Varnish is available in water-based, oil-based, high-gloss, satin or matte forms. When deciding on using varnish, it is important to ascertain if the type of varnish is formulated for floors or furniture. Use the varnish as specified for either interior or exterior applications, and never use an exterior finish varnish for interior use as it never dries completely. All varnishes are fairly easy to use, requiring several layers of coat for better application, and it can be applied using a brush for best result.

Shellac is an insect secretion that is formed on and collected from plants. Shellac is produced almost entirely in India and has been used since 1590. Shellac may be combined with Kauri gum and coloured by incorporation of pigments. It is spirit varnish and produces a tough film with a smooth finish resulting in a high polish with $GU_{60} = 80–90\%$. The addition of Kauri gum to shellac gives better water resistance and greater elasticity. The latter is important to avoid fine cracks and fissures by the flight of time. Fine furniture makers in Europe have widely used this surface finish (Politur). Today, some leading wood turners use the mix of linseed oil and shellac claiming that shellac will ensure the gloss for many years.

Shellac is available in a variety of colours, subject to its constituents. They are non-toxic, easy to apply, able to dry fast and makes a beautiful finish. It can be combined with varnish and provides a clear shine. However, on its own, it is not resistant to moisture, less durable and it can be easily damaged as it is dissolvable in water and alcohol. Thus, with these drawbacks, shellac is less preferred for surface finishing today but it is very much popular as a sealer before applying filler or stain, in order to prevent bleeding into the top finishes.

Lacquer is regarded as the best all round finish as it is the fastest drying finish, provides moderate to excellent level of durability and enhances the wood grain. Lacquers are available in matte, semi-gloss, high-gloss and stain finishes. They exhibit amazing depth and richness to wood and while it evaporates upon application, it also cures the wood. Although lacquer finishes are very hard and durable, yet it is not scratch-proof and gets damaged when in contact with water. Therefore, lacquers are commonly used by experts in spraying operations, as it involves several different types of lacquer, and exhibit varying performance characteristics.

Oil is a transparent finish that helps to nourish and provide protection for the wood. It is designed for both external and interior use. Despite its slow drying time, it has high durability compared to wax and is easily work with, and will not penetrate a sealed surface. Tung oil and linseed oil are the types of oil that require an addition of varnish, and consumes less application time. Linseed oil finish is not scratch-proof, but yields a shiny look, while Tung oil yields a slightly wet look and moreover, it is an eco-friendly finish.

Danish oil, made of either tung oil (*Vernicia fordii*) or polymerized linseed oil, is a slow drying oil which can polymerize into a solid form. It provides a hard-wearing, water-resistant satin finish. It can also be applied as a primer on bare wood, before

Table 5.2 Type of wood finish and curing method

Common finish	Type of cure
Shellac	Evaporative finish
Lacquer	Evaporative finish
Varnish	Reactive finish
Oil	Reactive finish
Wax	Evaporative finish

application of the varnish. It is a suitable finish also for food utensils. Danish oil provides a coverage of approximately 12 m^2/L.

Wax is a transparent or translucent decorative finish. It is available in both water-based and oil-based types. It is applied to feed and protect the wood; however, it will not penetrate a sealed surface. Few types of wax promote high gloss surface and is highly suitable for furniture finish as it is easily applied. Nevertheless, regular maintenance and several layers of coats are needed, as wax is not highly durable. Further, its application is limited to interior use only, and can be applied to bare wood or over unsealed finishes such as, dye.

In colour enhancement, the surface treatment with oils, lacquers, wax or their mix with shellac play an important role. The colour hue remains much the same, but the lightness decreases and the colour saturation considerably increases (see in Sect. 2.11.2). Especially the sanding operation makes a grey veil over the surface due to the broken cell walls having diffused reflection. At the same time, sanding produces an even mat surface which, using lacquer with a given gloss grade, yields an even surface glossiness.

For a better understanding of finishes and their differences, it is very useful to categorize them by the way they cure. Generally, they are three main types of curing method available:

- Evaporative finishes: in this type of finishes, when the solvent carrying the finish evaporates, they leave an actual resinous film.
- Reactive finishes: these types of finishes modify chemically as they cure and this is called "polymerization" and the resultant material is less readily dissolved in solvents.
- Coalescing finishes: this is a combination between evaporative and reactive finishes and is typically water-based. They are essentially emulsions with slow-evaporating thinners and the solvent is glycol–ether, while the thinner is water. Examples includes water-based finishes, pigment with water binder, etc. Table 5.2 summarizes the types of wood finishes commonly available.

5.5.4 Applying Coating Materials

The finish material can be applied by many different techniques, but the most common methods are dipping, roller-coating and spraying (Paul 1996).

Dipping can be categorized into two kinds. Firstly, the process involves immersion of the entire furniture piece into a tank, full of finish. By doing so, the finish will coat all the surfaces including all holes and cuts. An even application is not necessary in this process as long it covers all the side of the dipped furniture. During the dripping process, the excess dripping off is channelled into a different tank and is reused. However, the main drawback of this method is that the tank must be fully filled with a large capacity of finish and must be maintained until the last item is done. This disadvantage can be removed by applying the pouring method. Next, is dipping with controlled withdrawal where a furniture is immersed in the finish then withdrawn, and this process is controlled very carefully to avoid runs that will develop poor appearance.

The viscosity of coating fluids is a very important factor in dipping technology. The flow down velocity of excess layer is inversely proportional to the viscosity and the layer thickness adhering and remaining on the surface is proportional to the initial shear stress τ_0 required to initiate flow (see in Sect. 2.10).

Spraying method is one of the most widely used methods in finishing and is very cost effective. It is applied in order to shield large surfaces with an even layer of coating material.

There are three basic methods for spraying:

- high pressure jet atomization, airless,
- low pressure injection into a turbulent air jet transferring the coating droplets to the surface and
- air-mix system, which is a combination of the above two types.

Every spray system usually consists of several basic elements such as, an adequate source of compressed air, a reservoir or feed tank to hold a supply of the finishing material, and a device for controlling the combination of the air where the finishing material is emitted in an atomized cloud or spray against the surface to be coated. There are two main types of spray equipment available which are spraying gun to cover small areas and pressure-feed equipment to cover larger area.

Roller coating is a process of coating a surface with simple rotation of rollers in contact with the substrate, transferring the coat onto the surface. Roller coating is designed to be used on flat surfaces rather than three-dimensional elements or moulded panels. The main advantages of roller coating are its speed and efficiency on flat surfaces, especially for mass production of repeated parts, which provides a consistent coating layer by controlling tightly the roll gaps. This system can also be completely automated to achieve high productivity levels.

There are many types of industrial equipment utilizing rollers. These machines are equipped with one or two heads each consisting of an applicator roll, which applies the coating onto the substrate and a feeding (or dosing) roll, which proportions the coating to be applied. A detailed description of the coating processes and finish film defects is provided in the report by Ratnasingam et al. (1996).

5.6 Packaging of Finished Goods

Packaging is the activity undertaken to protect and preserve the finished product during their distribution, storage and handling until the product reaches the customer. Apart from that, a good packaging also provides instructions for their use, preserve the appearance and performance of the finished goods.

A well-packaged item must be able to maintain its important characteristics during transportation and storage for the described shelf life. The pack must be able to identify the furniture content and its amount, provide instruction on how to use and assemble it, as well as to impress upon any hazard warnings by print and pictures for a better understanding of the customer (Ratnasingam 2016).

Therefore, packaging a finished good is generally consistent with the quality of the product as well as the customer's preferences. In most cases, packaging has become an important instrument as an advertising medium; placing an important value in marketing and makes a competent tool for consumer attraction. In accordance with that, during the development of the packaging, the operators must keep in mind all the details, including colours, type of packages, information printed and innovation methods to generate a general emotional perception within the consumer segment.

A wide range of materials are used in the packaging process of furniture, such as cellophane (plastic film), polyethylene, corrugated cardboard, foam, polystyrene (Styrofoam) and plastic materials.

Corrugated cardboard is most popular form of outer protection, made from one or more sheets of arched paper known as "fluting" secured by an adhesive to two or more liners. Its construction may be single, double or triple wall and is made from different types of paper. The key raw material used in making corrugated cardboard involves bagasse, rice husk, bamboo, recycled paper, etc. Recycled includes waste paper and paper products made of long and short fibres. Other than providing a stable cushion for any product by decreasing damages via movement, corrugated packaging keeps the item safe for long-distance transportation as well keeps the moisture away from damaging the product.

Corrugated carton comes in different sizes and thickness depending on the amount of protection that is required to withstand long shipping times. Thus, as a fairly high-tech solution that utilizes advanced computer design and manufacturing systems corrugated cotton definitely become best choice for many packaging designers, Fig. 5.33.

Other packaging materials include cellophane, foam, polystyrene and polyethylene. However, the most commonly used are polyethylene and polystyrene which holds a good mechanical property of being less stiff, lower density, lower strength and with less hardness. Apart from that, plastic packaging materials and components

Fig. 5.33 Corrugated carton

are less affected by ambient environment, temperature and humidity fluctuations, and therefore is a good interior packaging material.

In packaging of furniture, there is no such thing as "*too much*" protection, as it is always better to have a thoroughly done packaging covering every square inch of the furniture part and even multiple coats of the wrapping to avoid any form of damages. For this, the use of paper, poly film, corrugated paper, bubble film and foam padding are done, whenever possible.

In the packaging process, there is no right formula which could be used as a standardized procedure. This is because, every item requires special attention, different types of packing materials and procedures to keep them protected safely during the transportation. However, a glimpse of the packaging process, as shown in Fig. 5.34, can give a general idea. The performance of the packaging is often ascertained through the multi-points drop test in accordance with established standards for product packaging (Ratnasingam 2016).

There are several types of packing machines used in the furniture industry. The criteria used to choose the right packaging machines depends on its technical capabilities, ergonomic values, labour requirements, worker safety, maintainability, serviceability, reliability, ability to integrate into the packaging line, capital cost, floorspace, flexibility, energy requirements, quality of outgoing packages, throughput, efficiency, productivity and return on investment. These machines can be purchased as standard, off-the-shelf equipment, custom made to specific operation or manufactured or modified by in-house engineers and maintenance staff. Figure 5.35 shows a type of packaging machine used in the industry.

5.7 Value-Adding Technologies

In the present, the amount and variety of precious timber resources, which would ensure the highest physical, mechanical and aesthetical properties in their natural

Fig. 5.34 Packaging processes

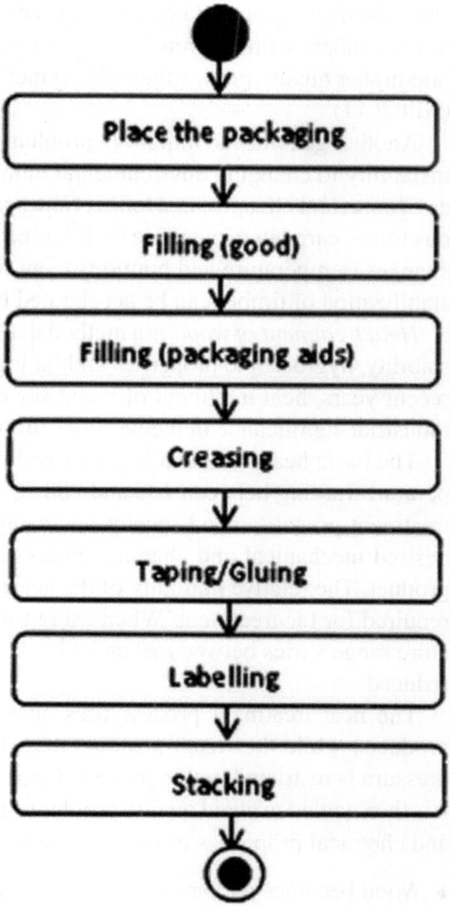

Fig. 5.35 Packaging machine for panels

state, are decreasing and becoming expensive worldwide. There is a need, therefore, to use timbers with inherently lower properties. In order to transform these timbers into higher quality ones, value-adding methods and techniques have been developed (Hill 2011).

Another general and important problem of timber materials is their "movement or instability to changing environmental humidity", i.e. swelling and shrinkage due to environmental changes, in a longer time span after they are cut into lumber. In previous times, carpenters put aside their lumbers for 5–6 years exposed to environmental changes (temperature and humidity), but not to rain and sun irradiation. Today, the stabilization of timber can be accelerated by short-term cyclic treatment.

Heat treatment of wood is a method that is undertaken to enhance the dimensional stability, hygroscopic properties and biological resistance of wood by using heat. In recent years, heat treatment of wood seems to be a novelty which has now gained industrial significance in the world of furniture production (Hill 2011).

The basic heat treatment is performed in a low-oxygen environment with a temperature varying between 180 and 280 °C for 15 min to 24 h depending on the heat treatment process, wood species and sample size, moisture content of the sample, desired mechanical and chemical properties and dimensional stability of the final product. The relative humidity of the heated air will considerably influence the time required for the treatment. When using saturated or superheated steam, the temperature range varies between 90 and 130 °C, and the time of treatment is significantly reduced.

The heat treatment process uses heat which is created from oil or wood by-products, while the steam prevents the wood from burning. Usage of chemical or pressure is restricted in this process. Once the wood is heated for a substantial time, it is then cooled to avoid quality problems. Heating permanently changes the physical and chemical properties of the wood as follows:

- Wood becomes darker in colour, changing to a brown or dark brown.
- Colour of heart and sapwood may be homogenized.
- Wood absorbs less moisture, so it becomes more dimensionally stable.
- Heat conductivity is reduced.
- Wood becomes more resistant to rot and decay.
- Wood becomes lighter weight (due to moisture loss).
- The wood weakens and becomes less bendable.

Heat-treated wood is generally used in indoors for wooden flooring, wall and ceiling panels as well as for outdoor furniture and other utensils. Heat-treated wood with its beautiful and pleasant colour after the treatment, becomes unique and provide more opportunity for the manufacturers to make specialized wood products.

As a value-adding activity in the furniture industry, heat treatment of wood has brought a new, ecological alternative to produce wood products, in which at the end of their life cycle, there is not any environmental hazard as well as conserving human health. This also increases the chances for manufacturers of such products to extend their market share into indoor and outdoor applications, providing quality wood products with highly desirable properties to customers.

For details concerning colour modification and colour homogenization, the interested reader is referred to Sects. 2.11 and 3.6.7. Applications of heat treatment with dry and humid air, and with steaming can also be found in another literature (Csanády et al. 2015).

Appendix

Ornamental Properties of Timbers

(grains, rays, figures, colours, irregular growth, infected with staining fungus)

Photographed and selected by G. Sitkei

See Figs. A.1, A.2, A.3, A.4, A.5, A.6, A.7, A.8, A.9, A.10, A.11, A.12, A.13, A.14, A.15, A.16, A.17, A.18, A.19, A.20, A.21, A.22, A.23, A.24, A.25, A.26, A.27, A.28, A.29, A.30, A.31, A.32, A.33, A.34, A.35, A.36, A.37, A.38, A.39, A.40, A.41, A.42, A.43, A.44, A.45, A.46, A.47, A.48, A.49, A.50, A.51, A.52 and A.53).

E. Csanády et al., *Optimum Design and Manufacture of Wood Products*,
https://doi.org/10.1007/978-3-030-16688-5

Fig. A.1 Oak veneer, tangential cut

Fig. A.2 American walnut, near tangential cut, lacquered

Fig. A.3 American walnut, near tangential cut, lacquered

Fig. A.4 Wenge, tangential cut

Fig. A.5 Wenge, intermingled cut

Fig. A.6 Sucupira, near radial cut

Fig. A.7 Angelim pedra, near radial cut

Fig. A.8 Rewa-rewa (*Knightia excelsa*)

Fig. A.9 Beefwood (*Grevillea striata*) Special cut

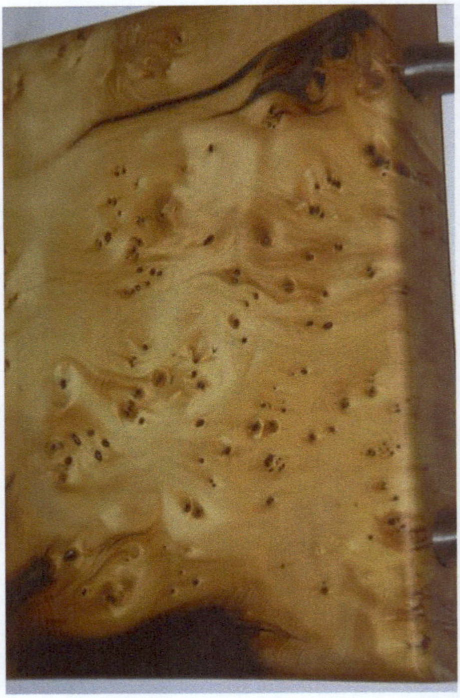

Fig. A.10 Bird's eye Huon pine

Fig. A.11 She-oak (*Casuarina*)

Fig. A.12 Sindora, reddish streaked

Fig. A.13 Veneer made of Carpathian Elm burl Ulmus minor (syn. *Urophora campestris*)

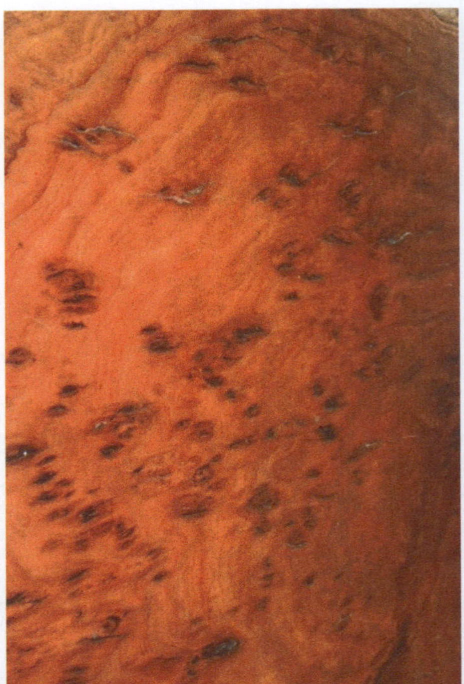

Fig. A.14 Karri bird's eye (fine)

Fig. A.15 Bird's eye Jarrah

Fig. A.16 Panther-stripped Jarrah

Fig. A.17 Tiger stripped Karri (*Eucalyptus diversicolor*)

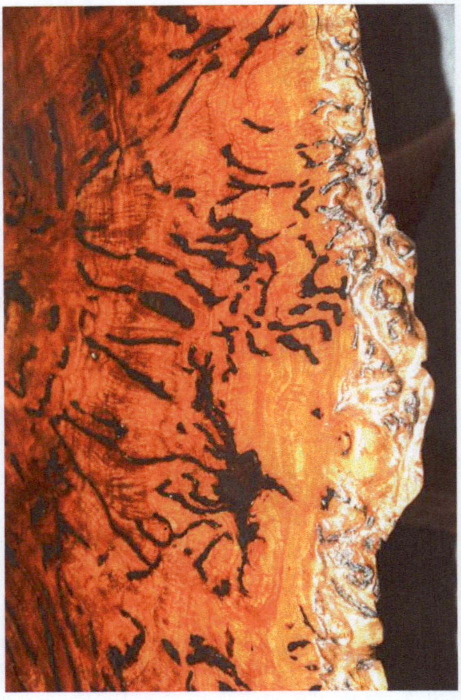

Fig. A.18 Karri table, natural edge cleaned with high pressure water jet (*Eucalyptus diversicolor*)

Fig. A.19 Burl of Tasmanian Myrtle (*Nothofagus cunninghamii*)

Fig. A.20 Walnut wood. Delicate colour pattern cut from the bottom part of the stem. Cabinet detail

Fig. A.21 Veneered door in four quadrangles perpendicular each other

Fig. A.22 Sugar box (*Tasmanian Sassafras*)

Fig. A.23 Turned table leg, Walnut

Fig. A.24 Door of a chest for bed clothes, Walnut

Fig. A.25 Machined ornament on a chest for bed clothes

Fig. A.26 Carved chair backrest, Stained

Fig. A.27 Upholstered chair with carved backrest frame Stink wood (*Ocotea bullata*)

Fig. A.28 Chair with carved backrest Stink wood (*Ocotea bullata*)

Fig. A.29 Carved backrest of a chair. Ebony, before 1700

Fig. A.30 Fully carved coffee-table, teak wood, detail

Fig. A.31 Mirror with carved frame

Fig. A.32 Carved mirror frame, detail

Fig. A.33 Drawer chest with inlay and carved ornaments

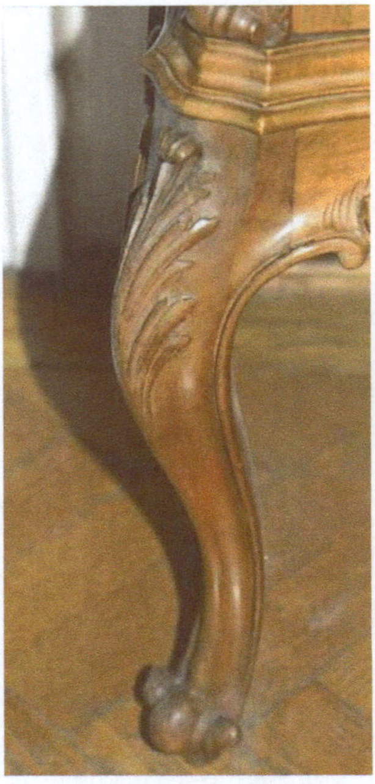

Fig. A.34 Drawer chest, leg detail

Fig. A.35 Carvings on a cabinet with central glass case

Fig. A.36 Inlay detail

Fig. A.37 Round table with inlay, 80 cm diameter

Fig. A.38 Leg of a round table

Fig. A.39 Drawer front of a cabinet with inlay. Around 1920

Fig. A.40 Biedermeier cabinet, upper half. Around 1860

Fig. A.41 Biedermeier cabinet, inlay pattern

Fig. A.42 Cabinet with upper drawers and carved doors. Early twentieth century. Yellow and Stink wood, S.A

Fig. A.43 Carving details to A42

Fig. A.44 Armrest of a sofa, Walnut

Fig. A.45 Relief carving for furniture Chinese motif

Fig. A.46 Open-work carving for cabinet door decoration

Fig. A.47 Open-work carving for wall decoration made of Jelutong, Malaysia

Fig. A.48 Carving for wall decoration made of Jelutong, Malaysia

Fig. A.49 Open-work carving on a jewellery box made of Jelutong, Malaysia

Fig. A.50 Jewellery box made of Chengal (brown), Rengas (red) and Jelutong (yellow). Temerloh, Malaysia

Fig. A.51 Jewellery box made of Bird's eye Huon pine, Tasmania

Fig. A.52 Deep bowl made of swamp Kauri (40.000 years old). Dargaville, NZ

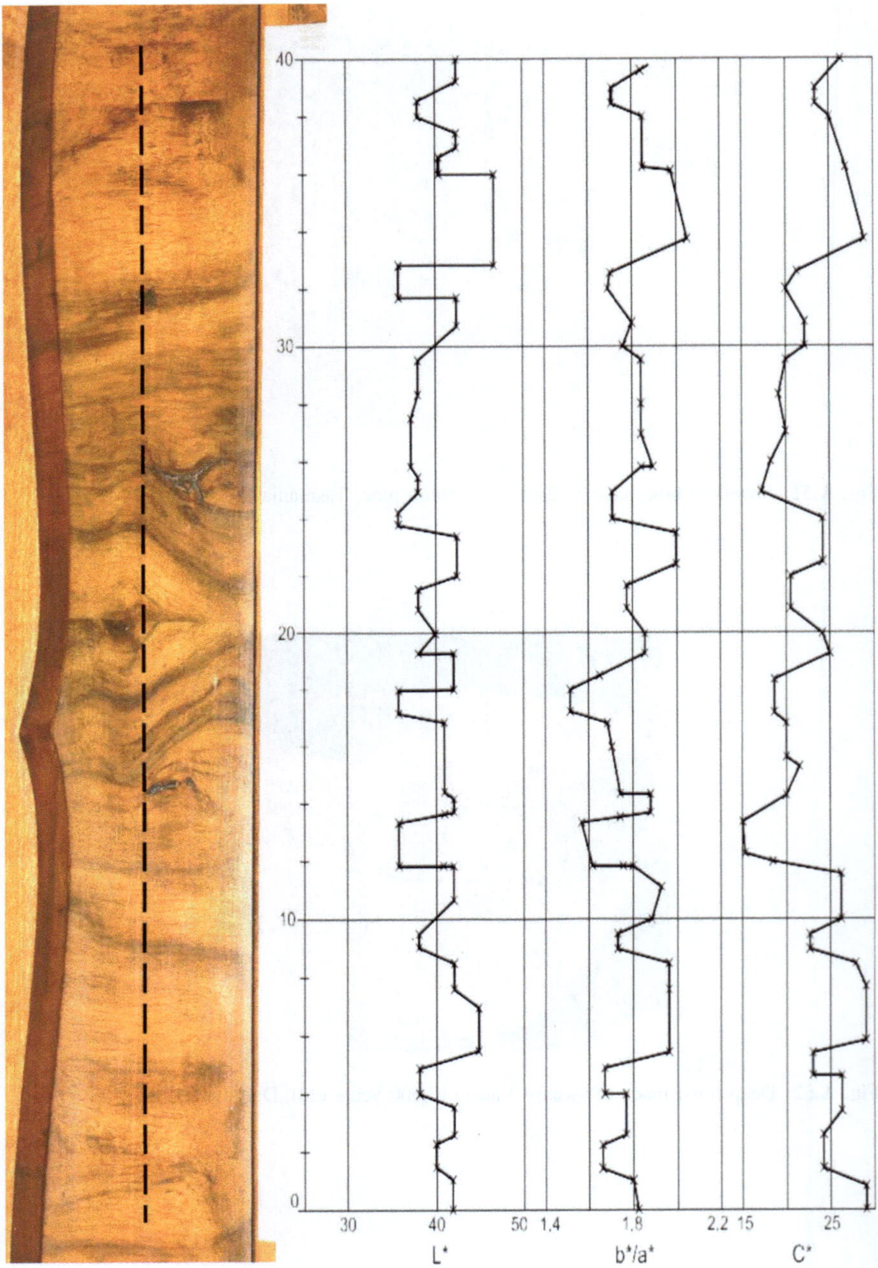

Fig. A.53 Representation of colour characteristics along a line

Literature

Abbott, E., Firestone, F.: Specifying surface quality. J. Mech. Eng., 569–572 (1933)

Acar, M., et al.: Withdrawal force capacity of T-type furniture joints constructed of various heat-treated wood species. In: Proceedings of the International Furniture Congress 2016, Mugla, Turkey (2017)

American Library Associations (ALA): The use of performance tests and quality assurance programs in the selection of library chairs. Libr. Technol. Rep. **18**(5), 483–571 (1982)

Anderson, E.: The Management of Manufacturing. Addison-Wesley (1994)

Andres, F.: Die Holzbiegerei und die Herstellung der Möbel aus gebiegenem Holz. Wien (1923)

Antal, M.R.: Analysis of the decisive form characteristics of exclusive furniture (in Hungarian). PhD theses, University of Western Hungary (2007)

Askenazi, E.K.: Anizotropia dreveszinü I dreveszнüh materilov. Izdatelsztvo Lesznaja Promüslennoszt, Moszkva (1978)

Atar, M., et al.: Impacts of the dovetail tooth angle on diagonal tensile strength in the leg-to-rail joints. Post conference paper, the xxviiith international conference research for furniture industry, Poznań, Poland, September 2017 (2017)

Bakar, E.: A practical method and apparatus for converting bamboo culms into flat sheets. In: Proceedings of 21st IWMS, pp. 171–179 (2013)

Bakar, E., et al.: Durability of phenolic-resin-treated oil palm wood. Int. Biodeterior. Biodegradation, 126–130 (2013)

Bandel, A.: Gluing Wood. CATAS Publication, Udine, Italy (1995)

Bardak, T., et al.: Maderas, Cienc. tecnol. vol. 19, no. 1. Concepción, Chile (2017)

Bertling, L.: Information-und Expert System für die Holzbearbeitung. 8. Holztechn. Koll. Braunschweig. S. 17–32 (1991)

Boadu, K.B., Antwi-Boasiako, C.: Assessment of the bending strength of mortise-tenon and dovetail joints in leg-and-rail construction using Klainedoxa gabonensis Pierre Ex Engl. and Entandrophragma cylindricum (Sprague) Sprague. Wood Mater. Sci. Eng. **12**(4), 242–250 (2017)

Bodig, J., Jayne, B.A.: Mechanics of Wood and Wood Composites. Van Nostrand Reinhold (1982)

Brush T, Arcan, M.: Modeling the body/chair interaction-an experimental numerical approach. Clin. Biomech., 217–219 (2000)

Candadai, R., Reddy, N.P.: Stress distribution in a physical buttock model: effect of simulated bone geometry. J. Biomech., 1403–1411 (1992)

Carter, W., Ragsdell, K.: The optimal column. Trans. ASME, 71–76 (1974)

E. Csanády et al., *Optimum Design and Manufacture of Wood Products*,
https://doi.org/10.1007/978-3-030-16688-5

Cescinsky, H., Gribble, E.: Early English Furniture and Woodwork, vol. I–II. Routledge, London (1922)
Chala, A., et al.: Duplex treatment based on the combination of ion nitriding and PVD process. In: Proceedings of 16th IWMS Matsue, pp. 471–479 (2003)
Chow, W., Odell, E.: Deformations and stresses in soft body tissues of sitting person. J. Biomech. Eng., 79–87 (1978)
Çolakoğlu, M.H., Apay, A.C.: Finite element analysis of wooden chair strength in free drop. Int. J. Phys. Sci. **7**(7), 1105–1114 (2012)
Cross, N.: Engineering Design Methods. Wiley (1996)
Csanády, E., Magoss, E.: Mechanics of Wood Machining. University of Wood technology Sopron (2011)
Csanády, E., Magoss, E.: Mechanics of Wood Machining. Springer, Heidelberg, New York, Dordrecht, London (2013)
Csanády, E., Magoss, E.: Einfluss der Rauhigkeit auf den Glanz von Oberflächen. Tagungsband des 17. Holztechnologischen Kolloquiums, Dresden, pp. 80–91 (2016a)
Csanády, E., Magoss, E.: Einfluss der Rauhigkeit auf den Glanz von Holzoberflächen. Holztechnologie, 57 6 S. 31–35 (2016b)
Csanády, Magoss, Tolvaj, : Quality of Machined Wood Surfaces. Springer, Heidelberg, New York, Dordrecht, London (2015)
CSIL: World Furniture Outlook 2018/2019 Report. Centre for Industrial Studies (CSIL, Milano, Italy) (2018)
Curtu, I., Ghelmeziu, N.: Mecanica lemnului şi materialelor pe bază de lemn. Editura Tecnică Bucureşti (1984)
Davidon, W.: Variable metric method for minimization. ARC Resources and Development Report, ANL-5990 (1959)
de Moura, L., Hernandez, R.: Evaluation of varnish coating performance. Wood Fiber Sci., 355–366 (2005)
de Moura, L., Hernandez, R.: Effects of abrasive mineral, grit size and feed speed on the quality of sanded surfaces of sugar maple wood. Wood Sci. Technol., 517–530 (2006)
Déry, J.: Spanunguntersuchungen an einer Blockbandsäge. Acta Fac. ligniensis Sopron, S. 43–52 (1985)
Diler, H., et al.: Determination of allowable design loads for wood chairs. In: XXVIIIth International Conference Research for Furniture Industry, Poznan, Poland (2017)
Dobler, K.: Der freie Schnitt beim Mähen von Halmgut. Hohenheimer Arbeiten No. 62 (1972)
Dobrint, P.: Optimiertes Schleifen von MDF und Spanplatten. Workshop Tagungsband, Braunschweig, S. 125–136 (1991)
Dong, W., et al.: Comprehensive study of parameters characterising three-dimensional surface topography. Wear, 29–43 (1994)
Dzięgelevski, S., Smardzewski, J.: Meblarstwo—Projekt I Konstrukcja. (Furniture—Design and Construction, in Polish) Państwowe Wydawnictvo Rolnice i Leśne Poznań (1995)
Dziuba, T.: Optimierung der Konstruktion von Stuhlseitenteilen. Holztechnologie **30**(6), 303–306 (1990)
Eckelman, C.: Furniture mechanics. Progress Report 274, Purdue University (1967)
Eckelman, C.A.: CODOFF Computer design of furniture frames. User's manual. Research Bulletin No. 857. Purdue University Agricultural Experiment Station. Lafayette, Indiana (1970a)
Eckelman, C.A.: CODOFF Computer design of furniture frames. Engineering concept. Research Bulletin No. 863. Purdue University Agricultural Experiment Station. Lafayette, Indiana (1970b)
Eckelman, C.A: Bending strength and moment-rotation characteristics of two pin moment-resisting dowel joints. Forest Prod. J. **21**(3), 35–40 (1971)
Eckelman, C.A.: Reasonable design stresses for woods used in furniture. Research Bulletin No. 916 Purdue University Agricultural Experiment Station. Lafayette, Indiana (1974)
Eckelman, C.A.: Strength Design of Furniture, 1st edn. Tim Tech Inc. (1978)

Eckelman, C.A.: Withdrawal strength of dowel joints. Forest Prod. J. **29**(1), 48–52 (1979)
Eckelman, C.A.: Out-of-plane strength and stiffness of dowel joints. Forest Prod. J. **29**(8), 32–38 (1979)
Eckelman, C.A.: Withdrawal strength of dowels from wood composites. FPJ **35**(5) (1985)
Eckelman, C.A.: Performance testing of furniture. Part I. Underlying Concepts. Forest Prod. J. **38** (3), 44–48 (1988)
Eckelman, C.A.: Performance testing of furniture. Part II. A multipurpose universal structural performance test method. Forest Prod. J. **38**(4), 13–18 (1988)
Eckelman, C.A.: An overview of the ALA test method with test reports on side chairs. Libr. Technol. Rep. **31**(2), 115–214 (1995)
Eckelman, C.A.: Textbook of Product Engineering and Strength Design of Furniture. Purdue University Agricultural Experiment Station. Lafayette, Indiana (2003)
Eckelman, C.A., Erdil, Y.Z.: General services administration upholstered furniture test method—FNAE 8–214: A Description of the method with drawings, p. 1159. Department of Forestry and Natural Resources, Purdue University, Extension Publication FNR 176 (2001a)
Eckelman, C.A., Erdil, Y.Z.: Performance test method for intensive use chairs-FNEW. A Description of the test method with drawings, vol. 20, pp. 83–269. FNR 176 (2001b)
Eckelman, C., Haviarova, E.: Performance tests of school chairs constructed with round mortise and tenon joints. Forest Prod. J. **56**(3), 51–57 (2006)
Eckelman, C.A., Munz, S.: Rational design of cases with front frames and semi-rigid joints. Forest Prod. J. **37**(6), 25–30 (1987)
Eckelman, C.A., Rabiej, R.: A comprehensive method of analysis of case furniture. Forest Prod. J. **35**(4), 62–68 (1985)
Eckelman, C.A., Resheidat, M.: Deflection analysis of shelves and case tops and bottoms. Forest Prod. J. **34**(6), 55–60 (1984)
Eckelman, C.A., Resheidat, M.: The analysis of five-sided cases. Purdue University Agriculture Research Bulletin No. 981 (1985)
Eckert, E.: Messung der Reflexion von Wärmestrahlen an technischen Oberflächen. Forsch. Ing.—Wes. pp. 265–270 (1936)
Elek, L.: Manufacturing parameters of hardwood furniture joints for attaining optimum strength. Industry—University Research Project, University of Sopron, Hungary, 2016–2018 (2018)
EN 14358:2016 Timber structures—calculation and verification of characteristic values
EN 16139:2013 Furniture—strength, durability and safety—requirements for non-domestic seating
EN 1728:2013 Furniture—seating–test methods for the determination of strength and durability
EN 1990:2001+A1. 2005, 2010. Eurocode—basis of structural design
EN 1991-1-1: 1995, 2009. Eurocode 1. Actions on structures Part 1-1. General actions—densities, self-weight, imposed loads for buildings
EN 1995 1-1: 2004+A1, 2008. Eurocode 5. Design of timber structures Part 1-1. General—common rules, rules for buildings
Erdil, Y.Z., Kasal, A., Eckelman, C.A.: Bending moment capacity of rectangular mortise and tenon furniture joints. Forest Prod. J. **55**(12), 209 (2005)
Exner, W.: Das Biegen des Holzes. Wien (1887)
Federhofer, K.: Zur Berechnung der dünnen Kreisplatte mit grosser Ausbiegung. Forsch. Ing.-Wesens, S. 148–151 (1936)
Feigenbaum, A.V.: Total Quality Control. McGraw-Hill Book Co., Singapore
Feld, H., et al.: Carbide selection for woodworking tooling. In: Proceedings of 17th IWMS, Rosenheim, 2005, pp. 520–527 (1991)
Feulner, A.: Kunstgeschichte des Möbels. Berlin (1927)
Fischer, R.: Beitrag zur Modellierung der Abstumpfung von Holzbearbeitungswerkzeugen. Holztechnologie, No. 2, S. 70–72 (1983)
Fischer, R., Schuster, C.: Qualitätsentstehungen spannend erzeugter Holzoberflächen. Mitteilungen aus dem Institute für Holztechnik der TU Dresden (1993)

Fletcher, R., Powell, M.: A rapidly convergent descent method for minimization. Comput. J., 163–168 (1963)

FPL 1973: Basic Engineering Properties of Particleboard. U.S.D.A. Forest Service Research Paper FPL 206 (1973)

Fritsch, A.: Application-optimized tools. In: Proceedings of 17th IWMS, pp. 164–173 (2005)

Füchter. K.: Experimentelle Untersuchungen zur Verschieblichkeit und Festigkeit von Flächeneckverbindungen in Korpusmöbelbau (Experimental study on flexibility and strength of corner joints in case furniture). Diploma thesis, Universität Hamburg (1984)

Ganowicz, R., Dziuba, T., Ożarska-Bergandy, B., Kwaiatkowski, K.: Experimentelle Prüfung der Theorie der Verformungen von Schrankkonstruktionen. Holztechnologie 4:202–206 (1978)

Garcia, J.: New developments in ultrafine hardmetals. In: Proceedings of 17th IWMS Rosenheim, pp. 534–542 (2005)

Gejdoš, P.: Continuous quality improvement by statistical process control. Procedia Econ. Finan. **34**, 565–572 (2015)

Gibson, P., Greenhalgh, G., Kerr, R.: Manufacturing Management: Principles and Concepts. Chapman & Hall, London (1995)

Gustaffson, S.I.: Furniture design by the use of finite element method. Holz als Roh-und Wekstoff **53**, 206–257 (1995)

Gustaffson, S.I.: Finite element modelling versus reality for birch chairs. Holz als Roh-und Wekstoff **54**, 355–359 (1996)

Gustaffson, S.I.: Optimising ash wood chair. Wood Sci. Technol. **31**, 291–301 (1997)

Hajdarević, S.: Ibrahim Busuladžić: stiffness analysis of wood chair frame. Elsevier, Procedia Eng. **100**, 746–755 (2015)

Hajdarević, S., Martinović, S.: Effect of tenon length on flexibility of mortise and tenon joint. Elsevier, Procedia Eng. **69**, 678–685 (2014)

Hegedűs, J., Kő, F.: Product development methodology based on value analysis (in Hungarian). Lecture notes, School of Engineering, Kecskemét, Hungary (2001)

Hermann, J.W., et al.: New directions in design for manufacturing. In: Proceedings of DETC'04 ASME 2004 Design Engineering Technical Conferences and Computers and Information in Engineering Conference, 28 September–2 October 2004, Salt Lake City, Utah USA DETC2004-57770 (2004)

Hill, C.A.S.: Wood Modification: Chemical, Thermal, and Other Processes. Wiley, Chichester, England (2011)

Hill, M.D., Eckelman, C.A.: Flexibility and Bending Strength of Mortise and Tenon Joints. Furniture Design and Manufacture, USA (1973)

Hinchman, M.: History of Furniture: A Global View. Fairchild Books, New York (2009)

Hinken, E.: Machining wood with coated abrasives. Progress Report No. 2. University of Michigan (1954)

Homeier, A.: Integral systems and process technologies to enhance productivity in CNC processing centres. In: Proceedings of 17th IWMS, pp. 320–331 (2005)

Horman, I., et al.: Numerical analysis of stress and strain in a wooden chair. Drvna Industrija **61**(3) (2010)

Horváth, R.: Kőris ollós-csapos kötések illesztési szorosságának vizsgálata, az eredmények integrálása a terméktervezésbe (Study on the fit tightness of open mortise joints made of ash; adaptation of the results in the product design, in Hungarian). Diploma thesis, University of Sopron (2017)

Horvath, G.P., Kovacs, Z., Denes, L.: Ergonomic design using quality function deployment and design of experiments. Forest Prod. J. **63**(7/8), 257–262 (2012)

Hüsing, L.: Rechnerische und Experimentelle Deformations- und Festigkeitsuntersuchungen an Korpusmöbeln unter Annahme Elasto-Mechanischer Stoffgesetzte und Verschieblicher Eckverbindubgen (Numerical and experimental study of deformations and strength of case furniture with supposed constitutive law and flexible joints. In German). Dissertation, Universität Hamburg (1986)

Imirzi, H.Ö., Efe, H.: Analysis of strength of corner joints in cabinet type furniture by using finite element method. In: Proceedings of the XXVIth International Conference Research for Furniture Industry, pp. 49–55 (2013)

Imirzi, H.Ö., Smardzewski, J., Döngel, N.: Method for substitute modulus determination of furniture frame construction joints. Turk. J. Agric. Forestry **39**, 775–785 (2015)

ISO 10303-236: STEP standard (Standard for the exchange of product data). AP236 Application protocol for furniture

ISO 3534-2: Statistics—vocabulary and symbols—statistical quality control, 1st edn. (1993)

Itaya, S., Tcuchiya, A.: Development of CrN coated tool to create smooth surface of solid wood. In: Proceedings of 16th IWMS Matsue, pp.74–81 (2003)

Janka, G.: Die Härte der Hölzer. Wien (1915)

Kamenicky, J., Paulenkova, M.: Rovnice na vypočet unosnosti kolikovych nabytkovych spojov zo smrekoveho dreva (Formulae for calculating the strength of dowel joints for furniture made of softwood, in Slovak). Drevarsky Vyskum **29**(2), 53–67 (1984)

Kamenicky, J.: Die Nachgiebigkeit von Zapfenverbindungen für Möbelkonstruktionen. Drev. Vysk. **20**(4), 197–214 (1975)

Kamperidou, V., Vasileiou, V.: Bending capacity of middle joints of upholstered furniture frames. Drvna Industrija **63**(4), 255–261 (2012a)

Kamperidou, V., Vasileiou, V.: Strength properties of the most frequent corner and middle joints of upholstered Furniture frames constructed with beech and coplar solid wood. Drvna industrija **63**(4), 255–261 (2012)

Kamperidou, V. et al.: Correlation between bending and tension strength of corner and middle joints constructed with beech and poplar wood. Annals of Warsaw University of Life Sciences —SGGW Forestry and Wood Technology No. 73, 44–50 (Ann. WULS-SGGW, Forestry and Wood Technology 73) (2011)

Kasal, A., et al.: Relationship between static front to back loading capacity of whole chair and the strength of individual joints. In: Proceedings of the 27th International Conference Research for Furniture Industry September (2015), Turkey (2015)

Kasal, A., Smardzewski, J., Kuskun, T., Erdil, Y.Z.: Numerical analysis of various sizes of mortise and tenon furniture joints. BioResources **11**(3), 6836–6853 (2016)

Kaygin, et al.: Simulating strength behaviors of corner joints of wood constructions by using finite element method. Drvna Industrija **67**(2), 133–140 (2016)

Kemény, S., Deák, A.: Design and Evaluation of Experiment (In Hungarian). Technical Publishing House, Budapest (2000)

Kiviniemi, A. et al.: Integration of Multiple Product Models: IFC Model Servers as a Potential Solution. Digital Library of Construction Informatics and Information Technology in Civil Engineering and Construction. w78-2005-kn-w2 (2005)

Koch, P.: Wood Machining Process. Ronald Press, New York (1964)

Kovács, Z.: A study on strength of windows with regards to the functional requirements (in Hungarian). CSc thesis, Scientific Academy of Hungary (1968)

Kovács, Z.: Strength behaviour of window and furniture frames. Acta Facultatis Ligniensis 1984/1. University of Forestry and Wood Technology Sopron, Hungary (1984)

Kovács, Z.: Strength design of furniture, doors and windows (in Hungarian). Lecture notes, hand scripts. University of Forestry and Wood Technology, Sopron, Hungary (1989)

Kovács, Z.: Engineering design of products (in Hungarian). Lecture notes, hand scripts. University of Forestry and Wood Technology, Sopron, Hungary (1995)

Kovács, Z.: Quality engineering. Lecture notes, University of West Hungary, Sopron (2002)

Kovács, et al.: Investigation of the possibilities of enhancing the effectiveness of furniture part manufacture. Industrial research, Sopron, Hungary (2011a)

Kovács, Z., et al.: Investigation of the process capabilities of solid wood part manufacture. Disciplinary research, University of Sopron, Hungary (2011b)

Kühne, G., Kröppelin, U.: Dimensionierung von Möbelbauteilen. Teil 1. (Dimensioning of Furniture Parts. Part 1. In German) Holztechnologie. Leipzig 28 (1987) 2. 67–69 (1987a)

Kühne G., Kröppelin, U.: Dimensionierung von Möbelbauteilen. Teil 2. (Dimensioning of Furniture Parts. Part 2. In German) Holztechnologie. Leipzig 28 (1987) 3. 135–139 (1987b)

Kündig, Fa. Quer, oder doch besser schräg? HOB die Holzbearbeitung: 7/8: 38–39 (2015)

Kungl, Á.: Kőris ollós-csapos kötések illesztési szorosságának vizsgálata, az eredmények integrálása a gyártásba (Study on the fit tightness of open mortise joints made of ash; adaptation of the results in the manufacture, in Hungarian). Diploma thesis. University of Sopron (2018)

Lang, M.: Untersuchungen zur Entwicklung eines kombinierten Plan- und Umfangfräsverfahrens mit kegelstumpförmigen Werkzeugen. PhD Dissertation, Dresden-Sopron (1989)

Langhaar, H.: Dimensional Analysis and Theory of Models. Wiley, New York (1951)

Lehmann, B., Hutton, S.: The kinematics of washboarding of bandsaws and circular saws. In: Proceedings of 13th IWMS Vancouver, pp. 205–216 (1997)

Licher, E.: Schnittwert-Datenbank für die Holzbearbeitung. 8. Holztechn, Kolloqium Braunschweig, S. 33–58 (1991)

Likos, et al.: Effect of tenon geometry, grain orientation, and shoulder on bending moment capacity and moment rotation characteristics of mortise and tenon joints. Wood Fiber Sci **44** (4), 462–469 (2012) #2012 by the Society of Wood Science and Technology (2012)

Lin, S.C., Eckelman, C.A.: Rigidity of furniture cases with various joint constructions. Forest Prod. J. **37**(1), 23–27 (1987)

Lindan, O. et al.: Pressure distribution on the surface of human body. Arch. Phys. Med. Rehabil., 378–385 (1965)

Magoss, E.: General regularities of wood surface roughness. Progress Report No. 1, DWE University of West-Hungary, Sopron (2008)

Magoss, E.: General regularities of the surface roughness sanding solid woods. In: Proceedings of 21st IWMS Conference, pp. 325–332 (2013)

Magoss, E.: Insight into the mechanism of wood sanding. Progress Report No. 3, DWE University of West Hungary, Sopron (2015)

Magoss, E., Sitkei, G.: Strukturbedingte Rauheit von mechanisch bearbeiteten Holzoberflächen. Möbeltage in Dresden, Tagungbericht S. 231–239 (2000)

Magoss, E., Sitkei, G.: Fundamental relationships of wood surface roughness at milling operations. In: Proceedings of 15th IWMS Conference Los Angeles, pp. 437–446 (2001)

Magoss, E., Sitkei, G.: Optimum surface roughness of solid woods. In: Proceedings of 16th IWMS Conference, Martsue, pp. 336–371 (2003)

Magoss, E., et al.: New approaches in the wood surface roughness evaluation. In: Proceedings of the 17th IWMS, pp. 251–257 (2005)

Márkus, Z.: Study of the rigidity of mortise and tenon joints in a chair frame with repeated and sustained loading. Diploma thesis, University of Forestry and Wood Technology, Sopron, Hungary (1990)

Martin, O.: Fenster glänzen mit erstklassigen Oberflächen. HOB die Holzbearbeitung: 3: 50–51 (2015)

Masuda, M.: Why human loves wood grain figures. In: Proceedings of ISWSF 2001, Ljubljana, pp. 11–23 (2001)

Mihailescu, T.: Finite element analysis of mortise and tenon joints in parametric form. Editure Universitatii "Transilvania" Brasov (2003)

Miles, L.D.: Techniques of Value Analysis and Engineering. McGraw Hill (1961)

Mistéth, E.: Többcélú létesítmények optimális biztonságának rendszer szemléletű vizsgálata (Examination of the optimal safety of multi-purpose systems. in Hungarian). Műszaki tudomány 52/1-2/ (1977)

Mistéth, E.: Az erőtani számítás szabályozásának méretezéselméleti alapjai (Theoretical bases of the standardisation of strength dimensioning, in Hungarian). Mélyépítéstudományi szemle XXI. 6, pp. 233–239 (1981)

Mistéth, E.: Teherhordó szerkezetek élettartam alatti biztonságának elemzése (Analysis of the reliability of load-carrying structures during their service life, in Hungarian). Study, Technical committee, Hungarian Institute of Standards, Budapest, Hungary (1986)

Molnár, S.: Faanyagismeret (Wood structure and properties, in Hungarian). Mezőgazdasági szaktudás kiadó (Agricultural Expertise Publishing House), Budapest (1999)

Molnár, Z.: Végmegmunkált természetes faanyagok felületi stabilitása nedvesítéskor (Stability of machined wood surfaces at wetting). PhD Dissertation, Sopron (2018)

Molnár, S., et al.: The World's Industrial Timbers (Földünk Ipari Fái, in Hungarian). ERFARET, Sopron (2016)

Nagy, G.: Strength anylysis of a chair by FEM. Diploma thesis. University of Sopron. Hungary. Advisers: Zsolt Kovács and László Elek (2017)

Najafi, A.: Investigation on the strengths of T-shape joints. Middle-East J. Sci. Res. **15**(3), 398–402 (2013) ISSN 1990-9233

Okay, R., et al.: What is relationship between tooth passage frequency and natural frequency of the bandsaw when washboarding induced. In: Proceedings of 12th IWMS Kyoto, pp. 267–280 (1995)

Orlowski, K., Wasielewski, R.: Wasboarding during cutting on frame sawing machines. In: Proceedings of 15th IWMS Los Angeles, pp. 219–228 (2001)

Pahl, G., et al.: Engineering Design a Systematic Approach, 3rd edn. Springer (2007)

Pallay, N.: Mechanikai technológia (Technology of timbers). University of Sopron (1951)

Palotai, P.: Study of the rigidity of dowel joints in a chair frame with repeated and sustained loading. Diploma thesis, University of Forestry and Wood Technology, Sopron, Hungary (1990)

Paul, S.: Surface Coatings: Science & Technology. Wiley, West Sussex, England (1996)

Peck, E.: Bending solid wood to form. Agr. Handbook No. 125 (1957)

Petrik, O.: Berendezések megbízhatóságra tervezésének néhány kérdése (Some Questions of Designing Systems for Reliability. In Hungarian). Communications of the Institute of Precision Mechanics and Optics of the Technical University of Budapest, Hungary (1985)

Prekrat, S., Smardzewski, J.: Effect of glueline shape on strength of mortise and tenon joint. Drvna Industrija **61**(4), 223–228 (2010)

Prodehl, A.: Untersuchung über das Biegen von gedämptem Holz. Diss. T.H. Dresden (1931)

Rabiej, R.: Untersuchungen zur Tragfähigkeit und Verformung geklebter Rahmeneck verbindungen bei Kurzzeitbelastung. Holztechnologie **22** 2, 89–96 (1981)

Ratnasingam, J., Ioras, F.: Effect of adhesive type and glue-line thickness on the fatigue strength of mortise and tenon furniture joints. Eur. J. Wood Wood Prod. **71**(6), 819–821 (2013)

Ratnasingam, J., Ioras, F.: The fatigue characteristics of two-pin moment resisting dowel furniture joints with different assembly time and glueline thickness. Eur. J. Wood Wood Prod. **73**(2), 279–281 (2015)

Ratnasingam, J., Scholz, F.: Wood sanding process. University Putra Malaysia (2004)

Ratnasingam, J., Scholz. F.: Optimal surface roughness for high-quality finish on Rubberwood (Hevea brasiliensis). Holz als Roh-und Werkstoff: 64: 343–345 (2006)

Ratnasingam, J., Tanaka, C.: Wood Machining Processes: A Managerial Perspective. The Tanabe Foundation Publication, Shimane, Japan (2002)

Ratnasingam, J., Hashim, I., Ganapathi, A.: The tearful woodworker. Universiti Pertanian Malaysia Press, Serdang, Malaysia (1996)

Ratnasingam, J., Reid, H., Perkins, M.: Furniture industry: regaining the competitive edge. J. Inst. Wood Sci. **14**(3), 115–120 (1997)

Ratnasingam, J., Ma, T.P., Perkins, M.: Productivity in wood machining processes: is it a matter of simple economics? Holz als Roh-und Werkstoff **57**, 51–56 (1999)

Ratnasingam, J., Scholz, F., Friedl, Riegel, A.: Wood Sanding Processes: An Optimization Perspective. The Shimbon Kogyo Foundation Publication, Kyoto, Japan (2004)

Ratnasingam, J., Scholz, F., Wong, D., Francis, E.: The Sanding of Wood-Based Panels. Swiss Industrial Abrasives Publication, Geneva, Switzerland (2005)

Ratnasingam, J.: The Malaysian Furniture Industry—Unravelling its Growth and Challenges to Innovation. Universiti Putra Malaysia Press, Serdang, Malaysia (2015)

Ratnasingam, J.: Furniture Manufacturing Processes—Principles & Practices. Centre for Industrial Development Publication, Singapore (2016)

Ratnasingam, J., Chin, K.A., Hazirah, A.L., Subramaniam, H., Khoo, A.: Innovations in the Malaysian furniture industry: drivers and challenges. Bioresources **13**(3), 5254–5270 (2018)

Ravindran, A., et al.: Engineering Optimization. Wiley, New York (2010)

Reddy, N.P., Patel, H., Cochran, G.V.B., Brunski, J.B.: Model experiments to study the stress distribution in a seated buttocks. J. Biomech., 493–504 (1982)

Richter, G.M.: Ancient Furniture: A History of Greek Etruscan and Roman Furniture. Oxford University Press, American Branch, New York (1926)

Richter, K., et al.: The effect of surface roughness on the performance of finishes. Forest Prod. J., 91–97 (1995)

Roosenburg, N., Eekels, J.: Product Design: Fundamentals and Methods. Wiley, Chister (1995)

Russel, R.S., Taylor, B.: Production and Operations Management. Prentice Hall (1995)

Sagmeister, J.: Study on strength of chair frames. Diploma thesis, University of Forestry and Wood Technology, Sopron, Hungary (1982)

Saljé, E., Dubenkropp, D.: Kantenfräsen von Holzwerkstoffplatten. Holz- und Kunstoffverarbeitung, No. 4, S. 307–311 und 490–494 (1983)

Sandberg, D., Johansson, J.: A new method for bending solid wood—high frequency heating of beech. Växjö University School of Technology and Design (2005)

Schmitz, H.: Das Möbelwerk. Berlin (1926)

Schmidt, E., Eckert, E.: Über die Richtungsverteilung der Wärmestrahlung von Oberflächen. Forsch. Ing.—Wes, pp. 175–183 (1935)

Scholz, F., Ratnasingam, J.: Optimization of sanding process. In: Proceedings of 17th IWMS Conference Rosenheim, pp. 422–429 (2005)

Sebastianelli, R., Tamimi, N.: Understanding the obstacles to TQM success. Qual. Manage. J. **10** (3), 45–56 (2003)

Semple, K.E., Smith, G.D.: Prediction of internal bond strength in particleboard from screw withdrawal resistance models. Wood Fiber Sci. **38**(2) (2006)

Siau, J.: Wood: Influence of moisture on physical properties. Virginia State University (1995)

Siklinka, M., Ockajova, A.: The study of selected parameters on wood sanding. In: Proceedings of 15th IWMS Conference Los Angeles, pp. 485–490 (2001)

Sitkei, G.: On the mechanics of oblique cutting of wood. In: Proceedings of 13th IWMS Conference, pp. 469–476 (1997)

Sitkei, G.: Die Natur als Ingenieur, oder welche Grössen können Bäume haben? Acta Fac. Ligniensis Sopron, S. 99–104 (2007)

Sitkei, G.: Similarity study on the energy requirement of saws. In: Proceedings of 21st IWMS, pp. 199–205 (2013a)

Sitkei, G.: Zur Beurteilung der Farbtöne von Holzarten, Fortschrittbericht No. 2, DWE Sopron (2013b)

Sitkei, G.: Gloss of wood surfaces. Progress report No. 2. DWE Sopron (2014)

Sitkei, G., et al.: Reducing apple fruit damage by using cushioning materials. Acta Techn. Hung., 353–362 (1979)

Sitkei, G., et al.: Schnittleistung und Energiebedarf von Gattersägen. Acta Fac. Ligniensis Sopron, S. 23–30 (1988)

Sitkei, G., et al.: Theorie des Spanens von Holz, Fortschrittberichte No. 1, Acta Fac. Ligniensis Sopron (1990)

Skaar, C.: Wood-Water Relations. Springer (1988)

Smardzewski, J.: Furniture Design. Springer (2015)

Smardzewski, J., Dzięgelevski, S.: Stability of cabinet furniture backing board. Wood Sci. Technol. **28**, 35–44 (1993)

Smardzewski, J., Papuga, T.: Stress distribution in angle joints of skeleton furniture. Electronic Journal of Polish Agricultural Universities EJPAU **7**(1), #05 (2004)

Soerianegara, I., Lemmens, R.: Timber Trees: Major Commercial Timbers. Prosea Bogor (1994)

Sparkes, A.J.: The Strength of Mortise and Tenon Joints. Furniture Industry Research Association, Great Britain (1968)

Stadler, W.: Multicriteria Optimization in Engineering and in the Sciences. Plenum Press, New York (1988)

Stonebraker, P.W., Leong, G.K.: Operations strategy: focusing competitive excellence. Allyn and Bacon Publication, Massachusetts. Strengths of mortise and tenon joints. Turk. J. Agric. Forestry **29**, 493–498 (2005)

Sugihara, H., et al.: Wear of tungsten carbide tipped circular saws in cutting particleboard. Wood Sci. Technol., 283–299 (1979)

Szalai, J.: A faszerkezetek méretezését és gyártását befolyásoló sajátosságok (Material properties influencing the dimensioning and manufacture of timber structures. In Hungarian) In: Mérnöki Faszerkezetek II (Engineering timber structures Volume II) edit Wittman, G. Chapter 18. pp 143–214. Mezőgazdasági Szaktudás Kiadó (Publisher Agricultural Expertise) Budapest (2001)

Szalai, J.: A faanyag és faalapú anyagok anizotrop rugalmasság- és szilárdságtana (Anisothropic elasticity and strength of wood and wood-based materials. In Hungarian). First edition 1994 Sopron, Hungary. Second, revised edition (2004) www.ssk.nyme.hu

Taguchi, G.: Introduction to Quality Engineering. Asia Productivity Organization, Tokyo (1986)

Tankut, A.N., Tankut, N.L.: The effects of joint forms (shape) and dimensions on the strengths of mortise and tenon joints. Turk. J. Agric. For. **29**, 493–498 (2005)

The Wood Database. https://www.wood-database.com/

Thomas, T.: Characterization of surface roughness. Precis. Eng., 97–103 (1981)

Thompson, R.: Manufacturing Processes for Design Professionals. Thames & Hudson, London (2007)

Timoshenko, S., Goodier, J.: Theory of Elasticity. New York (1951)

Tröger, J., Lang, M.: Hobelfräsen mit vermindertem Wellenschlag. HOB, 11/1990, S. 43–49 (1990)

Tröger, J., Läuter, G.: Zum Einfluss von Rundlaufabweichungen auf Bearbeitungsqualität und Standvorschubweg beim Fräsen. Holztechnologie, S. 24–29 (1983)

Ulrich, K.T., Eppinger, S.D.: Product Design and Development, 5th edn. McGraw Hill, New York (2014)

Upholstered Furniture Test Method. FNAE-80-214. General Services Administration. Washington, DC

Vanderbilt, M.N.: Matrix Structural Analysis. Quantum Publishers, Inc., New York (1974)

Vassiliou, V., et al.: Strength of corner and middle joints of upholstered furniture frames constructed with black locust and beech wood. Wood Res. **61**(3), 495–504 (2016)

Vorreiter, L.: Holztechnologosches Handbuch. Band 2, Wien (1958)

Wambier, K., Wilczyński, A.: Strength and stiffness of dowel corner joints—effect of joint dimensions. Folia Forestalia Polonica **31**(3), 29–41 (2000)

Wilczyński, A.: Axial load capacity of dowel joint in pine wood. Folia Forestalia Polonica **30**(4), 59–67 (1999)

Wilczyński, A., Warmbier, K.: Effect of joint dimensions on strength and stiffness of tenon joints. Folia Forestalia Polonica **34**(6), 53–66 (2003)

Wood Handbook: Wood as an engineering material. Forest Products Laboratory, United States Department of Agriculture Forest Service, Madison, Wisconsin (2010)

Zhou, Y., Chen, Y.: The methodology for business process optimized design. In: Proceedings of Industrial Electronics Conference (IECON), Institute of Electrical and Electronics Engineers Computer Society, vol. 2, pp. 1819–1824 (2003)

Ашкенази, Е., Анизотропия древесины. (Wood anisotropy) Лесная Пром. Москва (1978)

Бершадский, А. Н. Цветкова, Резание древесины. (Machining of wood) Минск (1975)

Демяновский, К., Износостойкостъ инструмента для фрезерования древесины, (Wear resistance of tools for milling wood) Изд. Лесная Пром. (1968)

Егоров, К., К вопросу деформации основания конечной толщини. (On the deformation of foundation of finite thickness). Сборник "Механика грунтов", 34, Гостройиздат (1958)

Егоров, К., Распределеные напраежений в основани конечной толщини. (Stress distribution in foundation with finite thickness). Сборник "Механика грунтов", 43, Гостройиздат (1961)

Королев, В. В., Основы рационалъного проектирования мебели. (Basics of rational design of furniture) Лесная Пром. Москва (1973)

Кряжев Н. А. Цилиндрическое и коническое фрезерование древесины. М., (1963) (Cylindrical and cone shaped tools for wood milling.)

Соболев, Ю., Древесина как конструкционный Материал. (Wood as construction material) Лесная Пром. Москва (1979)

Шубин.Г., Физические основы и расчет просессов сушки древесины. (Physical background of wood drying.) Лесная промышленностъ, Москва (1973)

Шуин.Г., Физические основы и расчет просессов сушки древесины. (Physical background of wood drying.) Лесная промышленностъ, Москва (1973)

Янов, В. и А. Белов, Художественное конструирование мебели. (Artistic design of furniture) Лесная Пром. Москва (1971)

Index

E. Csanády et al., *Optimum Design and Manufacture of Wood Products*,
https://doi.org/10.1007/978-3-030-16688-5

MIX
Papier aus verantwortungsvollen Quellen
Paper from responsible sources
FSC® C105338

If you have any concerns about our products,
you can contact us on
ProductSafety@springernature.com

In case Publisher is established outside the EU,
the EU authorized representative is:
Springer Nature Customer Service Center GmbH
Europaplatz 3, 69115 Heidelberg, Germany

Printed by Libri Plureos GmbH
in Hamburg, Germany